Creative Model Construction in Scientists and Students

The Role of Imagery, Analogy, and Mental Simulation

John J. Clement

Creative Model Construction in Scientists and Students

The Role of Imagery, Analogy, and Mental Simulation

 Springer

John J. Clement
University of Massachusetts
Amherst, MA 01003
USA

ISBN 978-1-4020-6711-2 e-ISBN 978-1-4020-6712-9

Library of Congress Control Number: 2007938452

© 2008 Springer Science+Business Media B.V.
No part of this work may be reproduced, stored in a retrieval system, or transmitted in any form or by any means, electronic, mechanical, photocopying, microfilming, recording or otherwise, without written permission from the Publisher, with the exception of any material supplied specifically for the purpose of being entered and executed on a computer system, for exclusive use by the purchaser of the work.

Printed on acid-free paper

9 8 7 6 5 4 3 2 1

springer.com

Acknowledgments

I would like to acknowledge the contributions of the following persons in preparing this book: First to my wife Barbara Morrell for all her support; and to the following for very valuable discussions: Ryan Tweney, Carol Smith, Lynn Stephens, Neil Stillings, David Brown, Melvin Steinberg, Tom Murray, William Barowy, and Jack Lochhead.

The research reported in this document was supported by the National Science Foundation under Grants MDR-8751398, DRL-0723709, and REC-0231808. Any opinions, findings, and conclusions or recommendations expressed in this book are those of the author and do not necessarily reflect the views of the National Science Foundation.

Contents

Acknowledgments v

1 Introduction: A "Hidden World" of Nonformal
  Expert Reasoning... 1

Part One Analogies, Models, and Creative Learning in Experts and Students

Section I Expert Reasoning and Learning via Analogy 19

2 Major Processes Involved in Spontaneous
  Analogical Reasoning... 21
3 Methods Experts Use to Generate Analogies 33
4 Methods Experts Use to Evaluate an Analogy Relation.............. 47
5 Expert Methods for Developing an Understanding
  of the Analogous Case and Applying Findings 57

Section II Expert Model Construction and Scientific
            Insight.. 65

6 Case Study of Model Construction and Criticism
  in Expert Reasoning.. 67
7 Creativity and Scientific Insight in the Case Study for S2 97

Section III Creative Nonformal Reasoning in Students
             and Implications for Instruction...................... 117

8 Spontaneous Analogies Generated by Students Solving
  Science Problems 119

9   Case Study of a Student Who Counters and Improves His Own Misconception by Generating a Chain of Analogies 127

10   Using Analogies and Models in Instruction to Deal with Students' Preconceptions 139
John J. Clement and David E. Brown

Part Two   Advanced Uses of Imagery and Investigation Methods in Science and Mathematics

Section IV   Transformations, Imagery, and Simulation in Experts and Students 159

11   Analogy, Extreme Cases, and Spatial Transformations in Mathematical Problem Solving by Experts 161

12   Depictive Gestures and Other Case Study Evidence for Use of Imagery by Experts and Students 171

13   Physical Intuition, Imagistic Simulation and Implicit Knowledge 205

Section V   Advanced Uses of Imagery in Analogies, Thought Experiments, and Model Construction 235

14   The Use of Analogies, Imagery, and Thought Experiments in Both Qualitative and Mathematical Model Construction 237

15   Thought Experiments and Imagistic Simulation in Plausible Reasoning 277

16   A Punctuated Evolution Model of Investigation and Model Construction Processes 325

17   Imagistic Processes in Analogical Reasoning: Transformations and Dual Simulations 383

18   How Grounding in Runnable Schemas Contributes to Producing Flexible Scientific Models in Experts and Students 409

Section VI   Conclusions 431

19   Summary of Findings on Plausible Reasoning and Learning in Experts I: Basic Findings 433

20	**Summary of Findings on Plausible Reasoning and Learning in Experts II: Advanced Topics** 457
21	**Creativity in Experts, Nonformal Reasoning, and Educational Applications** 507

References 575

Name Index 591

Subject Index 595

Detailed Table of Contents

Acknowledgments v

1 Introduction: A "Hidden World" of Nonformal Expert Reasoning. 1
    1.1 Why Study Nonformal Reasoning? 1
        1.1.1 The Need for a Theory of Learning with Understanding 1
        1.1.2 A Strong Parallel Between Expert and Student Learning Processes. 2
    1.2 The Background from Which I Approached This Work. 2
        1.2.1 Novice Problem Solving. 2
        1.2.2 Expert Studies. 3
        1.2.3 Background of Work on Expertise and Science Studies and Remaining Gaps in Our Understanding of Scientific Thinking 4
        1.2.4 Educational Applications of Expert Studies. 9
        1.2.5 Summary. 9
    1.3 Generative Methodology: Qualitative Nature of the Study 10
        1.3.1 Descriptive Case Studies. 10
        1.3.2 Exploratory Documentation of Imagery and Mental Simulation 11
        1.3.3 Instructional Applications. 11
    1.4 General Features of the Analysis Method Used: Contact Between Data and Theory. 12
    1.5 General Theoretical Framework 14
    1.6 Section Summaries and Approaches to Reading This Book 15
        1.6.1 Creativity, Imagery, and Natural Reasoning. 15

Part One Analogies, Models and Creative Learning in Experts and Students

Section I Expert Reasoning and Learning Via Analogy 19

2 Major Processes Involved in Spontaneous Analogical Reasoning. 21

   2.1 Some Major Issues in Analogical Reasoning 21
       2.1.1 Historic Recognition of Importance of Analogy 21
       2.1.2 Definitions of Analogy 22
       2.1.3 Theories of Analogical Reasoning 23
       2.1.4 Preview of Alternative Processes for Analogical Reasoning Identified in This Book 24
   2.2 Method of Study 26
       2.2.1 Data Collection 26
   2.3 Initial Observations 27
       2.3.1 Initial Results on Frequency of Analogy Use 27
       2.3.2 Observations from Transcripts 28
       2.3.3 Evaluating the Analogy Relation 29
   2.4 Major Processes Used in Direct Analogical Inference 29
       2.4.1 Analogies from a Second Subject 30
       2.4.2 Analysis of Major Events in S3's Transcript 31
   2.5 Conclusion 32

3 Methods Experts Use to Generate Analogies 33

   3.1 Introduction 33
   3.2 Definitions of Basic Concepts and Observations 34
       3.2.1 Definition of "Spontaneous Analogy" 34
       3.2.2 Observed Spontaneous Analogies 36
       3.2.3 Analogy Generation Methods 37
       3.2.4 Frequency of Different Analogy Generation Methods 40
       3.2.5 Summary of Observations with Respect to Analogy Generation 42
   3.3 Discussion 42
       3.3.1 The Presence of Analogies in the Solutions 42
       3.3.2 Generation Methods and Invention 44
       3.3.3 Summary 45

4 Methods Experts Use to Evaluate an Analogy Relation 47

   4.1 The Importance of Establishing the Validity of an Analogy Relation 47
   4.2 Examples from Case Studies 48
       4.2.1 Evaluating Analogies for the Sisyphus Problem 48
       4.2.2 Bridging Analogies 49

Detailed Table of Contents   xiii

		4.2.3 A Pulley as an Analogy for the Wheel	50
	4.3	Analogy Evaluation in the Doughnut Problem	52
		4.3.1 Bridging from Tori to Cylinders	52
	4.4	Discussion of Findings and Connections to History of Science	53
		4.4.1 Discussion of Findings on Bridging	53
		4.4.2 Analogies and Bridges in the History of Science	54
		4.4.3 Beyond Bridging	55
	4.5	Summary	56

5 Expert Methods for Developing an Understanding of the Analogous Case and Applying Findings 57

    5.1 Evaluating and Developing an Understanding of the Analogous Case 57
        5.1.1 Direct Methods 57
        5.1.2 Indirect Methods 58
        5.1.3 Summary: Developing Understanding of the Source Analogue 60
    5.2 Inference Projection 61
        5.2.1 Why Are Analogies Useful? 61
        5.2.2 Data on Inference Projection 62
    5.3 Section I Summary for Creative Analogy Generation 63

Section II Expert Model Construction and Scientific Insight 65

6 Case Study of Model Construction and Criticism in Expert Reasoning 67

    6.1 Issues Surrounding Theory Formation 67
    6.2 Background Questions from Philosophy of Science 68
        6.2.1 The Source and Pace of Theory Change 68
        6.2.2 Philosophical Positions: Empiricism vs. Rationalism 70
    6.3 How are Theoretical Hypotheses Formed in the Individual Scientist? 72
        6.3.1 Answer 1: Hypothetico-deductive Method Plus Induction 72
        6.3.2 Answer 2: "Creative Intuition" 73
        6.3.3 Answer 3: Analogies as a Source of Theoretical Hypotheses 73
        6.3.4 Definitions of "Model": A Thorny Issue 74
    6.4 Protocol Evidence on Construction Cycles That Use Analogies 76
        6.4.1 Purpose of Case Study 76
        6.4.2 S2's Protocol 76
        6.4.3 Analysis of Insight Episode 81

	6.5	Summary of Evidence for a Model Construction Cycle as a Noninductive Source for Hypotheses	84

- 6.5 Summary of Evidence for a Model Construction Cycle as a Noninductive Source for Hypotheses 84
  - 6.5.1 Model Construction Cycles 84
  - 6.5.2 Explanatory vs. Nonexplanatory ("Expedient") Models 88
- 6.6 Major Nonformal Reasoning Patterns in the Preceding Chapters ... 95
- 6.7 Appendix: Introduction to Concepts of Torque and Torsion 95

7 Creativity and Scientific Insight in the Case Study for S2 97

- 7.1 Eureka or Accretion? The Question of Insight in S2's Protocol... 97
  - 7.1.1 Defining a Pure Eureka Event 97
  - 7.1.2 Is There a Sudden Reorganizing Change in S2's Understanding?..................................... 98
  - 7.1.3 Does the Subject Use Extraordinary Reasoning Processes?............................... 100
  - 7.1.4 Defining "Insight".................................. 102
  - 7.1.5 Summary.. 104
- 7.2 Creative Mental Processes 104
  - 7.2.1 Anomalies and Persistence in Protocols and Paradigms..................................... 105
  - 7.2.2 Transformations, Invention, and Memory Provocation....................................... 108
  - 7.2.3 Productive Processes: Constrained Successive Refinement vs. Blind Variation....................... 110
- 7.3 Darwin's Theory of Natural Selection 112
- 7.4 Initial List of Features of Creative Thinking from This Case Study and Remaining Challenges.............................. 113
  - 7.4.1 Creative Thought 113
  - 7.4.2 Limitations of the Case Study........................ 115

Section III Creative Nonformal Reasoning in Students and Implications for Instruction........................ 117

8 Spontaneous Analogies Generated by Students Solving Science Problems 119

- 8.1 Use of Analogies by Students............................... 120
  - 8.1.1 Frequency 120
  - 8.1.2 Features of Spontaneously Generated Analogies.......... 120
- 8.2 Conclusion 123
  - 8.2.1 Similarities Between Experts and Students 123
  - 8.2.2 Implications 123
- 8.3 Appendix: Examples of Problems and Spontaneous Analogies 124
  - 8.3.1 Chariot Problem 124

Detailed Table of Contents xv

	8.3.2	Space Carts Problem	124
	8.3.3	Forces on a Stationary Cart Problem	124
	8.3.4	Rocket Problem	125
	8.3.5	Skaters Problem	125

9 Case Study of a Student Who Counters and Improves His Own Misconception by Generating a Chain of Analogies 127

- 9.1 Spontaneous Analogies in a Student's Problem Solution 127
  - 9.1.1 Protocol for S20 129
  - 9.1.2 Protocol Summary 130
  - 9.1.3 Protocol Observations: Creative Case Generation 132
  - 9.1.4 Developing Hypotheses about Cognitive Events that can Account for the Observations 133
- 9.2 Conclusion: Expert-Novice Similarities 136
  - 9.2.1 Instructional Implications 137

10 Using Analogies and Models in Instruction to Deal with Students' Preconceptions 139
John J. Clement and David E. Brown

- 10.1 Introduction 139
- 10.2 Teaching Strategy 140
  - 10.2.1 Introducing the Target 140
  - 10.2.2 Anchoring Case 141
  - 10.2.3 Bridging Strategy 141
- 10.3 Teaching Interviews 141
  - 10.3.1 Tutoring Session 142
  - 10.3.2 Discussion of First Case Study 144
  - 10.3.3 A Second Case Study 145
  - 10.3.4 Explanatory Models 148
  - 10.3.5 Abstract Transfer vs. Explanatory Model Construction 149
  - 10.3.6 Summary of Cases 150
- 10.4 Applications to Classroom Teaching 150
  - 10.4.1 Study of Classroom Lessons 150
- 10.5 Conclusion 153
  - 10.5.1 Persistent Misconceptions 153
  - 10.5.2 Explanatory Models vs. Specific Analogous Cases 153
  - 10.5.3 Two Roles for Anchors 153
  - 10.5.4 Plausible Reasoning vs. Logical Proof Processes in Learning 154
  - 10.5.5 Role of Thought Experiments vs. Observation Activities in Instruction 154

Part Two Advanced Uses of Imagery and Investigation Methods in Science and Mathematics

Section IV Transformations, Imagery and Simulation in Experts and Students............................. 159

11 **Analogy, Extreme Cases, and Spatial Transformations in Mathematical Problem Solving by Experts**.................... 161

    11.1 Introduction... 161
    11.2 Case Study of Analogical Reasoning in a Mathematics Problem 161
        11.2.1 Method 161
    11.3 Results on the Use of Analogies for Eight Subjects............ 163
        11.3.1 Analogy Generation Methods..................... 163
        11.3.2 Evaluating the Cylinder Conjecture 164
    11.4 Other Creative Nonformal Reasoning Processes 165
        11.4.1 Extreme Cases................................. 165
        11.4.2 Partitioning and Symmetry Arguments 165
        11.4.3 Reassembly of a Partition........................ 167
        11.4.4 Embedding 168
    11.5 Discussion... 168
        11.5.1 Imagistic Reasoning 169
        11.5.2 Conserving Transformations..................... 169
    11.6 Conclusion 170

12 **Depictive Gestures and Other Case Study Evidence for Use of Imagery by Experts and Students**..................... 171

    12.1 Introduction... 171
        12.1.1 Hand Motions 171
        12.1.2 Imagery Questions and Hypotheses 172
        12.1.3 Previous Research on Hand Motions 173
        12.1.4 Limitations of Previous Research................... 175
    12.2 Constructing Observational and Theoretical Descriptors........ 175
        12.2.1 Proposed Set of Hypotheses...................... 175
        12.2.2 Relations Between Observations and Hypotheses 177
    12.3 Case Studies 181
        12.3.1 An Expert Protocol.............................. 181
        12.3.2 Analysis of S15's Protocol 182
        12.3.3 Evidence Supporting the Use of Imagery in the Solution................................. 183
        12.3.4 Argument Structure............................. 184
        12.3.5 A Student Protocol 189
        12.3.6 Analysis of S20's Protocol 190
        12.3.7 Summary of S20 Analysis 192

Detailed Table of Contents   xvii

    12.4   Discussion 192
            12.4.1   Types of Processes Associated with Motions 192
            12.4.2   Can Depictive Hand Motions be a Direct
                      Product of Imagery? 193
            12.4.3   Summary of Relations Between Observations
                      and Hypotheses 194
    12.5   Relationship of These Findings to Others in the Literature 194
            12.5.1   The Existence of Kinesthetic Imagery 195
            12.5.2   Depictive Motions Are Not Simply Translated
                      from Sentences 195
            12.5.3   Movements Are a Partial Reflection of Core
                      Meaning or Reasoning 195
            12.5.4   Gestures Can Reflect Imagery 196
    12.6   Conclusion 197
            12.6.1   Sources of Information About Imagery
                      and Simulation 197
            12.6.2   Limitations 198
    12.7   Appendix 1 – Detailed Justification for using Evidence
            of Imagery from Hand Motions in S15's Protocol 199
            12.7.1   Motions Are Concurrent with
                      Solution Process 199
            12.7.2   Motions Can Be a Direct Product
                      of Solution Process 201
            12.7.3   Motions Not Translated from Verbal Sentences 201
            12.7.4   Evidence for Imagery 201
    12.8   Appendix 2 202

**13   Physical Intuition, Imagistic Simulation and Implicit
Knowledge** 205

    13.1   Introduction: Issues in the Area of Imagery, Simulation
            and Physical Intuition 205
            13.1.1   Abstract vs. Concrete Thinking in Experts 206
    13.2   Initial Examples of Physical Intuition 206
            13.2.1   Intuition Reports 207
            13.2.2   Defining Features and Observable Behaviors
                      Associated with Intuitions 208
            13.2.3   Physical Intuitions 209
    13.3   Imagery Reports and Imagistic Simulation 209
            13.3.1   Moving from the Findings in Chapter 12 to
                      Models of Imagistic Simulation 209
            13.3.2   Schema-driven Imagistic Simulation Processes 210
            13.3.3   Precedents in the Literature on
                      Perceptual/Motor Schemas 215
            13.3.4   Relations Between Observations and Hypotheses 218

		13.3.5	Importance of Concrete Intuitions and Imagistic Simulation................................... 219

13.4 Implicit Knowledge...................................... 221
    13.4.1 Distinguishing Different Levels of Implicit Knowledge 222
    13.4.2 Evidence for Unconscious Knowledge................ 225

13.5 Knowledge Can Be Dynamic 226
    13.5.1 Different Uses of the Term "Simulation" 226
    13.5.2 Knowledge Experienced in Imagistic Simulations Is Not Static 227

13.6 Conclusion: The Role of Concrete Physical Intuitions and Simulations in Embodied Thinking by Experts............ 229
    13.6.1 Summary of an Initial Framework for Modeling Physical Intuition and Mental Simulation via Perceptual/Motor Schemas and Imagery 229
    13.6.2 Imagery....................................... 229
    13.6.3 Intuitions and Imagistic Simulation 230
    13.6.4 How Is New Knowledge Generated from an Elemental Simulation?............................ 230
    13.6.5 Using Perceptual/Motor Schemas as an Initial Foothold for Understanding the Use of Intuitions and Imagistic Simulation 232
    13.6.6 Imagery, Intuitions, and Anchoring 233

Section V   Advanced Uses of Imagery in Analogies, Thought Experiments, and Model Construction 235

14   The Use of Analogies, Imagery, and Thought Experiments in Both Qualitative and Mathematical Model Construction 237

14.1 Introduction to Chapter 14–16 237
    14.1.1 Stages in Model Construction Leading up to Quantitative Modeling During the Solution.................................... 238
    14.1.2 Issues in the Field 240
    14.1.3 Ways to Read this Chapter 241

14.2 Composite Protocol Monologue for the Spring Problem 241
    14.2.1 I. Efforts to Develop an Initial Qualitative Description or Prediction for the Targeted Relationship 241
    14.2.2 II. Searching for and Evaluating Initial, Qualitative, Explanatory Model Elements 244
    14.2.3 III. Seeking a More Fully Imageable and Causally Connected (Integrated) Model: Attempts to Align and Elaborate the Model

			So as to Have Elements That Are Fully Connected Spatiotemporally.....................	255
	14.2.4	IV. Increasing the Geometric Level of Precision in the Spatial and Physical Relationships Projected from the Model into the Target Until They Are Ready to Support Quantitative Predictions............	258
	14.2.5	V. Developing a Quantitative Model on the Foundation of the New Qualitative and Geometric Models	260
14.3	Stages in the Solution	265
	14.3.1	Some Possible Precision Levels for Relationship R Between X and Y..................	265
	14.3.2	Transforms to "Close" Analogies in Later Stages of Solution	269
	14.3.3	Summary.......................................	270
14.4	Building a Theoretical Distinction: Explanatory Models vs. Expedient Analogies	270
	14.4.1	Expedient Analogies...............................	270
	14.4.2	Source Analogues.................................	271
	14.4.3	Triangular, Not Dual, Relation in Model Construction	272
	14.4.4	Source Analogues are Projected into the Composite Model, and Must Be Imagistically Aligned	273
14.5	Conclusion	274
	14.5.1	Plausible Reasoning and Stages of Investigation	274
	14.5.2	Parallels and Differences Between Qualitative and Mathematical Modeling......................	275

15 Thought Experiments and Imagistic Simulation in Plausible Reasoning 277

15.1	Nature of Thought Experiments	277
	15.1.1	Fundamental Paradox of Thought Experiments and Two Definitions	277
	15.1.2	Nersessian.......................................	278
	15.1.3	Focus of This Chapter.............................	279
	15.1.4	What are Some Major Functions of and Benefits from Untested Thought Experiments?	280
	15.1.5	Primary Function	280
	15.1.6	Secondary Functions...............................	281
	15.1.7	Can Schema-based Imagistic Simulation be Involved in Untested Thought Experiments with These Different Functions, and If So, What is Its Role?.......	281
	15.1.8	Summary.......................................	285

15.2 Addressing the Thought Experiment Paradox: How Can an Untested Thought Experiment Generate Findings with Conviction?.. 286
    15.2.1 Introduction................................... 286
    15.2.2 Sources of Conviction: Perceptual/Motor Schemas.... 287
    15.2.3 Sources of Conviction: Spatial Reasoning, Symmetry, and Compound Simulation............... 290
    15.2.4 Summary...................................... 293
15.3 Imagery Enhancement Phenomena Support the Proposed Answer to the Paradox 294
    15.3.1 Limitations on Simulation Ability 294
    15.3.2 Imagery Enhancement Focused on Enhancing the Application of a Schema in a Simulation......... 295
    15.3.3 Analysis of Transcripts........................... 297
    15.3.4 Sources of Conviction in Imagery Enhancement...... 298
    15.3.5 Implications of These Extreme Case Examples for a Theory of Thought Experiments...... 300
    15.3.6 Imagery Enhancement Focused on Enhancing Spatial Reasoning or Symmetry or Compound Simulations 301
    15.3.7 Enhancing Spatial Reasoning Via Image Size and Orientation................................ 302
    15.3.8 Symmetry Enhancement......................... 303
    15.3.9 Compound (or "Linearity") Enhancement 304
    15.3.10 The Effectiveness of Enhancement Can Be Explained Using the Present Theory of Conviction in Thought Experiments 304
15.4 How Are Imagistic Simulation and Thought Experiments Used Within More Complex Reasoning Modes?.............. 305
    15.4.1 Four Important Types of Plausible Reasoning 305
    15.4.2 Evaluative Gedanken Experiments as the Most Impressive Kind of Thought Experiment 312
    15.4.3 Multiple Types of Reasoning Processes that can Utilize Thought Experiments Run Via Imagistic Simulations................................... 315
15.5 Are Imagistic Simulations Operating in the Mathematical Part of the Solution? 316
15.6 How Thought Experiments Contribute to Model Evaluation... 317
    15.6.1 Evaluation Strategies 317
    15.6.2 Summary...................................... 319
    15.6.3 Combining Reasoning Processes into a Model Construction Process................. 319
15.7 Chapter Summary..................................... 321
    15.7.1 Addressing the Fundamental Paradox of Thought Experiments: Sources of Conviction 322

16 A Punctuated Evolution Model of Investigation and Model Construction Processes 325

- 16.1 Abductive Processes for Generating and Modifying Models 325
  - 16.1.1 Defining Abduction 325
  - 16.1.2 Construction Occured via Generative Abduction Rather than Induction or Deduction 327
  - 16.1.3 Generative Abduction: Basic Model................ 330
- 16.2 Qualitative Investigation Processes......................... 332
  - 16.2.1 Introduction to Three-part Model of Investigation Processes....................................... 332
  - 16.2.2 GEM Cycles 338
  - 16.2.3 The Explanatory Depth and Precision of Description Dimensions 338
  - 16.2.4 The Three Cycles in the Outlined Investigation Process can Generate the Five Major Observed Modes of Investigation in the Protocol.............. 342
  - 16.2.5 Separate Explanation and Description Processes 347
  - 16.2.6 Computational Model of Todd Griffith.............. 348
  - 16.2.7 Evaluation Functions can Guide Control 350
  - 16.2.8 Comparison to Griffith Study 352
  - 16.2.9 Explaining Insight: Unpredictable Spontaneous Accessing of Subprocesses....................... 354
  - 16.2.10 Generality 355
  - 16.2.11 Levels of Explanation and Precision................ 355
  - 16.2.12 Limitations of the Model Presented 357
- 16.3 Mathematical Modeling Processes 359
  - 16.3.1 Cycle III: Mathematical Modeling 359
  - 16.3.2 Untested Thought Experiments at Higher Levels of Precision than Qualitative Modeling......... 361
  - 16.3.3 Mathematics and Explanation..................... 362
- 16.4 Abduction II: How Evaluation Processes Complement Generative Abduction. 363
  - 16.4.1 Multiple Sources of Ideas and Constraints for the Generative Abduction Process 364
  - 16.4.2 Model Evaluation can Provide Inputs to the Next Abduction Cycle........................... 364
  - 16.4.3 Role of Transformations in Model Modification 367
  - 16.4.4 Distinctions Between Constructive Transformations, Running a Schema in an Imagistic Simulation, and Basic Spatial Reasoning Operators.................. 368
  - 16.4.5 Coherence and Competition Between Models 369
- 16.5 Seeking an Optimal Level of Divergence.................... 370
  - 16.5.1 The Problem of Accessing Relevant Prior Knowledge: An Ill-Structured Problem 370

		16.5.2	Need for an Effective Middle Road with Respect to Creative Divergence 371

        16.5.2 Need for an Effective Middle Road with Respect to Creative Divergence 371
        16.5.3 Analogies and Extreme Cases Appear to be a Fruitful Source of Divergence 371
        16.5.4 Dangers of Divergence: The Need for Optimal Divergence 372
        16.5.5 Some Methods for Reducing Einstellung Effects Via "Contained" Divergence 372
        16.5.6 Mechanisms for Modulating Divergence 374
        16.5.7 Summary for Section on Divergence 376
  16.6 Chapter Summary 377
        16.6.1 Diagrammatic Summary 377
        16.6.2 Multiple Cycles and Goals in the Overall Investigation Process 378
        16.6.3 Four Subprocesses at the Core of Complex Model Construction: Generative Abduction, Model Evaluation, Schema Alignment, and Mathematization. 380

17 Imagistic Processes in Analogical Reasoning: Transformations and Dual Simulations 383
  17.1 Two Precedents from the Literature 383
        17.1.1 Structural Mapping and Evaluation 383
        17.1.2 Wertheimer's Parallelogram 384
  17.2 Conserving Transformations 385
        17.2.1 Transformations are not Equivalent to Mapping Symbolic Relations 385
        17.2.2 Are Conserving Transformations Just Memorized Rules? 386
  17.3 Conserving Transformations in Science 386
        17.3.1 Wheel Problem 386
        17.3.2 Spring Problem 387
        17.3.3 Newton's Canon 389
  17.4 Dual Simulation 389
        17.4.1 Do Dual Simulations Differ from Transformations? 391
        17.4.2 Dual Simulation for the Square and Circular Coils 392
  17.5 Overlay Simulation 392
        17.5.1 Examples of Overlay Simulation 392
        17.5.2 Connection to Model Construction: Overlay Simulations and Model Projections May Involve Similar Processes 395
        17.5.3 Model Projection. 396

Detailed Table of Contents xxiii

		17.5.4	Imagistic Alignment Analogies	396
		17.5.5	Dual Simulation vs. Compound Simulation in Modeling..................................	397
	17.6	Summary and Discussion of Types of Evaluation Processes: Contrasting Mechanisms for Determining Similarity...........		398
		17.6.1	Mechanisms for Dual Simulation (Including Overlay Simulation).....................................	399
		17.6.2	Mechanisms for Conserving Transformations	400
		17.6.3	Bridging is a Higher-order Strategy Compared to Others...	401
		17.6.4	Combination of Evaluation Methods	401
		17.6.5	Comparison to Structural Mapping of Images	404
		17.6.6	Conclusion on Evaluation: Four Main Analogy Evaluation Methods, Not One.....................	405
	17.7	Use of Imagistic Transformations During the Generation of Partitions, Analogies, Extreme Cases, and Explanatory Models		405
	17.8	Conclusion		407
18	How Grounding in Runnable Schemas Contributes to Producing Flexible Scientific Models in Experts and Students......			409
	18.1	Introduction: Does Intuitive Anchoring Lead to Any Real Advantages?......................................		409
		18.1.1	Review of Findings on Imagistic Simulation and Runnable Schemas	410
		18.1.2	Transfer of Runnability Hypothesis	410
		18.1.3	Models Can Inherit the Capacity for Simulation from Anchors	412
		18.1.4	What, Exactly is Transferred?.....................	415
		18.1.5	Example of Transfer of Imagery and Runnability in Instruction....................................	416
	18.2	Cognitive Benefits of Anchoring and Runnability for Models....		418
		18.2.1	Traditional Benefits of Building on Prior Knowledge ...	419
		18.2.2	Benefits of Transferring Runnability from a Schema to an Explanatory Model........................	419
		18.2.3	Recursive Runnability of Models As Thought Experiments Explain Many of These Benefits	424
		18.2.4	Transfer of Conviction	424
	18.3	How Runnable Models Contribute Desirable Properties to Scientific Theories		425
		18.3.1	Scientific Theories and the Role of Runnability........	426
	18.4	Conclusion		428
		18.4.1	Initial Support for the Runnability Hypothesis........	429

Section VI Conclusions 431

19 Summary of Findings on Plausible Reasoning and Learning in Experts I: Basic Findings 433

- 19.1 Brief Overview of Theoretical Findings 433
  - 19.1.1 Model Construction in Experts 433
  - 19.1.2 Model Construction in Students 435
  - 19.1.3 Summary Table of Expert Subprocesses 435
- 19.2 Analogy Findings, Part One 435
  - 19.2.1 The Presence and Importance of Analogy in Expert Thinking: Significant Analogies 435
  - 19.2.2 Literal Similarity and the Problem of What Counts As an Analogy 438
  - 19.2.3 Analogy Subprocesses 438
  - 19.2.4 Initial New Distinctions and Findings on Analogy 439
- 19.3 Model Construction Findings, Part One and Initial Connections to General Issues in History/Philosophy of Science 440
  - 19.3.1 Extraordinary vs. Natural Reasoning 440
  - 19.3.2 Extraordinary Thinking? 441
  - 19.3.3 Eureka vs. Accretion Question 441
  - 19.3.4 A Case Study of Scientific Insight 442
  - 19.3.5 Initial Exploration of Mechanisms of Hypothesis Generation 444
  - 19.3.6 Section Summary 445
- 19.4 Imagistic Simulation Findings, Part One 446
  - 19.4.1 Imagery Indicators as Observational Concepts 446
  - 19.4.2 Mechanisms for Imagistic Simulation 447
  - 19.4.3 Terminology for Imagistic Simulations 448
  - 19.4.4 Imagery During Simulation Behavior 449
  - 19.4.5 Image-Generating Perceptual Motor Schemas as Embodied Knowledge 449
  - 19.4.6 Sources of New Knowledge in Imagistic Simulations 451
  - 19.4.7 How Perceptual Motor Schemas are Useful in Scientific Thinking 452
  - 19.4.8 Intuitive Anchors 453
  - 19.4.9 Role of Perceptual/Motor Schemas in the Construction of Model Assemblies 453
  - 19.4.10 Connection to Experiments and Situated Action 454
  - 19.4.11 Section Summary 454

20 Summary of Findings on Plausible Reasoning and Learning in Experts II: Advanced Topics 457

- 20.1 Analogy Findings, Part Two 457

	20.1.1	Comparison to Classical Views of Analogical Reasoning 457
	20.1.2	Analogies and Imagery 463
	20.1.3	Analogies and Model Construction. 468
20.2	Imagistic Simulation Findings Part Two: Thought Experiments and Their Uses in Plausible Reasoning 471	
	20.2.1	Overview 471
	20.2.2	Summary of Findings on Thought Experiments 471
	20.2.3	Broader and Narrower Categories of Thought Experiments. 473
	20.2.4	Can Thought Experiments Allow One to "Get the Physics for Free?" 474
	20.2.5	Section Conclusion 475
20.3	Model Construction Findings Part Two: An Evolutionary Model of Investigation Processes 476	
	20.3.1	Top Level of Scientific Investigation Process 476
	20.3.2	Process I: Description Cycle 478
	20.3.3	Process II: Explanatory Model Construction 478
	20.3.4	Process III: Mathematical Modeling. 484
20.4	The Important Role of Imagery in the Expert Investigations 485	
	20.4.1	Limitations of the Imagery and Simulation Systems 485
	20.4.2	Evidence for Imagery Involvement in a Wide Range of Reasoning Processes 486
	20.4.3	Evidence for the Importance to Subjects of Imagistic Simulation 488
	20.4.4	Possible Advantages of Imagistic Representations as Knowledge Structures 489
	20.4.5	Possible Advantages of Imagistic Representations for Creative Reasoning 492
20.5	Transfer of Runnability Leads to Outcomes of Flexible Model Application and Generativity 499	
	20.5.1	Example of Flexible Model Application. 499
	20.5.2	Role of Runnable Intuitions in Conceptual Understanding and Recursive Runnability 501
	20.5.3	Comparison to Lakoff and Nunez's Embodied Mathematics. 503
	20.5.4	Payoffs from Transfer of Runnability: 504
20.6	Comments on Methodology 504	
	20.6.1	Small Samples 504
	20.6.2	Links Between Data and Theory 505

21 Creativity in Experts, Nonformal Reasoning, and Educational Applications 507

- 21.1 Summary of the Overall Framework. 507
  - 21.1.1 View from Multiple Diagrams 507
  - 21.1.2 Central Role of Imagery 510
  - 21.1.3 Highlighted Findings. 510
  - 21.1.4 Larger Integrating Processes. 517
  - 21.1.5 Position on Concrete vs. Abstract Thinking 518
- 21.2 How Experts Used Creativity Effectively. 521
  - 21.2.1 Do Expert Discovery Processes in Science Always Have an Empirical Focus? 521
  - 21.2.2 How a Coalition of Weak, Nonformal Methods are Able to Overcome the Dilemma of Fostering both Creativity and Validity 523
  - 21.2.3 Overlap Between the Context of Discovery and Context of Evaluation. 530
  - 21.2.4 Section Conclusion 531
- 21.3 Educational Applications: Needed Additions to the Classical Theory of Conceptual Change in Education 532
  - 21.3.1 Uses and Criticisms of Kuhn. 532
  - 21.3.2 Criticisms of Classical Conceptual Change Theory 532
  - 21.3.3 Need for an Expanded Theory of Conceptual Change for Education. 533
- 21.4 Expert–Novice Similarities in Nonformal Reasoning and Learning. 533
  - 21.4.1 Similarities Concerning Resistance to Change 534
  - 21.4.2 Similarities in the Use of Intuition and Imagery. 536
  - 21.4.3 Use of Analogies by Students. 537
  - 21.4.4 Model Construction by Students. 538
  - 21.4.5 Summary: Expert–Novice Comparisons. 540
- 21.5 Implications for Instructional Strategies and Theory 540
  - 21.5.1 Strategies Suggested by Initial Studies of Analogy and Model Construction in Part One of the Book. 541
  - 21.5.2 Strategies and Implications Suggested by Findings on Imagistic Knowledge Representations in Part Two of the Book 548
  - 21.5.3 Educational Implications of Imagistic Learning Processes in Part Two of the Book 552
  - 21.5.4 Conclusion–Educational Applications 556
- 21.6 Are Creative Processes in Experts a Natural Extension of Everyday Thinking?. 559
  - 21.6.1 Expert–Novice Comparisons for Knowledge Structures: Science as an Extension of Intuition? 559
  - 21.6.2 Expert–Novice Differences in Reasoning. 560
  - 21.6.3 Expert–Novice Similarities in Reasoning 560

Detailed Table of Contents xxvii

	21.6.4	Some Expert Processes are Neither Extraordinary Nor Ordinary 560
	21.6.5	A Spectrum from Ordinary Thinking to Unusually Effective Creative Thinking to Extraordinary Thinking 562
	21.6.6	Summary: How Creative Expert Reasoning is not Ordinary 566
	21.6.7	Implications for Instruction: Utilizing Natural Reasoning Processes 567
21.7	Assessing the Potential for a Model of Creative Theory Construction in Science 568
	21.7.1	Expertise and Domain Specificity................... 568
	21.7.2	Can Creative Behavior be Explained?................ 569
21.8	Conclusion 572
	21.8.1	Creative Thinking 572
	21.8.2	The Model Construction Process Portrayed Here in Contrast with Oversimplified Models.............. 572
	21.8.3	Questions About Scientific Thinking 574

References 575

Name Index 591

Subject Index 595

Chapter 1
Introduction: A "Hidden World" of Nonformal Expert Reasoning

This book attempts to examine sources of creative theory formation in scientists. This interest leads to an investigation of the domain of nonformal reasoning in science, including analogical reasoning, mental model construction, imagistic simulation, applying physical intuition, and advanced techniques such as using Gedanken (thought) experiments. Some historians and philosophers of science believe that these nonformal, creative reasoning processes play a crucial role in original discoveries in science even though those processes can be well hidden from sight in published scientific articles and presentations. However, others are quite skeptical that such nonformal methods can play a role in scientific thinking. The book documents these methods actually being used, through the analysis of video tapes of scientists thinking aloud while attempting to solve problems and understand unfamiliar systems.

Transcripts from these tapes capture scientists in the act of generating creative analogies, extreme cases, mental models, and thought experiments, as well as mentally performing imaginative spatial transformations such as deforming, cutting, and reassembling objects in novel ways. They also allow the analysis of insight episodes where a subject makes a conceptual breakthrough accompanied by "Aha"-type exclamations. A major goal of this book is to better understand this "hidden world" of expert nonformal reasoning by describing subprocesses occurring in each of the above methods.

1.1 Why Study Nonformal Reasoning?

1.1.1 *The Need for a Theory of Learning with Understanding*

One section of the book deals with learning in science students. There has been a long history of struggle in studies of learning to explicate the idea of "meaningful learning" or "learning for understanding." The basic intuition underlying this struggle is that there appears to be a palpable difference between learning that leads to rote knowledge and learning that leads to deeper understanding. Students who achieve deeper understanding are able to do several important things, such as to give explanations in their own words, determine when it is appropriate or not to use a scientific principle, solve conceptual problems quickly on the basis of insightful, qualitative causal reasoning, and transfer knowledge flexibly to new situations. Deeper understanding of this kind can be very difficult to teach. So there is an important need to formulate better models of what "learning for understanding" entails.

By working toward a theory of three major sources for understanding, namely, analogical transfer, explanatory model construction, and intuitively grounded knowledge expressed in imagistic simulations, I believe that progress can be made on this question. (These correspond to three sections of the book.) I focus most on describing these sources in experts in order to examine the claim that they are legitimate sources of scientific knowledge. It is important to study these processes while experts are in the process of solving conceptual problems in order to develop descriptions of actual scientific understanding and reasoning *in practice* as opposed to the more formal versions on display in scientific articles. Whether these processes also occur naturally in students and the extent to which they can be utilized in instruction are topics of Section III of the book and part of the concluding Chapter 21.

1.1.2 A Strong Parallel Between Expert and Student Learning Processes

This book merges findings from what was originally two separate tracks of work. The expert track focused on creative nonformal reasoning. The applied track focused on methods for dealing with students' preconceptions in physics and mathematics. However, work on scientific reasoning in experts has many more implications for theories of instruction than I originally estimated. The simplest way to foreshadow this conclusion in one sentence is that *many of the powerful nonformal reasoning and learning processes used by experts to achieve scientific understandings are also useful in helping students learn scientific understandings.* This parallel between processes occurs in each of the three major categories of applying intuitively grounded knowledge, analogical reasoning, and model construction, and provides one motive for the study of nonformal reasoning.

1.2 The Background from Which I Approached This Work

1.2.1 Novice Problem Solving

A brief description of the path from two tracks of work which led to the studies described in this book may be in order, so as to say something about the particular point of view I bring to this work. The applied track began with a think-aloud study of conceptual physics problem solving with freshman science and engineering majors, leading to one of the first papers to identify misconceptions and reasoning

1.2 The Background from Which I Approached This Work

difficulties that impede learning in students (Clement, 1982b). However, a set of unexpected, more positive observations also emerged in this period. While students often had misconceptions (ideas that conflict with presently accepted theory) about abstract physical principles, they were sometimes able to solve the problems using more concrete methods involving analogies to familiar situations. Approximately 60 spontaneous analogies were generated by the students, and in some cases these tapped useful intuitions about familiar physical events that could be applied to the original problem (described in Chapter 8).

The most impressive protocol came from a freshman engineering student who produced a remarkable solution to a conceptual physics problem involving the concept of mass in space. In this protocol he generated a number of analogies, extreme cases, and even thought experiments, and used them to overcome a common misconception concerning inertial mass. (A case study analysis of this solution is given in Chapter 9.) I was amazed that one could observe such methods – heuristics that Polya (1954) and Wertheimer (1959) had identified as sophisticated and creative plausible reasoning strategies used by Archimedes and other great mathematicians – in a young student who had not been trained in their use. His use of an elaborate chain of analogies to familiar and invented cases, where he could apply his own physical intuitions about the behavior physical objects, was especially flexible and powerful. This convinced me that there might be some natural forms of nonformal reasoning that we knew very little about but that could be quite powerful, even for students. In addition, we eventually discovered what seemed to be several very different ways of generating and confirming analogies at work in his protocol. This provided an initial motive that work in this area might produce a more fine-grained, observation-based description of nonformal reasoning methods than had previously been offered. However, the reasoning used was unfamiliar and difficult to "decode." A coherent analysis of the protocol was not completed for 10 years.

I also eventually came to see an overlap between the processes this student was using spontaneously and processes I had seen operating in some successful tutoring interactions with other students who had asked for help in physics courses. Analogies to physical problems in the form of familiar examples that the students could relate to seemed to play a key role in tutoring in giving meaning to equations that were otherwise considered hopelessly abstract by the students. However, this tutoring strategy was based entirely on "teaching intuitions." At this time we had no theory whatsoever of how to use analogies effectively.

1.2.2 Expert Studies

In what started as a separate interest, I was intrigued by the plausible reasoning processes used by colleagues to solve unfamiliar problems. My colleagues and I would often challenge each other with qualitative physics problems. These were conceptual problems that were not very amenable to the immediate application of familiar mathematical principles from courses. In some cases, they drew out the use of more interesting problem-solving techniques such as analogy, and I became fascinated with the challenge of describing the form of those plausible reasoning methods. I decided to try capturing these processes in experts using think-aloud techniques. Polya's books are a large compendium of such plausible reasoning processes in mathematics, but they are not supported by systematically collected evidence on experts, and they do not treat reasoning about physical systems or physical intuition much at all. The challenge was to see whether we could actually document the form of plausible reasoning techniques in experts.

At this time in the 1970's some early efforts were being made by Papert, Gruber, and Cohen at MIT to collect qualitative problems for their effort to understand how basic knowledge and reasoning processes were used in physics. Like the problems I had developed, many of these problems could be solved in an elegant way without using mathematics; if one could only find the "right way to look at the problem," the underlying mechanism would suddenly, or not so suddenly, become obvious. But finding the "right way to look at the problem" was a deeply subtle and nontrivial task. The problem discussed in Chapters 4 and 17 of where to push on a heavy wheel to roll it up a hill was one of the tasks suggested in Cohen (1975). Their problems of this kind inspired me to invent other problems that were not amenable to homework-style solutions.

A series of grants from the National Science Foundation then allowed us to pursue the issue of plausible reasoning more seriously. This funded a study of experts that was done using standard think-aloud recording techniques. The initial focus was on analogical reasoning. An initial set of subjects were interviewed on a variety of problems. The subjects were advanced doctoral students (who had passed their comprehensive examination) or professors in technical fields. Subsequently I began to report results in the *Proceedings of the Cognitive Science Society* and the journal *Cognitive Science*. This set of solutions was expanded over a number of years and comprises the expert database for this book.

1.2.3 Background of Work on Expertise and Science Studies and Remaining Gaps in Our Understanding of Scientific Thinking

1.2.3.1 Higher-order Processes

In general, in the history of psychology, there has been a significant shortage of research on higher-order cognitive processing in proportion to that on other areas of human performance. For example, case studies of creative reasoning or hypothesis formation in experts were and continue to be rare. One cause of this was the tendency to restrict research to experimental rather than descriptive paradigms. It can be argued that this restriction is inappropriate for the initial study of higher-order reasoning and learning. There are several reasons that doing controlled experiments to analyze processes as complex as hypothesis construction are extremely difficult. First, the processes are so poorly understood that it is not clear what variables to try to vary or control. Second, there are likely to be many choices open to a creative problem solver, and many possibilities for feedback loops, processes that call each other, and recursion. For the goal of identifying what these processes are, these nonlinear properties make the use of a controlled experiment (usually designed to identify simple linear relationships between variables) very difficult to use at best. There are psychometric approaches to studying general characteristics of creative thinkers (Plucker and Renzulli, 1999), but authors such as Mayer (1999) have called for research that goes beyond these to study the nature of creative *processes*. One alternative to the experimental tradition came from the work of Piaget and his followers (e.g. Piaget, 1930, 1955; Inhelder and Piaget, 1958; Easley, 1979; Witz, 1975), who initiated the use of protocol analysis and opened up the descriptive study of intuitive knowledge structures and natural reasoning processes for different age groups.

1.2.3.2 Initial Studies of Expertise

Newell and Simon (1972) led the way in psychology in exploring alternatives to the predominant experimental tradition with their landmark think-aloud study of problem solving. This study was descriptive rather than experimental and attempted to propose hypothesized reasoning processes that were supported by observations from transcripts. This study led to a theory of general reasoning processes used by humans. Perhaps of necessity, this early study focused on problems that were much simpler and more well defined than scientific discovery. However, from that time, many expert studies continued to focus on puzzle problems or practiced skills rather than the process of learning for deeper understanding or conceptual change. Studies of typing skill and algebra symbol manipulation skills fall into the former category. Even physics problem solving, when approached as the use of a set of rules for solving problems in a certain category, can be treated as a highly practiced skill without much attention given to building conceptual understanding. There has been a tendency to study physics experts in the task context of first-year physics homework problems (Larkin and Simon, 1987) – not a very realistic domain for capturing creative processes experts use in their own research. However their research program did go beyond the study of algebraic symbol manipulation in identifying "working forward" as an alternative problem-solving strategy in which the subject first develops a (usually causal) model of different components in the problem system before attempting to solve equations. And Langley (1981) published important research using AI programs to model processes of scientific discovery. However, the analysis was limited to "empirical discovery" of inductive patterns in data sets, and did not address the question of how deeper theoretical concepts and hypotheses are invented to *explain* such data sets.

It can be argued that the focus of the above research was not at a high-enough level to be relevant to problems of theory formation in science; to develop models of *conceptual change or growth*, the focus in experts must be on processes at a higher level than homework-level problem solving or finding patterns in data sets. For the same reason, there is a significant problem in applying most studies of expertise to problems of conceptual change in education. Only when experts are dealing with new questions that force them to construct new models do we have a learning process that reflects significant conceptual change and the development of new understandings. This is the level that is most relevant to theory change in science and the kind of student learning that can overcome deep seated difficulties in science, and this is the level dealt with in this book. An example of more recent work in psychology that has used descriptive methods to study conceptual change is that of Chinn and Brewer (1998) who find large variations in student responses to anomalous data, from rejection, to theory accommodation.

1.2.3.3 More Recent Studies of Expertise in Science

The early focus on puzzle problems in experts is one reason that our own group began to study experts working on unfamiliar problems outside their own area of specialty. This creates the possibility for real learning and discovery situations because the expert is operating on the frontier of his or her own personal understanding, and may be able to expand that understanding during the interview. Thus, for the purposes of this study, by "expert" I mean a person who is an experienced problem solver in a technical field but not necessarily in the domain of the problems in this study. In other words the expertise studied is that of *adaptive expertise* in scientific problem solving in general (Hatano and Inagaki, 1986). In this way we began to collect and analyze expert protocols that exposed nonformal reasoning and qualitative model construction processes. These early studies included processes of analogy (Clement, 1982a), dissonance, and model criticism and revision (Clement, 1981).

A number of studies by cognitive historians of science have begun to expose the model based nature of theory development. Pioneering studies of Darwin (Gruber, 1974), Faraday (Tweney, 1985; Gooding, 1990), Kepler (Gentner et al, 1997) continental drift (Giere, 1988), biochemistry (Langley et al., 2006), and the history of genetics (Darden, 1983) have focused attention on the many steps necessary to create a new theory during model construction. Specific analogies are also cited as important in theory development, although the details of how they work as a process are difficult to infer from information at a broad historical timescale. For example, in a series of penetrating studies, Nersessian (1984,1991) has proposed the following findings, based on article drafts and other data from scientists such as James Maxwell. Although Maxwell's final theory was expressed compactly in the form of equations, he developed his initial theories of electromagnetic induction using qualitative models of fields, starting from hydrodynamic and mechanical analogies. He developed the qualitative model incrementally in cycles of improvement rather than all at once. Furthermore, these cycles were not cycles of theorizing alternating with empirical tests. Rather, Maxwell appeared to be occupied with thought experiments through which he would then evaluate and refine his model. Maxwell's theory was not constructed at the level of patterns in behavior of electrical apparatus. Rather, he was at pains to explain already established patterns in terms of theoretical entities such as fields and their interactions. Rather than behavior patterns induced from data, the models he proposed appeared to be abductive constructions that would explain behavior if they were true. His initial aim appeared to be to develop a set of coherent qualitative visualizable models. And he spent significant amounts of effort to achieve them. These then paved the way for quantitative modeling.

These fascinating findings document the importance of processes like analogy, thought experiments, and qualitative model construction in the thinking of preeminent scientists. The fact that Maxwell's early work was qualitative and used nonformal reasoning and concrete mechanisms is highly significant, since this work is foundational for one of the most successful mathematical abstractions ever achieved by a physicist: the E/M field equations. These findings add to our motives for better understanding of the processes of analogy, abductive model construction, and thought experiments, as well as of their relation to quantitative models – issues that I will attempt to contribute to in this book. To move further in the direction suggested by studies like Nersessian's, we need to address questions such as: do scientists use early analogies or thought experiments or models that are so conjectural that they are not even candidates for a written record of drafts on paper? How are these rejected or how do later versions grow from these? Do sudden insights or Aha events ever actually occur during theory development? What triggers these if and when they occur?

Thus, several historical studies using data from scientists' diaries and drafts of papers has opened an exciting window on nonformal, nondeductive processes in scientists. However, that kind of data has less to say about questions about short-term processing. To acquire evidence at a finer detail on these processes, we also need studies of experts "in process," working "live" on such problems.

1.2.3.4 Studies of Analogical Reasoning

It is true that some early psychological studies of analogical reasoning broke away from the "homework problem" mold to a certain extent. (Gentner, 1983, Holyoak and Thagard, 1989) These studies began to address issues about insight that were raised by the Gestalt psychologists. The use of analogies in problem solving cannot be fully characterized by the application of conscious rules, especially during the phase of accessing an analogous case. I will review structural mapping theories of analogy in Chapter 2. They are highly refined and have explained a large body of empirical evidence. However, the existing work in psychology on analogy has concentrated on analogies presented to the subject. Very little work has been done on spontaneous analogies suggested by the subject. Studies of analogy have also tended to be limited to novices and nonscientific adults, with too little work done on expert use of analogy. Therefore this book begins with studies of experts spontaneously generating analogies for unfamiliar problems and later looks at this process in students as well.

1.2.3.5 Discussions in Scientific Laboratories

An exception to the latter summary are the studies by Dunbar of biogeneticists in working laboratories by Dunbar (1997). These have focused on highly productive research teams and their interaction in large group research meetings. Thus Dunbar has crossed the threshold to studying "live" scientists in the act of reasoning, as was true of an early pioneering study by Tweney (1985). Dunbar also has been able to study the social interaction dimension in group discussions in real laboratory sessions as an important complement to studies of individual thinking. (Nersessian has also recently "moved in" to a bioengineering laboratory.) Dunbar was able to distinguish between three types of analogy used in these meetings. "Local" analogies usually involved mapping an idea for a procedure from a previous experiment to a current one and were quite common. "Regional" analogies mapped a system of relationships from one area to another (e.g. phage viruses to retroviruses). Their use was less common but still significant. More remote, "long distance" analogies were less common and usually used pedagogically or metaphorically rather than playing a role in discoveries. Thus, Dunbar has been able to study scientists as they generate, rather than respond to, analogies and finds that analogy use is quite common in research meetings. This provides further motivation to study the ways that scientific analogies and other kinds of generative processes work in detail.

I believe that individual thinking-aloud studies are complementary to these laboratory studies in an important way. A limitation of the research meeting studies is that the record cannot show the uninterrupted train of thought of any one scientist. This is because there cannot be a record of his or her thinking while others are speaking. Also, scientists think independently between the meetings and those processes are not recorded. Also, side issues may tend to be more common in meetings, complicating the flow of ideas and making it more difficult to study a particular type of reasoning.

For a finer-grained analysis of processes like analogy or model formation, there is a need to study the train of thought of individual scientists under think-aloud instructions. These may yield a more complete and continuous database that can allow one to examine issues such as: How exactly are the analogies generated and evaluated by a scientist? How are they then applied and perhaps modified in the service of model construction? Another reason for using individual think-aloud techniques is that videotapes of individual work allow documentation of the use of drawings by the subject. Data can be collected on when and where the subject is attending, pointing, or gesturing over part of a drawing. Beginning in Section IV, I work from data on drawings and depictive hand motions, among other observations, to begin modeling the role of spatial reasoning and imagery at an even finer-grain size than the earlier sections. This allows one to consider questions such as: How do thought experiments work; are there different kinds? What is the source of new knowledge in such thought experiments? Do they involve conclusions from "mental simulations"? If so, what cognitive processes could possibly be responsible for such "simulations"? What is the role of imagery in these processes? Can imagistic simulations play a role in analogies and model construction, and if so, how? Are subjects able to enhance (improve the quality of) their own imagery?

1.2.3.6 Studies of Imagery

Specific previous psychological studies of imagery are described in Chapter 12, but some of the general gaps I hope to speak to here are:
- The majority of previous studies of imagery have focused on static, two-dimensional images. There is a need to focus on dynamic, three-dimensional imagery in

order to deal with processes such as mental simulation. Our ability to model processes of spatial reasoning – especially the process of predicting the consequences of *actions in space* – is still in a very primitive state.
- A considerable amount of effort has gone into studying inferences along propositional chains and networks of causes (despite often being called mental simulation) beginning with de Kleer and Brown (1983); Forbus (1984). But the *elemental simulation* of a single causal relation is usually not studied and is not treated as an imagistic process with dynamic properties that takes place over time. Cognition involving the *coordination* of simultaneous actions is also avoided.
- In this regard, there is a need to consider the role of the motor system in dynamic imagery and kinesthetic imagery. In some contexts it may be important to include *perceptual motor images and knowledge structures* rather than just simply visual images. These may tap lower-level cognitive systems which take into account basic spatial and physical constraints affecting the design of any mechanism or any assembly of real-world objects. In Section IV of this book I treat perceptual motor schemas that originate from everyday practical actions on the world as a different type of knowledge than discrete linguistic encodings. Experts sometimes refer to these as "physical intuitions." Previous work has not tended to make connections between expert processes and this level of knowledge, although some beginnings have been made by diSessa (1985), and by others in laboratory studies (Schwartz and Black, 1996, 1999), situated learning theory (Clancey, 1997; Greeno, 1997) and history of science (Tweney, 1991; Ippolito and Tweney, 1995; Gooding, 1990).

1.2.4 Educational Applications of Expert Studies

Work on expertise typically assumes that novices display only *deficits* in comparison to experts. Novices are considered to be "experts with holes." However, novices may have prior knowledge conceptions that differ from the expert's. Some of these will be misconceptions, but others may be useful intuitions that can be molded into expert conceptions.

The traditional focus has been on expert–novice differences. As stated earlier, I will also focus on expert–novice *similarities*, where they occur, as a way of pointing out where education can tap into and build on unused reasoning abilities or prior knowledge intuitions in students. In particular, the use of physical intuition, imagery, simulation, and analogies to prior knowledge schemas will be examined.

1.2.5 Summary

In summary, science studies have provided initial evidence that nonformal processes like analogy and qualitative model construction were used by preeminent scientists and are actually used in formative research meetings in leading laboratories. These studies have provided strong motives for pursuing a fine-grained analysis of nonformal reasoning using think-aloud techniques. We appear to have entered an era in research on scientific thinking where historical, laboratory, sociological, and individual protocol studies can complement each other by focusing on processes at different time scales, contexts, and levels of thinking, appropriate to the character and grain size of their respective data collection methods.

1.3 Generative Methodology: Qualitative Nature of the Study

1.3.1 Descriptive Case Studies

I have chosen to use methodology appropriate to the embryonic state of the field of studying reasoning in scientists, namely descriptive case studies, in response to the need for studies which map out subprocesses used in nonformal reasoning. For example, the precise role of analogical reasoning in complex model construction is not at all clear, nor are the various subprocesses used in analogy generation, analogy evaluation, or imagistic simulation. Through the analysis of think-aloud case studies, models of cognitive processes in these areas will be proposed that will grounded in naturalistic observations. Because the processes are complex, these models will surely be incomplete, but the hope is that they can provide us with a starting point that is grounded in case study observations of real behavior.

The think-aloud procedures used with experts throughout this set of studies is described in detail early in Chapter 2. The goal was to use think-aloud instructions and occasional probing that encouraged a subject's verbalizations about their thought process but that avoided suggesting strategies or ways of describing thinking. I would therefore describe the data collected for the studies in this book as intensive fieldwork observations. As in Newell and Simon (1972) there was no manipulation of experimental variables in the study – simply the careful collection and analysis of think-aloud protocols, yielding evidence for processes that produced the protocol. Thus it is more like the methodology of naturalistic protocol used by Darwin in South America than an experimental or psychometric technique. There were a variety of methods used by different subjects on the same problem, so one might say that different "species" of problem solving and explanation behaviors were collected. The analysis of the data also shares some characteristics with Darwin's theoretical work, including its generative, abductive character, where the goal is to construct the simplest possible hypotheses about underlying cognitive process characteristics that are constrained by, and therefore still fit, as many of the observations in the database as possible. This often involved struggling to find a new way to look at some passages of transcripts that seemed hopelessly convoluted and tangled at first. Many of the reasoning "moves" the experts made were difficult to describe since there seemed to be no current vocabulary for their component parts, overall structure, or function. This was often further confused by sidetracks in their arguments and returns to previous lines of thought. Paradoxically, at some points the data seemed at once too rich and too incomplete to be tractable.

1.3 Generative Methodology: Qualitative Nature of the Study

However, certain patterns of reasoning were eventually discerned which do appear to form coherent strategies and which account for originally unexplained sections of transcripts from a number of different subjects. I have had to invent new vocabulary for naming some of these patterns, such as "extension analogies," "imagistic simulation," and "imagery enhancement." Certain other vocabulary is in common use but is used in too many ways, and these needed more precise definitions, such as "analogy," "extreme case," "intuition," and "thought experiment." Thus, part of the analysis task in an exploratory study of previously undocumented behavior is to formulate stable, independent, and useful concepts that appear to reflect or explain natural units or patterns in the behavior, and to formulate corresponding vocabulary. (See Glaser and Straus, 1967, for discussion of technique; and D. Campbell, 1979, for urging its importance.) When these new conceptual entities describe a cognitive structure or process, they also constitute what Harre (1972) calls an "existential hypothesis"; that is, a hypothesis that a certain process or structure exists in some form in the subject. These concepts then become the "atoms" out of which more elaborate theories of larger structures and processes can be built. Thus, the methodology used is that of concept and hypothesis construction from descriptive case studies.

1.3.2 Exploratory Documentation of Imagery and Mental Simulation

Another major difficulty adding to the rationale for a qualitative approach in this study was the lack of knowledge in the field about the role of mental imagery and mental simulation in higher-order thinking. Imagery has long been a controversial topic in psychology. Partly because it is much harder to collect evidence on than linguistic thought, the very existence of imagery and mental simulation, and especially their importance in thinking, have been challenged repeatedly in the history of psychology. As a consequence, very little work has been done on the possible roles imagery plays in cognition, especially for higher-level cognition (two exceptions are Finke (1990) and Shepard (1984). Therefore this book starts conservatively by postponing the discussion of imagery and mental simulation until Section IV. This allows the display of a variety of pertinent phenomena without making a premature theoretical commitment to an imagistic form of mental representation. Having this database in place subsequently allows one to outline a framework for thinking about higher-level uses of imagery and simulation that is grounded in a body of observations. This should allow readers a chance to try out alternative interpretations of the data in the first two sections of the book, if desired, as a way to comparatively evaluate the later interpretations using imagery concepts.

1.3.3 Instructional Applications

Eventually, the separate track of work we were conducting on ways to deal with students' preconceptions did converge with the expert findings. Although I had hoped this might happen, I am still surprised by the extent to which processes we saw experts use turned out to be useful as processes to encourage learning in students during instruction. Some of our planning diagrams for physics lessons (see Fig. 15.6), now use the same format as diagrams we used to describe the results of expert model construction processes. Thus, the present work is highly interdisciplinary, in the tradition of cognitive science, and makes use of concepts from various disciplines of cognitive science in order to describe model construction processes in both experts and students.

1.4 General Features of the Analysis Method Used: Contact Between Data and Theory

The theories of plausible reasoning processes developed here will consist of concepts and hypotheses constructed in order to account for observations of behavior in descriptive case studies. In most of this study I have tried to maintain close contact with detailed observations of human cognition. Contact with observations means that, the generation of these models has incorporated the constraint of providing coherent explanations for as many of the observed behaviors as possible. Whenever possible, the constructs are grounded by reference to multiple locations in the transcripts that support them.

This book provides evidence that the construction of scientific models in experts can proceed via a process in which an initial model is constructed and then successively refined through cycles of criticism and modification. One can also apply this view of model construction to the methodology of investigation used in this book. During their development, models of cognitive processes have been formulated to explain sections of transcript, then criticized and revised repeatedly in light of data in subsequent sections and patterns across the larger protocols. Unlike methods of statistical analyses, this process of early qualitative model evolution does not necessarily require large sample sizes. Thus Newell and Simon (1972), in their seminal work, *Human Problem Solving*, were able to develop a very influential theory for a particular kind of thinking on the basis of a small number of protocol analyses.

Much of this book takes the form of "existence proofs": it exhibits case studies of one or a few subjects using certain processes. Such small sample sizes may seem insignificant to experimental psychologists, but in unexplored areas of science where one does not even know what the relevant variables are, in-depth case studies can make an important early contribution (e.g. the study of the first identified case of a planet with a moon, or a new life form, or a new kind of astronomical object, will not be dismissed because of low sample size) (Cronbach, 1975; Newell and Simon, 1972; Clement, 2000). A model of mental processes is constructed that can explain behaviors in a protocol and that does not conflict with other events in the protocol.

In the paper "You Can't Play Twenty Questions with Nature and Win," Alan Newell (1973) advocates think-aloud case studies as an essential tool in the formation

1.4 General Features of the Analysis Method Used

of viable hypotheses in psychology. Also, Anzai and Simon (1979), in commenting on reasons for using this methodology in an article analyzing a single subject, remark:

> It may be objected that a general psychological theory cannot be supported by a single case. One swallow does not make a summer, but one swallow does prove the existence of swallows. And careful dissection of even one swallow may provide a great deal of reliable information about swallow anatomy. (p136)

From another point of view though, the sample size in their work and also this one is very large – if one counts each clause in the protocols as representing an act of cognition that should be consistent with the theory. The above construction process is more constrained than it may seem, at first sight, to be. Criticism of the theory in this case takes the form of finding inconsistencies, both between the theory and the numerous episodes within the transcripts, and between elements of the theory. One finds that it is not so easy to make a model that survives such criticisms. The theory is then continuously revised to remedy the inconsistencies. Numerous model generation, criticism and revision cycles of this kind foster the growth of more and more coherent models of the reasoning and learning processes used.

Within the realm of qualitative modeling, there are different stages of model development, from early exploratory work to later, more confirmatory work (Clement, 2000). A spectrum of methods from different stages has been used here. These methods include: analysis of individual short examples from problem solutions, used to provide "existence proofs" and introduce new observational and theoretical constructs by example, as in Chapters 2 and 4; extended individual case studies or microanalyses in Chapters 6 and 9, which examine a process in depth (Glaser and Straus, 1967; Easley, 1979); surveys across a sample, as in the analogy generation methods in experts and novices in Chapters 3 and 8, respectively, where frequency of occurrence data across 10 or more subjects are given; and expositions in Chapters 7, and the end of 18, where the findings were connected with issues in the history of science at a broader and more speculative level. Each method above is appropriate according to whether the purpose is to provide an existence proof for a phenomenon or process, to describe the frequency of a phenomenon in a sample, unpack a process into plausible hypothesized subprocesses, or to add to the coherence of a theory by connecting it to other areas.

The models put forward here tend to be somewhat more detailed than those from history of science since the source of data is much more fine grained. On the other hand, they tend to be less detailed than some AI models which actually run on a computer; that requirement often takes AI theories far beyond the level of detail currently available in data from humans. In this regard, first-order empirically grounded models at a low or intermediate level of detail are significant if there is little prior work in the field; they are a major advance over having no viable grounded models at all. They are appropriate to an early stage of development of the field. Here, for the most part, I have tried to stick to a level of detail in modeling corresponding to the level of detail in the data. This enables maintaining connections to empirical support. The resulting models of expert reasoning are then considered "grounded hypotheses" – models that have some initial support in the case study observations they explain.

1.5 General Theoretical Framework

The findings presented in this book support a theoretical and epistemological framework at a general level concerning the central role of explanatory models in conceptual understanding in science. The present study began with an interest in analogical reasoning, partly in response to the work of an important group of scholars in the philosophy of science, including Campbell (1920), Harre (1961), Hesse (1966), and Nagel (1961), who suggested that analogies may be a source of hypotheses in science. Their argument is that scientists not only find patterns of empirical observations in their work, but also think in terms of theoretical explanatory models such as molecules and black holes that constitute a different type of hypothesis than empirical laws. These models are not simply condensed summaries of empirical observations, but are actually theoretical mechanisms that scientists invent. On the other hand, these explanatory models do not consist simply of formulas and statements of abstract principles; they are concrete models, which *underlie* the comprehension of formulas. An example is the elastic particle model for gases. Scientific models such as this are built up from analogies to more primitive and familiar notions, so as to explain and fit with what is observed in nature. The above authors, as well as Black (1979), argued that scientific models involve analogies to a familiar system (such as a collection of colliding balls). In Nagel's (1961) terms, such analogue models help the scientist "make the unfamiliar familiar." This suggests that analogical reasoning may work as an important noninductive source for generating such hypothetical models by tapping into familiar prior knowledge schemas. In sum, these writers have put forward the thesis that conceptual understanding in science utilizes concrete mental models, which in turn depend on analogies to familiar experiences.

This book attempts to analyze protocol data in order to evaluate the view above and to develop more detailed theories about the relationships between intuitions, imagistic simulations, analogies, and explanatory models. Data to be presented here support the idea that many primitive components of models are refined intuitions based on one's personal experience in interacting with the world. For example, an analogy can be implicitly drawn between an intuition about the effect of a baseball hitting one's hand and the effect of molecules in an expanding gas on a piston. This kind of concrete grounding can provide the core meaning of a mechanistic explanation for the phenomenon of gas pressure that makes sense at an intuitive level and provides an important foundation for more formal theories.

Later in the book I will use protocol data to support and motivate the hypothesis that "concrete," as used above, should mean "imageable," and that the construction of such models involves nonformal reasoning processes such as analogy, spatial reasoning, abduction, and imagistic simulation. In this view, nonformal reasoning plays a key role in scientific thinking and learning for both experts and students, and central elements of science are based on an extension of intuitive knowledge structures via natural reasoning processes.

1.6 Section Summaries and Approaches to Reading This Book

Locations of definitions for most specialized terms are identified in the index. As seen in the table of contents, the book is divided into two parts of three sections each. Part One deals with the major topics of analogy and model construction, and Part Two analyzes these processes more deeply as depending on imagery and physical intuition. Within Part One, analogical reasoning is examined in Section I because it is the simplest to describe. In Section II model construction processes (which use analogical reasoning as a subprocess) are described. Applications of these ideas to problem solving and instructional situations with students are described in Section III. A discussion of spatial transformations, imagery, and intuition are postponed until Part Two, Section IV because the connections to behavioral evidence are more difficult to make. Section V uses the newly developed constructs for imagery and intuition to analyze analogical reasoning, thought experiments and model construction as heavily dependent on imagistic processes. This section also extends the scope of the investigation to mathematical model construction. Conclusions in Section VI summarize the theories of analogy, imagistic simulation, and model construction and tie them to larger questions about discovery in science.

Part One and to some extent Part Two can be read independently. Part One discusses similarities in expert and novice reasoning with respect to the role of analogies, and the educational applications of these findings. Section III is the most immediately relevant section for those who wish to apply this work to learning or teaching situations. The first two thirds of Chapter 19 provides a concluding summary of the findings in Part One. A summary of the implications for learning and teaching appears in the middle of Chapter 21. Part Two also examines analogies and model construction, but to show how they can depend on imagery, the analysis is done at a finer level of detail by examining hand motions and other indicators from transcripts. In addition, Part Two examines processes by which thought experiments work and the payoff properties of scientific models that are capable of generating mental simulations. Thus Part Two attempts to develop a more elaborated and detailed theory of imagistic learning. For the reader interested primarily in this part of the work, a review of the figures and tables in Part One, should prepare one to concentrate on Part Two.

1.6.1 Creativity, Imagery, and Natural Reasoning

In the final chapter the larger question of how experts used creativity effectively is discussed. Although nonformal reasoning modes are heuristic in the sense that they, unlike deductions, are not guaranteed to work or produce truths from given truths, they can combine in powerful ways to meet the challenge of fostering both creativity and validity at the same time during model construction. Imagistic transformations and simulations expand the potential for divergent creativity greatly. But demanding precisely coherent and connected imagistic models also deepens the capacity for repeated cycles of stringent criticism and revision. This system can be beautifully balanced and modulated, so as to provide varying degrees of divergence – volatile divergence at an early stage and focused convergence in a later stage of a solution.

Similarities between novice and expert reasoning processes documented elsewhere in the book are reviewed. Evidence indicating that these processes can be utilized in the classroom to foster the construction of qualitative models is summarized. In this view, qualitative, nonformal reasoning involving dynamic imagery plays a key role in both scientific thinking and student learning. Finally, findings are reviewed that indicate that although the best expert thinking is remarkable rather than ordinary, major aspects of creative processes in experts are neither unnatural nor unexplainable. This suggests that scientific thinking is an extension of natural forms of thinking. This means that there is a large potential for engaging students in these processes.

Part One
Analogies, Models, and Creative Learning in Experts and Students

Section I
Expert Reasoning and Learning Via Analogy

It has often happened in physics that an essential advance was achieved by carrying out a consistent analogy between apparently unrelated phenomena.... The association of solved problems with those unsolved may throw new light on our difficulties by suggesting new ideas. It is easy to find a superficial analogy which really expresses nothing. But to discover some essential common features, hidden beneath a surface of external differences, to form, on this basis, a new successful theory, is important creative work.

A. Einstein and L. Infeld (1967, p. 270)

Chapter 2
Major Processes Involved in Spontaneous Analogical Reasoning*

2.1 Some Major Issues in Analogical Reasoning

2.1.1 Historic Recognition of Importance of Analogy

When one examines the literature on creativity in science, one of the first topics one encounters is analogical reasoning. The quote from Einstein and Infeld on the preceding page captures the respect that some prominent scientists have for the role of analogies. Investigators such as Campbell (1920); Dreistadt (1969); Gentner (1982); Hesse (1966); Einstein and Infeld (1967) have argued that analogies can play an important role in the creation of new theoretical hypotheses. In some cases these hypotheses can become established analogue models, such as the "billiard ball" model for gases. Most of this work has been at a philosophical level or is based on retrospective reports of scientists. However, the present study aims to provide an initial body of more direct evidence from "live" think-aloud protocols that capture scientists in the act of analogical reasoning as it occurs. Other literature on the role of analogy in constructing scientific models will be reviewed in Chapter 6. Historically, in psychology and education, analogical reasoning has long been suspected of being important in both the learning of scientific models and in the transfer of this learned knowledge to new, unfamiliar problems (diSessa, 1983, 1985; Rumelhart and Norman, 1981; VanLehn and Brown, 1980; Vosniadou and Ortony, 1989). A full issue on the role of analogy in science teaching appeared in the *Journal of Research in Science Teaching* (vol. 30, issue 10). Again historically, investigators have long ascribed an important role to analogical reasoning in problem solving (Dunker, 1945; Gick and Holyoak, 1980; Polya, 1954; Schon, 1981; Wertheimer, 1959), measures of intelligence (Sternberg, 1977), and the development of concepts (Lakoff and Johnson, 1980). Thus there has been recognition for some time of the importance of analogical reasoning in advanced cognition.

---

*Some portions of this chapter are based on findings reported in: Clement, J. (1991). Nonformal reasoning in experts and in science students: The use of analogies, extreme cases, and physical intuition. In J. Voss, D. Perkins, and J. Siegel (Eds.), *Informal reasoning and education* (pp. 345–362). Hillsdale, NJ: Lawrence Erlbaum.

This chapter first foreshadows some contrasts between findings from previous work on analogy and findings that will emerge in this book. It then provides some initial examples from think-aloud protocols of the use of spontaneous analogies in expert problem solving. Four major processes involved in using spontaneous analogies will be identified: generating the analogy, understanding the analogous case, determining whether the analogy relation is valid, and applying findings from the analogy. These processes provide an initial view of analogy as a rational but nondeductive type of creative reasoning with both generative and evaluative components.

2.1.2 *Definitions of Analogy*

By "analogy" most investigators have meant the following. Given a problem in the context of an original situation called the *target*, the analogy is a connection based on structural similarity between the target and a different case called the *base* or source. Once the similarity is deemed valid or sound, useful additional information can sometimes be inferred in the target.

2.1.2.1 Presented vs. Spontaneous Analogies

Of the existing psychological studies of analogy, almost all have focused on presented analogies, in which at least part of the analogy is presented to the subject for completion. Here I will focus on spontaneous analogies, where the subject initiates and forms the entire analogy. Roughly, these occur when a subject, in thinking about a target situation A, shifts, without being prompted, to consider a situation B (the base) which differs in some significant way from A, and hopes to apply findings from B to A. In successful solutions by analogy the two contexts being compared are often perceptually different but are seen by the scientist to be functionally or structurally similar in some way. (In Chapter 3, I will give a more detailed definition.)

2.1.2.2 Analogy vs. Literal Similarity

Dedre Gentner (1983) proposed a pioneering theory of analogical reasoning based on what she terms *structural mapping*. This theory represents both the base and target as propositional semantic network representations. This allows her to talk about knowledge within each representation as symbols used in predicates at different levels of abstraction, from higher-order relations to lower-order relations to concrete attributes. For Gentner, analogies can be contrasted with what she terms a literal similarity between two cases. In a literal similarity, the two cases share many concrete surface attributes as well as abstract relations, whereas in an analogy, the

2.1 Some Major Issues in Analogical Reasoning

cases predominantly share abstract relations and very few surface attributes. I will develop an alternative view to this syntactically defined view of analogy. I will consider the possibility that some "close" analogies that Gentner would call "literal similarities" are difficult, powerful, creative achievements that deserve to be called analogies (Clement, 1988).

2.1.3 Theories of Analogical Reasoning

Gentner (1983) and Forbus et al. (1997) described the processes involved in analogical reasoning as follows:

1. The analogous case is *accessed* by being activated associatively and retrieved from permanent memory.
2. A *mapping* is generated between corresponding entities in the base and the target and the *soundness* of the analogy is assessed.
3. One or more key elements are *inferred* in the target.

In this theory access is a process that relies predominantly on surface attributes, whereas mapping[1] and soundness assessment tend to focus more on abstract relations.

Holyoak adopted a similar point of view on the essential elements above. His analysis has focused most strongly on the mapping process as the most difficult process to account for (Holyoak and Thagard, 1989). This reflects the view, which I share, that it is easier to find a *possibly* analogous case but not so easy to find a good or meaningful analogy. What are the properties of a *good* analogy? This turns out to be a deep and difficult question. The two cases should be alike in important ways – but what is the best meaning for "important"? One property of good analogies according to Gentner is the presence of interconnected higher order relationships in the base that can be placed in correspondence with interconnected higher order relationships in the target (a syntactical criterion called systematicity).

Holyoak has argued that evaluating soundness is a more complex and goal oriented process, and has argued for the strongly weighted influence of *corresponding problem contexts* (such as problem goals) and of *particular* corresponding *semantic relations such as causal relations*. He has also proposed connectionist models for implementing associative access processes and for doing the weighted calculations of soundness evaluation. Nevertheless, for both authors, the processes are assumed to operate on sets of explicit, discrete propositional descriptions of the base and

---

[1] One difficulty is that different papers in the literature use different meanings for the term "mapping." Possible subprocesses as referents are: identifying identical relations in base and target; identifying other corresponding relations and elements in base and target; scoring the strength of the above matches for evaluating soundness; and identifying candidate inferences in the target. Because I see evidence for the fourth process sometimes taking place separately, I will use the term "mapping" to refer to the first three subprocesses

target in the classical theory. Although Gentner and Holyoak have disagreed on details concerning the mechanisms of access and evaluation, these details are not so germane to the focus of this book. As an advanced organizer, the following section foreshadows a comparison between findings to be presented in this book and the common features of the classical theory as espoused by Gentner and Holyoak.

2.1.4 Preview of Alternative Processes for Analogical Reasoning Identified in This Book

In the studies to be described in this book, it was found that when experts use spontaneous analogies, they exhibit a wider variety of creative behavior than those in the classical theory. For example, the chapters that follow argue that the main processes mentioned above are more richly varied in the following ways:

A. *Access*:
  1. Analogies are not just accessed in permanent memory by association from the target; they can be generated by transforming the target case.
  2. Many of the cases generated in this way turn out to be newly invented, rather than accessed cases; this leads to several implications that differ with the basic theory above.
B. *Evaluating soundness*: In previous theories, mapping connections between discrete symbols in the representations of the target and base is the lion's share of the work of determining soundness; once the mapping process is complete, soundness is calculated from a weighted scoring system for the soundness of the mapping.
  The present study suggests that there are other important methods for evaluating the soundness or validity of an analogy besides mapping of discrete symbols, including generating secondary bridging analogies (Chapter 4), finding conserving transformations, or conducting dual imagistic simulations for the base and the target (Chapter 17). These methods may all rely heavily on imagery.
C. *Comprehension of the source* or analogous case: Most previous studies assume that the subject has an adequate understanding of the knowledge in a source conception that is pertinent to the target (although the subject may not yet see the relevance of the source to the target). The present study finds that subjects may also need to develop and refine their understanding of a source analogue before they are able to apply results to a target problem.
D. *Application via direct inference*: It also finds that analogies do not always lead to a direct inference from base to target. They can also play other roles; one of these is a more provocative role in activating an essential schema that has never been applied before to either the target or the base. In the classical theory of analogy, Gentner and Holyoak emphasized the roles that analogies can play as:
  1. An aid to problem solving or predicting behavior, via direct inference as mentioned above

2.1 Some Major Issues in Analogical Reasoning

2. An aid to learning via an inductive generalization from the common features seen in a target and one or more analogous cases

In contrast, this book discusses evidence from transcripts for the following additional roles:

1. As mentioned above, an analogy can work through indirect provocative triggering of a new principle, schema, or method, as the analogous case itself is analyzed.
2. Analogies can be applied recursively to evaluate earlier analogies. These may contribute by helping to confirm the previous analogy relation, for example, rather than via direct inference.
3. There are differences in the types of tasks used in the literature on analogy in problem solving. Much of the previous literature emphasizes "insight" problems (Holyoak) or story comprehension (Gentner) where the task is usually to plan an action or to make a prediction. Part of the present work focuses on conceptual *explanation* tasks in science. In these the task is to make a prediction and explain why the physical system behaves that way. The analogical reasoning processes used in explanation tasks may be somewhat different than in simpler tasks. In particular, analogies may serve as proto-models, that is, as a starting point or source analogue for the development of a scientific explanatory model. Clement (1981, 1989); Falkenhainer (1989), Holland et al. (1986), and Gentner et al. (1997) have proposed that advanced model construction in science may begin with an analogy that is modified as it is criticized and revised to eventually become a useful model. A related but expanded role, and its connection to previous literature in history and philosophy of science, will be discussed in Chapters 6, 7, and 16.

2.1.4.1 Frequency of Use

Catrambone and Holyoak (1989) paint a discouraging picture in finding that the use of cross-domain analogies is quite infrequent for laymen solving insight problems. Frequencies rise only when the experimenters prompt the solver to use a previously described analogous case or ask the solver to describe what several previous analogous cases have in common. This provides a rather discouraging prognosis–since these results are for presented analogies, they suggest that non-presented, spontaneous analogies might occur very rarely. However, Chapter 3 questions this prognosis in general by examining scientific problems where analogies occurred fairly frequently among a sample of experts, and Chapter 8 does the same for a sample of engineering students. Studies undertaken by Dunbar (1997) document ongoing group problem-solving processes in genetics laboratories, and these also show more frequent uses of analogical reasoning. On the other hand, the present study will show that successful uses of analogy in expert problem solving can also be difficult and time consuming, rather than being quick shortcuts to a solution, as they are sometimes portrayed.

In summary, several groups in cognitive science have focused on analogical reasoning processes over the last 15 years, but there has been very little prior research done on the spontaneous use of analogies in experts. Chapters 3–6 focus on this type of analogy in order to describe any new processes that may be involved. The remainder of this chapter simply introduces the method used to study spontaneous analogies, and develops some initial hypotheses about them. I will examine two brief examples from think-aloud protocols and develop a notation for analyzing four major processes involved in analogical reasoning.

2.2 Method of Study

2.2.1 Data Collection

Subjects and tasks. All expert subjects in this book were professors in technical fields or advanced doctoral students who had passed their comprehensive examination. The database for much of the present section of the book comes from videotapes of subjects' working on the "Spring Problem" shown in Fig. 2.1. An example of an analogy for this problem would be to think about the weights hung vertically from long and short elastic bands of the same thickness instead of from wide and narrow springs. Knowing that the larger band will stretch more might suggest that the larger spring will stretch more. In fact, the correct answer to the spring problem is that the wide spring will stretch farther. This seems to correspond to many people's initial intuition about the problem. However, giving a full explanation for why this is correct is a much more difficult task.

A second problem about whether it would take more force to push a wheel up a hill by pushing parallel to the slope at the top or at the rear of the wheel (the "Sisyphus" problem) will be presented and discussed primarily in Chapters 4 and 17. The name refers to the character in Greek mythology who was condemned to pushing a large stone up a hill forever. A third problem called the Torus Problem

A weight is hung on a spring. The original spring is replaced with a spring:
   --Made of the same kind of wire,
   --With the same number of coils,
   --But with coils that are twice as wide in diameter.
Will the spring stretch from its natural length, more, less or the same amount under the same weight? (Assume the mass of the spring is negligible compared to the mass of the weight.)
Why do you think so?

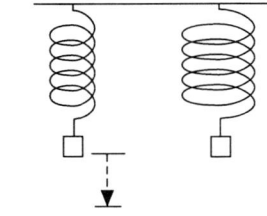

Fig. 2.1 Spring problem

2.3 Initial Observations   27

involved calculating the volume of a torus without taking an integral. Data collection was complicated by the fact that solution times on the problems varied tremendously (e.g. from 6.4 to 52 min on the Spring Problem). Since subjects were at a considerably higher level professionally than undergraduates and were not able to offer more than an hour or two of time, this meant that subjects completed different numbers of problems, depending on how long it took them to do the first problem. These numbers for book sections I–IV are: Spring Problem (10 subjects), the Torus Problem (8 subjects), and the Sisyphus Problem (7 subjects).

Smaller sections of chapters that report on specific issues also use protocols from several other simpler problems that are similar in character to the Spring and Sisyphus problems.

Think-aloud instructions. Before solving a set of problems, subjects in the studies throughout this book were told that the purpose of the interview was to study problem solving methods and were asked to think aloud as much as possible during their solution attempt, including reporting preliminary thoughts. Subjects were given instructions to solve the problem "in any way that you can," and were asked to give a rough estimate of confidence in their answer. No suggestions were made to encourage the use of analogies, imagery, or any other method and in fact the interviewer avoided introducing those terms altogether. Probing by the interviewer was kept to a minimum, usually consisting of a reminder to keep talking. Occasionally the interviewer would ask for brief clarification of an ambiguous report. At the end of the problem solution more extensive clarification was sought on other ambiguous reports. Subjects were able to use a large pad of paper and magic marker if needed. Almost all sessions were videotaped. Exceptions to this procedure occur in the instructional interviews in Chapters 10 and 18, where the interviewer as tutor was also introducing new questions and ideas in addition to encouraging thinking aloud.

2.3 Initial Observations

2.3.1 *Initial Results on Frequency of Analogy Use*

The spring problem solutions took from 6.4 up to 52 min, and the average length was 23.7 min. All subjects favored the (correct) answer that the wide spring would stretch farther. But the subjects varied considerably in the types of explanations they gave for their prediction. The frequency of analogy usage was:

Total number of spontaneous analogies generated 38
Number of subjects generating at least one analogy 8

Thus the proportion of subjects using analogies was high, indicating that analogy generation among experts is not limited to a few special individuals. These data will be discussed in more detail in Chapter 3.

2.3.2 Observations from Transcripts

The main purpose of this chapter is to introduce some initial examples of the phenomena of spontaneous analogical reasoning by examining excerpts from the protocols of two subjects solving the spring problem. A broader survey of the solutions will be given in Chapter 3.

First, consider an excerpt from the solution of a research physicist, S1. To counter the idea that analogies are used only by those who lack more formal reasoning methods, it can be noted that this subject was a Nobel laureate in physics. This subject had actually wound springs in the lab, and after stating with confidence, on the basis of experience, that the wider spring will stretch more, he proceeds to consider the harder quantitative question of determining how much more.

> S1: The equivalent problem that might have the same answer is – suppose I gave you the problem in a way instead of being a coiled spring, it's a long U spring like that, just like a hairpin. (draws Fig. 2.2). And now I hang a weight on the hairpin, and see how far it bends down. Now I make the hairpin twice as long with the same wire and see how far it bends down. Now that [deflection] goes with the cube [of the wire length]. That's the deflection in the length of the cantilever beam. Heh, heh – and maybe it comes out that way with the spring. So my – I would bet about, about 2 to 1, I would bet that the answer to this [the wider spring] is that it goes down 8 times as far. (*In this book comments in brackets are the author's for clarity*).

S1 has generated an analogous case, that of a hairpin, and made a quantitative prediction about its behavior: its deflection will increase with the cube of its length. Although he is confident of his prediction for the analogous case, he is not positive that he can transfer this conclusion to the original problem. In fact, he feels that his conclusion warrants a bet with only "2–1" odds. Although the process of deductive reasoning starting from assumed certain principles can produce certain conclusions, reasoning by analogy starting from assumptions can only produce varying degrees of confidence. Although S1 eventually used more formal methods in his solution, his starting point in attacking this problem was to generate an analogy. His subsequent work served to confirm that the wider spring will deform more. This supports the view that spontaneous analogies can be used by even the most sophisticated experts during problem solving.

Other subjects in the study had had less experience with springs than S1, and for them, the qualitative question of whether the wide spring stretches more was much

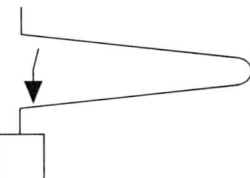

Fig. 2.2 Hairpin analogous case

more challenging, and occupied the entire session. Therefore in this chapter, I will be concerned only with the qualitative aspects of the problem. (Arguments about whether the deformation varies with the cube of the spring's width are discussed in Chapter 14.)

2.3.3   Evaluating the Analogy Relation

The term *analogy* is not used consistently across the various disciplines; in some instances it is used strictly to refer to a relationship, while in others it refers to the base case in the relationship. I am interested here in the process of analogical thinking as much as in the result of this process. Therefore, in this book, unless a more restricted meaning is clear from the context, I mean the term *analogy* to include all of the above; it includes the *target case* and the *analogous case*, together with their *analogy relation*. The target case in the preceding example was the spring, the analogous case was the hairpin, and the subject proposed an analogy relation of a partial equivalence between the spring and the hairpin with respect to the relationship between width and stretch. This subject appeared to have very *high confidence* in his belief that the large hairpin would stretch eight times as much as the small hairpin, that is, in his understanding of the analogous case. But he had only *moderate confidence* in his answer for the original target case of the spring. This leads us to infer that he has only moderate confidence in the equivalence of the hairpin to the spring problem, that is, in the validity of the analogy relation. This supports the view that there are *two different evaluation processes* involved here: one for the analogous case and one for the analogy relation. In other instances, subjects were observed to reject the validity of an analogy relation completely, that is, they decided that the analogous case was not similar enough to the original problem to draw any conclusions from it whatsoever. They then moved on to another analogy or another method.

2.4   Major Processes Used in Direct Analogical Inference

Observations of this kind suggest the hypothesis that the processes listed in Table 2.1 are fundamental in making an inference by analogy (Clement, 1982a, 1988). However, subjects can initiate P2, P3, and P4 in any order and go back and forth

Table 2.1 Major Processes in Direct Analogical Inference

P1. Generation of the analogous case
P2. Evaluation of the analogy relation
P3. Evaluation (and if necessary, development) of understanding of the analogous case
P4. Application via inference projection

between them. Thus, at least at this level, the subjects do not appear to use a simple, well-ordered procedure for controlling their solution processes. In the case of S1 there was an initial process of accessing or generating the idea of a hairpin (process P1). This was followed by his analyzing details about the hairpin as a cantilever beam, a standard model in physics (process P3). In the quotation, he also appears to briefly assess the validity of the analogy relation (process P2) in order to give his confidence estimate. He also indicates that one may be able to apply the finding about the width–stretch relationship from the hairpin to the spring (process P4).

2.4.1 Analogies from a Second Subject

I will next examine the solution of a subject, S3, who was an advanced Ph.D. candidate in computer science, and who had worked as an electrical engineer. The full protocols for difficult problems are quite long, therefore I present verbatim segments of protocols.

> 008 S3: (Reads Spring Problem)...Umm... I have no idea. Umm, and my first thought is that the length...of the coil spring being greater (traces circles in air with finger spiraling downward) and the strength of the metal being the same means that there's going to be kind of more leverage for bending [in the wider spring].
>
> 009 S3: And that therefore it's going to hang farther down. And that's pretty much strictly an intuition based on my familiarity with metal and with working with metal.... Let me just think through that.
>
> 010 S3: (Draws horizontal wires in Fig. 2.3) And my intuition about that is that if you took the same wire that was fastened on the left here [short horizontal wire] and doubled the length and hung some weight on it, that the same material uh, with some weight on it, would bend considerably further....
>
> 019 S3: It would seem that that means that um, that back in the original problem, the spring in picture 2 [the wider spring] is going to hang farther; it's going to be stretched more.
>
> 021 S3:and I have a confidence of about 75%....
>
> 022 S3:I have a great deal of confidence that Da [the displacement of the long wire] is greater than Db [the displacement of the short wire] in any case. I would say 100% confidence.

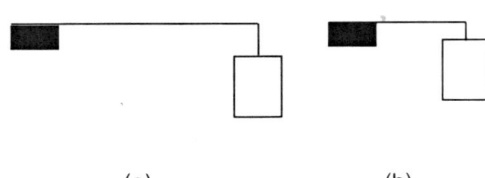

Fig. 2.3 Analogous case of bending rods   (a)   (b)

2.4.2 Analysis of Major Events in S3's Transcript

1. The subject refers to an "intuition" that predicts the larger spring will stretch farther (line 009).
2. He draws the picture of a new problem, a straight, horizontal wire bending under a weight (line 010). His discussion indicates that he has spontaneously generated a case he considers analogous to the spring. His belief is that, in this new case, the long wire would bend more than the short wire, again on the basis of "intuition." (Later, in line 022, he reports a 100% level of confidence in this prediction.) This is evidence that that, by the end of line 010, he has generated and comprehended the analogous case (steps P1 and P3 in Table 2.1).
3. He applies his findings from the new case to the original case (step P4) inferring that his analogy predicts the larger spring in the original problem will stretch further (line 019). However, he reports only 75% confidence in his answer to the original problem (line 021). A plausible explanation for this lack of confidence is that he was not fully satisfied with requirement P2 (evaluating the analogy relation between A and B).

It is interesting that step P4, *inference projection*, was initiated *before* the subject had completed step P2: the subject was able to make a tentative prediction about the original case before fully evaluating the analogy relation. The prediction at this stage is of the form "*If* the analogy is valid, then the wider spring will stretch more." Figure 2.4 shows another way to represent the four major processes. In this notation, the dotted squares represent poorly understood or uncertain predictions about

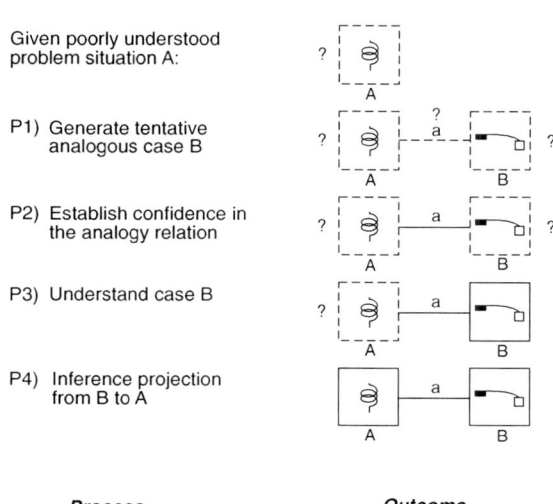

Given poorly understood problem situation A:

P1) Generate tentative analogous case B

P2) Establish confidence in the analogy relation

P3) Understand case B

P4) Inference projection from B to A

Process   Outcome

Fig. 2.4 Four major processes in a direct analogical inference

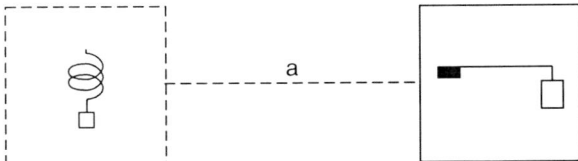

Fig. 2.5 Final status of S3's bending-rod analogy

cases. When a prediction for a case is strong, the dotted square is changed to a solid square to represent a confident prediction. Dotted and solid lines between squares represent unconfirmed and confirmed analogy *relations* between cases, respectively. Process P1 results in setting up a tentative analogy, but there are three question marks indicating uncertainties to be resolved before it becomes a useable analogy. Each of the subsequent steps then eliminates one of these question marks. The diagram shows an idealized example in which steps 2 and 3 are fully confirmed (assigned a high level of confidence). Unlike the previous examples, the diagram depicts a situation where the analogy relation is questioned and confirmed first, followed by questioning and confirming a prediction for the analogous case; thus the order of steps P2 and P3 is not fixed. This leads to the original case A about springs being confidently predicted in step 4, as symbolized by the solid box around A. This can only happen when the previous three steps have all been completed. However, in a nonidealized situation with real problem solvers, the confidence in each of these steps may be only partially rather than fully confirmed.

The final state of the analogy at the end of the section of S3's protocol, above, is shown in the diagram of Fig. 2.5. A poorly understood conception of the spring is linked by analogy to a well-understood conception of the wire, with the analogy relation not yet confirmed. That is, even though the subject was sure that he understood how to make a prediction for the bending-rod situation, he was still unsure whether it was sufficiently similar to the spring to use to predict its behavior. Therefore, I describe him as having a tentative or unconfirmed analogy relation at this point.

2.5 Conclusion

In summary, with regard to the four processes in Fig. 2.4, the most basic initial observation is that subjects appear to speak separately at times about each of these processes, suggesting that they exist as separate cognitive activities. Further evidence for these processes will be provided by the other case studies of analogy use examined in this book. The chapters of Section I and II that follow will examine each of the four processes in more detail, revealing a richer variety of processes than has been described previously.

Chapter 3
Methods Experts Use to Generate Analogies*

3.1 Introduction

The first step in using a spontaneous analogy is to generate the analogous case. By "generation" here I mean accessing or constructing an analogous case and raising the question of whether there is a valid analogy relation between it and the target problem. This is often considered to be the most creative part and by some even an unconscious part of using an analogy, and therefore it may be the one which is least well understood. The classical view of analogy generation is that a related case is accessed in permanent memory by association and brought into conscious attention in working memory (Gentner, 1983; Holyoak and Thagard, 1989). The purpose of this chapter is to examine think-aloud evidence on spontaneous analogies generated by experts. I will present evidence that some analogies are indeed accessed by association but that many are generated via a transformation or via a principle and that the classical view misses certain methods for generating analogies, some of which are quite powerful. Thus this chapter focuses on subprocesses used for process P1 in Table 2.1

The first part of this chapter elaborates the definitions used in referring to analogies, and the criteria for recognizing them in transcripts in more detail than was possible in Chapter 2. A variety of analogies from the study of ten expert subjects solving the spring problem are then described. The number of significant analogies and examples of the different types of analogies generated, including several creative invented cases, are presented. Finally different methods of analogy generation are identified and discussed.

---

*Some segments of this chapter, along with Figs. 3.1 and 3.2 are reproduced with kind permission of The Cognitive Science Society as copyright holder from Clement, J. (1988). Observed methods for generating analogies in scientific problem solving. *Cognitive Science*, 12, 563–586.

3.2 Definitions of Basic Concepts and Observations

3.2.1 Definition of "Spontaneous Analogy"

In order to describe in more detail the frequency with which analogies were generated, one must begin with a more careful definition of the concept of a spontaneous analogy. In defining criteria for recognizing a "spontaneous analogy," it is desirable for the definition: (1) to include attempts to produce cases that are similar to but different from the original problem situation; (2) to include such attempts whether or not they ultimately yielded an answer to the problem; (3) to rule out trivial cases that involve only a surface similarity without a structural or functional similarity; and (4) where appropriate, to separate analogy generation from other problem-solving processes such as breaking a solution into independent parts or analyzing the problem in terms of a theoretical principle.

The following observation criteria were used to code for the generation of a *spontaneous analogy*: (1) the subject, without provocation, refers to another situation B where one or more features ordinarily assumed fixed in the original problem situation A are different, that is, the analogous case B violates a "fixed feature" of A (to be defined below); (2) the subject indicates that certain structural or functional relationships (as opposed to surface attributes alone) may be equivalent in A and B; and (3) the related case B is described at approximately the same level of abstraction as A.

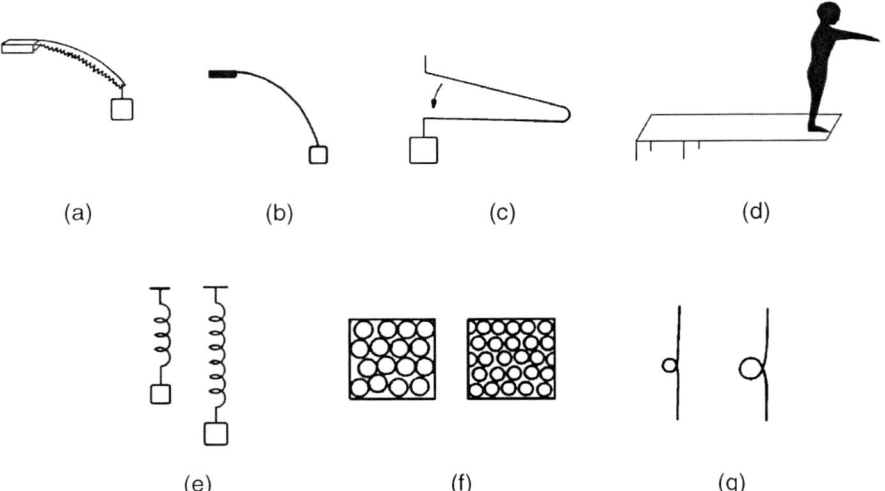

Fig. 3.1 Some analogous cases generated for the spring problem: (a) Longer sawblade bends more. (b) Longer rod bends more. (c) Longer hairpin bends more. (d) Longer diving board bends more. (e) Longer spring stretches more. (f) Foam rubber with larger air holes compresses more. (g) Larger kinks in a wire easier to remove

3.2 Definitions of Basic Concepts and Observations

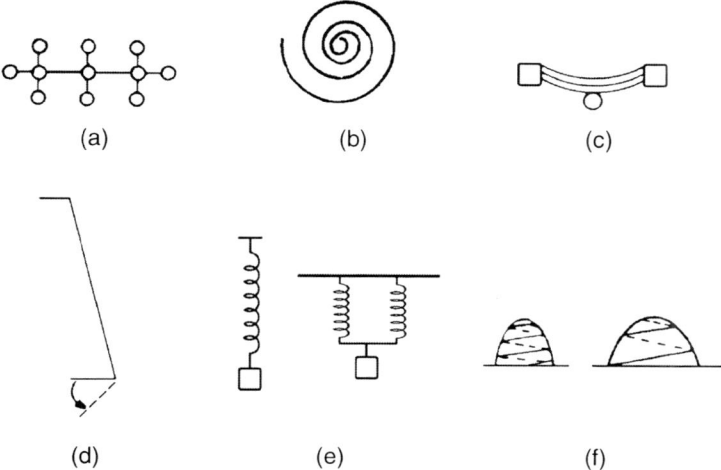

Fig. 3.2 Six more analogous cases generated for the spring problem: (a) Polyesters. (b) Spiral spring in two dimensions. (c) Car spring. (d) Longer rod twists more under same torque. (e) Parallel springs stretch less. (f) Car climbs farther per circuit on wider mountain, given the same incline angle (so wide spring stretches more)

We have already seen an example of spontaneous analogy generation in Chapter 2 where the subject attempted to relate the spring problem to the analogous problem of comparing long and short horizontal wires or rods bent by the same weight, as shown in Fig. 2.3. (The saw blade in Fig. 3.1a is another variation of this analogy from another subject.) Several of the subjects generated similar analogies and had a strong intuition that a long straight object would bend more than a short one. They reasoned that since the longer object would bend more, the wider spring would probably stretch more. This analogy in fact leads to the correct prediction, and provides a plausible initial justification for it. In some instances, a more complicated analogy was constructed (such as a spring with square coils) which led to a more accurate justification of the answer.

As used here, *fixed features* are those features of the problem situation that are commonly assumed to be given which are not subject to change; and *problem variables* are features that are assumed to be changeable or manipulable. Two aspects that are assumed to be fixed features in the spring problem are the equal thickness of the wire in the two springs and the helical shape of the springs. Aspects that are assumed to be problem variables are coil diameter and amount of stretch. Effectively, the subject's assumptions about which aspects of the situation are fixed and which are variables determine a stable context that affects the problem representation within which he or she works on the problem. Considering the problem of a horizontal rod, then, represents a change in what was originally a fixed feature (the shape of the spring) in the subject's initial comprehension of the problem. Thus the bending rod can be treated as an analogous case. An analogy, then, changes the problem representation being considered.

The above definition excludes several types of related cases that were not counted as analogous. First, when subjects used a simple partition such as looking at a single coil of the spring, it was not counted as an analogy if it consisted simply of thinking about a part of the original system (without changing the shape or other characteristics of the parts). Second, the indication of a mere surface similarity, such as one subject's comment that the drawing of springs in the original problem "reminded him of eels," was not counted as an analogy. Third, certain extreme cases, such as considering a very narrow or very wide spring, were not counted as analogies, because width is considered to be a problem variable, not a fixed feature. Fourth, the use of the term "analogy" was confined to a related case B at approximately the same level of abstraction as A. This criterion rules out saying that a robin is analogous to a bird, or that a spring is analogous to the general notion of a harmonic oscillator. Thus, when one subject thought about the behavior of a door spring as a particular example of a helical spring, this was not counted as an analogy.

3.2.2 Observed Spontaneous Analogies

Instances of spontaneous analogies were coded from the transcripts and videotapes using the definition given above. In addition, an analogy was only classified as significant if it appeared to be part of a serious attempt to generate or evaluate a solution, and as nonsignificant if it was simply mentioned as an aside or commentary. As an example of a nonsignificant analogy, one subject was reminded of another problem he had seen involving the deflection of piano strings of different lengths, but apparently mentioned this as an aside without the intention of applying findings back to the spring problem. Since the primary focus here is on processes involved in *attempts* to use analogies, the significance of an analogy did not depend on whether the solution generated was correct. Two independent coders analyzed transcripts for 80 related cases not identical to the original problem, with 84% agreement on identifying significant analogies, and with resolution by concensus. The results were as follows (Clement, 1988).

Number of subjects	10
Total number of spontaneous analogies generated	38
Total number of significant analogies generated	31
Number of subjects generating at least one analogy	8
Number of subjects generating a significant analogy	7

Thirty-one of the analogies were significant according to the criterion above, and a number of these are illustrated in Figs. 3.1–3.3. The 31 significant analogies include three generated by the Nobel laureate discussed in Chapter 2. The most common species of analogy was the bending rod and variations thereof, such as a bending saw blade, a bending wire, and a diving board. Six of the subjects generated an analogy of this type. Thus the number of significant analogies for this problem was quite high, indicating that analogical reasoning is a process that is available to many expert subjects.

3.2 Definitions of Basic Concepts and Observations

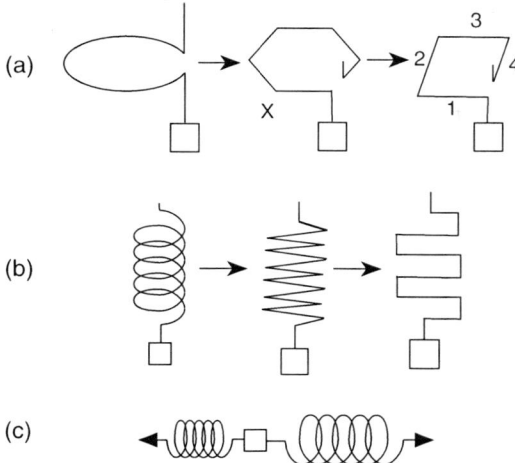

Fig. 3.3 Three novel analogous cases generated for the spring problem: (a) Circular, hexagonal, and square coils (b) Two-dimensional zigzag spring and modified zigzag with stiff joints. (c) Pitting the wide spring against the narrow spring

3.2.3 Analogy Generation Methods

Analysis of the transcripts indicated that there were at least three types of analogy generation methods. "Generation method" here refers to the way in which the analogous case B first comes to the attention of the subject during the solution. Examples of each type are discussed below.

3.2.3.1 Generation from a Formal Principle

A plausible hypothesis to explain how analogies are generated in science derives from the situation where a single equation or formal abstract principle (such as conservation of energy) applies to two or more different contexts. This suggests that analogies may be formed by first recognizing that the original problem situation, A, is an example of an established equation or principle, P, as shown in Fig. 3.4a. The analogous situation, B, is then retrieved or generated as a second example of principle P. For example, after S1 referred to the fact that bending is proportional to the cube of the length in the engineer's model of a cantilever beam, he immediately thought about a person standing on the end of a diving board (an example of this principle). If this turns out to be the main method used by subjects, it will support the hypothesis that analogy generation can be reduced to the processes of assimilation by a formal principle, followed by accessing an example of the principle.

Fig. 3.4 Three types of analogy generation: (a) Generation via a principle. (b) Generation via a transformation. (c) Generation via an association

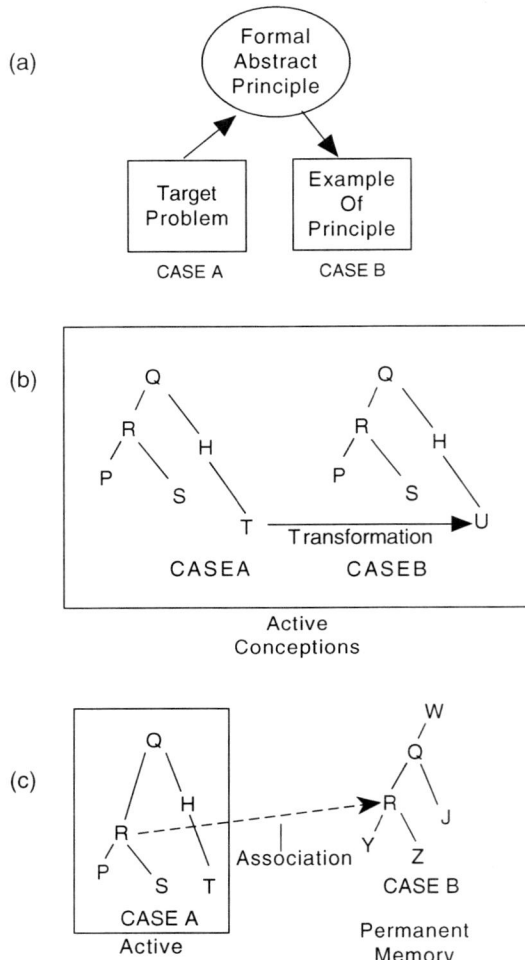

3.2.3.2 Generation via a Transformation

This occurs when a subject creates an analogous situation B by modifying the original situation A and thereby changing one or more features that were previously assumed to be fixed. In these instances there is no mention of a formal principle or equation. Consider the following example from subject S9:

> 041 S: I'm going to unroll these things [the two springs] and see if that helps my intuition any. Um. if I essentially, uh, uncoil or project the spring into a wire...the wire will actually go from here to here (draws horizontal line.) That's if I actually unroll the wire.

3.2 Definitions of Basic Concepts and Observations

The subject proceeds to consider the effects of hanging weights on the ends of long and short horizontal wires. Unrolling the spring into a straight wire is an example of a transformation. It is hypothesized that such a transformation occurs when the subject focuses on an internal representation of the problem situation A and modifies one or more aspects of it to change it into a representation of situation B, as shown in Fig. 3.4b. If the changed elements are not causally important to the behavior in question, such a transformation should produce an analogous case.

The transformation cited above involved a single continuous action but other examples can involve discrete acts of modification. S2 uses a discrete transformation to generate the idea of a square spring early on in his solution, while debating about whether a horizontal bending rod (which has a changing slope) works in the same way as the stretched spring (which has a constant slope)

> 023 S: I still don't see why coiling the spring [from a horizontal rod] should make any difference. ... Why does it have to be a [circular] coil? Surely you could coil a spring in squares, let's say, and it...would still behave more or less the same.

3.2.3.3 Generation via an Association

In contrast to generating an analogy via a transformation, the subject generating an analogy via an association is "reminded" of an analogous case B in memory, rather than transforming A into B. Such an analogous case may differ in many ways from the original problem but still have important features in common with the original situation. For example, S2 produced evidence for several analogies generated via an association in his protocol when he said: "I feel as though I'm reasoning in circles and I think I'll make a deliberate effort to break out of the circle somehow. ... What else stretches?. ... Like rubber bands, molecules, polyesters." Intuitively, it is as if the subject were "letting his mind wander" in a divergent process that allows him to retrieve similar situations. However, the focus on the concept of stretching here appears to play a role in constraining and guiding the activation process.

In another example, subject S6 compared the wide and narrow springs with two blocks of foam rubber, one made with large air bubbles and one made with small bubbles in the foam, respectively (Fig. 3.1f). He had a strong intuition that the foam with large air bubbles would be easier to compress. Another subject, S5, examined the relationships between coil width, coiling angle, and wire length by thinking about mountain roads winding up narrow and wide mountains (Fig. 3.2f).

Comparing the Generation Processes. As shown in Fig. 3.4c, the fact that cases generated by association differ in many ways from the problem situation suggests that an established schema B is being activated associatively in permanent memory, as opposed to being constructed via a transformation of A in working memory. A stretched rubber band, for example, does not appear to be a construction created by modifying the original spring situation; rather, it appears to be a familiar idea that has been activated as a whole. Thus associative analogies would tend to be more "distant" from the original situation conceptually than those produced by transformations in the sense that they share fewer features with A.

It may be difficult for some to see how generating an analogy via a principle could aid in problem solving, since if the principle is already known, there should be no need for an analogy. However, if the subject is unsure of the applicability of the principle to the target problem, then an analogy might help to evaluate this. In the use of both analogies and principles, evaluation of appropriateness is all important.[1]

Figure 3.4c is undoubtedly oversimplified, since it portrays an association as a single connection, whereas in some cases a much more complicated process of weighted activation from multiple sources may be involved. Similarly, Fig. 3.4b gives the impression that a transformation is always a simple, discrete replacement, whereas in the case of spatial transformations of entire shapes (such as "unbending") the process may be a much more distributed and continuous one: Given the present data, it seems premature to make the assumption (as is done in most of the existing literature) that activation caused by association to discrete features is the only method for initiating analogical reasoning processes.

3.2.4 Frequency of Different Analogy Generation Methods

3.2.4.1 Observation Criteria and Results

The 31 significant analogies in the ten solutions to the spring problem were classified according to their method of generation. Observation criteria used to provide evidence that a certain generation method was used are given below, along with the number of analogies in that category. Generation methods were coded jointly by a team of two transcript analysts until a consensus was reached.

Generation via a Formal Principle

Number of Significant Analogies 1

Characteristics used as indicators of an analogy generated via a principle were: (1) the subject refers to an abstract formal principle (mathematical or verbal) near the first reference to he analogous case B; or (2) the subject may refer to case B as an "example" of a principle.

---

[1] Two of the four steps involved in using an analogy discussed in Chapter 2 were: (P1) access or create an analogous case and generate a tentative analogy relation between it and the original problem; and (P3) understand the analogous case B. In stating that generation via a transformation takes place in working memory rather than accessing permanent memory, I am referring to step (P1), not step (P3). However, in all cases one way (P3) can be achieved is by accessing other familiar schemas in permanent memory which can interpret or analyze B.

3.2 Definitions of Basic Concepts and Observations

Generation via a Transformation

Number of Significant Analogies: 18

Indicators, in order of importance, were: (1) the subject refers to modifying an aspect of situation A to create situation B; (2) the subject states that B is an invented situation he has not encountered before; (3) the novelty of the analogous case suggests that it has just been invented; or (4) there exist a small number of transformations which can change A into B since the analogous case is not different in many ways from the original problem.

Generation via an Association

Number of Significant Analogies: 8

Indicators, in order of importance, were: (1) the subject mentions "being reminded of" or "remembering" case B; (2) B is different in many ways from the original problem; (3) the subject refers to B as a "familiar" situation; (4) B is a situation which obviously should be familiar to S (but may not necessarily be well understood by S).

Method Unclear

Number of Significant Analogies: 4

An analogous case was placed in the category "method unclear" when there was not enough data in the protocol to make a confident classification of the generation method.

Note that the largest number of significant analogies was *generated via a transformation* and that evidence was observed for generation via a principle in only one case.[2]

3.2.4.2 Novel Cases

There were five analogous cases observed that were clearly novel, shown in Fig. 3.3. (Novelty is not a fourth type of analogy generation method but rather a descriptive characteristic. Each of these five cases was classified as having been generated via a transformation.) They include springs with polygonal coils, two-dimensional zigzag springs, and an experiment where the subject pits the narrow spring against

---

[2] At one point consideration was given to splitting the generated via a transformation category into two parts: those cases generated by a simple modification or transformation of the original

the wide spring by attaching them to opposite sides of the weight along a simple line. The significance of these cases is that their novelty suggests that they have been invented rather than retrieved directly from memory. They are thought experiments in the broad sense of being situations where the subject attempts to predict the behavior of a new system without making new empirical observations.

3.2.5 *Summary of Observations with Respect to Analogy Generation*

In summary, spontaneously generated analogies were observed to play a significant role in the problem solutions of scientifically trained subjects. Generation via a transformation and via an association were the two primary analogy generation methods for which evidence was observed. Evidence for analogies generated via a formal principle occurred only rarely. This result certainly does not rule out the possibility that the latter method may be used in scientific problem solving, but it does suggest that it may not be the most common method for generating analogies, and that the other two methods may play a significant role. In addition, several novel analogous cases were generated that can be described as invented thought experiments.

3.3 Discussion

3.3.1 *The Presence of Analogies in the Solutions*

3.3.1.1 Analogy Generation as a "Horizontal" Change in Representation

From the point of view of problem-solving theory, an analogy can be said to involve a shift in the subject's problem representation. One way, then, to view analogy generation is as a meta-operator which operates on the initial problem representation rather than within it. However, it is a shift of a special kind. Other instances of shifts in problem representation can occur when the subject engages in abstract planning or in using symbolic representations, such as equations. However, in the latter two instances the subject moves "vertically" to a more abstract representation whereas in moving to an analogous case, the subject moves "horizontally" to

---

problem A; and those *constructed* by combining and assembling several schemas into one mechanism (items in Fig. 3.3 were among the candidates for the latter.) It might prove theoretically useful to distinguish the latter process, but this proved difficult at an observational level for this data base since all of the cases in question resembled the spring in some way. Therefore only the single category (which might be more aptly labeled "transformation or construction") was used.

another problem representation at roughly the same level of abstraction. Using an analogy can be thought of as the most creative of these three strategies in the sense that one is shifting one's attention to a different problem, not just to an abstract version of the same problem.

3.3.1.2 Developing Useful Boundaries for the Concept of "Spontaneous Analogy"

This idea of a horizontal change in representation leads to another motive for the definition of spontaneous analogy presented earlier. The definition is consistent with the idea that analogy generation is a creative and divergent process. The condition that the analogous case be one where "features ordinarily assumed fixed in the original situation are different" means that the subject must somehow break away from the original problem and shift his or her attention to a significantly different problem. This may be difficult for some people to do, probably because of the difficulty involved in breaking set – breaking out of the assumptions built up in considering the original problem.

To some, analogies such as a zigzag spring or a hexagonal coil (Fig. 3.3a) may seem too similar to the original spring to be counted as "real" analogies. Instead they might be seen as what Gentner terms a "literal similarity." The important issues here are: "What is the form of the basic reasoning patterns being used?" and, "What are the most useful and fundamental distinctions to emphasize in constructing definitions for terms like 'analogy'?" Certainly much data have been collected on problem solving where no spontaneous analogies occur. What seems to distinguish spontaneous analogies when they happen, more than anything else, is the fact that the subject is somehow bold enough to break away from the previous assumptions about the problem context. Just because an analogous case appears to be "close" to the original problem from hindsight does not mean that the assumption- breaking act of generating it was easy, by any means.

For example, the hexagonal coil case cited earlier is quite close in shape to the circular coil case, and yet it was only generated after many frustrating attempts to develop other analogies. Furthermore, its "closeness" does not imply "weakness." It is a powerful idea that led to a genuine scientific insight. As we shall see in Chapter 6, it was generated by only one of the ten subjects and was used by this subject to discover the major mechanism underlying resistance to stretching in a helical spring. Its identification in the hexagonal coil constitutes a scientific insight involving the discovery of a new variable and the discovery of a new causal relation in the system Clement (1981).

Thus the action of "considering a situation B which violates one or more fixed features of A" is taken as central to the definition of a spontaneous analogy. I consider this a more important criterion than requiring case B to have many surface features that are different from A's features, and so cases like the hexagonal coil are included as examples of analogies. Such "close" analogies appear to be one of the most fruitful and powerful types of analogies observed. The definition of analogy

3.3.2 Generation Methods and Invention

3.3.2.1 Formal vs. Informal Methods

Of the three methods, generation via a principle is a less direct method where thinking of the principle serves as an intermediate step on the way to producing the analogous case. If this had been the only method used, it would argue that the analogy generation process reduces to the process of assimilation of the problem by an abstract principle followed by accessing a familiar example of the principle. Of the three generation methods, it is also the most formal. However, generation via a principle was observed in only one case. The fact that less formal methods were observed in all but one case provides further support for the idea that experts can use nonformal as well as formal methods in solving problems.

The protocols indicated that more analogies were generated via a transformation than via an association. Although an association process is usually cited as the first step in using an analogy and as an important source of creativity in scientific problem solving, it may be that transformation processes are just as important, if not more important, in scientific work.

3.3.2.2 Invented Analogies

The method of generation via a transformation (or constructed) is of interest because of its potential for creating new cases. When we ask the general question of what it means to think of an analogous case, the standard view is that the analogous case is a familiar knowledge structure residing in memory which is at some point activated or retrieved as being related to the current problem. However, in using this kind of model, it is difficult to account for the production of the four novel cases shown in Fig. 3.3. The occurrence of these novel cases supports the hypothesis that some analogous cases are actually invented, not retrieved or reconstructed from memory. That is, in addition to creating a new analogy relation between cases A and B (which is assumed to occur in all of the spontaneous analogies discussed in this book) the subject also creates the analogous case B itself.

Inventing a novel analogous case is something like inventing a new machine or composing a piece of music in that some very major aspects of the case have never been experienced by the subject before. As in composition, although individual elements used in the invention of a novel analogous case originate in permanent memory, it makes little sense to say that the case as a whole was retrieved from memory, since the case has never been in mind before. It makes more sense to say that this kind of analogous case was *created* (see Clement, 1988).

3.3.3 Summary

In summary, the study described in this chapter provides evidence for some experts using spontaneous analogies to a significant degree during problem solving. Scanning the collection of analogous cases produced in Figs. 3.1–3.3, one is impressed by the sheer variety of possible approaches generated for the simply stated spring problem. In my mind, Figs. 3.1–3.3 represents a kind of "botanical garden" of species of problem solutions. In subsequent chapters I will try to dissect some of these species in more detail. Experts are unlikely to produce such a variety of forms when given a standard textbook problem for which they have preestablished procedures of solution. But when given an unfamiliar problem for which no standard method is available, their behavior becomes much more flexible and inventive, and new forms can evolve.

In addition, rather than one method, evidence was found for at least three different methods of analogy generation in the protocols: generation via a principle (1 case), generation via an association (8 cases), and generation via a transformation (18 cases). Although the mechanism underlying analogy generation is usually described as an association process, transformation processes, where the subject modifies or transforms some aspect of the original problem, may be just as important if not more important. In contrast to the usual view of an analogous case as already residing in memory, several of the analogous cases were quite novel, indicating that they were newly invented. This changes one's view of analogy from a process that always accesses knowledge of familiar situations to one that is capable of creating new situations (Clement, 1988). The presence of inventive processes would seem to be necessary in order to explain the emergence of novel analogue models in science. This theme will be developed further in Chapter 6.

The wide variety of cases produced by the subjects described in this chapter is symptomatic of the creative possibilities in the processes of advanced problem solvers and reflects the divergent side of expert thought. This can be a powerful mode of thinking when combined with convergent, critical processes for evaluating validity, and those processes are the topic of the next chapter.

Chapter 4
Methods Experts Use to Evaluate an Analogy Relation*

4.1 The Importance of Establishing the Validity of an Analogy Relation

We have seen that expert subjects can generate and use analogies spontaneously during problem solving. However, since not all analogies are valid, it is important for the subject to have a way to evaluate their validity. In a solution to the problem of whether a wide spring stretches more than a narrow spring, S2 generated the following analogy:

> 039 S2: a spring that's twice as long…stretches more…now if this is the same as a spring that's twice as wide, then that [the wide spring] should stretch more [than the narrow spring]…uhhhh, but IS it the same as a spring that's twice as wide?

This last question is the topic of this chapter: how can a subject tell whether a proposed analogy is trustworthy? That is, how can one evaluate the soundness or validity of the analogy relation between an analogous case and a target? If one cannot establish confidence in the validity of the analogy relation, then the analogy is useless. In fact, the subject above was not able to confirm the validity of this analogy in his protocol and was not able to make confident inferences from the analogy. "Validity" is used here in a sense outside the context of deductive certainty. Since conclusions reached by analogy are viewed as always having a certainty level of less than 100%, establishing validity here means "raising confidence in the appropriateness of the analogy relation to a high level." One view of the validity evaluation process described by Forbus et al. (1997) is based on the identification of structural correspondences through the mapping of discrete symbolic relations identified in the source and target.

In this chapter and Chapter 17, I will present some evidence from expert protocols, which seems to fit the above hypothesis: experts do at times refer to names of features that correspond in the source and the target. But I will also present evidence for other processes that can be involved in validity evaluation that tend not to be

---

*Some portions of this chapter are based on findings reported in Clement, J. (1998). Expert novice similarities and instruction using analogies. *International Journal of Science Education, 20*(10), 1271–1286.

discussed in the literature. In this chapter, I discuss one of these: analogical bridging. Several examples of analogical bridging will be presented from expert solutions to different problems, followed by two examples from the history of science. We shall see that bridging analogies can be quite imaginative: as in Chapter 2, we are confronted here with evidence that experts are not just followers of simple algorithms. Rather they are capable of inventing new forms and representations that can lead to novel and imaginative solutions.

4.2 Examples from Case Studies

4.2.1 Evaluating Analogies for the Sisyphus Problem

The "Wheel Problem" also called the "Sisyphus Problem" below, illustrated in Fig. 4.1, poses a question about whether one can exert a more effective uphill force parallel to the slope at the top of a wheel or at the level of the axle (e.g. as in pushing on the wheel of a covered wagon).

Wheel or "Sisyphus" Problem You are given the task of rolling a heavy wheel up a hill. Does it take more, less, or the same amount of force to roll the wheel when you push at x, rather than at y?

Assume that you apply a force parallel to the slope at one of the two points shown. And that there are no problems with positioning or gripping the wheel. Assume that the wheel can be rolled without slipping by pushing it at either point.

4.2.1.1 Subjects

The expert subjects discussed in this book solved a variety of problems in their interviews. Because their solution times on different problems had a very large range (e.g. from 6.4 to 52 min on the spring problem), not all subjects could be scheduled to work on all of the same problems in the time allotted. The wheel problem was given to seven expert subjects who were advanced doctoral students or professors in technical fields. Subjects were given instructions to solve the problem "in any way that you can," and no suggestions were made to encourage the use of analogies or any other method.

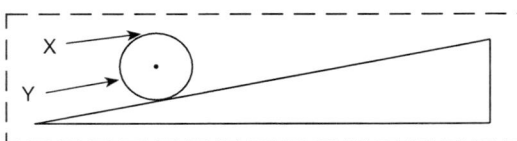

Fig. 4.1 Wheel problem

4.2.1.2 Case Study of a Solution

Subject S2 compared the wheel to the analogous case of raising a heavy 5 ft long pole or lever hinged to the hill (Fig. 4.2b). He reasoned that pushing at the point higher up on the lever would require less force. If the lever's fulcrum is the hinge where it is attached to the hill, one should have more leverage for raising the lever by pushing at the top as opposed to the middle, and therefore it should take less force to push at the top. (This can be confirmed by thinking about the extreme case of pushing at the bottom of the lever where, intuitively, it should take much more force to raise the lever.)

Having made a confident prediction for the lever, he then made an inference by analogy that the wheel would be easier to push at the top (the correct answer). However, the subject seriously questioned whether there was a valid analogy relation between the wheel and the lever. (As with the spring problem I will refer to the lever situation as the *analogous case* and to the possible structural similarity relationship between the lever and the wheel as the *analogy relation*.) Where is the "fulcrum" for the wheel? Is it at the center or at the bottom? S2 leans toward the view that the fulcrum corresponds to the bottom of the wheel (correctly). But can one really view the wheel as a lever, given that the "fulcrum" at the bottom of the wheel is always moving and never fixed? This is the question of whether the analogy relation is valid.

As an initial example of mapping discrete symbolic features, S2 had found a potential mismatch between his descriptions of the stationary fulcrum at the bottom of the lever and the moving fulcrum at the point of contact of the wheel. This led him to doubt the analogy. He apparently felt that this was an important relation that should probably be the same in the base and the target for the analogy to be valid. If we focus on his linguistic description of these relations, we can think of this as an example of mapping discrete symbolic features as one method for evaluating the validity of analogies.

4.2.2 Bridging Analogies

In order to evaluate the validity of the lever analogy, S2 generated an intermediate analogy or "bridging" case in the form of the spoked wheel without a rim shown in Fig. 4.3C. This allows one to see the wheel as a collection of many levers. One can

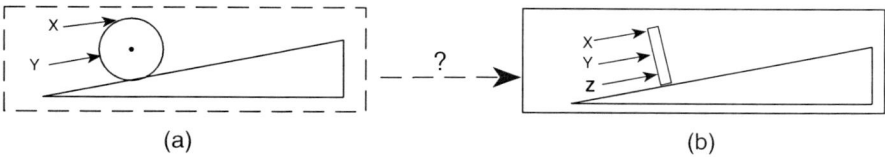

Fig. 4.2 Analogous case of lever (b)

think of it as an intermediate case that is "in between" the source and the target and that shares features with both. This bridging analogy raised the subject's confidence in the validity of the original analogy by reducing the subject's concern about the moving fulcrum issue.

An initial rough hypothesis for why this method works is that it is easier to evaluate a "close" analogy than a "distant" one. The bridge divides the analogy into two small steps which are easier to evaluate than one large step, as shown in Fig. 4.3. It is easier to see that the real wheel should behave like a rimless spoked wheel, and that the rimless spoked wheel should behave like a lever, than to make this inference in one step. This is an example of a bridging case used as a method for evaluating – and in this case confirming – the validity of an analogy relation.

We can represent this process using the notation of dotted boxes and relations developed in Chapter 2. In Fig. 4.2, the subject has become quite confident about his prediction for the analogous case of the heavy lever, so the box around this situation is shown as solid. He is trying to transfer his understanding of the lever to the case of the wheel, but he is unsure of the validity of this analogical relationship and this tentative but unconfirmed relation is shown as the dotted line and question mark between the cases. The bridging case of the spoked wheel without a rim shown in Fig. 4.3 then provides a pathway that does allow him to transfer his understanding to the wheel in two steps, from the well-understood base case B to the bridging case C and from there to the original target problem A.

4.2.3 A Pulley as an Analogy for the Wheel

Subject S7 dealt with the wheel in a different way. He first thought about the extreme case of rolling the wheel up an extremely steep hill that was almost vertical and was trying to decide intuitively whether it would be harder to push on the edge of the wheel or on the middle. While thinking about this extreme case he then goes on to propose an innovative analogy. He attempts to recast the problem in the rather

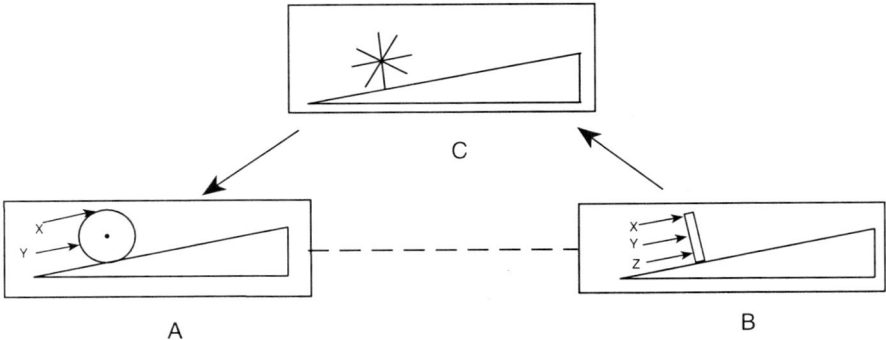

Fig. 4.3 Bridging analogy for the wheel problem: spoked wheel without a rim (C)

4.2 Examples from Case Studies

different context of a pulley system, as shown in Fig. 4.4B. This allows him to predict that it would be easier to push the wheel at x, since he knows that the pulley would cut the required force in half. In the following section, he generates a bridge in order to help him to evaluate the analogy relation. He imagines the pulley laid on the ramp as shown in Fig. 4.5B. The bridge in this case takes the form of a rope tied to the wheel, as shown in Fig. 4.5C.

> 162 S: Assuming...we attach a rope.... Now it becomes more like the pulley problem which I was thinking before (draws Fig. 4.5C).

> 163 S: Seems a lot easier than getting down here behind it [at "Y"] and pushing. Why? because of that coupling pulley effect. It seems like it would be a lot easier to hold it here [at "X"] for a few minutes than it would be to get behind it or even to attach a rope here [at "Y"] and; yeah, my confidence here is much higher now, that it's right. [to push at "X"]

In line 163 above, we have evidence that the bridge has increased the subject's confidence in the pulley as a valid analogy for the problem situation. There is also evidence in line 162 that the bridge of the rope tied to the wheel is an intermediate case for the subject in that he says that it is more like the pulley problem than the original problem was. The subject may be worried about the fact that in the original problem one was applying a force only at one point of the wheel whereas in the pulley, the rope may be applying force to the wheel at every point where it touches

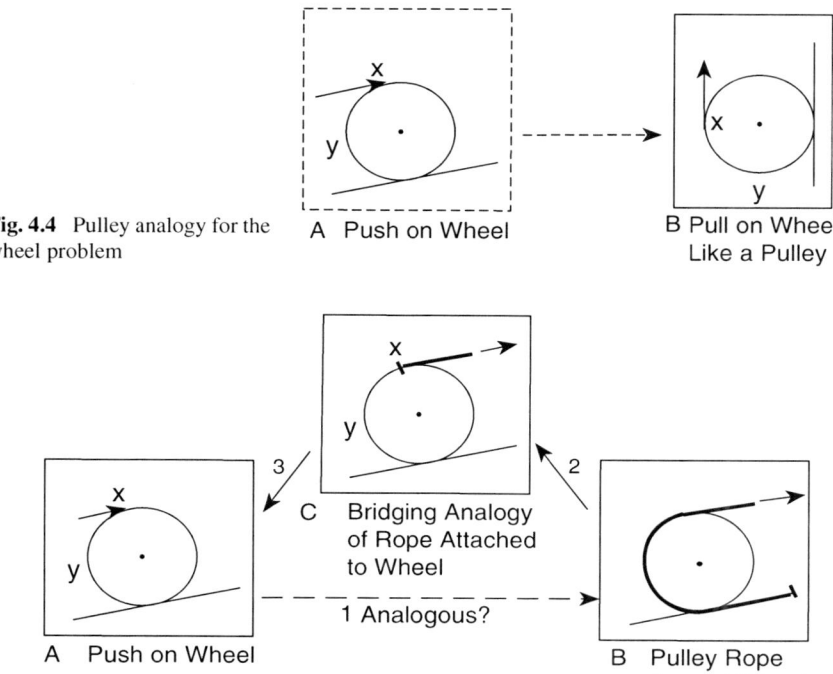

Fig. 4.4 Pulley analogy for the wheel problem A Push on Wheel   B Pull on Wheel, Like a Pulley

Fig. 4.5 Bridging analogy to confirm pulley analogy

the wheel on its circumference. The bridge appears to help him resolve this criticism.

Rather than letting the problem dictate the space in which he will operate, in the above sections the subject diverges into some innovative and unique spaces of his own construction. Methods like these contrast sharply with a more standard approach where the problem triggers a well-known procedure for solving the problem in a series of familiar steps.

Overall on the wheel problem 3 of the 7 subjects generated analogies. Two subjects generated bridging analogies (described above).

4.3 Analogy Evaluation in the Doughnut Problem

4.3.1 Bridging from Tori to Cylinders

Another example of a bridge occurred in a solution to the mathematics problem (shown in Fig. 4.6) of finding the volume of a torus without using calculus. This problem was given to eight subjects, six of whom referred to the analogy of a cylinder. More details on their solutions are given in Chapter 11.

Subject S13 at first thought that the answer might be the same as the answer to the analogous case of the volume of a cylinder. He guessed that the appropriate length for the cylinder would be equal to the central or "average" circumference of the torus ($2\pi (r_1 - r_2)$). However he was only "70% sure" of this. Thus he needed to evaluate the plausibility of an analogy between the torus problem and the base problem of finding the volume of a cylinder of a particular length.

A bridging analogy which allowed him to accomplish this evaluation is the case of the square-shaped torus shown in Fig. 4.7. By showing that the four sides of the square torus can be joined to form a single long cylinder with slanted ends, he

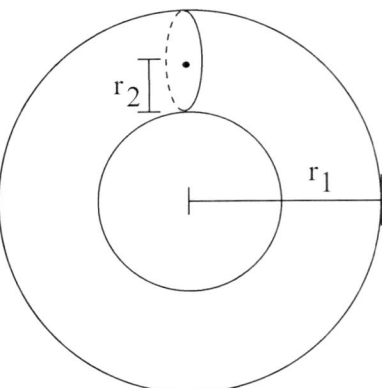

Fig. 4.6 Torus or "Doughnut" problem: compute the volume of the torus (doughnut) without taking an integral. Give an approximate answer if you cannot determine an exact one

Fig. 4.7 Bridging analogy: square torus

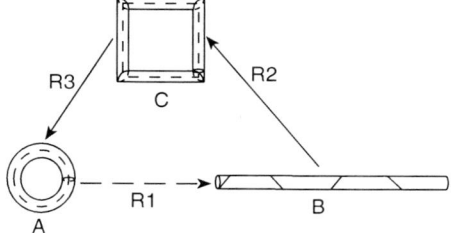

established the validity of the analogy between the cylinder and the square torus. He inferred that the volume of this long cylinder would be exactly equal to its central length times its circular cross section. From this he concluded that the appropriate length to use in the square torus was the average of its inner and outer perimeters, and this raised his confidence in his original solution to "85%." He was able to use the same reasoning for the hexagonal torus, and this raised his confidence to "100%" for the problem. This is an example of a *multiple bridge*. Thus the bridging cases of a square and hexagonal torus helped the subject change his original conjecture about the equivalence of the torus and the cylinder into a firm conviction.

Many sequences of mathematical models, especially in applications of the calculus, have the form shown in Fig. 4.7. In this view, mathematical limit arguments, which examine properties as one passes from an analyzable simpler base case to approaching the limit of the target case, are a sophisticated methods used to justify the intuitive validity of a series of bridging analogies between the base and the target situation.

4.4 Discussion of Findings and Connections to History of Science

4.4.1 Discussion of Findings on Bridging

Bridging is indicative of the recursive possibilities inherent in reasoning processes like the use of analogies. Since a bridging case is itself an analogous case, it can be described as an analogy used to evaluate a previous analogy (or, more precisely, as a second analogous case used to evaluate the analogy relation between a first analogous case and the original problem). We can summarize a view of bridging as one method for evaluating the analogy relation between a problem situation, A, and an analogous case, B, as follows:

1. The subject constructs a representation for an intermediate bridging situation C which shares features with both A and B.
2. The subject asks whether the analogy relation between A and C is valid with respect to certain important features.

3. The subject also asks whether the analogy relation between C and B is valid with respect to the same important features as in step 2.
4. If the subject can answer yes to both questions with high confidence, this constitutes evidence for the validity of the original analogy. (The subject may also use bridging recursively by bridging again between C and B or C and A as in the case of the square and the hexagonal tori.)

Here A being analogous to C and C being analogous to B means that A is analogous to B. We can refer to this type of inference as *analogical transitivity*. However, it should be noted that analogical transitivity is considered a form of plausible reasoning which does not carry the force of a logical implication.

As discussed in the previous chapters, it is clear that many of the bridges discussed here are novel situations that the subject is unlikely to have studied or worked with before. This suggests that they are invented representations that have been constructed by the subject, not simply retrieved from memory. In theories of scientific discovery, hypothesis generation is ordinarily seen as a more creative process than hypothesis evaluation. However, in the case of bridging we are faced with a creative, nonempirical evaluation method which generates novel constructions. Thus there appears to be evidence here for a type of "creative hypothesis evaluation" process.

4.4.2 Analogies and Bridges in the History of Science

4.4.2.1 A Bridge Used by Galileo's Predecessor, Benedetti

Legend has it that Galileo investigated the question of whether light bodies accelerate as rapidly as heavy bodies in an empirical manner by dropping objects from the tower of Pisa, but this legend has come under doubt. However, it is known that Galileo and his predecessor, Benedetti, did use thought experiments like the following one to argue their side in this issue. Figure 4.8A shows two equal objects of one unit each being dropped while Fig. 4.8B shows a heavier object being dropped that is equal to the two smaller objects combined. According to Aristotle the one-unit objects will fall much more slowly than the larger object. Galileo claimed that they will reach the ground at nearly the same time. In saying this he was effectively proposing an analogy between cases A and B in Fig. 4.8 to the effect that each body falls according to the same rule irrespective of its weight.

A marvelous bridging case used to support this analogy is the case shown in Fig. 4.8C. The argument was first published by Benedetti (Drake and Drabkin, 1969) and a related argument was given by Galileo (1954). Imagine the two-unit objects in A to be connected by a thin line or gossamer thread. Does the mere addition of this tiny thread, which makes the two objects become one, cause their rate of fall to increase by a large amount? Because this is implausible, the bridge argues that A and B are indeed equivalent with respect to rates of fall. In this ingenious thought experiment, we again appear to have a form of creative rational hypothesis evaluation.

4.4 Discussion of Findings and Connections to History of Science

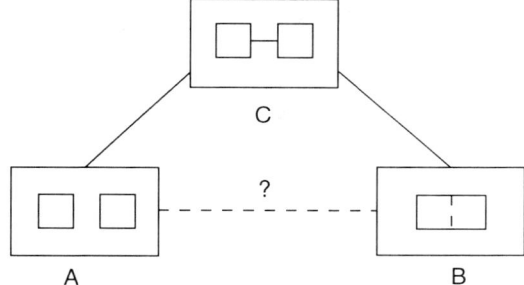

Fig. 4.8 Benedetti's thought experiment

4.4.2.2 A Case of Multiple Bridges Used by Newton

One of the most far-reaching scientific analogies of all time was proposed by Robert Hooke and Isaac Newton in the 17th century. They claimed that the moon is attracted by and falls toward the earth as it travels through its orbit just as an everyday object (such as an apple) does. To a modern physicist, this may seem more like an obvious fact than a creative analogy, but to advocate such an idea in Newton's time was not an obvious step at all. One has only to imagine the disbelief that would be produced by telling someone ignorant of science that the moon is falling. The horizontal dotted line in Fig. 4.9 shows the proposed analogy relation. Newton's conjecture was that the moon revolving around the earth and the apple falling both have the same causal mechanism. Newton appeared to use a multiple bridge shown in Fig. 4.9c to support this analogy in *The Principia*. A cannonball is fired farther and farther until it finally goes into orbit around the earth. Those cases lie between the case of the moon in orbit and a cannonball falling straight down under the influence of gravity. Assuming one attributes the vertical fall of a ball to gravity, the bridging cases help one see how the dropped object's motion and the motion of the moon can both be caused by the gravitational pull of the earth.

4.4.3 Beyond Bridging

An enigmatic aspect of bridging strategies is that they "create more work" – more analogy relations to evaluate – so how can they be advantageous? And it is difficult to explain their exact purpose or advantage over other more direct strategies. Perhaps they make the "gap" between cases smaller, but what does this mean exactly? We still lack a fully satisfying explanation for the usefulness of bridging analogies. In fact the presence of a series of many bridging cases in Newton's cannon experiment suggests the possibility of *smooth transition* from the vertical drop to the orbiting object. In the case of multiple bridges we are approaching what can be considered a third method of analogy evaluation, namely, finding a *conserving transformation*. Such a transformation changes case A into case B while conserving

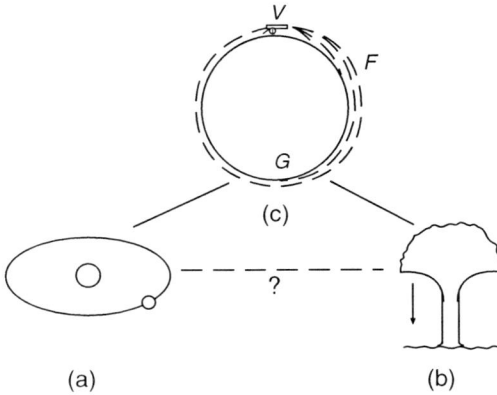

Fig. 4.9 Newton's cannon

important relationships that make A analogous to B. Thinking about continuously increasing the horizontal speed of launch in this case would constitute a conserving transformation since the major relationship of gravity causing the acceleration of the object remains unchanged and is conserved.

Protocol observations of conserving transformations and "dual simulations" as alternatives to traditional methods of analogy evaluation will be discussed in Chapter 17. There I will propose the hypothesis that these methods may not depend on the mapping of discrete symbolic features, and that bridging may be a way of making these methods more feasible. The spontaneous generation of a bridging analogy by a physics student will be discussed at the end of Chapter 19. The use of bridging analogies as an important teaching strategy for use in dealing with students' preconceptions in science will be discussed in Chapter 10.

4.5 Summary

Since not all analogies are valid, it is important for the subject to have a way to evaluate their validity. The ability to *evaluate* the validity of analogies appears to play as important a role in insightful problem solutions as the ability to *generate* analogies. Previous research has identified processes that involve the mapping of discrete features for evaluating analogy relations. An additional strategy identified here is the construction of bridging analogies (Clement, 1986). The bridging pattern was observed to occur in different problem contexts in both science and mathematics.

In constructing a bridge, the subject finds an intermediate case that is seen as "in between" the analogous case and the problem situation because it shares important aspects of both. Many of the bridges observed appeared to be novel inventions created by the subject. They can therefore be considered a kind of creative, nonempirical, hypothesis *evaluation* strategy, even though it is hypothesis *generation* strategies that are usually thought of as the creative side of scientific thought.

Chapter 5
Expert Methods for Developing an Understanding of the Analogous Case and Applying Findings

Chapter 2 outlined four major processes involved in analogical reasoning: generation, evaluation of the analogy relation, evaluation of understanding of the analogous case, and applying findings via inference projection, as illustrated in Fig. 2.4. As an advanced organizer, Table 5.1 summarizes the hypothesized subprocesses involved in each of these four processes. Processes in part (a) and (b) there have already been discussed. This chapter discusses subprocesses involved in the last two processes, understanding the analogous case, and applying findings.

5.1 Evaluating and Developing an Understanding of the Analogous Case

There are several plausible methods by which a subject might evaluate, and if necessary, develop their understanding of an analogous or source case, shown in Table 5.1c. In the simplest instances the subject may have direct knowledge about the source case that is immediately accessible and sufficient to understand it. In other instances, however, the understanding of the source case itself must be actively developed by the subject in order to make progress (Clement, 1981, 1982a; Burstein, 1983).

5.1.1 Direct Methods

The first two methods in Table 5.1c, factual knowledge and mental simulation, are the most direct ones if applicable. If the predictions they provide are adequate to the task, then no further development of the analogous case is necessary. For example, S2 was convinced that a long rod would bend more than a short one, based on an "intuition" accompanied by moving his hands as if he were bending rods with his hands (this will be discussed as an example of *mental simulation* in Chapter 13). For him, no further development of this analogous case was needed. A third method, application of a known principle, is exemplified by S1's analysis of the hairpin as a "cantilever beam" with known characteristics. Such methods are direct in the sense that they stay "within" the analogous case – the subject does not modify the case under consideration.

Table 5.1 Analogical reasoning processes

(a) Generation
    1. Via association
    2. Via transformation
    3. Via a principle

(b) Evaluation of the analogy relation
    1. Via mapping of discrete symbolic features
    2. Via a bridging analogy

(c) Evaluation (and if necessary, development) of understanding of the analogous case
    1. Direct methods
        Factual knowledge of solution
        Mental simulation
        Application of known principle(s)
    2. Indirect methods
        Break problem into parts
        Use extreme case
        Use extension analogy

(d) Application of findings, e.g. via direct inference of a prediction or transfer of a method of attack

Qualitative analysis techniques for understanding more complex cases have also been described, including chaining causes together in Driver (1983) and Clement (1979), and inferences about feedback loops, and control of rates of flow in Forbus (1984) and de Kleer and Brown (1983).

5.1.2 Indirect Methods

5.1.2.1 Breaking the Problem into Parts

Other methods are less direct. For example, the understanding of an analogous case might be developed via standard analysis techniques such as breaking the system into parts. S2, for example, analyzed single sides of his hexagonal coil analogy (see Chapter 6).

5.1.2.2 An Extreme Case Can be Used to Confirm the Source Analogue

I consider that an Extreme Case has been generated when, in order to facilitate reasoning about a situation A (the target), a situation E (the extreme case) is suggested, in which some feature from situation A has been maximized or minimized. In the rod example, S2 referred to an extreme version of his source analogue: the case of a very short rod.

5.1 Evaluating and Developing an Understanding of the Analogous Case

025 S: I have a strong intuition – a physical intuition that this [longer rod] will bend a lot more more than will. ... In fact, the intuition is confirmed by taking it to the limiting case; it becomes intuitively obvious to me that as one moves the weight closer and closer to the fulcrum that the thing will not bend at all...

This method served to further boost his confidence in his ability to predict the behavior of the system. It is used here as a method for evaluating and developing his understanding of the analogous case. A number of other extreme cases will be examined in upcoming chapters.

5.1.2.3 Extension Analogies

In still other cases the development of the analogous case may be facillitated by generating a second analogous case to aid in understanding the first analogy, as in the following excerpt from S3 working on the spring problem (I examined the first part of S3's solution in Chapter 2). At this point S3 had generated the analogy of a bending rod, which helped him predict the correct qualitative answer to the spring problem. S3 was hoping to go beyond this to predict from his analogy whether the radius/stretch relationship in the spring was linear or quadratic or cubic, but his understanding of the bending of the rod was not sufficient to give him a quantitative relationship between applied weight and bending in the rod. So he generated an "extension" analogy in order to help him understand the original rod analogy. The diagram in Fig. 5.1c shows this extension analogy in the form of two parallel pipes. The two pipes are fixed at the left side and held together in such a way that when the weight is applied to the right side, the upper pipe is stretched and the lower pipe is compressed. This allows one to model tension and compression in the corresponding upper and lower parts of the rod as a way of understanding the rod's resistance to bending. A simplified version of the analysis similar to one given by Galileo (1967) is as follows: one can imagine for an enlarged view of a bar, slicing it lengthwise by making a horizontal cut along the center of the bar. This transforms it into two thinner bars, one lying on top of the

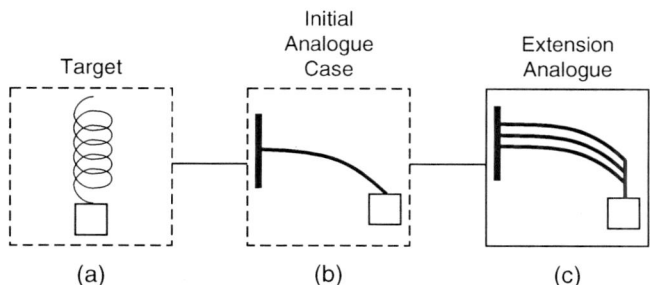

Fig. 5.1 Extension analogy of parallel pipes

other. With the bars glued together at each end but free to slide over one another elsewhere, when one applies weight on one end of this "sandwich," one sees that the bottom half will compress and the upper half will stretch. S3's analysis of his analogy with pipes was part of a similar attempt to model the bending rod in more detail and determine its length/deflection relationship so that this information could in turn be used in analyzing the spring.

Here the pipes analogy is effectively a second analogy used to understand the previous analogous case of the bending rod. I call this an extension analogy and take it to be an illustration of the recursive application of analogical reasoning as an approach to understanding an earlier analogous case (Clement, 1982a). It is similar to the bridging analogies discussed in Chapter 4 in that it is a second analogy used to help make an initial analogy successful. However, it is different than a bridging case in the following ways. Whereas a bridging case is used to evaluate the validity of the original analogy relation, an extension analogy is used to develop the understanding of the original analogous case. Also, whereas a bridging case falls "in between" a base and a target in terms of being similar to both, an extension analogy falls on the "other side" of the base from the target, as shown in Fig. 5.1c. Thus bridging analogies and extension analogies each lead to forming a chain of analogies, but each serves rather different purposes.

5.1.3 Summary: Developing Understanding of the Source Analogue

Table 5.1c summarizes plausible methods by which one's predictions and/or understanding for a source analogue could be developed. Of particular interest here are the last three more indirect methods: breaking the problem into parts, extreme cases, and generating an extension analogy. These illustrate the creative potential of expert reasoning.[1]

The present findings contrast with previous research which has seldom included development of the source analogue as a major process in analogical reasoning. Perhaps this difference is due to the present focus on spontaneous analogies, which allows one to observe analogous cases that are constructed by the subject. As mentioned above, previous studies have tended to concentrate on analogous cases that were presented to the subject or recalled from memory. It stands to reason that constructed analogies would more often require the process of developing understanding. The necessity to develop understanding of the analogous case also has

---

[1] In fact, Table 5.1c can be viewed as an entry point or top-level process for prediction and understanding problems in general, instead of being subservient to understanding analogous cases. This illustrates the potentially recursive nature of using an analogy to understand the target, since using a first analogy can have the subprocess of calling up this set of methods again in order to understand the analogous case. I will elaborate on these points in Chapter 16.

educational implications, because too often teachers make an unwarranted assumption that students have a sufficient understanding of a presented analogous case. This implication will be discussed in Chapter 10.

5.2 Inference Projection

It remains to discuss the fourth step in Table 5.1, by asking what is involved in applying findings from the source to the target. Answers to this question interact with our view of why analogies are useful at all, and I begin with a discussion of that question.

5.2.1 *Why Are Analogies Useful?*

An interesting characteristic of analogical reasoning lies in the paradox that by seeming to *move away* from a problem the subject can actually *come closer* to a solution. In order to use an analogy effectively one must be able to postpone working directly on the original problem and be willing to take an "investigatory side-trip" with the faith that it may pay off in the end. This is a risky thing to do (especially while being recorded); there is no guarantee the side-trip will make any contribution to the solution at all. Why do scientists bother to attempt analogies?

In the view of Ernest Nagel (1961), scientists use established analogies in the form of models in science (such as a "billiard ball" model for gases) in order to "make the unfamiliar familiar." This is one of the major functions of scientific models, in Nagel's view. Nagel is referring to established analogue models in science, whereas in the present study the analogous cases are usually based on familiar ideas from personal experience. But the function Nagel describes for established analogies in science could be taken as equivalent to the function of the analogies in this section in the following sense. The resolution of the "moving closer by moving away" paradox would seem to lie in the idea that humans often appear to *build up new knowledge by starting from old knowledge.* Moving closer to the answer by moving away from the problem via an analogy can work because one is moving to a more familiar area one knows more about, and one may then be able to *apply* part of this knowledge back in the original problem. Consideration of the analogies discussed in this section of the book suggests that the knowledge inferred from an analogy could be useful in at least three possible ways in that it may:

1. Predict an answer to a specific problem
2. Provide a suggested method of attack
3. Provide a principle that applies to the target

For example, S1's analogy of the hairpin (Chapter 2) seemed to imply both *inferring a prediction* and perhaps transfer *of a principle* because he mentions the possibility that the cubic relation between width and stretch will apply to the spring. Second, the "mountain roads" analogy (Fig. 3.2f) suggests a *method of attack* since it assumes that the roads descend at the same angle, suggesting that the subject examine whether the springs of different widths have the same helix angle under the same weight. Third, in Chapter 6, it will be argued that after generating the bending-rod analogy, S2 comes to take the bending seriously as a *possible model* for explaining what is happening to the spring wire, which goes beyond the function of simply inferring a prediction as an answer.

5.2.2 Data on Inference Projection

Our discussion from data will be short, because although sometimes we see data indicating an inference process, often we do not see data for an inference process that is separate from other processes such as evaluation of the analogy relation.

 This leads one to suspect that in many cases the process of confirming the analogy relation is where most of the work was done as preparation for inference, after which the act of inference projection itself was already completed or trivial. This makes sense in the case of a problem asking for a requested relation such as the relation of width to stretch in the spring. The subject's examination of the similarity of the source and the target in sufficient detail to evaluate an analogy relation could in some cases include a mapping of the implied solution to the target problem. Thus, the tentative proposal of an inferred prediction could occur early on in the solution process, even before the confirmation of the analogy relation or of the prediction for the base. But how can such an inference occur with any certainty until other confirmation processes are completed? The confusion here can be attributed to three possible choices for the meaning of the word "inference." First, it could refer to a tentatively proposed inference projected from the base to the target that yields the prediction asked for by the problem. Second, inference could mean establishing this particular correspondence with some confidence. And third, it could mean having confidence in the whole analogy – that is, confident transfer based on an overall assessment of the soundness of the analogy and understanding of the base. It appears that inference of the first and second types can sometimes take place early on in the development of an analogy, often during the process of confirming the analogy relation. Confident inferences of the third type can only happen after all other processes have been completed.

 If the inference process in the second sense above can be taken care of in conjunction with the evaluation of the analogy relation, should it be listed as a separate process at all? I believe it should because there is evidence in some cases that it can take place as a separate process. S2 spoke of it separately in line 39 above, apparently before he had evaluated the soundness of the analogy relation. And in other cases to be examined in this book, inference happens after evaluation as a separate consideration. But it is true that the inference process is difficult to study because it is often hidden by being intertwined with other processes. Since in this study I have tried to focus analysis on areas and levels that can be grounded in protocol data, I have devoted the least amount of time to an analysis of inference projection, simply because many subjects did not provide enough evidence on it.

5.2.2.1 Inference via an Inverse Transformation

Transfer via an inverse transformation can occur when an analogy is generated or confirmed via a transformation like unrolling the spring into a wire. Inference can then involve the process of using the inverse transformation to convert the answer for the base into the answer for the target. Rolling the straight wire back up to make the spring would be an example of such a process. This example is discussed in Chapter 17.

5.3 Section I Summary for Creative Analogy Generation

In Section I of the book so far, evidence has been presented indicating that scientifically trained individuals can generate analogies spontaneously during problem solving. Such methods are not often observed in expert solutions to standard lower-level textbook problems where more straightforward and familiar techniques can be used. But when given a problem like the spring problem, where most subjects have no preestablished, readymade procedures to apply, creative processes like analogy generation do come into play, and a wide variety of "species" of problem representations emerge, as indicated in Figs. 3.1–3.3.

Table 5.2 Roles played by analogy in expert thinking

Applying findings via inference projection of: an answer or prediction; a method of attack for problem solving; or an abstract principle
"Bridging analogy" used to evaluate a previous analogy relation (Chapter 4)
Extension analogy used to help develop understanding of previous analogous case

Drawing on the cases in the previous chapters, an initial list of roles that analogy can play in expert thinking is shown in Table 5.2. The second and third roles are non-traditional and not discussed in the classical literature on analogy. This list will be expanded in later chapters as other roles are identified in the case study examples. In particular, an additional role of a source analogue for developing a new explanatory model will be discussed in Chapter 14.

Table 5.1 summarizes the spontaneous analogical reasoning processes identified so far in this book. An attempt was made to propose processes that are tied to protocol observations. Previous work on analogical reasoning has been largely based on philosophical grounds, proposals for sufficient information processing strategies, or empirical studies of provoked rather than spontaneous analogies. These have emphasized the ideas of generation via an association, access to a retrieved analogous case which itself does not require development, evaluation via mapping, and application via direct inference. New observations in the present study suggest the presence a number of other important processes as well, including generation via transformations, generation of novel, invented analogous cases, efforts to improve or develop greater understanding of the analogous case, and the use of bridging analogies for evaluation of an analogy relation. (Additional methods for evaluating analogy relations will be discussed in Chapter 17.) This breakdown of subprocesses for analogical reasoning provides the beginning of a typology of plausible reasoning processes that will be expanded in future chapters.

Section II
Expert Model Construction and Scientific Insight

The mind is no more a 'windowless monad' with a 'pre-established harmony' (Leibniz, 1714) than it is an unharmonized monad with a picture window. More nearly is it a community of pretuned monads that come into harmonious action, with each other and with the world outside, through many glasses darkly.

R. Shepard (1984, p. 439)

Chapter 6
Case Study of Model Construction and Criticism in Expert Reasoning*

6.1 Issues Surrounding Theory Formation

The preceding section of the book has focused on an analysis of spontaneous analogical reasoning processes observed in experts. This provides a foundation for examining the more complex process of scientific theory formation and insight, and this chapter uses evidence from a think-aloud case study to examine whether analogies can play an important role in that process. I begin by reviewing some longstanding controversies in philosophy of science over the sources of new theories and insights in science. I will focus here mostly on "classic" work from 20th-century authors, and refer to more modern approaches later in the book.

Galileo's theory of motion, Faraday's concept of the magnetic field, Darwin's theory of natural selection, and Einstein's theory of relativity are commonly cited examples of creative achievements in science. Each is a major event in the history of scientific ideas, and in each case something very new emerged that affected the entire scientific community and subsequently affected civilization as a whole. Analyzing how such achievements take place is a worthwhile goal, but achieving this goal has unfortunately proven to be surprisingly difficult. In Darwin's case, for example, it is possible to argue that the theory of natural selection was built up gradually through a large number of detailed empirical observations. But on the other hand, it also is possible to argue that the theory was the result of a mental breakthrough well after the Beagle's voyage in the form of an insight that constituted a sudden reorganization of Darwin's ideas. An intermediate position is also possible. Thus, even with respect to specific historical examples, disagreement emerges as to the basic sources and pace of theory change in science.

At issue here is an important question concerning the nature of science. Cast in its most global and extreme form, the question is: "Does science change in an incremental manner, with a series of many small empirical observations inching it forward, or do occasional large breakthroughs occur in the mind of the scientist in

---

*Portions of this chapter are reproduced with kind permission of Springer Science and Business Media from: Clement, J. (1989) Learning via model construction and criticism: protocol evidence on sources of creativity in science. In J. Glover, R. Ronning, and C. Reynolds, *Handbook of creativity: Assessment, theory, and research*, New York: Plenum.

the absence of new data, each causing a great leap forward in the field?" One purpose of this chapter and the next is to determine whether the methodology of protocol analysis has the potential to illuminate some aspects of this question by using data from transcripts of scientists solving problems aloud. I will concentrate most on an example of a breakthrough episode in a thinking-aloud case study and discuss the senses in which it is and is not an example of a scientific insight. In particular, the case study is used to address elements of the following questions:

1. What processes are involved in the generation of a scientific theory? In particular, are theories always generated as inductions from data?
2. What role do analogies play in creative scientific thinking?
3. What is a scientific insight? Can one identify "insight events" in thinking-aloud protocols? Why do insights occur? Do they indicate that deep scientific thinking is more "revolutionary" than "evolutionary" in character?

I will attempt to show that empirical evidence from protocols can be collected which can speak to certain aspects of these questions. Discussion of questions 1 and 2 is initiated in Chapter 6, while question 3 is taken up in Chapter 7.

6.2 Background Questions from Philosophy of Science

6.2.1 The Source and Pace of Theory Change

6.2.1.1 Eurekaism vs. Accretionism

It is useful to separate out two major issues involved in the controversy over theory formation, the *pace* of scientific theory change (question 3 above) and the *source* of new theories (questions 1 and 2 above). With respect to the pace of theory change, one can contrast Eurekaist and accretionist positions. A Eurekaist claims that a theory can be changed at a very fast pace by an insight that reorganizes its structure. In its strongest form, Eurekaism is associated with sudden flashes of inspiration, possibly following a period of incubation, or where some ideas may form in and arrive suddenly from the unconscious mind.

An "accretionist" or incremental view of the pace of scientific theory change holds that a scientist gains knowledge in small pieces and puts them together deliberately at a slow and even pace. This process can lead to a smooth progression in the attainment of knowledge – an incremental "march of progress" without large-scale reorganizations.

The idea that analogies can be involved in hypothesis formation might be used to support a Eurekaist view of scientific discovery. If analogy generation is a fast, creative process, and if it is important in hypothesis formation, then it is an interesting candidate for a cognitive process underlying insight or Eureka events. However there are still real questions facing us about how central analogies are in discoveries and how fast the analogical reasoning process is. This issue will be examined more closely in Chapter 7.

6.2.1.2 Rationalism vs. Inductivism

A second major issue is the *source* of new theoretical knowledge. The question of the sources of and justification for new knowledge is a central point of controversy between the rationalist and empiricist traditions in Western thought. The rationalist tradition emphasizes the power of reasoning from prior knowledge and greatly values the consistency and beauty of the resulting theories. Reasoning power, coupled with the prior beliefs and intuitions of the learner are emphasized as sources of knowledge. On the other hand, the empiricist tradition emphasizes the importance of careful observation and greatly values the reliability of repeatable experimental procedures. Here the term induction will denote a process by which a more general principle is abstracted from a set of empirical observations as the source. I will use the term inductivism to refer to the belief that induction is the primary, if not exclusive, source of hypotheses in science. Stated most simply, scientists gradually gather facts, use inductive reasoning to organize them into general statements, and finally build up a pyramid of general empirical laws that summarize all of the gathered data. Theory-driven and data-driven approaches in artificial intelligence can to some extent be thought of as modern inheritors of the rationalist and inductivist viewpoints.

Although they refer to different issues, the Eurekaist vs. accretionist and rationalist vs. inductivist controversies are not independent historically, but tend to interact. Eurekaism tends to be associated with rationalism, while accretionism tends to be associated with inductivism. Thus it is sometimes useful to refer to an individual position as "rationalist–Eurekaist" or "inductivist–accretionist." A rationalist–Eurekaist view of theory change is associated with the idea that scientists at times must be very creative, whereas the inductivist–accretionist view suggests that scientists can make progress by relying on small changes without large creative breakthroughs. This simplified picture of two opposing camps can then be used as a starting point for introducing some important issues concerning the nature of science.

Writers on both sides of this controversy have tried to claim Darwin's theory of evolution as an example. Historically, inductivist–accretionists claimed that it was a prime example of the power of induction, as facts gathered by Darwin during the voyage of the Beagle were slowly pieced together into a grand theory. Rationalist–eurekaists claimed that Darwin had a sudden, crucial insight after reading Malthus' theory of human population constraints. But both of these positions run the risk of being oversimplified. As Gould (1980) puts it: "Inductivism reduces genius to dull, rote operations. Eurekaism grants it an inaccessible status more in the domain of intrinsic mystery than in a realm where we might understand and learn from it." The implied challenge here is to find a less simplistic view that helps to explain creative behavior in a nontrivial way. In this and the following chapter accounting for the data from the case study in this chapter leads to a more complex view of scientific discovery than any of the extreme Eurekaist, accretionist, rationalist, or inductivist positions can provide. In Chapter 7, I will also cite some historical studies of Darwin's insights which point to the same conclusion.

6.2.2 Philosophical Positions: Empiricism vs. Rationalism

I give a brief introductory outline here of how these two broad questions concerning the source and pace of scientific theories interacted with some of the major 20th-century philosophical positions on the nature of the scientific enterprise.

6.2.2.1 Empiricism

Prior to this century, empiricists focused on observation as the primary source of knowledge in science, and the 20th-century logical positivists built on their tradition by attempting to show that scientific knowledge could be grounded firmly in sense experience. In their view careful observations, and the assumptions of a common scientific observation language and the applicability of the laws of logic and probability, could provide science with knowledge of the utmost reliability, if not certainty. Although the logical positivists concentrated on issues surrounding the justification of theories rather than their origin, their empiricism also affected views of the origins of scientific knowledge. Science was described in an accretionist manner as building and extending theories incrementally, approaching truth in a monotonic way. For example, Carnap held the inductivist belief that science advances upwards from particular empirical facts to generalizations which summarize or provide an abbreviation for a body of such facts (Suppe, 1974, p. 15). Certainly positivism has influenced the methodology of other disciplines (e.g. behaviorism in psychology) in this direction.

6.2.2.2 Attacks on Empiricism

Important attacks on the positivist position, such as Popper's success in showing that induction cannot fully confirm the truth of theories, Hanson's claim that observations are "theory laden," and Kuhn's (1970) claim that theoretical advances often precede the empirical findings used to support them in science, have raised serious problems by arguing against the empiricist emphasis on sense experience as the preeminent basis for knowledge. Popper (1959) held that the proper role for data is in the criticism rather than the confirmation of hypotheses. Hypotheses are conjectures made by scientists rather than certainties abstracted from data. All theories are hypothetical and so it is appropriate to refer to them as hypotheses. But these conjectures can be reliably criticized and falsified by collecting data. This allows science to make progress via a series of conjectural hypotheses and reliable criticisms. Popper's work provided support for the model shown in Fig. 6.1, the hypothetico-deductive method. There are three main stages shown here: (1) a hypothesis is formed by conjecture; (2) predictions deduced from the hypothesis are tested empirically; (3) if the prediction is incorrect, the hypothesis is rejected

6.2 Background Questions from Philosophy of Science

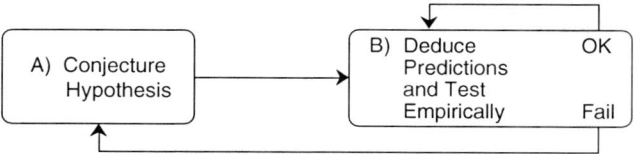

Fig. 6.1 Hypothetico-deductive method

and the cycle restarted; if correct, further testing is pursued. Popper maintained, contrary to the logical positivists, that a single successful empirical test did little to confirm a hypothesis, but that failing such a test was grounds for rejecting a hypothesis. Those hypotheses that survive the gauntlet of repeated testing become accepted laws. Favored laws emerge through the survival of the fittest conjectures, so to speak. However, Popper's allowance for conjecture also opens up the possibility that a noninductive, non-accretionist process, or even a Eureka event, could be involved in hypothesis formation.

6.2.2.3 Attacks on the Hypothetico-deductive Method

Popper's views have in turn been criticized in a number of ways. The most relevant shortcoming for the purposes of the present study is that his classic work does not specify mechanisms for generating hypotheses; he relegates this task to psychology. Also, Hanson's notion that observations can be "theory-laden" implies that empirical testing in the hypothetico-deductive method may not be fully reliable and sufficient on its own as a means of hypothesis evaluation.

6.2.2.4 Kuhn

With regard to the pace of theory change, Kuhn's (1970) ideas of revolution within a scientific discipline and the creative "gestalt switch" required for individual scientists to move outside of their own paradigm argued against an accretionist view of theory change. In his view, normal science may be accretionist in character, but revolutionary periods in science involve crisis and reconstruction, implying that science progresses at an uneven pace with periods of slow and fast change. On the other hand, critics of Kuhn, such as Toulmin (1972), have in turn questioned the reality of scientific revolutions, arguing for a more evolutionary view of theory change.[1]

---

[1] Placing different scholars on these two broad spectra ignores many other differences between them and requires a number of simplifications. For example, some scholars (e.g. positivists and Popper) tend to concern themselves with the formal justification of theories while others (e.g. Hanson, Kuhn) also focus on their psychological origin; arguments also vary as to whether they refer to science as a whole or to the individual scientist.

6.2.2.5 Summary

An inductivist–accretionist view of science sees it as compiling facts and generalizations in a piece by piece fashion. Put most simply, induction is the primary process of hypothesis generation, with a one-directional flow of knowledge from data upward to theories. In a rationalist–Eurekaist view, on the other hand, significant theoretical developments can occur when a scientist formulates mental constructions at some distance from existing data and can actually develop new ways of looking at old data and of looking for new data. Thus knowledge can flow from a newly invented, general theory downward to influence the formation of new specific theories, to reorganize one's view of existing data, and to suggest new places to collect important data. Such reorganizations presumably would require a large degree of creativity, perhaps even extraordinary "Eureka" episodes of insight.

6.3 How are Theoretical Hypotheses Formed in the Individual Scientist?

In this section, I will touch on some of the literature addressing the question that Popper chose not to deal with: "What are the mental processes by which theoretical hypotheses are formed in an individual?" Discussion of this narrower question about individuals may be of some interest to those investigating the broader question about science as a whole, even though the latter issue is more complex. In fact, surprisingly little work has been addressed to this question, especially in comparison to the complementary question: "How are scientific hypotheses tested?" Here I give a brief overview of several possible positions that can be taken on the first question concerning formation. Popper's position and the hypothetico-deductive method shown in Fig. 6.1 can be taken as a starting point here in the form of a nonanswer. The method shows one way in which hypotheses might be tested but does not show how they are generated.

6.3.1 Answer 1: Hypothetico-deductive Method Plus Induction

Popper argued convincingly that induction cannot be used to confirm the truth of scientific theories. However, some modern scholars retain some form of induction in their model of scientific method as a way to *suggest* hypotheses, rather than to confirm them unequivocally. This can be represented by the model shown in Fig. 6.2 – combining the hypothetico-deductive method with induction as a source of hypotheses. Here there is no claim for a sure "logic" of discovery, but only for a fallible method for generating hypotheses. Further experiments must be performed in order to evaluate the inductions. Such a diagram is commonly implied in everyday characterizations of scientific method as a combination of induction and deduction. Scholars such as Harre (1983), Achinstein (1970), and Gregory (1981) argue that induction can play

6.3 How are Theoretical Hypotheses Formed in the Individual Scientist?

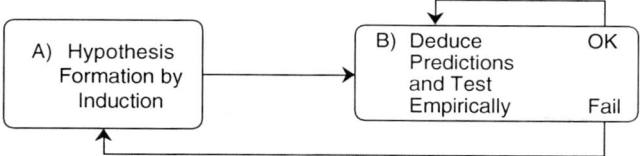

Fig. 6.2 Hypothetico-deductive method with induction as a source of hypotheses

a major role in hypothesis formation. However, they believe that other processes can be involved as well. Langley (1981) developed simulation models of data-driven inductive processes for generating certain scientific laws.

6.3.2 Answer 2: "Creative Intuition"

Is some form of induction or guessing the only source of scientific hypotheses? A number of authors have answered "no" to this question by pointing to the role of creativity, intuition, and the unconscious in generating hypotheses (Koestler, 1964; Polanyi, 1966; Rothenberg, 1979). Their views can be roughly characterized as replacing the "Hypothesis Formation by Induction" step in Box A of Fig. 6.2 with a process labeled "Hypothesis Formation by Creative Intuition." For example, Polanyi emphasizes the role of intuition and tacit knowledge in science. Rothenberg proposes a process of "Janusian thinking," whereby a person is able to juxtapose seemingly contradictory ideas, as a common element in creative thinking. Koestler points to "bisociative thought" – the ability to connect normally independent frames of reference – and to the role of the unconscious in accounting for creativity.

An interesting controversy has emerged in this area. Perkins (1981) and Weisberg (1993) argued that all of these descriptions attempt to point to *extraordinary* thinking processes; they attempt to supplement ordinary reasoning with something more powerful. The opposite view would claim that most creative acts can be explained plausibly by a model where a person uses certain ordinary thinking processes more intensively, or with special goals in mind. In that view, the difference between a creative and an uncreative person is a difference of degree and purpose, not a difference of kind. Perkins also describes authors like Koestler as contributing mainly to the description of the *products* of creative thinking; a remaining problem is to specify the *processes* of creative thinking in more detail.

6.3.3 Answer 3: Analogies as a Source of Theoretical Hypotheses

The work of another group of scholars in philosophy of science, including Campbell (1920), Harre (1961), Nagel (1961), and Hesse (1966), suggests that *analogies* may be a source of hypotheses. They argue that scientists often think in terms of qualitative visualizable models as a major locus of meaning for a scientific theory. The above authors, as well as Black (1979), argued that models are often constructed from analogies to familiar situations (e.g. gases are analogous to a collection of colliding balls; light is analogous to water waves). In Nagel's terms, such visualizable analogue models could help scientists "make the unfamiliar familiar." In that view, analogies would replace induction in the role of the most common source of theoretical hypotheses in the upper box in Fig. 6.2. This marks a large shift to a rationalistic as opposed to empiricist view of the origin of scientific theories.

6.3.4 Definitions of "Model": A Thorny Issue

In summary, philosophers and historians of science have proposed several possibilities for the nature of theory formation processes in science, including guessing, induction, creative leaps of intuition, and analogies. However, in historical studies it is always difficult to find objective records saying much in detail about the actual process of theory formation in the individual scientist. In this chapter I examine whether analyzing an extended think-aloud protocol of an expert trying to understand a physical system can speak to this issue.

As might be expected, real thinking is messier than the reasoning that shows up in published papers. One theme that will become apparent in the present data is that a scientific theory does not spring full blown into a subject's head from an initial analogy. The development of a theory is more gradual. It may start from a very rough or fanciful analogy that is then gradually improved. One thorny issue that will arise is how to define the transition point where an initial fanciful analogy changes into something we can call a scientific theory. This is a very difficult issue requiring examples, so I am going to postpone it until later in the chapter, and it will need to be revisited beyond that point as well. For now I will refer to all relevant analogies, explanations, and theoretical models collectively as *scientific models* construed broadly. This is not inappropriate in that I will end up arguing that many of the analogies do appear to be a powerful generative force contributing to the development of a theory. Then once we have a variety of examples of scientific models, in the broad sense, on the table, I can revisit the issue and define a much narrower concept of "explanatory model" as a special kind of scientific model that appears to lie at the core of a theory. Meanwhile, an important immediate purpose of the present protocol analysis will be to understand the kinds of analogies and improvement cycles that make progress toward a theory possible.

6.3.4.1 Mental Models

To get a working definition of the broad concept of "scientific model" I first need to consider the even broader concept of a "mental model." I will use the term

6.3 How are Theoretical Hypotheses Formed in the Individual Scientist?

(mental) model in the broadest sense to mean a (mental) representation of a system that can predict or account for aspects of its structure or behavior. I will make some minimal assumptions about models. Models are often idealized; one might say they are always simplified since we cannot comprehend every microscopic detail of entities in the world. Presumably models are useful when they represent the important interrelationships in a system, as opposed to being a collection of isolated facts. Then a single model can account for many events, making it an efficient way to store knowledge. Models can be developed at different levels of detail. For example, people can have a mental model for, say an old style three-speed bicycle at many different levels. My own model includes a chain and pedal gear drive, and cables for brakes and gear shifting, as well as some ideas about how the spinning wheels stabilize the bike, but many adults do not understand or represent the cable system nor the gyroscopic action of the wheels that aids in balance. Piaget showed that many children do not comprehend the role of the chain or have it in their model. And I myself do not understand how the gear shift system works that is hidden in the rear hub, so I have almost no representation for it. My model is a simplified, schematic, and somewhat general one that applies to many bikes, not just one. External representations, such as schematic diagrams, may serve to record features of a mental model, and may allow one to develop a more complex model than can be stored or envisioned at once in working memory.

6.3.4.2 Scientific Models

Recall that in the "Sisyphus problem" shown in Fig. 4.1 subject S2 was confident that it would be easiest to move the heavy lever in Fig. 4.2b by pushing at point X. Is it appropriate to refer to the analogous case of the lever as a "qualitative scientific model" for the wheel? I will use the term scientific model to mean a particular kind of mental model – thus it will again be a simplified (mental) representation of interrelationships in a system that can predict or account for its structure or behavior. Minimal additional criteria for consideration as a scientific model include having a certain level of precision as opposed to vagueness, a basic level of plausibility that rules out, for example, occult properties, and a requirement that the model be internally consistent (not self-contradictory) if possible. On this broad definition, analogies like the lever for the wheel, or water wave reflection for light reflection, or a mechanical thermostat for the body's temperature regulation system, can be scientific models when they are used in an attempt to predict or account for the behavior or structure of the system. (A hypothesis to be developed in Chapters 12–16 is that images of moving mechanisms can play a central role in scientific models, with mental simulations via dynamic imagery providing a major piece of the inferencing process. In this chapter, however, I postpone the discussion of imagistic processes in order to concentrate first on the form of the higher-order reasoning processes (such as analogy) being used to form models in investigations.)

6.4 Protocol Evidence on Construction Cycles That Use Analogies

In the remainder of this chapter, I will focus on the case study of subject S2 working on the spring problem. Part of this solution will also be examined in Chapter 7 as an example of a scientific insight and a case where the subject must "break out of" a previous set of assumptions (or "Einstellung" or "set") for the problem. Thus this is an in-depth case study of insight and analogy use by a single subject. For those interested in an overview of many types of strategies used by many subjects on the spring problem, see Chapter 14.

6.4.1 Purpose of Case Study

One of the main reasons for doing an in-depth case study is to develop and refine a basic vocabulary of concepts for describing psychological observations and theories. The initial challenge of such a study is to develop and describe the "units" of behavior to be used in observation and to propose an initial cognitive model in the form of a set of cognitive structures and processes that can account for the behavior and that is both plausible and consistent. For simpler types of behavior, such modeling can be fairly detailed, and in some cases can be expressed as a computer simulation. For more complex or poorly understood phenomena, an initial step in modeling can be achieved by formulating a general description of the basic units or cognitive objects to be used in the model, a general outline of the model, and a set of "design criteria" that a more detailed model would need to fulfill. In this chapter, I am aiming at the latter level.

6.4.2 S2's Protocol

In the spring problem, subject S2 first generated the model of a horizontal bending rod, comparing a long rod with a short one (a weight is attached to the end of each rod) and inferring that segments of the wider spring would bend more and therefore stretch more. However, he was concerned about the appropriateness of this model because of the apparent lack of a match between seeing bending in the rod and not seeing bending in the wire in a stretched spring. One can visualize this discrepancy here by thinking of the increasing slope a bug would experience walking down a bending rod as opposed to the constant slope the bug would experience walking down the helix of a stretched spring. This discrepancy led him to question whether the bending rod was an appropriate model for the spring. (Another way for the reader to see this problem is to note that the bending model predicts that the slope of the wire and the distance between coils will increase as one goes down the spring, as shown in Fig. 6.3. Yet this does not happen in real springs). The full

Fig. 6.3 Asymmetrically stretched spring predicted by bending model

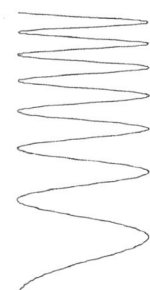

Bend Model
Predicts
Slope of Wire
and Distance
Between Coils
Increases

transcript is quite long; therefore verbatim excerpts are presented here. (Brackets in transcript indicate my comments.)

> 5 S2: I have one good idea to start with; it occurs to me that a spring is nothing but a rod wound up uh, and therefore maybe I could answer the question for a rod. But then it occurs to me that there's something clearly wrong with that metaphor because if I actually took spring wire and it was straight instead, it certainly wouldn't hang down like a spring does...It would droop...and its slope would steadily increase as you...went away from the point of attachment, whereas in a spring, the slope of the spiral is constant.
>
> 7 S2: Why does a spring stretch?... I'm still led back to this notion...of the spring straightened out [a bending rod]...(e) I'm bothered by the fact that the slope doesn't remain constant as you go along it. It seems as though it ought to be a good analogy, but somehow, somehow it doesn't seem to hold up....
>
> 23 S2: I feel I want to reject the straightened spring model – as a bad model of what a spring is like. I feel I need to understand the nature of a spring in order to answer the question. Here's a good idea. It occurs to me that a single coil of a spring wrapped once around is the same as a whole spring...In the one-coil case, I find myself being tempted back to the straightened spring [bending rod] model again....

Certainly an important positive feature of the above section is the subject's ability to criticize his own initial model. Several other subjects who thought of the bending rod model did not make this interesting criticism of it. Still, the model appears to be his best idea and he has real difficulty in rejecting it, as follows.

> I still don't see why coiling the spring should make any difference. ... Surely you could coil a spring in squares, let's say, and it...would still behave more or less the same. Ah! from squares, visually I suddenly get the idea of a zig-zag spring rather than a coiled spring; that strikes me as an interesting idea (draws Fig. 6.4a).... Might there be something in that idea....
> I see a problem with this idea. The problem...is that...if I assume these bars to be rigid... the stretch...has to do with the joint. But the springiness of the...real spring is a distributed springiness. ... So...I wonder if I can make the [zig-zag] spring...where the action...isn't at the angles...it's distributed along the length. ... And I'm going to do that; I-I have a visualization. ... Here's a stretchable bar (draws modified zig-zag spring in Fig. 6.4b) a bendable bar, and then we have a rigid connector. ... And when we do this what bends...is the bendable bars...and that would behave like a spring. I can imagine that it would....

Here there is evidence that the subject is generating a *series of analogue models* for the spring – from the rod, to the square, to the angular zigzag spring, to the rectangular zigzag spring with stiff joints. The zigzag spring is eventually

Fig. 6.4 (a) Zigzag spring #1 (b) Zigzag spring #2

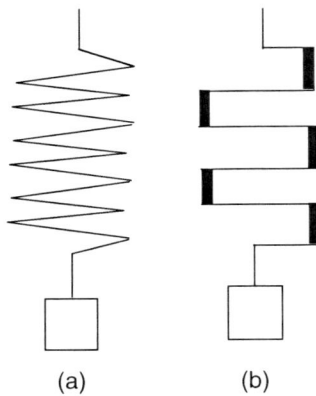

dropped, presumably because he was still critical of this model and could not reconcile the bending going on in sections of the zigzag spring with the lack of change in slope in the original helical spring. However, these attempts do provide evidence for an important thought pattern in the form of a *repeated successive refinement process* of *model construction, criticism, and modification.*

Next, he considers the analogy of a double-length spring instead of the double-width spring appearing in the original problem.

> 37 S2: This rod here: as the weight moves along, it bends more and more the further out the weight is.... Hmmm, what if I imagined moving the weight along the spring...would that tell me anything? Would that? I don't know. I don't see why it should. What if the spring were twice as long...instead of twice as wide? .... It seems to me pretty clear that the spring that's twice as long is going to stretch more. ... Now if this is the same as a spring that's twice as wide, then that should stretch more. ... Uhh, but is it the same as a spring that's twice as wide? Again, I just don't see why...the coiling should make any difference. It just seems geometrically irrelevant to me somehow. ... But I...can't – I have trouble...bring that into consonance with the behavior of an actually stretched out spring...the slope problem anomaly [increasing slope in the rod, but not in the spring] – If I could resolve that anomaly...then I would feel confident of my answer...but this anomaly bothers me a lot.

He seems critical of the appropriateness of the double-length spring analogy and reverts again to the bending rod.

At this point, the bending rod, double-length spring, and zigzag spring analogies have each pointed S2 to the correct answer to the problem, yet he remains unsatisfied with his understanding. In line 57, he continues to search unsuccessfully for a more satisfactory analogous case.

> 57 S2: I feel as though I'm reasoning in circles. I think I'll make a deliberate effort to break out of the circle somehow. What else could I use that stretches...like rubber bands...what else stretches...molecules, polyesters, car springs [leaf springs]...what about a...two-dimensional spiral spring? That doesn't seem to help.

6.4.2.2 Insight Episode

Subsequently, this subject produces an extremely productive analogy when he generates the idea of the hexagonally shaped coil in Fig. 6.5a and moves from there to the idea of the square-shaped coil in Fig. 6.5b. Imagining stretching these polygonal coils apparently led him to a major breakthrough in the solution which corresponds to the way in which engineering specialists view springs. Much of the remainder of this chapter will focus on this insight below.

The impressiveness of the reasoning displayed by different subjects in solving the spring problem depends on the depth of understanding sought by the subject and on the subject's prior knowledge. The first level of depth in understanding is simply to state an intuition that the wide spring will stretch more; a second level is to give some plausible justification or explanation for this. The following portion of transcript shows a subject going significantly deeper into the second level process. This part of the protocol is reported in sections as follows:

1. Subject is still in conflict about whether spring wire is bending
2. Generates a series of polygonal coil analogies
3. Subject has an insight
4. Evaluates and adapts square coil as a preferred model of the spring
5. Comments on his increased understanding

Section 1: Subject is still in conflict about whether spring wire is bending

> 57 S2: I just...have the intuition that a...straight rod ought to in some sense be a good model for a spring. But there are these anomalies that won't go away. And yet I can't see...a better model.
>
> 79 S2: I'm just trying to imagine the coil... (traces circle about 7 inches in diameter in air in front of self) a circle with a break in it....
>
> 81 S2: (has just drawn Fig. 6.6a)...you could just hold it there...and apply a force there, and the spring stretches.... I'll be damned if I see why it [the coil] should be any different from that case [the rod]....
>
> 87 S2: ...if you start with a helix and unwind it...you should get a bow [bend], but you don't. I mean visually imagining it, you don't. I don't see how you could make the bow go away – just to wind it up – Damn it!
>
> 111 S2: Darn it, darn it, darn it...why should that [the difference between a rod and a coil] matter?.... I'm visualizing what will happen when you just take this single coil and pull down on it and it stretches; and it stretches....

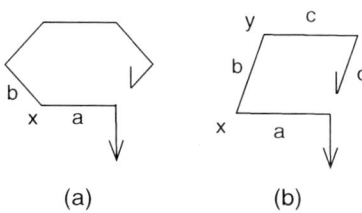

Fig. 6.5 (a) Hexagonal coil (b) Square coil

Fig. 6.6 (a) Single coil (b) Bending rod

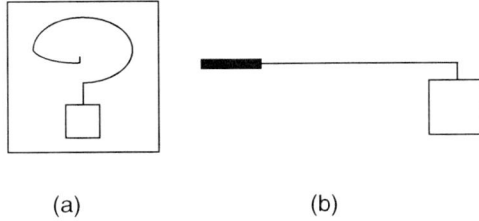

(a)   (b)

(The subject spends a considerable amount of time trying to resolve this issue without making progress.)

Section 2: Generates a series of polygonal coil analogies

117 S2: (40 minutes into the protocol) I keep circling back to these *same issues* without getting anywhere with them.... I need to...think about it in some radically different way, somehow. Let me just generate ideas about circularity. What could the circularity [in contrast to the rod] do? Why should it matter? How would it change the way the force is transmitted from increment to increment of the spring? **Aha!** Now let me think about; **Aha!** Now this is interesting. I imagined; I recalled my idea of the square spring and the square is sort of like a circle and I wonder...what if I start with a rod and bend it once (places hands at each end of rod in Fig. 6.6b and motions as if bending a wire) and then I bend it again.

119 S2: What if I produce a series of successive approximations to...the circle by producing a series of polygons! Maybe that would clarify because maybe that, that's constructing a continuous bridge, or sort of a continuous bridge, between the two cases [the rod and the coil]. Clearly there can't be a hell of a lot of difference between the circle and say, a hexagon...

121 S2: ...or even a triangle...square...(draws hexagon in Fig. 6.5a).... Now that a [hexagon] is essentially a circle. I mean, surely springwise that [hexagon] would behave pretty much like a circle does.

Section 3: Subject has an insight

121 S2: Now that's interesting. Just looking at this it occurs to me that when force is applied here, you not only get a bend on this segment, but because there's a pivot here (points to x in Fig. 6.5a), you get a torsion effect....

122 S2: **Aha!** Maybe the behavior of the spring has something to do with twist (moves hands as if twisting an object) forces as well as bend forces (moves hands as if bending an object). That's a real interesting idea. That might be the key difference between this [bending rod] which involves no torsion forces, and this [hexagon]. Let me accentuate the torsion force by making a square where there's a right angle. (Draws a Fig. 6.5b). I like that. A right angle...that unmixes the bend from the torsion.

123 S2: Now...I have two forces introducing a stretch. I have the force that bends this... segment [a] and in addition I have a torsion force which twists [segment b] at vertex, um, X...[in Fig. 6.5b] (makes motion like turning a door knob with one hand).

Section 4: Evaluates and adopts square coil as a preferred model of the spring

129 S2: (b)...Does this (points to square-shaped coil) gain in slope – toward the bottom?....

130 S2: (c)...indeed we have a structure here which does not have this increasing slope as you get to the bottom...(e)...it's only if one looks at the fine structure; the rod between the Y and the X, that one sees the flop effect [downward curvature].

132 S2: (b)...Now I feel I have a good model of sp- of a spring.... Now I realize the reason the spring doesn't flop is because a lot of the springiness of the spring comes from torsion effects rather than from bendy effects....

133 S2: And now I think I can answer the stretch question firmly by using this...square model of the spring. What does it mean, in terms of the square model to increase the diameter of the spring? .... Now making the sides longer certainly would make the [square] spring stretch more.

135 I: How can you tell?

136 S2: (a) Physical intuition...and also recollection... the longer the segment (moves hands apart) the more the bendability (moves hands as if bending a rod)...(b) Now the same thing would happen to the torsion I think, because if I have a longer rod (moves hands apart), and I put a twist on it (moves hands as if twisting a rod), it seems to me – again physical intuition – that it will twist more....

143 S2: So...doubling the length of the sides...it will clearly stretch more. Both for reasons of torsion and for reasons of the segment [bending].

Section 5: Comments on his increased understanding

144 S2: And my confidence is now 99%. ... I now feel pretty good about my understanding about the way a spring works although I realize at the same time I could be quite wrong. Still, there seems to be something to this torsion business; I feel a lot better about it.

178 S2: Before this torsion insight, my confidence in the answer was 95% but my confidence in my understanding of the situation was way way down, zero. I felt that I did not really understand what was happening; now my confidence in the answer is near 100% and my confidence in my understanding is like 80%.

6.4.3 Analysis of Insight Episode

6.4.3.1 Torsion Insight

Probably the most difficult achievement occurs when the subject, not knowing about the invisible twisting in the wire, is somehow able to construct that (correct) hypothesis. (S2 was the only subject interviewed who achieved this.) To see why his square spring model is helpful, note that it can in turn be understood in terms of two simpler cases, the twisting rod and the bending rod, as shown in Fig. 6.5b. *That is, pulling the end of the lever "a" down not only bends rod "a," but it also twists rod "b."* (One way to comprehend this idea is to view rod "a" as a wrench that is twisting rod "b.") The same is assumed to be true for all other adjacent rod pairs. Thus, twisting becomes an important type of deformation in the spring wire in this model; it is referred to by engineers as torsion. For those unfamiliar with the concepts of torsion and torque, an introduction is provided in the Appendix to this chapter.

The short transcript excerpts displayed here do not fully convey the frustration involved while the subject spent a considerable period of time (over 25 min) alternately questioning and trying to justify the initial bending rod model of the spring. Coming after this struggle, the invention of the polygonal coil with the subsequent

82   6 Case Study of Model Construction in Expert Reasoning

torsion discovery is a candidate for being termed a significant scientific insight for several reasons.[2] First, the idea is productive in the sense that it leads immediately to a considerable amount of cognitive activity. In fact one is given the impression of a "flood" of ideas occurring immediately afterward. Progress is made rapidly, as if the polygonal coil idea were a "trigger" that stimulates a series of further ideas. Second, the torsion idea appears fairly quickly, with little warning. Third, the subject makes a major change in his original model of stretching – by considering torsion the subject introduces a new causal factor into the system. Torsion constitutes a very different mechanism from bending for explaining how the spring resists stretching. Fourth, the subject says that he is now able to resolve the paradox of the apparent lack of bending in a helical spring and states that he feels he has achieved an increase in his understanding of the system (lines 129–130). This is true because he sees that in the square coil, any bending in the wires starts anew at each corner of the square and does not accumulate as one travels down the spring wire. Of course, his "theory of springs" could be developed further beyond the polygonal coil idea, but the fact remains that this model is a significant advance over the single bending rod models. Fifth, the subject reacts emotionally to his ideas, calling them "interesting" and exposing a "key difference," as well as producing some emphatic "Aha" expressions with a raised tone of voice. In Chapter 7, I will formulate a more careful definition for the term "insight" that is motivated by these factors.

6.4.3.2 Analogies Used by S2

How did the subject produce this new way of viewing the mechanism underlying the spring system? A hypothesized outline of the cognitive events producing S2's new understanding is shown in Fig. 6.7. The figure shows hypothesized "snapshots" of a series of S2's analogue models during the last insight episode as they develop over time, with solid lines showing confirmed analogy relations, and dotted lines showing tentative analogy relations. Poorly understood situations are shown in dotted boxes with well-understood situations shown in solid boxes.

Figure 6.7a (line 81): S2 has already reduced the spring situation to the equivalent single circular coil situation as shown by the solid line labeled (1) in the diagram. Also there is a tentative analogy relation shown as a dotted line labeled (2), from the single coil to the well-understood bending rod model.

Figure 6.7b (line 117): S2 then recalls his idea of a square spring and generates the model of a hexagonal coil. In his words, this is "constructing a continuous bridge or sort of a continuous bridge, between the two cases [of the circular loop and bending rod]." Figure 6.7b shows the form of a bridging analogy here of the kind discussed in Chapter 4. This quotation did in fact influence my original selection of the term "bridging" for the bridging analogies phenomenon described here and earlier in articles on experts and students (Clement, 1986, 1993b). Here the hexagon can be interpreted as a case constructed to be an intermediate bridge between the circular

---

[2] In fact, twisting is the predominant source of stretching in a helical spring. See Chapter 14 for a more complete analysis of the spring problem.

6.4 Protocol Evidence on Construction Cycles That Use Nalogies

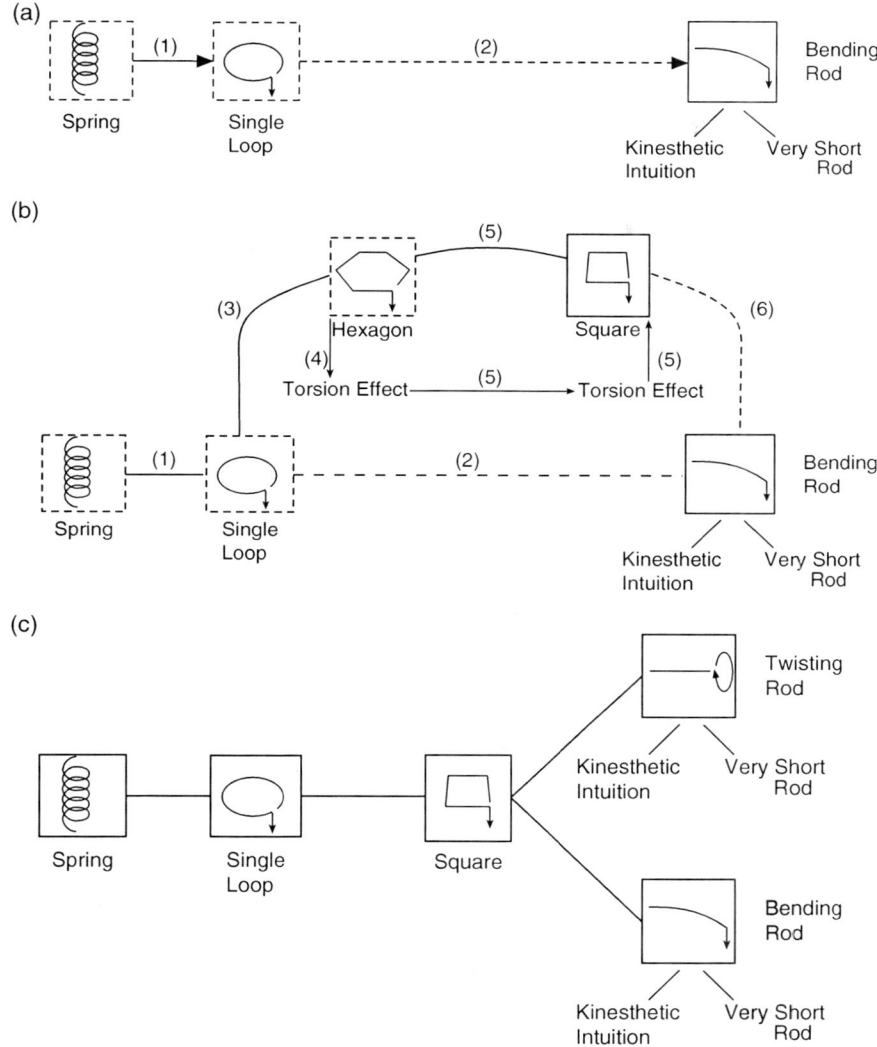

Fig. 6.7 Final argument structure in S2's spring protocol. (a) Initial rod analogy (b) bridging analogies and torsion insight (c) cases supporting S2's final understanding in section

loop and bending rod. Its function as a bridge, however, is quickly superceded by the unexpected insight that is triggered by the hexagonal coil.

(Line 121): While analyzing the hexagon in terms of bending effects, it occurs to him ("Aha!") that there will also be twisting effects. At this point he shifts to the simpler square model.

Figure 6.7c (line 123): By the final stage, S2's understanding of the underlying structure which makes a spring work has changed significantly. He now appears to have a mental model of the spring as working like a square coil which contains elements that bend and twist. His physical intuitions about the difficulty of (1) bending, and (2) twisting a long vs. short rod seem to play a role similar to axioms; they are basic assumptions on which the rest of his conclusions are founded. (I will discuss the nature of these intuitions in Chapter 13.)

These diagrams portray the solution as depending strongly on analogy relations. When torsion is recognized in the hexagonal coil, it is quickly transferred to the square coil. Since the square coil is considered analogous to the circular spring coil, torsion is transferred there and hence to the full helical spring. Intuitions about bending – that long rods are easier to bend than short rods – are transferred down the line in Fig. 6.7c from right to left in a similar way. In this view analogy plays a central role in forming the subject's new understanding of the spring system. In what follows I will refer to the square, hexagonal, and many-sided coil models collectively as polygonal coil models, since they are scientific models in the broad sense.

6.5 Summary of Evidence for a Model Construction Cycle as a Noninductive Source for Hypotheses

6.5.1 Model Construction Cycles

As a more general description of a construction process, one source of the growth in S2's ideas appears to have been the cyclical process shown in Fig. 6.8. Essentially, the diagram depicts a cyclical process of model generation, evaluation, and modification (I refer to this as a GEM cycle). It is difficult to describe so complex a process in such a simple diagram, but a simplified picture will aid in the present analysis. In particular, after problems arose with the bending rod case, the zigzag spring was proposed, and after problems with it were detected, it was in turn modified into a "better" zigzag case in line 23. And the hexagonal coil case was criticized and changed to the "better" square coil case in line 122. Thus these models were evaluated and improved in the manner shown in Fig. 6.8. Table 6.1 summarizes evidence from the protocol that S2's progress is a result of this kind of

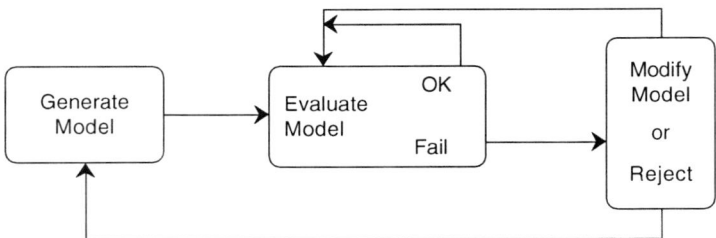

Fig. 6.8 Cyclical process of generation, evaluation, and modification (or rejection)

6.5 Summary of Evidence for a Model Construction Cycle

Table 6.1 Location of evidence for model construction cycle of hypothesis generation, criticism, and modification or rejection

Line	Process	Hypothetical model	Comments
5	G	Horizontal bending rod	Initial analogy
5	C	Horizontal bending rod	Bending in rod, but not in helix
23	G	Square coil	
23	M	Zigzag #1	Modifies square to produce zigzag model
23	C	Zigzag #1	Joints confounding
23	M	Zigzag #2 with stiff joints	Modifies zigzag #1 to produce #2
	[C*]		Bending in zigzag, but not in helix*
	Rj	Drops zigzag models	
57	Rc	Rod model	
87	C	Rod model	Bending in rod, but not in helix
117	Rc	Square coil	
119	M	Hexagonal coil	
121			Makes torsion discovery in hexagon
122	C	Hexagonal coil	Hexagon geometry too complex
122	Rc	Square coil	(Leads to successful prediction of restoring forces without cumulative bending in spring wire)

Key G = Generates hypothesized model; C = Criticizes model; M = Modifies model; Rc = Reconsiders model; Rj = Drops or rejects model.
* Inferred in absence of direct evidence in protocol.

cyclical process rather than being a result of either a convergent series of deduction or an induction from observations.

Thus it appears that real-time protocol evidence can be gathered to evaluate the plausibility of such models of scientific reasoning.

6.5.1.1 Comparison to Prior Research

In contrast to the hypothetico-deductive scheme and its derivatives in Figs. 6.1 and 6.2, in Fig. 6.8, models are not just generated and then evaluated in an attempt to reject them. Instead, some can be revised and reevaluated repeatedly in an improvement cycle. This can lead to the rejection of some models, but also to a series of progressive refinements in others. Theory formation and assessment cycles using analogies have also been discussed by Nersessian (1992), Holland et al. (1986), Millman and Smith (1997), Gorman (2006), Gentner et al. (1997), and Darden and Rada (1988) in historical or AI (Falkenhainer, 1990) contexts; and by Miyake (1986) for novice protocols.

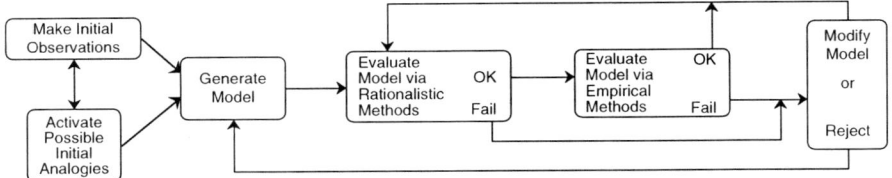

Fig. 6.9 GEM cycle with initialization processes and both empirical and rationalistic evaluation

Secondly, the schemes in Figs. 6.1 and 6.2 place great emphasis on empirical evaluation of theories. But subjects in the present study did not have access to empirical data, and yet they appeared to evaluate certain models – followed by improving some and rejecting others (e.g. in Table 6.1). Therefore there are nonempirical evaluation methods being used of some kind. Thirdly, a prominent role for analogies has been implicated in this analysis as a source of hypotheses, and that could replace raw conjecture or hypothesis by induction (or intuition) in Figs. 6.1 and 6.2 as a source.

This representation of investigation processes is extended further in Fig. 6.9. The subprocess of "activating possible initial analogies" reflects the observation that subjects may cast about for a variety of analogies when they are attempting to generate an initial model, as was implicated in the rapid search for analogies such as "molecules, polyesters, and car [leaf] springs" in line 57 of the transcript. Then when an analogy starts to be taken seriously as a model it can enter the GEM cycle shown.

Going beyond analysis of the present data, in order to connect to the empirical side of science, the double ended arrows between *Make Initial Observations* and *Construct Initial Model* represent the idea that not only does model construction respond to observation but that one's focus of attention during observation can be guided by one's initial model. This and other double ended arrows indicate that the initial model generation process can be highly interactive and complex. It is still poorly understood. Essentially, the scientist must construct a conjectured picture of a structure or process which can predict or account for what happens or explain why the phenomenon occurred.

Figure 6.9 also adds an empirical evaluation processes that complements a rationalistic process. Hypothesis evaluation or criticism can then take place in two major ways. Empirical testing can add support to or disconfirm a hypothesized model. Rationalistic evaluation can also support or disconfirm a model, depending, for example, on whether it is found to be externally consistent or inconsistent with other established theories as well as internally consistent with itself. Evaluation processes cannot provide full confirmation, but can lead a scientist to have increased or reduced confidence in a theory. Once generated, a hypothesis undergoes repeated cycles of rationalistic and empirical testing, and modifications as needed. A limitation of the diagram that is not intended to be part of the cognitive model is the order in which rationalistic and empirical evaluation occur; tests can occur in different orders on different cycles. The endless loops in Fig. 6.9 indicate that ideally, theories

6.5 Summary of Evidence for a Model Construction Cycle

in science are always open to new criticisms. However, as Kuhn (1970) points out, scientists will sometimes ignore or discount some criticisms in order to protect a favored theory. In practice, research groups may adopt a "protected core" of theories which they take as givens (Lakatos, 1978).

A missing element in the figure is the influence of the subject's prior theoretical framework. It is difficult to depict, since it could affect so many of the processes shown. Since the scientist operates from a background of broader theoretical assumptions, these may have an early influence on the model elements and analogies which come to mind, and sometimes even (according to Hanson and Kuhn), on what is observed. In sum, a possible synthesis of ideas is proposed in Fig. 6.9. It allows for the possibility that the hypothetico-deductive method, induction, abduction, analogy, rationalistic evaluation, and model modification may all play important roles at different times in scientific thought.

6.5.1.2 Hypothesis Generation Processes that are Neither Inductive nor Deductive

Let us examine more carefully the claim that S2's final model is neither the result of a convergent series of deductions nor an induction from observations. Consider his statements during the torsion insight:

> Maybe the behavior of the spring has something to do with twist forces as well as bend forces. That's a real interesting idea. ... That might be the key difference....

When S2 generates analogue model hypotheses, they appear not to be deduced logically from prior principles – they appear to be reasoned conjectures as to what might be a fruitful representation for analyzing how a spring coil works. The reasoning involved does not appear to carry the certainty associated with deduction.

Nor, apparently, are they built up inductively as abstract generalizations from observations. S2 is unable to collect new data during the interview, and consequently his reasoning is independent of new empirical processes. One can also consider whether he might be making new inductions on perceptual memories of *prior* observations, but he does not appear to recall observing bending, twisting, zigzags, or squares in springs; instead these appear to be newly imagined models. The novelty and nonobservability of the polygonal coil with torsion model, and its evolution from criticisms of the earlier horizontal rod model argue that the hypothesis generation process in this case was an imaginative construction and criticism process rather than one of induction from observations. Quite possibly, S2 would have made some new observations of springs as well, had they been available (although it is extremely doubtful that he would have observed torsion effects). But the present case study documents the possibility that impressive progress in model construction can be made via noninductive processes.

Of course, it is highly likely that empirical information was involved in the original development of some of the prior knowledge he uses. In attempting to speak to the rationalism vs. inductivism issue it is important to identify the time period of focus.

For the purposes of this analysis, the focus is on the new knowledge developed during the hour or so of reasoning in the interview rather than on the origins of the prior knowledge he uses. For example, he uses prior knowledge in the form of the concept of twisting. One assumes his earlier learning of the concept of twisting involved empirical experiences with wrenches, cranks, knobs, etc. His new model of the polygonal spring with torsion uses his old concept of twisting as one of its elements, but the total structure of the model is a larger new construction. The point is that the new knowledge developed by S2 – the construction of a new model for how a spring works – was apparently formed by nonempirical processes during the protocol.

6.5.1.3 Abduction

Peirce (1958) and Hanson (1958) used the term *abduction* to describe the process of formulating a hypothesis which, if it were true, would account for the phenomenon in question. The hypothesis can be a guess as long as it accounts for the observations collected so far. Such a process might include using the knowledge structure from an analogous case to form the starting point or core of a new model. Or it might integrate several related model elements – constructing a new model by combining several existing knowledge structures previously known to the subject. Empirical law hypotheses which consist only of a recognized regularity or repeated pattern in the variables might be formed via a more data-driven inductive process. This is possible on those occasions when one has the prior advantage of possessing the right variables, or components of compound variables, to look for. But there is also the possibility that hypotheses can be formed by a less data-driven abductive process, possibly for just a single instance of the phenomenon, or perhaps even before any observations have been made.

I do not wish to say here that patterns perceived in data cannot be involved in some types of scientific hypothesis formation; after all they provide the thing or pattern to be explained. Rather, the present case study acts more like an "existence proof" in showing the possibility that noninductive construction processes can be very important in the formation of a new theoretical model. Thus the power of these models does not seem to be inherited deductively from some prior principle or axiom, nor does it come from generalizing on many observed instances; rather it appears to come from an analogical generation process, plus a criticism and revision cycle leading to a series of successively better mental models for how the situation works.

6.5.2 *Explanatory vs. Nonexplanatory ("Expedient") Models*

6.5.2.1 Four Types of Knowledge Science

I now return to an issue mentioned in the chapter introduction to give some further details in the description of scientific models and to discuss differences between fanciful analogies and scientific theories. Campbell (1920), Hesse (1966), and

6.5 Summary of Evidence for a Model Construction Cycle

Harre (1972) developed important distinctions between qualitative theoretical models, empirical law hypotheses, and formal principles, as shown in Fig. 6.10. They believed that hypothesized, theoretical, qualitative models (I will call these "explanatory models") such as molecules, waves, and fields, are a separate kind of hypothesis from empirical laws. Such explanatory models are not simply condensed summaries of empirical observations but rather are inventions that contribute new theoretical terms and images which are part of the scientist's view of the world, and which are neither "given" in nor implied by the data. Campbell's (1920) oft-cited example is that merely being able to make predictions from the empirical gas law stating that PV is proportional to RT, is not equivalent to understanding the explanation for gas behavior in terms of an imageable model of billiard-ball-like molecules in motion. The model provides a description of a hidden, nonobservable process which explains how the gas works and answers "why" questions about where observable changes in temperature and pressure come from. (Summaries of these views are given in Harre, 1967; Hesse, 1967). Beyond these basic requirements, scientists often prefer explanatory models which are general, visualizable, simple, causal, and which contain familiar entities (Nagel, 1961).

The above considerations motivate a distinction between two types of scientific mental models: expedient analogies and explanatory models. Recall the proposal to use the term qualitative model to refer to a (mental) representation M of a target situation T that a subject can use to predict or account for T's structure or behavior. One kind of model then is merely an expedient and often temporary analogy which predicts some aspects of the target's behavior. M may happen to behave like T, and therefore provide a way of predicting what T will do. For example, a closed cylinder with a piston under certain conditions will behave analogously to a spring: the distance the piston moves is roughly proportional to the force one exerts on the piston (for small displacements) and the same is true for the spring. However, such an expedient analogy may say nothing about the underlying process which explains the gas's behavior. An explanatory model, on the other hand, should explain how T works, leading to a feeling of "understanding" T. For example, the elastic particle model for gasses explains why volume decreases with force or why pressure increases with temperature in an enclosed gas in terms of the collisions of the particles with the walls.

THEORETICAL KNOWLEDGE

-Formal Principle

-Explanatory Model

-Empirical law hypothesis: mathematical or verbal descriptions of patterns in observations

-Observations

Fig. 6.10 Four types of knowledge in science

EMPIRICAL KNOWLEDGE

Hesse (1967) and Harre (1961) describe the distinction this way: (1) a model which shares only its abstract form with the target (Hesse cites hydraulic models of economic systems as one example); I call this an "expedient analogy"; and (2) a model that has become in Harre's terms a "candidate for reality," where a set of material features, instead of only the abstract form, is also hypothesized to be the same in the model and the target situations. I will refer to the latter type of model, M, as an *explanatory model*, M_e, if some of the basic objects, attributes, and concrete relations in M are hypothesized by the subject to be a hidden part of T and to underlie the behavior of interest in T. For example, the gas is thought of, not just as something that happens to have the same behavior as a collection of particles, but as something that *is* a collection of particles. This ordinarily means that the subject can attain some degree of ontological commitment to (belief in the reality of aspects of) M_e if empirical and rationalistic support are obtained for it.[3] M_e is thought of as a hidden structure within T which provides an explanation for T's behavior – M_e contains some entities that are initially not directly observable or obvious in T at that point in time. This concept is designed partly to account for the remarkable ability of scientists to formulate and propose hidden structure and processes in nature before they are observed more directly, such as atoms, black holes, and the "bending" of light rays. An explanatory model can allow the scientist to see a phenomenon in a new way via a hypothesized underlying structure that is considered to be hidden in the target situation to be explained. This is something that empirical law hypotheses cannot do (see Fig. 6.10).[4]

6.5.2.2 Are S2's Models Explanatory?

S2 makes a clear distinction between *confidence in his answer* to the problem and *confidence in his understanding* of the spring:

144 S: There seems to be something to this torsion business; I feel a lot better about it....

178 S: Before this torsion insight, my confidence in the answer was 95%, but my confidence in my understanding of the situation was way, way, down, zero. I felt that I did not really understand what was happening; now my confidence in the answer is near 100%, and my confidence in my understanding is like 80%.

---

[3] This need not imply absolute certainty about the correctness and completeness of one's theoretical model, something that cannot be attained in this author's view.

[4] In one sense I am appropriating the term "explanatory" here since, as Kuhn (1977c) points out, what counts as explanatory is different for Aristotle, Newton, and quantum physics. I am proposing that what counts for S2 in this problem fits the definition given – an analogue model that has material elements which are hypothesized as "candidates for reality." The sharing of material elements between model and target can be termed material correspondence. Whether a satisfying explanation is actually attained, however, will also depend on other factors such as the support for and comprehensibility of the model.

6.5 Summary of Evidence for a Model Construction Cycle   91

This perceived increase in understanding is one indication that the idea of twisting and torsion in the polygonal coil has become an explanatory model for the subject, not just an expedient analogy for generating the answer to the problem. (Karmiloff-Smith and Inhelder (1975) have documented a related distinction in children's thinking.)

In the case of the present protocol the twisting and bending ideas qualify as explanatory, since the subject now believes that twisting and bending effects may actually be operating in the spring wire to produce its behavior. Twisting and bending are features that are not ordinarily observed in springs and so they are theoretical in this context. To be sure, they are purely qualitative hypotheses that will require more evaluation to be confirmed and they lack detail. But they qualify as the beginning of an explanatory model because they express for the subject a hypothesis concerning the hidden structure underlying the way stretching produces deformation and restoring forces in the spring wire. Furthermore, the square coil example appears to remove the anomaly of a potentially critical dissimilarity in the original bending rod model – that of the lack of cumulative bending in the spring. All of these factors presumably increase S2's feeling of understanding and of having a satisfying explanation for the behavior of a spring, as expressed in lines 144 and 178 quoted above. For the above reasons, the ideas of torsion and bending effects occurring in the spring wire qualify as two strands of an initial explanatory model which provides a hypothesis about the nature of springs. In this sense S2's protocol is an example of learning via the construction of a new explanatory model.

The double-length spring analogy in line 37 on the other hand, does not seem to contribute to a theoretical explanation for why the spring stretches. It is seen as an expedient analogy, as is the comparison of foam rubber with large holes and small holes portrayed in Fig. 3.1f. Both of these analogies predict the correct qualitative answer for the problem, but say nothing about why it happens.

6.5.2.3 An Explanatory Model Can Develop from an Initial Nonexplanatory Analogy

A further hypothesis is suggested by S2's problem solution: an expert can develop an explanatory model via the modification and refinement of an initial model that is merely expedient or has low explanatory status. In this view, whether a model is explanatory is a matter of degree. The explanatory status of a model depends on the degree to which one believes that the model contains elements that are like elements hidden in the target to be explained.

It is reasonable that when an analogous case is first proposed, it will often be unclear whether it has potential as an explanatory model – whether its elements could be something like the hidden elements in the target or not. Its explanatory status may grow gradually rather than in one decisive jump. Improvements in the model may also raise its explanatory status. Indeed, this seems to be what occurred in S2's case. He used the bending rod early on as a model, which gave him a prediction in which he was highly confident. However, he said his resulting understanding was very low. The recognition of the cumulative bending anomaly appeared to prevent him from accepting bending as an explanatory model. Cumulative bending is a centrally important material property which was present in the model, but not in the targeted spring system. This led him to generate and evaluate a number of alternative models, culminating with the polygonal coil model. The identification of torsion in the polygonal coil raised his feeling of understanding significantly. This is consistent with the interpretation that S2 had then acquired some confidence that torsion is a real but hidden mechanism operating in the spring. Thus, S2 appears to take an initial, nonexplanatory analogy (the bending rod) and develop it, via criticisms and modifications, into a model that in fact does have some explanatory status for him. Later, in Chapter 14, we will examine other more advanced protocol sections, in which bending in the vertical plane is rejected fully, leaving twisting as the "last model standing" as the main source of deformation in the spring. Nevertheless, the consideration of the bending model was an extremely important stepping stone toward this result. Thus an analogy does not have to be correct to play an important, useful role.

6.5.2.4 Simplifying Function of Models

Toward the end of the session S2 considers a multisided coil, but is unable to make further progress in his analysis before the end of the interview. Figure 6.11 shows the set of polygonal models referred to by S2, placed in order of increasing simplicity or analyzability from left to right. Note that these models attain a higher degree of perceptual resemblance to the spring in the opposite direction from right to left. Of the models shown, the bending and twisting rod models on the right are the simplest to understand, but appear to be least like the spring coil. One might be tempted to call the multigon in (b) the only "really" explanatory model in the sense that it is seen as actually present in the spring, while the others are not. But even in the multigon, there are material elements which are not present in the spring, such as corners and straight line segments. Apparently even the multigon model is not a full candidate for the mechanism in the spring.

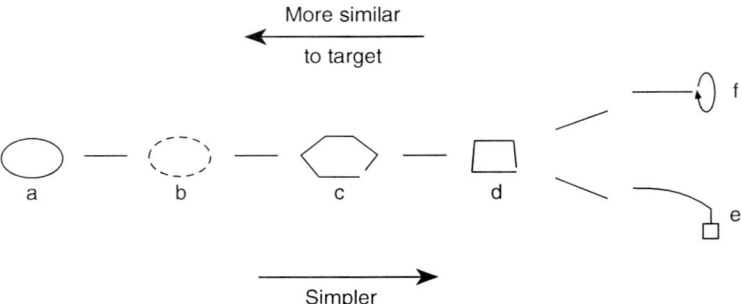

Fig. 6.11 Cases become simpler but less similar to target

6.5 Summary of Evidence for a Model Construction Cycle

Hesse (1967) and Harre (1972) describe some models in science as simplifying models where the scientist intentionally uses a model with features that are known to be somewhat different from those in T in order to make M simple enough to analyze. Although all models can be thought of as simpler than Nature in full detail, I will reserve the use of the term "simplifying model" for intentionally simpler cases such as those described above. S2's polygonal spring models appear to be simplifying models which are partially explanatory; he sees the spring as probably really twisting, as in the square coil, but not as really square. The square provides a simplifying geometry. In summary, this appears to be a case where modifications of an initial analogy with low explanatory status led to the development of a model with considerably higher explanatory status. Here I assume that the polygonal coil and zigzag spring models are simplifying models, that the extent to which they are explanatory is unknown to S2 at the time he proposes them, and that they are part of his attempts to develop an explanatory model which culminates in the torsion idea.

6.5.2.5 Can One Separate the Context of Discovery and the Context of Justification?

I consider this question in light of the model construction cycle shown in Fig. 6.9. It is traditional in philosophy to separate the contexts of discovery or theory formation (translated as model generation or modification in Fig. 6.9) from the context of justification (translated as model evaluation in the figure) in science. However, there is evidence in this protocol that loops in the cycle can at times be traversed extremely rapidly. For example, S2's criticism of the bending rod in line 5 indicates that the time interval between model generation and criticism can be as small as 15 s. In addition, his modification of the zigzag spring model in line 23 indicates that an entire generation, criticism, and modification cycle can take place within 90 s. This might suggest to some that one cannot separate the contexts of hypothesis generation and hypothesis evaluation. In fact the very idea of model evolution of this kind could be taken to challenge the separation of theory formation and justification.

I believe one's answer here should depend on the size and duration of the change process, the timescale perspective being used, and the grain size of the available evidence. While an evaluation in the form of a carefully designed laboratory experiment can take days or even years, other evaluation processes such as certain nonempirical checks for consistency can at times take place much more rapidly. When examined in a protocol on a second by second basis as was done here, it may be possible to separate the "context of model generation" from the "context of model evaluation." However, history of science tends to look at developments over a timescale of years or at best weeks. From this perspective it may be impossible to separate these two contexts for certain processes in the early stages of development. However, the subject's think-aloud information here does seem to shift between generation and criticism in the above examples. In such cases, one can separate out generation and evaluation events in the protocol, but one can also see a rapid, dialectic interplay between generation and evaluation processes.

94   6 Case Study of Model Construction in Expert Reasoning

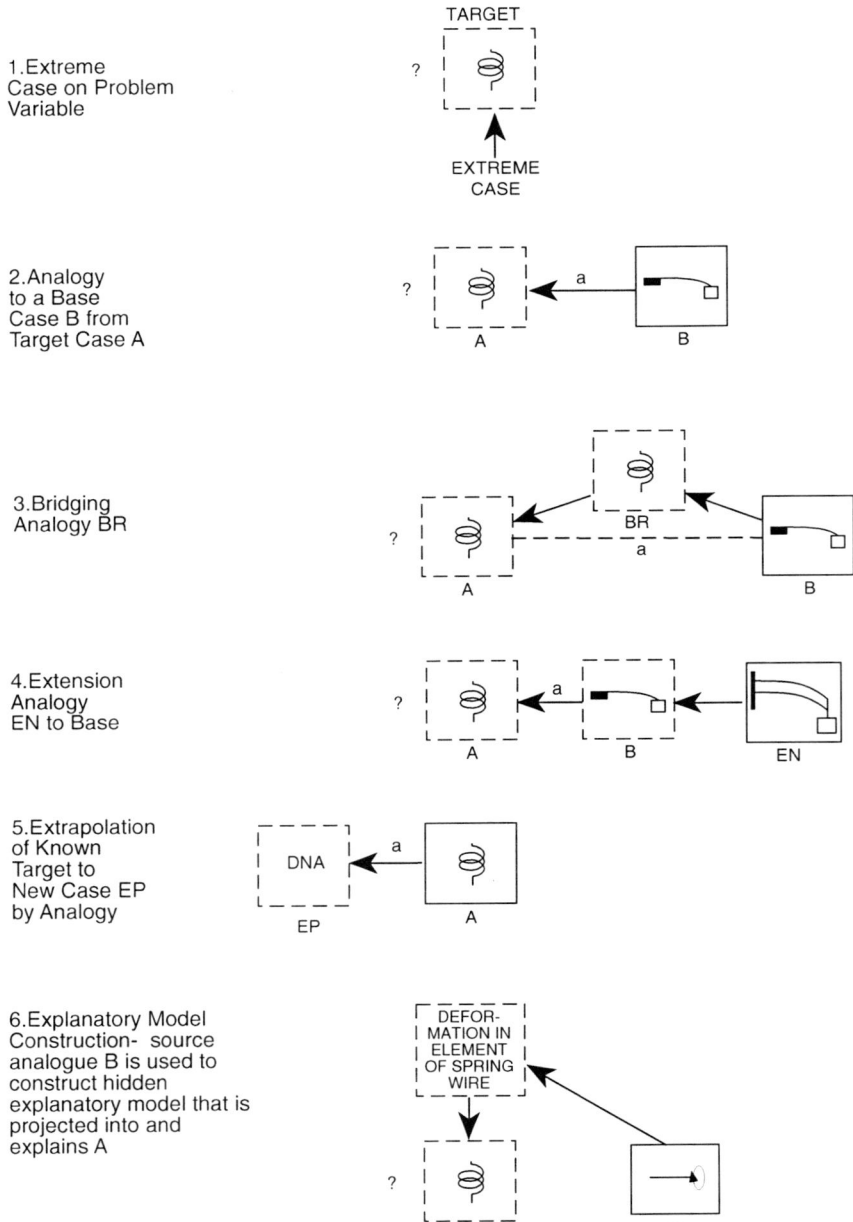

1. Extreme Case on Problem Variable

2. Analogy to a Base Case B from Target Case A

3. Bridging Analogy BR

4. Extension Analogy EN to Base

5. Extrapolation of Known Target to New Case EP by Analogy

6. Explanatory Model Construction- source analogue B is used to construct hidden explanatory model that is projected into and explains A

Fig. 6.12 Some species of plausible nonformal reasoning

6.6 Major Nonformal Reasoning Patterns in the Preceding Chapters

Figure 6.12 collects together some of the most important nonformal reasoning patterns evidenced in the preceding chapters, using the notation system developed so far. In this diagram arrows show the primary direction of information flow. One can then highlight contrasts such as the following. An extreme case on a problem variable is represented just below a target case A as a close variation, whereas an analogous case is shown horizontally apart to signify the change in a fixed feature of the problem. A bridging analogy (Chapter 4) moves closer to the target from the base in order to connect them, whereas an extension analogy (Chapter 5) may move farther away in order to anchor the base. In (6) explanatory model construction, a source analogue like the twisting rod is used to construct a hidden model that can then be projected into the target to explain it. This foreshadows a more detailed three-part view of the relations between a source analogue, explanatory model, and target that will be developed starting in Chapter 14. This diagram represents the "major species of reasoning patterns" collected so far. A species that has not been discussed but that occurs in history of science is (5) the extrapolation of a concept to a case outside its normal domain of application. For example, once the behavior or mechanisms for a spring are understood, one may consider attempting to apply this knowledge to a DNA molecule by treating it as a spring, even though it is not ordinarily viewed that way. Here, instead of having a problem and looking for a source of information, one has a source of information and looks for a problem to apply it to, a strategy that has sometimes been used by inventors and scientists with great success.

The present chapter has focused most heavily on gathering evidence that successive refinements in a scientific model via a model evaluation and revision cycle can be important in scientific thinking. This evidence argues for an *evolutionary view of gradual theory formation in small steps*. In the spirit of a dialectic, in Chapter 7, I will consider the opposite side of this augment: S2 also exhibited a *sudden insight* after a long and frustrating delay, *arguing for a revolutionary view of theory formation in large sudden steps*. Thus there is the potential to argue in both directions from this protocol, and a position on this issue will be formulated at the end of Chapter 7.

6.7 Appendix: Introduction to Concepts of Torque and Torsion

The concepts of torque and torsion can be introduced by looking ahead to Figure 14.11. If we ignore segment hb there and think of segment *ab* as a pipe and segment *ga* as a pipe wrench we are using to turn the pipe clockwise so that the pipe goes into a tight, threaded socket at b, then **torque** can be thought of roughly as the "twisting force" applied by the wrench to the end of the pipe at 'a' to turn it. The torque will be greater in proportion to the length of the wrench, r, since longer wrenches provide more leverage and more "twisting force". When F is perpendicular to r, the torque applied to the end of the pipe is equal to the force applied, F, times r.

$$T = F \times r$$

To define **torsion**, we need a different scenario. Imagine that ab is a steel rod only 1/8" thick with the end at b fixed in concrete so that the far end of the rod at 'b' cannot turn. Then if we clamp a vise grip wrench to the near end of the rod at 'a', applying the same torque will end up only twisting (deforming) the metal in every element of the entire rod somewhat, so that the near end at 'a' turns through the angle β shown in Figure 14.11 (called the angular displacement, or, informally, total amount of twist in the rod) and stops. If the rod is made of resilient metal, it will be elastic, meaning that if we remove the force F, the metal in the rod will untwist and spring back to its original orientation where β was zero. **Torsion** refers to an action that twists a material resulting in stresses and strains that makes the rod want to spring back to its original shape. If the rod is twice as long, but r and F are the same, the angle β will double. That is because the torque and resulting torsion stress will be the same as before, but there will be twice as much metal to deform under that stress, producing twice the total twist. In the protocols, subjects sometimes use the word "torsion" as defined above, but also sometimes misuse the term torsion slightly to mean torque, so they must be read in context.

Chapter 7
Creativity and Scientific Insight in the Case Study for S2*

7.1 Eureka or Accretion? The Question of Insight in S2's Protocol

I can now move to the second issue raised in the introduction to Chapter 6 – the pace of theory change. This chapter continues the analysis of the case study protocol in Chapter 6 by asking: "Does S2 make progress via Eureka events that involve sudden reorganizations, or does he progress smoothly in an incremental manner?" The answer to this question is not obvious. It seems to be possible to argue in either direction from this protocol. One can point to what appear to be sudden insights, but on the other hand, sections precede these insights in which the subject prepares the context and groundwork for having them. Some of his work fits the pattern of the improvement (GEM) cycle shown in Fig. 6.8, and that would appear to be symptomatic of a more gradual, evolutionary approach. Sometimes his methods appear to be systematic, but at other times ideas arrive in a rush, as if they are partly outside of his control. Thus, there seem to be mixed signals in the protocol on this issue.

7.1.1 Defining a Pure Eureka Event

In order to say something useful about the Eureka question, one needs to become more precise about the meaning of a Eureka event. Here I will propose an initial definition of a pure Eureka event as an extremely sudden, reorganizing, extraordinary break away from the subject's previous ideas. I use the term "extraordinary" here to refer to processes such as unconscious or supernormal reasoning that are different from those used in normal thinking. If the appearance of a new

---

*Portions of this chapter are reproduced with kind permission of Springer Science and Business Media permission from: Clement, J. (1989) Learning via model construction and criticism: protocol evidence on sources of creativity in science. J. Glover, R. Ronning, and C. Reynolds, *Handbook of creativity: assessment, theory, and research*, New York: Plenum.

hypothesis constitutes a break in the train of thought – if the hypothesis comes "out of the blue" and does not appear to be connected to the subjects' previous ideas in the protocol – this would constitute one kind of evidence for an extraordinary and probably unconscious thought process. The accretion vs. Eureka question in extreme form then becomes: Is the subject's accomplishment the result of a smooth, incremental, controlled, buildup from previous ideas? Or is it a sudden, reorganizing, extraordinary break with his previous ideas? I will consider two subissues of this question expressed by the two pairs of keywords in it: sudden reorganizing, and extraordinary break. In this section I would like to use the analysis of the protocol as an initial test bed for concepts developed to describe the quality and pace of structural change during creative hypothesis generation activities. I use the terms "structural change" and "conceptual change," as synonyms throughout.

7.1.2 Is There a Sudden Reorganizing Change in S2's Understanding?

This subquestion itself can be broken down into two parts: Is there a significant structural change? and is it a sudden change? For the latter part, a pertinent time period must be identified over which the change takes place, and a pertinent concept of "rate of hypothesis formation or modification" must be defined. I will conclude that although the torsion discovery was not a "blinding Eureka event" – an instantaneous reorganization of his ideas – it certainly was an impressive and relatively sudden breakthrough. The problem is to develop a more precise language for saying this.

7.1.2.1 Is There a Significant Structural Change?

One first needs to ask about the size of the change in representation or understanding produced by the torsion insight. Does it simply add on a small new fact? Is it a complete reorganization? The type of change in understanding to be discussed here is a structural change (change in relational structure as opposed to surface attributes) in a currently assumed mental model.

It is clear that the polygonal coil with torsion insight does not constitute a reorganization in his understanding of any domain larger than the "theory of springs" (such as the "theory of elastic materials"). However the insight does appear to add more than a simple fact; it appears to constitute the addition of a significant set of structural relations to the subject's hypothesized model of the spring system, including the new causal chain of weight causing twisting and torsion, which in turn causes resistance to stretching; and the new global effect of finding no cumulative effect of bending throughout the square spring.

7.1.2.2 Can the Insight Be Characterized as a Reorganization of the Subject's Mental Model?

In some senses it can. Torsion is a completely different geometric deformation than bending and constitutes a significantly different hypothesis. The case here would be even stronger, though, if the subject had switched completely by replacing the view of spring forces coming from bending with the (engineer's) view that spring forces come primarily from torsion. He did not go this far; instead he switched from using bending alone to using bending and torsion together in his explanations. But he did raise the question of which of these two effects predominates. Later, by the end of his interviews he believes that the stretch could be due completely to torsion, but this change of view does not occur during the insight episode.

What one can say then is that the subject achieved a major breakthrough in adding a major chain of casual factors to his model of the spring. This can be considered to be a reorganization in the sense that a new system of relationships was created. And not long afterwards the subject is considering the possibility that the old causal mechanism associated with bending may have to be rejected rather than augmented. Thus the structural change in this subject's model of the spring appears to be of intermediate size. The change process was characterized by imaginative attempts to switch to different problem representations, most of which failed. When a productive representation is found (the polygonal coil), it leads to the recognition of a completely new system of new relationships involving force, torsion, and twisting that threaten the previous view, and this is the sense in which it is "revolutionary." Of course it is on a much smaller scale than what Kuhn called revolutions in the history of science.

7.1.2.3 The Pace of Change in Understanding

I have taken a high rate of change in the subject's model as one defining characteristic of a pure Eureka event. This rate of change could be conceptualized as the ratio of the size of the change in the model's structure to the time interval over which the change takes place. The latter concept may not be easy to operationalize as an observable variable, depending on the comparisons being made and the complexity of the protocol, but it can at least play a role at the theoretical level.

It is a challenging task to point to a specific time interval in the protocol representing the "period of insight" because of the difficulty in defining the latter. As an upper limit, the time for this session was 52 min. Thus, it is certain that the subject changed from the bending rod model of the spring to the square coil with torsion and bending in a period smaller than 52 min. Viewed on a large timescale appropriate to the history of science, this would certainly be considered a tiny interval that indicates a relatively sudden conceptual change.

But much of this time was spent testing the simpler rod model and trying out other analogies, most of which were blind alleys. Can one identify a shorter period of insight within the protocol? The bending rod model was proposed within 1 min after reading the problem. Then a long period without lasting progress in model development of about 40 min ensues as the rod model is questioned, the "zigzag" models are proposed and rejected, and other analogies are tried. Finally, there is a breakthrough in a 4 min period during which the subject refers to the square hexagonal coils, makes the torsion discovery, and incorporates it into his final square coil model of the spring. When the subject finally generates the hexagonal coil toward the end of the protocol, it takes less than 80 s for him to recognize the torsion effect, and less than another 2 min to settle on the square coil as his final model of the spring. This 4 min period is therefore a candidate for the period of insight.

However, the square coil idea is considered very briefly and quite early on, only about 6 min into the protocol. But it is quickly dropped in order to consider the zigzag spring. Thirty-four minutes later, it is taken up again and leads to the torsion insight. Should this 34 min between the dropping and reemergence of an idea be counted as part of the period of insight? I will assume not, since the subject was following separate ideas during this time which turned out to be blind alleys. If one makes this assumption, one can point to this 4 min segment as a relatively sudden "period of insight." But the difficulties involved in defining the period of insight here are clear. The benefit of this exercise, however, is that it forces one to develop some useful distinctions between concepts such as structural change in a model, the period of insight, and the rate of structural change in a model.

On the other hand, the insight was not instantaneous: criticism and modification processes did occur during this 4 min period as shown in Fig. 6.8 and Table 6.1. This means that from a microscopic perspective which looks at fine-grained elements in the data, the insight appears to be "unpackable" into potentially understandable subprocesses. This leads me to describe it as "fairly sudden," rather than as an extremely sudden "bolt from the blue." This is the first sense in which the insight fails to qualify as a "pure Eureka event."

In summary, there appear to be periods in the protocol where progress is made slowly or not at all and others where progress is quite rapid. Those periods where little progress is made are frustrating to the subject but they in fact may provide necessary preparation for the later insight. The pace of structural change is uneven rather than consistent, and progress comes intermittently. When it does come, it is in the form of a relatively sudden breakthrough that involves a significant structural change in the subjects' hypothesized model.

7.1.3 Does the Subject Use Extraordinary Reasoning Processes?

The second major subquestion to the main question of whether there is a pure Eureka event in the protocol is whether S2 used extraordinary thought processes during his breakthrough. If the processes are found not to be extraordinary, one can also ask the opposite question of whether the subject's thinking is highly controlled in the sense that he always pursues a series of well defined, conscious plans and procedures.

7.1.3.1 Extraordinary Thinking

By extraordinary thinking, I mean the use of special processes which are outside of the set of normal reasoning processes used in everyday learning and problem solving. From a psychological point of view, this means I cannot imagine a plausible explanation for a particular thought process based on a sequence of inferences, associations, estimates, and criticisms. Two ways extraordinary thinking could occur during a problem solution, then, are: if the subject performs some supernormal feat of synthesis without preparation; or, more generally, if there is a break in the train of thought – a jump into a new train of thought that has no apparent connection to any previous thought. This last kind of event might be evidence for unconscious processing.

7.1.3.2 Two Types of "Breaks"

However, it is important to distinguish between a break away from the subject's currently assumed model, and a break in the train of thought. Clearly, S2 "breaks away from his initial model" of the problem. The torsion insight represents a real break (in the sense of "breakthrough") with his previous bending-rod model for understanding the problem.

On the other hand, S2's work does not contain an obvious "break in the train of thought." It does seem possible to construct a believable psychological account of his thought process as a series of connected conscious ideas. The growing series may actually look more like a branching tree or network than a single chain, and there may be jumps of attention from the end of one branch to the end of another, but the essential point is that a new idea does not appear from nowhere; it is always plausible that it was an outgrowth of the subject's previous conscious ideas.

Two major parts of S2's insight in the solution are the generation of the square coil analogy and the discovery of torsion. A plausible explanation for the torsion discovery can be given as follows. As S2 was examining adjacent sides in the newly constructed hexagonal coil model, an existing mental schema for dealing with twisting situations was activated. Such a recognition process is a common event in everyday problem solving and should not be considered extraordinary. It does happen to be a key event in the solution to this problem. He was not certain about this conjectured recognition at first, and needed to examine it critically, which led him to consider a square coil as an easier case.

In the case of the original square coil analogy, recall that it was generated while S2 was thinking about whether there was a difference between a bending rod and a single spring coil:

> 23 S2: Why should the coil have anything to do with- ? it's just so arbitrary. Why does it have to be a [circular coil]? Surely you could coil a spring in squares, let's say, and it... would still behave more or less the same.

This is a highly creative idea but not one that necessarily involves extraordinary reasoning. Here the subject appears to be imagining ways to bend a piece of wire into a spring. The plausible normal reasoning process is one of imagining a simple transformation one could perform with one's hands.

7.1.3.3 The Worth of This Idea was Not Recognized Immediately

Only after thinking hard about and confirming the lack-of-bending anomaly in the spring does S2 return to the square coil idea in line 117 and use it productively. Here there is a branch or sidetrack in the train of thought, but the return to the square coil idea can be seen as connected to its earlier appearance.

In some cases, the connection to a previous idea may be a weak one – a loose association or conjectured recognition or playful transformation rather than deductive inference or a precise subquestion. Associations, transformations, and recognitions in this light are divergent, unpredictable, and sometimes highly creative processes, but not extraordinary ones in the sense of being unconnected to the network of current representations. I consider S2's overall achievement – the marshalling and orchestration of a large number of reasoning processes to produce the invention of new explanatory model elements – to be remarkable in the sense of being unusually productive and creative. However, I can see no evidence that the reasoning processes he uses, taken individually, are extraordinary. The train of thoughts S2 reports weaves a "coherent story" in the sense that each new idea appears connected to previous ideas.

S2's ideas are also connected by the specific relationships implied in Fig. 6.9 in which new ideas can grow out of modifications of or reactions to past ideas. This is an even more specific sense in which his insight did not emerge from "out of the blue," and it will be discussed further in the section on creative processes below.

It should be noted that Tweney (1985) cites evidence to discredit the idea that Faraday's discovery of induction was a "bolt from the blue," as some have thought; and Perkins (1981) came to the conclusion, after reviewing the literature on insight in creative thinking, that there is no convincing body of evidence that insights occur via special or extraordinary processes. This does not eliminate the possibility that such special processes might exist, but it does indicate that it has been difficult so far to find convincing evidence for them.

7.1.4 Defining "Insight"

I have discussed some senses in which S2's protocol does not provide evidence for a pure Eureka event. In this section I will propose some criteria for a less extreme, but still very impressive kind of event I will term a "scientific insight." In order to sort out the different senses in which S2's solution is and is not an example of insight behavior, it will be useful to refer to the following list of the features of his polygon with torsion breakthrough which are insight-like.

7.1 Eureka or Accretion? The Question of Insight in S2's Protocol

1. The breakthrough is an important idea:
   (a) It is a key idea – an important component of a solution.
   (b) It overcomes a barrier that blocked progress; it comes after a frustrating series of false leads and blind alleys – after a period where little progress has taken place; it resolves an anomaly.
2. The breakthrough adds significantly to the subject's knowledge. It produces a large structural change in the subject's model where he:
   (a) Identifies new variables or causal factors in the system
   (b) Identifies a new hypothesized mechanism in the form of an explanatory model element
   (c) States that it increased his understanding
3. The subject's ideas are generated fairly quickly during the breakthrough, and he achieves rapid subsequent progress towards a solution.
4. The breakthrough is accompanied by more complex phenomena:
   (a) It is accompanied by indicators of emotional response – surprise, joy, satisfaction.
   (b) The subject realizes immediately that something important has been discovered.

On the other hand, the following are senses in which S2's breakthrough, however insightful, was not a Eureka event:

1. The breakthrough idea was not generated extremely suddenly without preparation.
2. It did not involve the total replacement of one hypothesized model with another.
3. There is little evidence that it was:
   (a) An extraordinary thought process
   (b) An unconscious process
   (c) A break with all previous trains of thought

One can now use the criteria developed in the above list to define three categories of insight behavior. The categories (designed to refer to hypothesis development activities) are "breakthrough," "scientific insight," and "pure Eureka event," defined in increasing order of specificity and unusualness so that the "breakthrough" category includes "scientific insight," and the "scientific insight" category includes "pure Eureka events."

A breakthrough is a process that produces a key idea – an important component of a solution – and that overcomes a barrier that can block progress toward a solution.

A scientific insight is a breakthrough occurring over a reasonably short period of time leading to a significant structural improvement in one's model of a phenomenon. That is, it constitutes a shift from the subject's previous way of representing the phenomenon and leads to an increase in understanding of the phenomenon, as determined by the evaluation process in Fig. 6.9. This is the descriptor that appears to me most appropriate for S2's breakthrough.

A pure Eureka event is a scientific insight where: (1) there is an extremely fast emergence of a new idea with little evidence of preparation; (2) the new idea is a whole structure replacing the subject's previous model or understanding of a situation; (3) the process is not explainable via normal reasoning processes; extraordinary thought processes that are unconscious or different from normal thought processes are involved.

This recasts the earlier initial definition of a pure Eureka event (an extremely sudden, reorganizing, extraordinary break from the subjects' previous ideas) in a way that relates it to other types of insight behavior. For some purposes, reducing everything to these three categories may be less important than having something like the above list of features for describing different ways in which an idea can be insightful. But the three terms may provide a useful shorthand.

7.1.5 Summary

This section has attempted to answer the question: "Was the polygonal coil with torsion more like a sudden Eureka event or an example of steady accretion?" Near the beginning of Chapter 6 I conjectured that analogy had developed a reputation for being a creative process and that perhaps it could be a mechanism underlying Eureka events. Indeed analogy was shown to play a central role in S2's discovery. But when one examines the thinking aloud case study microscopically over minutes on a small timescale, one sees an arduous dialectic process of conjecture, evaluation, and rejection or modification of analogue models that precedes the breakthrough, as opposed to an event that takes place instantaneously and effortlessly.

However, there was also a case to be made against accretion. After a long and sometimes frustrating period, a single analogy generated by the subject led to a fairly sudden insight which led to the formation of a new hypothesized model. Thus, insight processes were found which are not accretionist in character and which support a view of scientists as capable of significant reorganizations in a relatively short period of time.

However I argued that it should not be considered to be a product of extraordinary thinking that cannot be explained – the processes do not appear to be supernormal or unconscious ones. The upshot of the present analysis, then, is that rather than being an example of a pure accretion or pure Eureka process, the pace of progress is uneven, with "more revolutionary" and "less revolutionary" periods of work. S2's breakthrough can be characterized in the above terms as an impressive scientific insight triggered by a series of analogies, but not as a pure Eureka event.

7.2 Creative Mental Processes

In theory, the processes in the model construction cycle shown in Fig. 6.9 can be divided into two main categories, the productive processes of generation and modification and the evaluative processes of empirical testing and rationalistic evaluation. In this section, I examine questions about these individual processes and how they interact. First, Evaluative processes will be discussed with respect to the role that anomalies play within them. This leads to the suggestion that a tension condition indicated in the protocol is partially analogous to the motivating tension between an anomaly and a persistent paradigm in science. In the following section I will discuss the roles of transformation and invention in the generation of analogue hypotheses. These processes can provoke the recognition of a new principle revealed within a novel construction. In a final section, I discuss the role of divergence and constraint in productive processes, leading to the view that these processes are less constrained and convergent than established procedures, but more constrained and "intelligent" than a blind selection and variation process.

7.2.1 Anomalies and Persistence in Protocols and Paradigms

In this section, I take a more detailed look at the dialectic view of model construction as a cyclical process of generation, evaluation, and modification in an attempt to provide a deeper level of explanation for the phenomenon of extended periods of little progress between insights in the protocol. Table 6.1 outlines evidence in the protocol for the presence of such a dialectic process. A striking feature in watching the tape is the strenuous activity that S2 poured into this process. Even for those who admit that analogies can play a role in scientific discovery, a common view is that a subject may be passively reminded of an analogous situation C, and be able to transfer or infer a prediction from C back to the problem. The image is of the insight "coming to the subject" as a passive receiver. In the present case, the subject is much more active: inventing tentative analogies, rejecting a number of them, pursuing those that have promise by criticizing them, and modifying them until he is satisfied he has a valid model. A more apt informal image here is a constructivist one of the subject "aggressively constructing and testing different models in an effort to capture an understanding of the phenomenon."

What drives all this strenuous activity? In particular, why does the subject persist in criticizing his understanding when he is already 90% sure that the wider spring will stretch more? Why is there a period of very little progress followed by a period of insight? In this section, I attempt to speak to these questions in terms of conflict between a persistent model and a perceived anomaly.

7.2.1.1 Dialectic Tension

There is a palpable tension obvious in the first section of the video tape that is conveyed only to a limited extent by the transcript: a frustration with not being able to resolve the anomaly of the lack of bending in a helical spring. For example, in lines 87 and 111, he says:

> 87 S2: [I]f you start with a [stretched] helix and unwind it...you should get a bow [bend], but you don't. I mean visually imagining it, you don't. I don't see how you could make the bow go away – just to wind it up – Damn it!

> 111 S2: Darn it, darn it, darn it...why should that [the difference between a rod and a coil] matter?

The tension apparently occurs between the rod model, and the lack-of-bending anomaly. It bothers him enough to drive him to keep searching for a way to modify the rod model or replace it. This search takes up the better part of the 52 min interview which is peppered with expressions of frustration. Line 178 provides evidence that the reason for his dissatisfaction has to do with an important difference between having a confident prediction and having a feeling of understanding. I take this as an interesting example of a situation where good performance is not equivalent to deep understanding, and, because of the subsequent events which raise his confidence, I take the important difference to be the lack of a satisfying explanatory model.

7.2.1.2 Persistence of the Initial Model

The persistence of the bending rod model, with its image of the spring coil made of segments, each of which are bending, appears to be an example of an Einstellung effect (Maier, 1931; Luchins, 1942); a problem space or method dominates the subject's thinking, and prevents him from generating necessary new ideas. In order to make progress, the subject must redescribe the problem using new descriptors; he needs a new problem representation. But even though he proposes rejecting the model several times, he is repeatedly tempted to return to it. It is as if the idea has an autonomous "life of its own."

Surprisingly, this is related to an observation made in preschool children by Karmiloff-Smith and Inhelder. They found that children, in modeling a (simpler) system, would often switch from believing in one cause to believing in a different cause after experimenting with the system. However, before changing to a completely new hypothesis, the children often went through an interim period where they thought there were two causes (the old one plus the new one). They did not see that the new one had the potential to explain all the data. This suggests that this type of perseverance may be a very deep property of our thinking.

7.2.1.3 A Powerful Anomaly

Pitted against this persistent model is a powerful anomaly. Bending in the vertical plane is central to the rod model, but he cannot imagine a way for bending to take place in a helical spring without producing an abnormally shaped spring. I follow S2 in using the term "anomaly" here. Precisely how we view the cause of dissonance in S2 depends on a cognitive interpretation: If the increasing slope idea conflicts with a memory of prior observations of uniformly stretched springs, then that memory can be thought of as an empirical anomaly for the bending model. But if the model conflicts with a sense a symmetry about how a spring should stretch, then it is a theoretical inconsistency. Since the choice is not clear from the transcript, I have been using the term anomaly here in the broad sense of a new finding which conflicts with previous ideas, whereas in some narrower usages, its referent is limited to a conflicting observation. This distinction is not very important here however, since we are simply interested in identifying a source of dissonance that motivates further work.

7.2.1.4 Analogy to the Persistence of a Paradigm

There may be a partial analogy between S2's persisent Einstellung effect and Kuhn's idea of the persistence of a paradigm in science (Kuhn, 1970). Even when anomalies are known to exist, it is difficult to reject a paradigm until something better is found to replace it. But this is very difficult to do since it requires breaking out of the current, stable point of view. As discussed in Chapter 21, several of Kuhn's theses have become controversial. However, the present findings suggest an intermediate position that is compatible with at least some important Kuhnian ideas including resistance to change and the role of anomalies.

When the polygonal coil with torsion model is found, it appears to finally break the tension. Here the bending rod model is hard to reject until the better model is found, and this requires a great deal of imaginative effort. Compared to a problem on the frontier of science, the scale here is, of course, very much smaller and easier. For example, there are no social forces to reinforce the cognitive stability of the subject's initial model. Nevertheless, this tension between a persistent initial model and a recognized anomaly, which helps to explain the long period of slow progress followed by a period of scientific insight in the protocol, has interesting similarities to the persistence of a paradigm in Kuhn's descriptions.

7.2.1.5 Anomaly as a Source of Motivation; Analogy as a Source of Divergence

Furthermore, the tension associated with his dissatisfaction with his understanding apparently drives him to keep reattacking the problem repeatedly until he makes a breakthrough. Here it appears to require something as divergent as analogy generation to break out of the Einstellung effect formed by a persistent inadequate model. In the present situation the generation of a new or sharply modified model is required in order to break the deadlock; and it is in such cases that analogies should prove to be particularly useful, since they help the subject break away from his current model. When they are successful, they apparently can lead to fairly large and rapid changes in a mental model. S2 considers no less than 12 analogous cases during the protocol, including some that do not appear in the transcript excerpts given here, and this high degree of generative activity can be seen largely as a response to the tension urging him to find a more satisfying model. Thus, this example suggests that the tension between a previously established model and a prominent anomaly can be a major driving force behind hypothesis generation. This would seem to occur for individuals who have a high standard for the degree of coherence between their new models and other knowledge.

7.2.1.6 Evolution or Revolution? Or Punctuated Evolution?

Thus the phenomenon of intermittent progress involving periods of little progress punctuated by occasional insights can be seen as a natural outcome of psychological processes that may also operate in real science. We can refer to this metaphorically by using the concept of *punctuated evolution* from biology, which describes relatively sudden intermittent events as well as smooth periods of evolution in nature (Gould and Eldredge, 1977). In short, instead of evolution, *or* revolution, this case study exhibits evolution *and* revolution. The evidence in Table 6.1 pointed to a construction cycle of model generation, evaluation, and modification that can drive a gradual evolution of better and better models. However, the cycle can apparently "get stuck" due to Einstellung effects, leading to periods of inertia and stagnation of ideas. At these times, one needs a more revolutionary event in the form of an insight, to break out of the stalled pattern. His eventual insight accomplishing this constitutes a "mini-revolution" that resembles the kind of "Gestalt switch" Kuhn identified in scientific revolutions, although here it is much smaller in scale.

The GEM cycle in Figure 6.8 can be used to provide a first-order account for this punctuated evolution pattern by identifying smaller evolutionary changes with the uppermost cycle on the right hand side, and larger ["mini-revolution"] changes with the lowermost loop going to the left hand side, in which the modeling process is "restarted" anew as the result of the subject becoming discouraged with the prospects for the present model. Something like this dual loop feature is needed to account for evolutionary model improvement *and* radical shift behavior.

While this provides a very basic first order model of a process that could produce evolution and (mini)revolutions, it does not answer the question of how S2 is able to break the powerful Einstellung effect he was stuck in by generating a scientific insight. We will revisit and speak to this question in Chapter 16, by considering voluntary strategies for modulating the level of divergence in investigations, and an involuntary process of volatile activation during mental simulation.

7.2.2 Transformations, Invention, and Memory Provocation

7.2.2.1 Transformations as a Source of Creativity

From this protocol can one point to any processes that are particularly important for creativity? For one, transformations appeared to be a very important source of creative or divergent ideas in this protocol. For example, after considering the bending-rod case, in line 23, S2 says: "Surely you could coil a spring in squares, let's say, and it would still...." Here the subject seems to be constructing a new case by transforming the rod into a square coil rather than making an association to an existing idea in memory. Also, in line 37 the double length spring analogy originates from the transformation of sliding a weight along a wire. And in Chapter 3, it was found that of the analogies generated by ten subjects in solving the spring problem, more were generated via a transformation than were generated via an association (Clement, 1988). More broadly in this book, a transformation in the physical world is defined as an action that modifies one or more features of a system to produce an altered system. Correspondingly, a mental transformation is an action that modifies one or more features of a representation of a system to produce an altered representation. Using this definition, one can say that the modification process referred to in Fig. 6.9 is a mental transformation applied to the previously hypothesized scientific model. Although association often is cited as a primary source of creativity, it may be that transformations are just as important, if not more important, in scientific problem solving. This book will concern itself only with mental transformations, and that will be the intended meaning when "transformation" is used without an adjective.

7.2.2.2 Invention of Analogous Cases

The novelty of the zigzag and polygonal springs supports the claim that they are invented cases. Although analogous cases are traditionally thought of as schemas already in long-term memory which are activated or retrieved during problem solving, it can also happen that the analogous case is invented along with the analogy relation. For example, the square coil was apparently invented via a transformation, not recalled from memory. Models generated by inventing an analogous case are in this sense even more creative than those generated by being reminded of an analogous case.

The polygonal coil is a new problem representation amenable to a new method of analysis (torsion). In such an instance, the knowledge that one gains from an analogous case C need not be "stored in" C. Thinking about C may activate a useful schema (such as torsion) which has not previously been applied either to the original situation to be explained or to C. This instance provides some support for Black's view that the interaction between the original and analogous cases can produce knowledge in the form of an insight that was not residing beforehand in either the original or the analogous cases: "It would be more illuminating in some of these cases to say that the metaphor creates the similarity than to say that it formulates some similarity antecedently existing" (Black, 1979, p. 37). In the present case study, in contrast to the usual view of analogy generation, the recognition of the key relationship (torsion) in the analogous case occurs well after the generation of the analogy. The analogy plays a *provocative* role in activating a principle whose applicability was previously unrecognized, rather than a "direct source of transferred information" role (Clement, 1988). In Chapter 20, I will discuss the role of *provocative starting point for developing a solution* as one of several new roles for analogy to be added to those already listed in Table 5.3. Thus some analogies are invented rather than recalled, and some play a "provocative" role in accessing new information rather than a "direct inference" role.

7.2.3 Productive Processes: Constrained Successive Refinement vs. Blind Variation

Hypothesis generation and modification processes are sources of creativity within the model construction cycle. To what extent are these processes random or constrained? Some of the divergent processes in the protocol that precede the torsion insight are associations, transformations, the activation of analogous cases in memory, and the invention of new analogous cases. This leads to the following question: "Are S2's processes so divergent as to constitute a random trial and error process"? Certainly S2's divergent thinking seems to be less systematic or formal than either logical deduction or methodical procedures of induction. And yet this less formal method allows the subject to make impressive progress in his understanding. In this process, it does not matter so much if one makes a faulty conjecture; it may still be possible to transform it into a successful conjecture by carrying out a series of criticisms and modifications. In this section, I discuss the sense in which the subject's successive refinement process goes beyond a random trial and error strategy.

The cycle in Fig. 6.9 may constitute a random trial and error process in those instances when the old hypothesis is discarded and a totally new random hypothesis is tried on each cycle, without learning or attempts at modification between cycles. A somewhat less divergent strategy would be to randomly modify a *part* of the previous hypothesis and keep the other part in the next cycle. This is analogous to a random variation theory of evolution. (See Campbell, 1960, for an exposition of this analogy.) However, one can argue that the generation and modification processes are not random ones in the case of S2.

First, this subject uses the generation of spontaneous analogies as a strategy for generating hypotheses. Analogous cases are generated primarily by association or transformation processes which means that they are *connected in some way to the target*. The connection may not be a strong one, but this is probably better than no connection at all. The second type of evidence indicates that at times, a *conscious constraint* is held in mind while a new association or transformation is being generated. For example, in line 57, S2 appears to focus on the idea of stretching as a constraint as he generates several tentative analogies by association after asking himself, "what else stretches"? In a second example in line 117, he generates polygonal coils after attempting to "generate ideas about circularity...why should it matter? How would it change the way the force is transmitted," in the spring? The use of constraints during generation is one sense in which the model construction cycle can go beyond a random variation and selection process.

A further kind of evidence is the observation of an intelligent modification process in the cycle. Most of the analogies generated by S2 were rejected in the end. But several did clearly serve as stepping stones by preparing the way for suggesting better ideas later on. Criticisms of the earlier model are used to suggest modifications that can repair them. This gives the cycle the property of successive

7.2 Creative Mental Processes

refinement, in which one can learn from the mistakes of the past. For example, the first zigzag spring in line 23 is criticized as a model because of the contaminating effect of bending at the joints. This is then modified into a second zigzag model with stiff joints which is aimed at removing the criticism. As a second example, the hexagonal coil model is changed to the square coil because in S2's words, it "unmixes the bend from the torsion" in a simplifying way. In these instances the subject seems to generate or search for modifications which remove particular difficulties that the evaluation process has identified in an existing model. Thus the cycle involves nonrandom, intelligent modification based on information about prior difficulties. This is a particularly powerful way in which modification processes can be constrained. (See Darden (1983), Rada (1985), and Darden and Rada (1988) for a further discussion of nonblind hypothesis generation, including the use of interrelations between scientific fields as a heuristic. Also, Holland et al. (1986) discuss goal-weighted summation of activation as a possible mechanism for guiding retrieval of relevant information, while Lenat (1977, 1983) discusses heuristics for learning by discovery in mathematics.

7.2.3.1 Less Constrained Methods

Not all generation methods are highly systematic or constrained however. The generation of the double-length spring analogy in line 37 provides an interesting example. Here the analogy originates from the idea of sliding a weight along a rod. He then imagines this transformation happening on the spring itself, as if it were simply an "interesting thing to try." There is some evidence here that the subject is exploring new and uncertain directions rather than trying to achieve a specific goal using a conscious strategy of generation under constraints. Although the analogy in this case does not lead to a breakthrough, one cannot rule out the possibility that the ability to think playfully in a relatively unconstrained manner can at times be a powerful method.

7.2.3.2 Summary

Thus I arrive at an intermediate position concerning the nature of the subject's hypothesis generation and modification processes. Compared to a pure Eureka event, they form a more ordinary and connected train of thoughts. Compared to a problem-solving process governed by established procedures, they include processes that are relatively ungoverned and divergent. Rather than simply accessing prior knowledge about systems, they can produce new systems that are novel inventions like the polygonal coil as well as a presumably infinite variety of other representations. As used here within the model construction process however, they often appear to be more constrained and "intelligent" than a blind variation and selection process, exhibiting an intermediate level of control.

7.3 Darwin's Theory of Natural Selection

Having reviewed some philosophical views of hypothesis formation processes in science and having presented some findings from expert protocols, I will briefly consider a third approach to the study of creativity in science: the analysis of notebooks and other historical documents produced by innovative scientists. I return to the example of Darwin's theory of natural selection mentioned at the beginning of Chapter 6. Early writers had described the origin of this discovery as the net result of a gradual buildup of information – a process of accretion that occurred during Darwin's voyage on the Beagle, principally in South America. However, Gruber (1974) countered this view by pointing to evidence in Darwin's notebooks indicating that after the Beagle's voyage, he, like a number of other naturalists, believed in the existence of evolution (gradual change in species) but still had no model to explain it. He lacked the theory of natural selection. It was only after a year and a half of conceptual struggle after his return to England that Darwin was able to formulate a satisfactory theory. A particularly famous piece of evidence arguing against the accretion view is the role of an analogy that occurred to Darwin when he read Malthus. In his autobiography (written much later) he wrote:

> I happened to read for amusement Malthus on population, and being well prepared to appreciate the struggle for existence which everywhere goes on from long-continued observation of animals and plants, it at once struck me that under these circumstances favorable variations would tend to be preserved, and unfavorable ones to be destroyed. (Darwin 1892, p. 42–43)

Darwin saw that factors similar to those that limited population growth in man (such as a limited food supply) might be a source of a selection factor in a survival of the fittest model for animals. Thus, the accretion by induction view is hard to maintain in Darwin's case.

Does the Malthus episode then provide evidence for a Eurekaist view of Darwin's achievement? The analyses of Darwin's private notebooks carried out by Gruber (1974) and Schweber (1977) argue against this opposite extreme as well. They show that Darwin struggled long and hard after returning from the voyage of the Beagle, considering and rejecting or modifying several hypotheses and gradually fitting a large number of pieces together into the theory of natural selection. The notebooks indicate the analogy from Malthus was only one event in a complicated process of model generation, evaluation, and modification.

Darwin read widely in fields outside of biology, and apparently drew analogies from several of these fields in constructing his theory, including the ideas of variation and selection (from breeding in domestic husbandry), and the idea of natural competition (from Malthus as discussed earlier) (Darden, 1983; Millman and Smith, 1997). Some believe Darwin also was influenced by the laissez-faire economics of Adam Smith which showed that an ordered and efficient economy could emerge from free competition. In addition, Gruber (1974) cited Darwin's early geological theories on the growth of Pacific barrier reefs over tens of thousands of years as fertile preparation for the idea that small individual forces acting over long periods of time could effect vast changes in nature.

Thus historical evidence in Darwin's case now supports a more complex view than either accretionism or Eurekaism. *Both* selected sets of many individual observations and the nonempirical insights formed by key analogies to other fields were apparently very important in Darwin's case. We have less direct evidence on how large Darwin's moments of insight episodes were, although it is clear that the overall process took a long time, with a large number of revisions, and he himself claims to have had moments of insight. It seems reasonable that the pace of change could have been uneven, with intermittent progress and a number of important, but not all encompassing insights. These studies suggest that a more realistic hallmark of genius than large and pure Eureka episodes is the ability to generate a variety of tentative analogue models and to carry out the long struggle of repeated conjectures, criticisms, rejections, and modifications necessary to produce a successful new theory. Both model evolution and the occasional insight are important. Although the timescale is much longer in Darwin's case, it is interesting that these are the same distinguishing criteria that emerge from the most impressive cases of model construction in the protocols discussed so far.

7.4 Initial List of Features of Creative Thinking from This Case Study and Remaining Challenges

7.4.1 Creative Thought

To the extent that an extended analysis can remove the initial subjective impressiveness of an event, perhaps I am in danger here of seeming to trivialize the processes of analogy generation, model construction, and insight as hypothesis development activities, and I would like to avoid giving that impression. Once one has thought through the answer to a problem, the solution process can appear to be less impressive or even obvious from hindsight. While one is actually solving a problem, however, creative reasoning such as that exhibited by S2 is impressive in a number of ways:

1. First, there is the *insight* in the protocol which seems to lead to a "flood" of ideas. The speed of progress during this episode is impressive, and it argues against an accretionist view of the pace of change.
2. S2's central achievement is the generation of a new structural hypothesis – the *invention of a new model of hidden mechanisms* in the spring that he has never observed. This involves the identification of new causal variables in the system (such as torsion) and new causal chains, as well as the identification of a new global effect (lack of cumulative bending).
3. An important factor in producing this achievement is the subject's *desire to ask "why" questions* and to seek a deep level of understanding beyond what is required for the solution of the immediate problem. Presumably, this urge to penetrate surface features and conceptualize an underlying explanatory model at the core of a phenomenon is a basic drive underlying creative theory formation in science.

4. He exhibits a remarkable *persistence* in this quest in the face of recognized internal inconsistencies and repeated failures. There is something of an existential twist here: although the problem has no practical significance for the subject, he puts enormous energy into the problem of understanding as a challenge for its own sake.
5. Scientists can get stuck. His playful and uninhibited inventiveness in producing conjectures and modifications of the problem counters this. The analogous cases he generated in searching for a better way to represent the problem included the bending rod, polyester molecules, leaf springs, watch springs, two types of zigzag springs, two or more types of polygonal springs, and double-length springs. He displays an ability to think *divergently* and the *flexibility* to modify thought forms in novel ways.
6. There is a *willingness to vigorously criticize and attack the validity of his own conjectures*. S2 is able to engage in a dialectic conversation with himself, proposing new ideas on the one hand, and criticizing them on the other. This seems to require viewing the failure of any single idea as not very important; although as has been shown, the apparent failure of five or six ideas does lead to some degree of frustration for S2.
7. Since the subject does not have access to experimental apparatus, it is remarkable how far he takes the development of his model without new empirical input, and therefore one reason for my interest in this protocol has been as evidence for the *power of nonempirical processes*. (I will discuss and qualify this position in Chapter 21).
8. With respect to Fig. 6.9, one can contrast the productive function of the generation and modification processes with the evaluative function of the rationalistic and empirical testing processes. The divergent and creative generation processes (such as the use of analogies) represent a significant departure from the more systematic, rule-governed processes of theory growth envisioned by inductionists, who would tend to see them as much too unrestrained to be part of the disciplined scientific enterprise. However, the generation processes are not entirely unconstrained, as has been discussed, and the evaluation cycle in Fig. 6.9 provides some strong constraints which can in fact act to control the enterprise of model construction. Thus, *alternating between generative and evaluative modes in scientific thinking is seen as a powerful method*, even when new empirical tests cannot be performed.
9. I concluded that S2's protocol contained examples of both evolution *and* a revolution in the form of an insight (of more modest size than a scientific revolution) and that both were important. A simple initial model for this "punctuated evolution" pattern was provided by the uppermost and lowermost loops in Fig. 6.8. More detailed models will be examined in Chapter 16.
10. Moving beyond data-based inferences, I conjecture that perhaps his awareness of his own ability to criticize ideas, and the resulting faith in himself as a self-correcting system, allows him a freer hand – allows him to be more uninhibited in generating conjectures and considering directions to pursue. It may be

that generative ability and critical ability are mutually supporting. Critical ability gives one the freedom to be unusually associative or inventive. Generative inventiveness, or the ability to replace and repair what one removes, gives one the confidence or assurance to be critical of and to at times tear down existing ideas. S2 seems willing to consider "risky" analogies such as the double-length spring and the bending rod that appear to be very different from the original problem. However, it has been shown that even when a risky initial analogy does not turn out to be explanatory, modifications of it may lead to an explanatory model. Realization of *this potential for debugging or redesign via criticism and modification may allow one to feel freer to explore more imaginative models* or a wider range of models. This freedom in turn would appear to be an important tool in the difficult job of breaking out of previous conceptions of the target situation. *Again, rather than the ability to hit on the best possible idea in one stroke, it may be that it is the ability to engage in a cycle of hypothesis construction and improvement that is the most viable form of scientific thinking.*

The above qualities appear to be some of the most impressive characteristics of creative thinking visible in the case study.

7.4.2 *Limitations of the Case Study*

Two areas which the present chapter does not address are question formulation and empirical investigation. Also S2's strong drive to ask "why" questions mentioned earlier, a kind of curiosity, has not been explained. Even though such gaps remain, the conclusions reached here suggest that creative hypothesis formation processes are not outside the realm of possible study.

Some topics that I will take up later in this book include the presence of multiple interruptible goals in such solutions, and the balancing of divergent and convergent processes. These speak to unanswered problems of complexity in S2's thought processes, including returns to previously attempted solution paths, and the resolution of competing influences. This is part of the general problem of insight as well as how "guided" conjecture is – why one person's initial conjectures are much more fruitful in the long run than those of others. In addition, each of the subprocesses shown in Fig. 6.9 is in need of more detailed study. Finally, S2's flexibility in inventing new problem representations is hard to model. His image of the spring appears to be malleable; he appears capable of modifying it into an infinite number of forms and variations. This suggests that spatial reasoning and imagery may be involved. In fact there are a number of spontaneous imagery reports in the protocol which suggest that certain forms of spatial reasoning on spatial representations may be central to S2's thinking here. This opens up a large and important question of the nature of these imagery processes and the role they play in scientific thinking, and this topic will be taken up in Sections IV and V.

Section III
Creative Nonformal Reasoning in Students and Implications for Instruction

The principle goal of education in the schools should be creating men and women who are capable of doing new things, not simply repeating what other generations have done; men and women who are creative, inventive and discoverers, who can be critical and verify, and not accept, everything they are offered.

Jean Piaget

(Quoted in Education for Democracy, Proceedings from the Cambridge School Conference on Progressive Education, Kathe Jervis and Arthur Tobier (Eds.) (1988).)

Chapter 8
Spontaneous Analogies Generated by Students Solving Science Problems*

So far this book has concentrated on expert reasoning on unfamiliar explanation problems. In this section we begin to ask whether there are any educational implications of these findings. However, it is possible that students are so different from experts that it is difficult to apply findings on expert reasoning and learning strategies to the problem of student learning. There is an existing literature on expert/novice differences. In addition to differences in amount of content knowledge possessed by experts and novices, it is believed that the structure of this knowledge is different (Chi et al., 1981; diSessa, 1985). Other research indicates that students harbor persistent preconceptions in all of the major science areas and that these can prevent the assimilation of new material. This makes teaching much more difficult and Chapters 9 and 10 will in fact look at learning processes which address this problem.

Another possibility is that, in addition to possessing different content knowledge, concepts, and beliefs, experts may also use a different set of reasoning processes than the naive student uses. Having presented evidence that experts use analogical reasoning, this chapter asks whether students do as well.

Textbooks and teachers often use analogies, and they are often recommended as a teaching tool in science education. But we suspect that in order for analogies to help students with their conceptual difficulties, the students need to do the reasoning, not just memorize the analogous cases. Thus, it is relevant to ask whether there is evidence that students can reason analogically when solving problems. This chapter attempts to examine whether novices generate analogies spontaneously, and, if so, to determine the characteristics of these analogies.

---

*Portions of this chapter are based on findings reported in: Clement, J. (1989). Generation of spontaneous analogies by students solving science problems. Topping, D., Crowell, D., and Kobayashi, V. (Eds.), *Thinking across cultures* (pp. 303–308). Hillsdale, NJ: Lawrence Erlbaum

8.1 Use of Analogies by Students

8.1.1 Frequency

To do this I will discuss the results of a set of think-aloud interviews. Sixteen freshmen engineering majors who had not taken college physics were each given a set of six qualitative physics problems and asked to think aloud in the presence of the interviewer (the author). Tapes of the interviews were examined in order to determine whether the students had spontaneously generated any analogies during their solution processes. Recall the definition from Chapter 2 that a spontaneous analogy occurs when the subject, without provocation, refers to a different situation B that he believes may be structurally similar to the original problem situation A. In fact, the students generated a large number of analogies in solving the problems, 59 in all. Of the 96 problem solutions, 24 of them (or 25%) contained analogies, as shown in Table 8.1.

By *significant, articulated analogies*, we mean analogies that were fairly clearly articulated and were used by the students to generate or add support to their problem solutions. This category excludes analogies that were vague or not pursued at length. In view of the fact that informal arguments and divergent thinking on the part of students are rarely encouraged in secondary schools, I was surprised at the relatively high number of analogies spontaneously invented by these students and used during their solution processes.

8.1.2 Features of Spontaneously Generated Analogies

We were particularly interested in the 34 significant, clearly articulated analogies, which occurred in 18 (or 19%) of the solutions. As a result of analyzing observable differences in these analogies, we proposed some basic categories of features of analogies. Table 8.2 shows how many of the significant analogies shared each feature. (Comparable percentages could not be determined for the entire sample of 59 analogies since a number of them were not sufficiently clearly articulated.)

Table 8.1 Spontaneous analogies generated by students

Subjects	N = 16
Problems solved by each	6
Problem solutions	96
Solutions containing one or more analogies	24 (25%)
Solutions containing one or more significant, articulated analogies	18 (19%)
Total number of analogies generated	59
Number of significant, articulated analogies	34

8.1 Use of Analogies by Students

Table 8.2 Types of analogies generated

N = 34 Articulated analogies	
Correct prediction	28 (82%)
Personal (vs. Physical)	18 (53%)
Invented (vs. Factual)	6 (18%)
Evaluated validity	5 (15%)
Successively refined series	3 Groups involving 11 Analogies

Several kinds of observations can be made about this collection of spontaneous analogies.

1. *Correctness.* Only six (18%) of the 34 significant analogies were incorrect in the sense that they led to an incorrect prediction.
2. *Personal vs. physical analogies.* The problems in the set were qualitative physics problems, thus one might assume that students would tend to generate analogous situations that were physical rather than personal in nature. However, as shown in Table 8.2, 18 (53%) of the analogies referred to body actions, indicating a preference for anthropomorphic explanations. For example, one problem asked students to consider a chariot moving forward at 60 miles an hour and were asked what would happen to an arrow shot backwards if it left the bow at 60 miles an hour. S9 responded: "If you were in a train that was starting up...and you run to the back of the train, the train's running underneath you, but if you run at the same speed as the train, then, uh, you're going nowhere." Here the student has solved the problem using a beautiful, simplifying analogy. (In fact we have successfully adapted this analogy for use in large group instruction.) Presumably part of the intuitive appeal and familiarity of the analogy comes from its injection of personal action into the problem situation (the act of running).
3. *Invented vs. factual analogies.* Most of the analogies generated by these students appear to be based on the students' own experience (what they believed to be fact) or on information from authority. Nevertheless, at least six (18%) of the cases were so novel that they were clearly new inventions. This demonstrates that students are sometimes capable of spontaneously producing "custom-designed," untested thought experiments. One problem concerned a rocket that was initially moving sideways. Students were asked to predict the final trajectory of this rocket after forward thrust had been added. S5 replied, "If somebody threw me out of a cannonball (*or cannon*)...and I pushed the wall right here, I wouldn't go down like that 'cause I'd still be moving this way." Many students will predict that the rocket will simply go forward in response to the thrust. In order to understand the addition of velocities, this student imagined his own arm providing the new thrust. He playfully imagined

that this push could save him from the trajectory on which the cannon had launched him. This invented thought experiment helped to convince him that the initial movement, this time due to a powerful cannon rather than to some unknown cause as in the original problem, would not simply disappear in response to a new thrust.
4. *Evaluated validity.* In five cases (15.7%), students gave evidence of criticizing or evaluating an analogy after it was constructed. For example, in response to the chariot problem, S3 said, "I'm kinda trying to think of what happens when I throw stuff out of cars...well, that's cause of wind, too." Here, he initially considers the car and object as analogous to the chariot and arrow but then reconsiders, observing that wind resistance could be a primary factor in the behavior of objects thrown from cars. However, many other students did not give evidence for evaluating the appropriateness of their analogies. This may be an area where spontaneous analogical reasoning needs to be improved by instruction.
5. *Successively refined series.* Several students (at least 3) generated a sequence of analogies to solve a problem. These students demonstrated an ability to refine their explanations progressively by criticizing and improving the first analogies they produced. S10 was trying to explain to another student how several forces acting on a stationary cart can cancel. His first analogy was to a case they had both studied in a mathematics class: "You know ... if you have one point ... there's an infinite number of, like, lines you can put through that point? Like, if you make a sphere, almost, like, out of 'em? Well, if you cut 'em out, you know, like, if you make 'em all, as long as they're the same on the top and on the bottom, they're gonna cancel each other out." He continued to try to refine his explanation for several more moments, at first simply by altering his language. Then he altered the substance of his analogy, making it more tangible: "Air. Yeah, air is always gonna be pushing, pushing down, isn't it?...but air is always, but because, because it's the same from all over, it cancels out. It's like on a still day, there's always air around. But nothing moves." Note that in the last two sentences he has improved his new, air analogy by providing a concrete example from daily life.
6. *Differences between subjects.* There was considerable variation in the number of analogies generated by individual students. One student generated 13 of the 34 significant analogies, while almost a third (5) of the 16 students generated no articulate analogies (although several of these students hinted at unclearly articulated analogies.)

A few analogies appeared to serve a powerful function in student reasoning: beyond helping the student solve the problem at hand, they led to new generalizations. This indicates a degree of conceptual change. For example, S5, who generated several thought experiments, gave evidence at the end of his interview of having overcome a misconception commonly held among students. His solution for the space cart problem will be examined in detail in Chapter 9.

8.2 Conclusion

8.2.1 *Similarities Between Experts and Students*

Rather than examining differences between experts and novices as a number of recent research studies have done, I have focused here on one way in which experts and novices are alike: they can both arrive at creative solutions during problem solving through a process of generating and tailoring analogies. We would not expect the analogies of novices to be appropriate and successful as often as are the analogies of experts; it is also probable that students are less likely than experts to criticize an analogy. However, concerning the size of the gap between experts and students, we can still conclude that many experts and students share the ability to generate spontaneous analogies, at least of the kind studied here. This is more interesting when we consider that analogies can be one of the most sophisticated tools of scientific problem solving. Unlike most previous studies, we did not examine the uses students can make of analogies generated by educators or experimenters. Rather, we were able to study students in the act of forming spontaneous and novel analogies while thinking aloud about physics problems. A few of the students in our sample even generated chains of analogies and constructed custom-designed thought experiments.

The fact that these problem-solving processes are creative and that they are also observed in the solution processes of expert scientists and mathematicians supports our position that many creative reasoning processes are ordinary thinking processes, not unanalyzable acts of "genius." This suggests that analogies are an intuitive form of reasoning. If so, it would seem profitable to encourage analogy *generation* during instruction in addition to using prepared analogies.

8.2.2 *Implications*

Several suggestions for pedagogical directions: (1) Teachers can attempt to generate compelling analogies when introducing an abstract principle. (2) Catalogues of analogous cases generated by students in studies like the present one can be assembled and organized by topic areas. These are intuitive examples that have made sense to students, and the best ones are candidates for adoption in teaching and new curricula. (3) A beginning has been made in cataloguing more general intuitive knowledge structures that are in rough agreement with scientific theory and that are possessed by students. An example is the intuitive idea that a greater force can produce more motion in an object starting from rest (Clement et al., 1989). Such intuitions can serve as anchors for grounding more complex ideas if they can be extended to other situations by analogy. (Chapter 10 gives descriptions of teaching experiments using this approach.) (See Camp and Clement et al., 1994)

This study suggests that students such as these possess a rich store of practical knowledge from concrete experiences, and that they possess the ability to relate this store of experiences fairly flexibly to new situations. Thus, there is reason to be optimistic that analogical reasoning, although not a perfect, reliable tool, can be tapped as a resource to at least initiate active thinking in instruction.

8.3 Appendix: Examples of Problems and Spontaneous Analogies

8.3.1 Chariot Problem

A man is in a chariot that is traveling at 60 miles per hour. The man fires an arrow backwards toward a target that he has just passed. The arrow leaves the bow at 60 miles per hour.
How fast will the arrow hit the target?

> #11 S25: Like air craft flying at 60 mph into a 60 mph wind goes nowhere in relation to the ground

How fast will the arrow hit the target?

> #12 S25: Like an aircraft shooting itself down

8.3.2 Space Carts Problem

Two carts, one with a weight in it and one without, are shot from elastic bands out from a rocket floating in space. The same stretch is used for both bands.
Does one cart travel faster, or are they both the same?

> #10 S26: Like a 10-lb. weight and a 1-lb. weight dropped from a building go the same speed

8.3.3 Forces on a Stationary Cart Problem

How do several forces acting on a stationary cart cancel out?

> #13 S28: Like when air [pressure] is the same from all over, it is a still day; vs. when the wind is from one direction, leaves move

8.3.4 Rocket Problem

A rocket is moving sideways with the engine off out in space. The engine is turned on and burns for 2 s.
What is the rocket's path during the burn?

> #14 S20: Like shooting me out of a cannon and, while moving, I push on a wall to the side

8.3.5 Skaters Problem

Two skaters of equal weight are facing each other on a frozen pond. The ice is very smooth, and practically frictionless. Skater "A" is stronger than skater "B".
What happens if "A" pushes "B"?

> #15 S25: They both slide the same distance in opposite directions, like a guy pushing off from a boat, diving into the water

What happens if they both push at the same time?

> #16 S30: Amount each moved when just one pushed would add, like both of them pushing on a solid wall

Chapter 9
Case Study of a Student Who Counters and Improves His Own Misconception by Generating a Chain of Analogies

Research in science education has identified a large variety of student conceptions that can conflict with currently accepted theories in science. These are variously called alternative conceptions or misconceptions. This chapter is a case study of an 18-year-old freshman in college who is able to counter one of his own misconceptions in physics and make considerable progress toward constructing a new conception. The self-correction occurs while the student is solving a qualitative problem aloud. The problem describes a situation that is unfamiliar to him, but he eventually solves it by relating it to several analogous situations that are more familiar. He also exhibits other types of behavior that Polya has described as important in creative problem solving: generating thought experiments, generalizing and specializing, and generating extreme cases. In addition he produces body movements that parallel his arguments, generates personal as well as physical analogies, and generates a bridging analogy. In fact, his method of using analogies and a number of his other reasoning patterns have the same form as those we have observed in experts solving unfamiliar problems. Thus, although the student has not taken college science courses, he spontaneously uses thought processes similar to those of creative scientists and mathematicians. However, since the student's arguments are simple ones based on common physical intuitions, there is reason to believe that such reasoning strategies could be useful in instruction with other students. This leads to the recommendation that qualitative arguments and chains of analogies be tried as techniques for helping students overcome misconceptions in the classroom.

9.1 Spontaneous Analogies in a Student's Problem Solution

The subject S20 was a freshman just entering a School of Engineering. He had taken a high school course in physics, but in the interview, he seems to use little knowledge from the course, relying instead on intuitive arguments. He solves a problem about the behavior of a metal cart being launched by an elastic band. The problem has two related parts. In part A of the problem he is asked to launch the cart across the top of a table using the elastic band (see Fig. 9.1a) and watch it roll to a stop. He correctly predicts that the cart will attain its maximum speed near the point of release from the

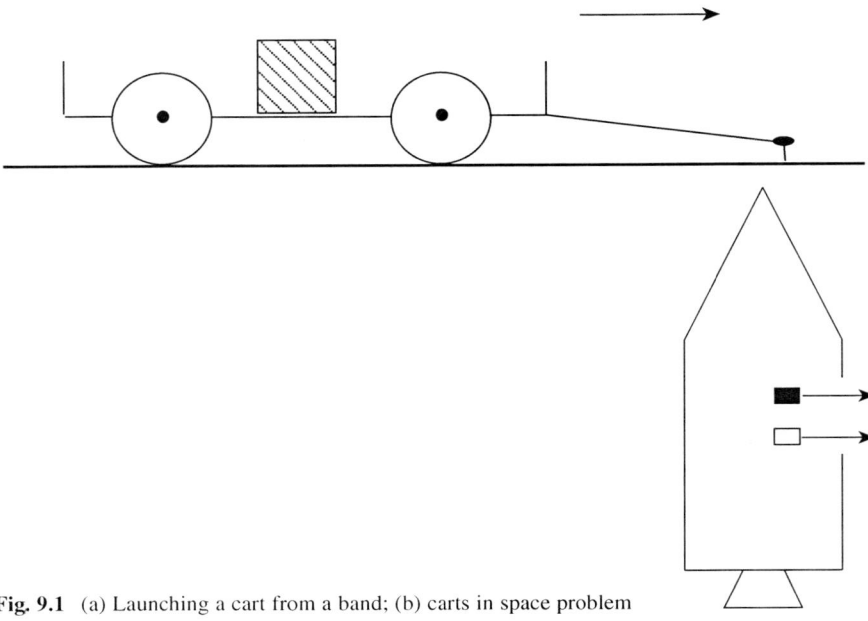

Fig. 9.1 (a) Launching a cart from a band; (b) carts in space problem

band, and that its maximum speed will be lower if pieces of steel are placed in the cart, because "you're having a larger mass to pull – I know in chemistry that a bigger atom goes slower – if you applied the same amount of force, it would go slower." Also, it will not go as far with the extra mass, "because now it's pushing down on this table – it would just be weighted down – it's like a shopping cart – you push a shopping cart a lot further without groceries than you could with it." Thus the student refers to two preliminary analogies. His two main conceptions here appear to be summarizable as: "The more you stretch the band, the faster the cart will go over the table. The more weight you put in the cart, the slower it will go over the table."

A more interesting series of analogies appears in his solution to part B of the problem. The experimental situation is the same, except that the subject is now asked what will happen when the force of gravity is absent. The student is asked whether there will be a winner in a race between the two small carts launched sideways from a rocket floating in outer space (see Fig. 9.1b). Both are launched with an elastic band of the same strength and with the same stretch, but the upper cart has more mass. The correct answer is that the more massive cart's speed should be lower. This is because the stretched bands store an equal amount of energy, and because it takes more energy to accelerate the more massive cart to a given speed.

However, many students will predict a tie here, saying that since the carts are both "weightless" in space, there will be no difference between their speeds. Thus the answer given to this question often reveals whether the student lacks a distinction between the concepts of mass and weight, where mass is a measure of an object's inertial resistance to acceleration and weight is the gravitational force exerted by a planet on the object.

9.1.1   Protocol for S20

Section A

24 S: Oh, um, one's heavier. Uh, all right, I'll start with the weight one [the upper cart in drawing]. Um, if it's heavier – uh – it's heavier – let's see. What makes heavy is gravity, so heaviness wouldn't matter if you're in space.

25 I: Uh huh.

26 S: (a) – 'cause there's no gravity. But, I know they try and make the rocket as light as possible – but that's only to get away from earth, so it doesn't matter. OK, uh, if it doesn't matter how heavy it is – (b) but it does (nods), 'cause the equation says E goes $1/2 mv^2$, so –

27 I: That's the equation?

28 S: Yeah.

29 I: For what?

30 S: Energy. Uh, if you pull this out with a force (points to upper cart and slides finger to right on paper) of, let's say, one, all right, the E = 1 – no, the E doesn't equal one.

Section B

31 S: Uh, I'd say that – it would still go slower (points to upper cart and slides finger to right). If you pulled this out with the weight on it, it would go slower than if you pulled this out (points to lower cart and slides finger to right) without the weight.
32 S: Without the weight you'd go faster and with the weight you'd go slower, because I can still think of, uh, you pulling (holds right hand up and moves it toward himself) something very heavy and pulling something very light. (Repeats hand motion.)

33 I: In space?

34 S: All right, in space. I've never been in space – Yeah, in space, too, I guess, in space. If you push (moves right hand away from chest) something heavier than you, you would go back (moves head and right hand back) uh, more than it would go that-a-way (points forward).

35 I: Mmm.

36 S: I just think this. I figure if you pushed (moves hand forward) the [large] rocket, you'd go back (moves hand back) more than the rocket would....

38 S: So if, uh, so if you pulled this (slides finger to right of upper cart) with a heavier [weight in it], it would go slower....

46 S: (a) So (looks back at drawing) if you had the rubber band here (slides finger to right) it would still pull the lighter rocket [cart] faster than it would pull the heavier rocket. The heavier rocket would stay slower. b) Oh, I really, I – the only thing I can think of is that it's still harder to push (moves clenched fist away from chest) that heavy rocket, c) 'cause I could throw a pen (makes flipping motion with hand) out in space, it's really light, and it would go (repeats hand motion) away. I (points thumb back) would go away, too, but it would go away (moves closed hand forward) more than I would (moves hand backward)....

66 S: a) Right. Uh, and so the same thing for this one. If, even though there's no weight, but still, I can just think of me trying to hit (punches air with fist) a rocket and trying to make that rocket go away (moves open hand away). I figure if I hit (punches) that rocket, I'd go away more (points back) than the rocket would because it's just so big (spreads hands apart). b) Oh! If, uh, if a meteor comes down and hits the earth (raises hand and moves it down) the earth is just so big, it's not gonna move out of its orbit. But the meteor sure gets splattered. and the meteor wouldn't – hm, I wonder if that's the same thing.

67 I: The earth doesn't get pushed?

68 S: No, not by a little thing. (Holds up hand with fingers closed) You'd have to have something bigger (opens fingers as if holding a ball) than the earth or –

69 I: Even though the meteor is weightless?

70 S: (a) Well, ooh, well, the meteor comes down, hits the earth (raises hand and moves down) the earth just sits there, doesn't, I mean, it's just so little, it doesn't matter about the orbit. (Sweeps hand in circle) It wouldn't knock it [the earth] out of the orbit. (b) But, if that meteor could bounce back, the meteor would bounce right back (moves hand down, then up over shoulder), if it could, if it didn't splat (moves hand down) (sound effects), it would bounce back. (c) And so, that means that if you try to hit (punches air) something, if I was little (points to self) and I was that meteor and I tried to hit (punches air) that rocket, I'd bounce back (points back) and the rocket wouldn't really move that much. (d) So, that means, uh, to move a big object (hands open) is harder than to move a little object (hand closes) 'cause if I –

71 I: With a rubber band, too.

72 S: With a rubber band, too. No, see, it's the same amount of force, like the force pulling (points to upper cart and slides finger to right) on this, uh, it's a big object and it's hard to move, so the rubber band, the force of the rubber band would be, uh, like, one, and this is such a hard object to move that it, it would go "slow" (moves hand away slowly) and it wouldn't go as fast. and if you had a really light thing, it would just (moves hand away quickly) zip along with the rubber band. It would go faster.

9.1.2 *Protocol Summary*

As evidenced by line 24 in the transcript, this subject exhibits the common misconception, saying: "What makes "heavy" is gravity, so heaviness wouldn't matter if you're in space." However, he also seems to question this argument. During the rest of the session, his comments all seem to relate to the question of whether differences in heaviness (as measured on earth) "matter" when one is trying to change the speed

9.1 Spontaneous Analogies in a Student's Problem Solution

of an object in space. The student makes no clear distinction at first between the concepts of mass and weight, and indeed, whether such a distinction is necessary is the deep issue implied by this question. Nevertheless, by the end of the protocol he does appear to reach a strongly held, correct conclusion, based on intuitive arguments. He concludes in line 70d that "to move a big object is harder than to move a little object," even in space. Although the student has not given a rigorous proof for why the lighter cart will reach a higher speed, he has succeeded in finding an intuitive way to counter the common misconception that the size of an object does not affect accelerated motion in space. That is, S20 appears to make substantial progress in overcoming his own misconception, and this raises the question of what the reasoning/learning processes were that achieved this.

It is convenient to divide the protocol into two parts. In the first part, in lines 24–30, there are three main arguments:

1. (Line 24) "what makes heavy is gravity, so heaviness wouldn't matter if you're in space." (He eventually refutes this argument.)
2. (26a) "But I know they try and make the rocket (a rocket launched from earth) as light as possible – but that's only to get away from the earth, so it doesn't matter." (He seems to decide that this fact is not relevant.)
3. (26b) "But it (heaviness) does (matter) because the equation says E goes $1/2mv^2$." (He tries to assign values to the variables in this equation for launching the cart but seems to be unsure of this argument and does not develop it.)

With these three preliminary arguments, S20 appeared to give reasons, pros and cons, for whether "heaviness matters" when one is trying to get something moving in space. In doing this he displays a pattern of proposing a conjecture and then evaluating it from more than one point of view – a pattern that he continues to display during the remainder of the interview. Thus he emulates the fundamental scientific method of generating conjectures and evaluating them, a pattern described for an expert problem solver in Chapter 6.

In the second section beginning in line 31, he continues to address the above question. He refers to the four new situations listed below and appears to consider whether or not each one is analogous to the original "launching carts in space" problem.

1. *Pulling objects.* (32) "You can still think of pulling something very heavy and something very light."
2. *Recoil from pushing.* (34) "In space – if you push something heavier than you, you would go back more than it would go that-a-way (points forward)."
3. *Colliding meteor.* (66b) "Oh! If a meteor hits the earth, the earth is so big it's not going to move out of its orbit. But the meteor sure gets splattered!"
4. *Bouncing meteor.* (70b) "But, if that meteor could bounce back, the meteor would bounce right back...if it didn't splat."

These four situations are referred to several times in lines 31–72 in mixed order. They constitute spontaneous analogous cases in the sense of being related situations that S20 thinks about to help him solve the original problem. After considering these analogies S20 appears to feel confident that he has solved the original problem correctly, and so the analysis will concentrate on them. A first possible hypothesis for how this happens is the following: each analogous case is somehow generated from the original problem and used to suggest a prediction for the original problem. He does this for the four cases separately, making his prediction stronger with each case. We will evaluate this hypothesis in a later section. How the analogies are evaluated and why they have the explanatory power they do for S20 are two major questions raised by the protocol.

9.1.3 *Protocol Observations: Creative Case Generation*

We first consider some observations that can be made from the interview transcript.

1. S20's arguments are primarily *qualitative* with the exception of the equation he mentions in line 26b (which he apparently discards).
2. He refers to a number of related cases – situations that are not the same as the situation in the given problem. They are potential *analogies* in the sense that the subject seems to believe that a structural relationship in the related case may also prevail in the original problem.
3. The "bouncing meteor" episode is distinguished by its *novelty* as a situation. This suggests that this analogous case was *generated* or *constructed* by S20 – not simply recalled from previous experiences or factual knowledge. It is an example of a thought experiment where he invents a new situation. This is reminiscent of the invented thought experiments documented in experts in Chapter 3.
4. He refers to extreme cases, such as his statement: (32) "You can think of pulling something very heavy and something very light"; and his opposition of the tiny meteor and the enormous earth.
5. The solution as a whole has a "wandering" character – he seems to *explore* various aspects of the problem, and these will sometimes remind him suddenly of new situations – as opposed to solving the problem in a planned way with a series of carefully organized steps. However, it is important to stay watchful for initially hidden patterns in such a solution that may expose more underlying structure.
6. In many of his statements he talks about putting himself in the place of one of the objects. At times, he appears to sort out the way forces are acting by pretending to push or pull on one of the objects himself. These will be called *personal analogies*; examples are the "pulling objects" and "recoil from pushing" themes. Nonpersonal analogies, such as the "colliding meteor" theme, will be called *physical analogies*.
7. As will be discussed in Chapter 12 many of S20's statements are accompanied by *body motions*, and these parallel the kind of actions he is describing at the time. These occur in both the personal and physical analogies.
8. There are several *levels of generality* in his statements varying from the very specific: (46c) "I could throw a pen out in space and it would go away more than I would (go back)," to the more general (70d) "that means to move a big object is harder than to move a little object." That is, there are differences in the level of generality of the sets of actions that he refers to.

9. As an analogy, the "colliding meteor" situation seems somewhat far afield and difficult to account for. The same is even more true for the "bouncing meteor" situation. Why S20 invents this fanciful situation is not at all clear. These *anomalous events* provide a major obstacle to achieving a coherent analysis of the reasoning in the protocol.

9.1.4 Developing Hypotheses about Cognitive Events that can Account for the Observations

Observations 2, 3, and 4 above point to S20's generation of analogous cases, extreme cases, and thought experiments, and to his connection of the problem situation to previously known facts. S20's arguments here are examples of what Polya (1954) calls plausible reasoning rather than being logical arguments in the strict sense. These reasoning patterns of S20's parallel those recognized by Polya as highly important for scientific and mathematical problem solving. When these methods are consciously employed as strategies they are often referred to as heuristics. However, in S20's case it is not at all clear that he makes a deliberate decision to use any of the strategies that he does use or that he has labels for the strategies. In his case they appear more to be a natural way of operating and to be a spontaneous reaction on his part to the dilemmas he finds himself in. We call this the use of *implicit heuristics*. The fact that S20 has had so little previous training in science suggests that we may be able to study certain types of heuristics in untrained subjects as spontaneous reasoning patterns rather than products of specific training in science.

With regard to a model of how the subject used analogical reasoning, the initial hypothesis given earlier was that each case was generated separately in connection only to the original problem and used to suggest an answer to it. However, Fig. 9.2 summarizes the *sequence* in which he refers to these cases, plus the "bouncing meteor." The numbers in the diagram indicate the order in which transitions

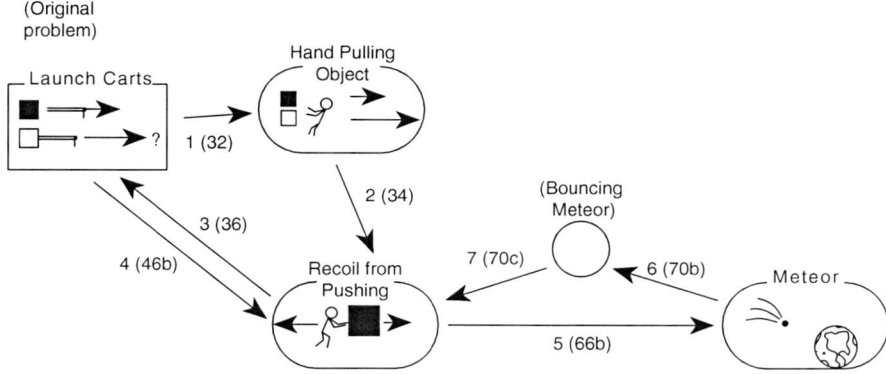

Fig. 9.2 Analogy theme sequence

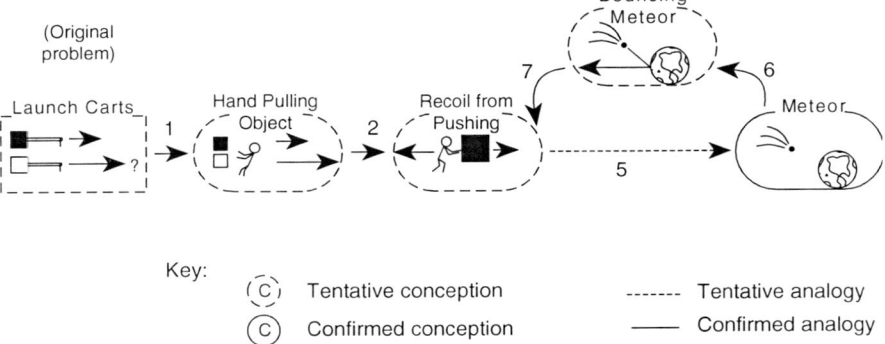

Fig. 9.3 Formation of analogies between S20's conceptions

between cases occur. This suggests that some of the related cases S20 talks about are secondary analogies, in the sense that they are generated in relation to a previous analogy, rather than in relation to the original problem. For example, S20 seems to relate the "colliding meteor" theme directly to the intermediate "recoil from pushing" theme, rather than to the original cart launch problem.

The diagram in Fig. 9.3 uses the notation developed in Chapters 2, 3, and 4 to outline a new hypothesis for how S20 generates his analogies. This diagram assumes that each of the major themes identified in his output is produced by one of these internal conceptual frameworks. The solid arrows signify pairs of conceptions that S20 considers to be analogous. Using this diagram we can account for much of S20's behavior during the interview as follows:

1. Arrows indicate the hypothesized sources from which the analogies form or from which they come to mind for the first time in the numbered order. This is supported by the observed order of transitions between analogous cases shown in Fig. 9.2.
2. One can also hypothesize the presence of two types of analogy formation processes that were identified in Chapter 3: (a) *formation by association* where the subject jumps to a more familiar situation that is different in many ways; and (b) *formation via a transformation* where the subject generates an analogy by modifying aspects of the current problem to produce a second situation.

    S20's sudden shift to the "colliding meteor" theme in line 66 is an example of the first type. The main evidence for this is the fact that the meteor situation differs with respect to a large number of features from the immediately preceding "hitting a rocket" example. An example of formation via a transformation is S20's "hand pulling a cart" theme, where he appears to have modified only one feature of the original situation, namely, he thinks of himself rather than the elastic band pulling the cart.
3. The conceptions shown as regions with a dotted border in Fig. 9.3 are *tentative conceptions* or beliefs. That is, they are each not fully understood and S20 is unsure of their internal validity. For example, he is unsure that it is harder to pull the heavier cart in space with his hand.

9.1 Spontaneous Analogies in a Student's Problem Solution

He does appear to be confident of the analogy relation in this case – that is, he seems to believe that *if* it is true that it would be harder for him to pull a heavier cart in space, then it would be harder for the elastic band to pull one as well. Presumably, he believes this because he feels that the change from pulling with the band to pulling with the hand does not affect the critical aspects of the experiment. In other words he believes that the critical factors determining the behavior of the system are conserved when there is a change in "what's doing the pulling." This will be described as a *conserving transformation* in the expert protocols discussed in Chapter 17. It leads to a confirmed analogy relation as opposed to a tentatively proposed analogy relation for analogy number 1 in Fig. 9.3.

4. This figure also shows a chain of tentative analogous conceptions formed by S20. Near the end of the transcript a breakthrough comes when this chain connects the original cart-launching problem that he is not sure of to a confirmed conception of meteor behavior that he is sure of. This allows him to make a chain of inferences back to a prediction for the original problem, A. The final inference chain takes the form: E–D–C–B–A, where the presence of a single situation E of high certainty, and the presence of confident analogy relation links, creates a domino effect to make confirmation of all the other situations plausible.

5. By line 69 (prior to the "bouncing meteor") the one remaining obstacle that has prevented the above domino effect from taking place is the unconfirmed analogy relation between the "recoil from pushing" and "colliding meteor" cases (shown as dotted line number 5 in Fig. 9.3, and recalling the distinction between an unconfirmed analogous case and an unconfirmed (or tentative) analogy relation). The unconfirmed status of the relation is supported by S20's statement in line 66 after proposing the analogous case of the colliding meteor: "Hmm, I wonder if that's the same thing [as the 'The Recoil from Pushing' case]?" I distinguish between the formation of a tentative analogy relation such as number 5 where the connection is initially uncertain, and a confirmed analogy relation, such as 1 and 2 where the analogy relation is established with confidence immediately.

6. One can now provide a clearer theory for the role of the somewhat anomalous and enigmatic "bouncing meteor" conception that S20 generates. In line 70b, S20 says, "If that meteor could bounce back the meteor would bounce right back." This is an intermediate case that can be described in terms of a new combination of two previous themes that S20 has referred to, namely: his previous first reference to a meteor colliding with the earth and his previous reference to him hitting something heavier than himself and having himself go back more than the heavy object moves forward. As shown in Fig. 9.3, this thought experiment provides a link by which S20 can replace the tentative analogy relation between the "recoil from pushing" conception and the first "colliding meteor" conception with a chain of two confirmed analogy relations. As such, the "bouncing meteor" fits the expert pattern of *bridging analogy* described in Chapter 4. The earlier "colliding meteor" conception provides an anchoring conception, as a known fact. We can then imagine in Fig. 9.3, a chain of inferences or "domino effect", as the analogies are confirmed proceeding from right

to left in the figure. This has the effect of linking all of the previous tentative conceptions with the confirmed "colliding meteor" conception. Thus the "bouncing meteor" is explained in this theory as a bridging analogy that constitutes a successful attempt on the part of the subject to confirm a tentative analogy relation.
7. The "bouncing meteor" scenario satisfies all of the important criteria for a *spontaneous thought experiment*: it is a novel case indicating that the subject has not previously observed or been informed about its behavior; and it has the character of a specific, concrete experiment but is not actually performed. His other analogies have these same features as well, since they are conducted in a strange context, outer space.
8. In Chapter 12, I will examine the idea that the subject's hand motions give us some tangible clues concerning the form and dynamics of his imagery during these thought experiments. In fact, the first two thought experiments seem to be anchored in intuitions with motor components as well as perceptual ones. It appears to be fairly natural for him to attribute these motoric conceptions to an energy-containing object like an elastic band.

In summary, we have been able to account for several of the observations listed at the beginning of this chapter by positing a first-order theory which states that S20 generates a chain of analogous but tentative conceptions. Many of these thought experiments seem to carry some weight on their own as arguments in favor of the view that it is harder to move a big object than a small object in space. However, an important confirmation is achieved when these experiments are linked analogically to the known fact that meteors do not move the earth significantly. This analysis makes the solution less anomalous and supports the hypothesis (proposed in Chapter 2 for experts) that four processes are important in reasoning by analogy: (1) given the initial conception A, the analogous conception B must "come to mind"; (2) the analogy relation must be "confirmed"; (3) conception B must be "confirmed"; and (4) findings must be applied to A.

9.2 Conclusion: Expert-Novice Similarities

The main finding from this case study is that many of the impressive creative reasoning strategies observed in experts were also observed in the problem-solving behavior of this naive student who had not been trained in these techniques. He *critically evaluates* his preconception by using thought experiments (e.g. "I can still think of you pulling something very heavy and something very light"). Second, he spontaneously uses *analogical reasoning*, including: (1) the generation of a *chain of multiple analogies*; (2) the specific strategies of generating an analogy via a *transformation* and via an *association*; (3) the use of *bridges* as a relatively advanced strategy for confirming an analogy relation. Third, he uses several other intuitive heuristics observed in expert problem solvers, including *generalizing and*

9.2 Conclusion: Expert-Novice Similarities

specializing, generating extreme cases, and generating *thought experiments*. Finally, he generates and confirms the idea (opposite to his original conception) that "to move a big object is harder than to move a little object even in space."

Thus there is some evidence that the subject is able to overcome one of his own preconceptions. The subject described in this chapter is an "extreme case" himself with respect to the frequency with which he generates and evaluates creative arguments by analogy. Many other freshmen subjects were observed to reason by analogy, but not at so high a frequency (see Chapter 8). We cannot make normative predictions about the population from the behavior of one subject, and this subject is probably unusually creative. Nevertheless, the case study provides an "existence proof" showing that such natural reasoning processes exist. This suggests the possibility that such reasoning patterns can be utilized to foster conceptual change in students. And it should be noted that the subject was truly a novice – he did not have a strong background in science. It is extremely unlikely that he had been trained in the use of analogical reasoning. Indeed, the subjective impression from the video tape is that of a cheerful, naive person taking an almost playful approach to exploring the problem.

9.2.1 Instructional Implications

In short, S20 is entering college with some very impressive creative reasoning skills that could be tapped, exercised, developed, and refined. Unfortunately, standard science courses do not appear to be tapping and developing these skills. This defines an important and challenging pedagogical problem for course improvement. This subject spontaneously corrected one of his own misconceptions and made considerable progress toward constructing a new conception. He did this using a series of analogous thought experiments which provided counter arguments to the initial alternative conception. After considering these specific cases, he was able to construct a generalization "So that means that, uh, to move a big object is harder than to move a little object" (even in space). This would seem to be the ideal point at which to introduce the mass/weight distinction instructionally. The student has prepared himself to give meaning to the distinction by grappling with the qualitative consequences of such a distinction. At this point he is lacking the appropriate labels that name the two quantities in the distinction. The meaning of the distinction can now be meaningfully grounded at an intuitive level by relating it to his kinesthetic thought experiments involving his own body. The subject's arguments provide an example of the qualitative arguments students may need to go through before considering more formal definitions. This suggests that rather than starting instruction in this area with operational definitions, one might do better to first raise the problem within specific examples and consider intuitive analogies and arguments at a qualitative level. This is the strategy used in the tutoring and classroom experiments described in the next chapter.

Chapter 10
Using Analogies and Models in Instruction to Deal with Students' Preconceptions*

John J. Clement and David E. Brown

10.1 Introduction

This chapter discusses an approach to teaching basic physics concepts that utilizes some of the learning strategies that have been identified for experts in this book. Research has shown that physics students can harbor persistent preconceptions which can constitute difficult barriers to learning and are often quite resistant to instruction (for reviews of research on students' alternative conceptions, see Driver and Easley, 1978; Driver and Erickson, 1983; Duit, 1987; McDermott and Redish, 1999; Confrey (1990)). The teaching interviews to be discussed illustrate how strategies such as the use of analogies, bridging analogies, and explanatory models can help students deal with such preconceptions. We have seen in the last two chapters that under certain conditions students can use analogical reasoning and other plausible reasoning processes used by experts. The question is whether these natural reasoning resources can be tapped and organized in such a way as to produce conceptual change during instruction.

Such strategies would take advantage of *positive* elements of prior knowledge by building on students' existing valid physical intuitions. It is perhaps confusing that we are attempting to build on students conceptions in order to change their conceptions. However, there is evidence that students have both useful and competing intuitive conceptions (from the perspective of the scientific theory being taught). (See Clement et al. (1989) for documentation of such useful "anchoring" intuitions.) In this teaching approach we attempt to increase the range of application of the useful intuitions and decrease the range of application of the competing intuitions. We hope to present evidence that such instruction can lead to conceptual change in that the student can make intuitive sense of aspects of the scientific theory that were counterintuitive before.

It is not unusual to see an experienced teacher using an analogy on occasion in instruction. However, the strategy described here differs in several ways from common

---

*Portions of this chapter are reproduced with kind permission of Springer Science and Business Media from: Brown, D., and Clement, J. (1989). Overcoming misconceptions via analogical reasoning: Factors influencing understanding in a teaching experiment, *Instructional Science*, 18:237–261.

uses of analogy (for a description of such uses see Thiele and Treagust, 1994). First, empirical research is done to search for an analogous case and establish that it is intuitively understood by most students. If it is, we refer to it as an "anchoring example." Second, considerably more effort than normal is put into establishing the analogy *relation* between the anchoring case and the target case. One of the strategies used to do this is the bridging strategy observed in experts, as described in Chapter 4.

By establishing analogical connections between anchoring situations and more difficult ones, students may be able to extend the range of their valid intuitions to initially troublesome target situations. This strategy has been used in classroom instruction, with apparent success (Clement, et al., 1987). In this chapter, we examine two case studies of students tutored with this strategy in order to examine their learning processes in more detail than is possible in a classroom study.

10.2   Teaching Strategy

10.2.1   *Introducing the Target*

As an example of anchoring and bridging strategies, consider the following hypothetical tutoring interaction that would represent the fastest possible use of bridging with a student who was in a state of maximum readiness for it. The numbers below refer to those noted in Fig. 10.1 which represent the situations considered by the student.

The first step in the teaching strategy is to make the preconception explicit by means of a target question. For example, a question which draws out a preconception for a majority of introductory physics students concerns the existence of an upward force on a book resting on a table. Students typically view the table as passive and unable to exert an upward force.

> 1. 1. *Book on a table*. In response to a question about the forces acting on a book at rest on a table, the student indicates that the table is not exerting an upward force on a book. (The physicist would say that the table is exerting an upward force on the book balancing the downward force of gravity.)

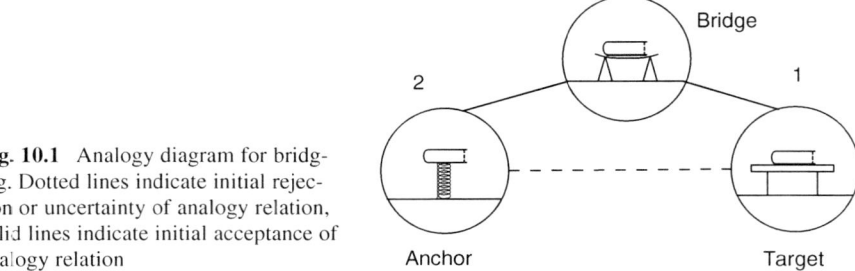

Fig. 10.1 Analogy diagram for bridging. Dotted lines indicate initial rejection or uncertainty of analogy relation, solid lines indicate initial acceptance of analogy relation

10.2.2 Anchoring Case

The next step is to suggest a case in which students tend to give a correct answer on the basis of intuition. We call such a situation an anchoring example (or, more briefly, an anchor). However, even though students may reason appropriately about the anchoring example, they may still be unconvinced of a valid analogy relation to the target case.

> 2. 2. *Book on a spring.* As a potential analogy, the tutor asks the student to consider the situation of a book resting on a spring. In this case the student indicates that the spring would be exerting an upward force since the spring is compressed. However, he/she rejects the analogy relation to the case of the book resting on the table, since the table is rigid and does not need to return to its original position.

10.2.3 Bridging Strategy

When this occurs the instructor attempts to strengthen the analogy relation. The instructor first asks the student to make an explicit comparison between the anchor and the target. If the student still does not accept the analogy relation, the instructor then attempts to find a bridging analogy (or series of bridging analogies) conceptually intermediate between the target and the anchor.

> 3. 3. *Book on a flexible board.* At this point the tutor introduces the "bridging" situation of a book resting on a flexible board between two supports. Upon reflection the student accepts that this situation is analogous to the book on a spring situation, since in both situations there is compression or bending and an accompanying tendency to return to an equilibrium position. After some discussion and internal struggle, he/she also accepts that the situation of the book on the flexible board is analogous to the situation of the book on the table since the table can be viewed as a thick board which would still bend, although imperceptibly.

10.3 Teaching Interviews

We now turn to the two learning case studies. Although these interviews could be called tutoring interviews, the tutor provided relatively little information. Rather, the students were primarily asked to think about a series of qualitative explanation questions and were informed that the tutor would take a "devil's advocate" stance in order to foster discussion. In this way students were encouraged to play an active role in the learning and to adopt only those ideas that seemed reasonable to them, as they would be unsure whether the arguments the tutor was advancing were "correct" or simply made to encourage discussion.

The first subject was a college freshman we will call "Mark" who had not taken physics in high school or college. The interviews were conducted by David Brown. In both case studies the target problem is the book on the table problem.

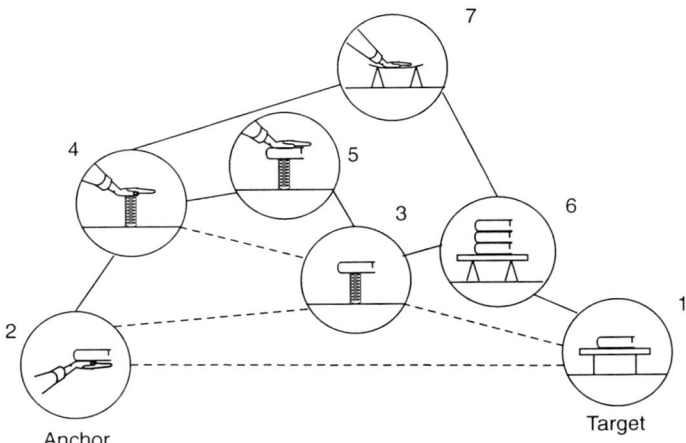

Fig. 10.2 Cases and analogy relations for first student. Dotted lines indicate initial rejection or uncertainty of analogy relation, solid lines indicate initial acceptance of analogy relation

Although this interview dealt with the same target problem as the hypothetical case above, many more potential analogies needed to be introduced and discussed as shown in Fig. 10.2. Some of the examples are what could be called "sub-bridges," that is, a situation which is intermediate between two situations, at least one of which is a previous bridge. For example, node 5 is a bridge between situations 3 and 4 and both of these are earlier bridges. The numbered sections in the analysis below correspond to the numbers noted in the diagram in Fig. 10.2.

10.3.1 Tutoring Session

1. *Book on the table.* The target question asked whether a table exerts an upward force on a book resting on the table. In response to this question, Mark replied:

   042 S: No, it's just, ah, a barrier between the floor and the, um, the position the book is at right now.

2. *Books on the hand.* Although he said the table would not exert an upward force, he indicated he would definitely have to exert a force upward in the case of several books resting on his hand. However, he did not view these situations (book on the table and books on the hand) as analogous. When asked why he answered differently in the two situations, he replied that his arm has muscles.

3. *Book on a spring.* The first bridging analogy introduced was that of a book resting on a spring (S stands for student, T for tutor).

   070 S: Ah, the book is on the spring and, um, this spring is absorbing, ah the force caused by the mass of the book and the gravity. But wouldn't say that the spring is, ah, pushing up on the book. That's just my sense.

071 T: Uh huh.
072 S: The spring itself doesn't initiate any movement.
Mark apparently views the spring as a passive entity, one that can absorb force but cannot "initiate any movement" itself.
073 T: What's the difference then between the book on the spring and the book on the hand?
074 S: Uh, muscles in the arm.
075 T: And the spring doesn't have any muscles?
076 S: Right. The spring is just, ah, a piece of metal and it'll absorb, ah, as much as it can until the point where it's completely contracted and then it will probably, ah, not absorb more energy.

4. *Hand on a spring*. The tutor proposed a hand pressing down on a spring as a bridge between the books on the hand and the book on the spring. Mark believed that the spring would push up against his hand. However, he views the book vs. the hand on the spring as not analogous. When asked why, he replied:

> 088 S: Ah, because the force, ah being exerted on the spring by the book is only the mass of the book and the gravity. But the, ah, the force of the hand, um, could be all kinds of, is you know, your muscles in your hand.

5. Hand *on a book on a spring*. As a bridge between the hand on the spring and the book on the spring, the tutor suggested a hand pressing down on a book on a spring. Mark said the spring would definitely be pushing up against the book in this case. When asked to compare this to the situation of the book resting on the spring, he responded:

> 106 S: Because now with your hand off [the book] the, no downward pressure is really being exerted. Actually now I see the point you're trying to make, it's, ah, it's only the amount of the force being, push being exerted on the spring is varying. It just seems to me that there's no force being exerted on the spring when the book is on there, the gravity's almost invisible, we don't even think about it. But now I realize that it, there is no difference between the two [situations] that you just asked me.

6. *Books on a flexible board*. Now that Mark believed that the spring exerts an upward force on the book, the tutor attempts to establish the case of several books resting on a flexible board between two supports as a bridge between the book on the spring and the book on the table.

7. *Hand on a flexible board*. Initially, Mark said that the board would be simply a barrier, but then he generates his own bridge between the earlier situation of the hand on the spring and the book on the flexible board.

> 123 T: Would you say the board is pushing up against the books?

> 124 S: Ah, no I would say the board is, ah, just a barrier between the books and the area underneath the board.

> 125 T: Uh huh.

> 126 S: I don't think the, ah, well now that I think about it a little more, ah, the spring, ah, this board might have some of the properties similar to the, ah, spring, because the, ah, if you push down on the middle of the, right at the point where the books are located....

> 127 T: Uh huh.

> 128 S: [T]he, it would probably come back up depending on the, whether the board was flexible.

129 T: I'm assuming that is [flexible].

130 S: Or it could break if it weren't flexible, but since it is, ah, I suppose you could say that the board is pushing up the books.

Now that Mark believed the flexible board exerted an upward force on the books, the tutor asked him to compare this situation with the situation of the books on the table.

135 S: Uh, the board is flexible and, yeah I guess that's, that's essentially it, the board is flexible and, ah, it probable isn't different, um, I'm starting to realize how technically it probably isn't different, it just appears different. Ah, you know, because it's a thin board, it's flexible and you can see easier that it's, um, the board is pushing up on the books. Especially after talking about the springs previously and, uh, the table is, ah rigid, it doesn't appear flexible though it is in the, ah, you know, in a really, really, really small microscopic, ah, sense. And, ah, so there probably, scientifically there probably is no difference, it's just a matter of, ah, numbers, you know, the board is very flexible and the table is immeasurably, ah, flexible.

Later in this interview, Mark said that the idea of the table exerting an upward force now made "complete sense." Thus there is some indication in this case that the bridging strategy was successful in bringing about conceptual change.

10.3.2 *Discussion of First Case Study*

It is interesting to note several differences between this use of analogy and a more standard approach, such as presenting an analogy in a text passage and noting the points of correspondence to the target. First, Mark felt he *already understood* the target situation, that the table was simply a barrier preventing the book from falling to the ground, but not exerting a force on the book. Second, as a result of this self-evaluated understanding, he strongly *resisted accepting the aptness of several proposed analogies*. Third, whereas the traditional use of analogy would involve presenting the base explicitly as an analogous situation, in this case the tutor simply suggested situations *without stating that the situations were analogous*. The purpose of the interview was to engage the student in a *process* of analogical reasoning, and not simply to present an analogy. Fourth, a very extensive amount of time was spent on establishing the validity of the analogy between the anchoring case and the target; in this case five intermediate bridging cases were introduced in an attempt to form this link. (This can be called a recursive bridging strategy.) In the end, three of these bridging cases appeared to enable the student to confirm the validity of the analogy in a way that made sense to him. Fifth, the result of the process of analogical reasoning was to *change rather than add to his existing understanding* of the book on the table situation. Finally, one could argue that the result of the process was that Mark came to *view the table as springy*, and not simply *as like* a spring, in that it too exerts an upward force.

10.3.3 A Second Case Study

The second subject, whom we will call "Ellen," a humanities graduate student with no background in physics, was also instructed that if she expressed a view on a question, the tutor might take the opposite view in order to generate discussion. She was asked to maintain her views unless it seemed reasonable to her to change her views. After these instructions she was asked the following question about a drawing of a book on a table: "What forces are there on a book resting on a table?"

> 10 S: I don't want to put an arrow up but I feel like this (circle in drawing representing the earth) is forcing that (book) to come down.
> T: Like the earth is forcing the book to come down?
> S: Yeah, and the table gets in the way. So that's why the book stays.
> T: But the table isn't pushing back on the book?
> S: How can a table push back on a book? (Laughter)

Ellen's laughter here indicates that she finds the idea of a motionless table exerting a force ludicrous. The tutor now introduces the analogous case of the book on the hand.

> 18 T: If I were to put the book on your hand, if you were to hold out your hand, and you just held it there, would you be pushing back on the book?
> S: Yeah
> T: You would be?
> S: I'm taking the book. I'm putting it on my [hand] I'm, yeah, I'm pushing against the book 'cause if I don't the book is heavy enough that I'd drop it if I don't push against it.

Ellen believes that the hand pushes up. However, she is unsure that this is analogous to the case of the table, as shown below.

> 26 T: The case with your hand is different than with the table?
> S: Well, the table just doesn't have a choice on what it does, where I have the choice about how I move my hand.
> S: I mean in a way I can understand how you can say the table is pushing against the book except it's not the same type of push... I can relax my hand, while the table is, it's just there...it cannot relax itself to allow the book to fall any further...this (table) is immobile...so I guess that's how I see pushing because I'm actively pushing.

Notice Ellen's conceptions about force appear to be different from the physicist's: (1) volition is involved: "the table just doesn't have a choice;" and, "I can relax my hand"; and (2) the source of force is an active source of power: "I'm actively pushing."

Here the book on Ellen's hand has been proposed as an analogy to the book on the table in an attempt to help her see the table as pushing back on the book. That is, one hopes that Ellen will believe in a force up from the hand on the book, and that this will make a force up from the table more plausible. There was a glimmer of success; "I mean in a way I can understand how you can say the table is pushing against the book." However, Ellen does not appear to be convinced.

Now another analogy is attempted which is an intermediate bridging case between the table and the hand in hopes that this will make the first analogy relation more plausible. The bridging case consists of placing the book on an imaginary spring of about the "springiness" of a bedspring. The spring shares with the table the features of being inanimate and non-volitional. It shares with the hand the feature of being obviously capable of motion.

> 34 T: Would you say that the spring is pushing on the book?
> S: I suppose you could say that in a reverse manner, but not – so (anticipating the next question!) why doesn't the table push on the book? (laughs) Umm, again it just seems like the spring is being acted upon, I mean, I guess in a way you could say it's pushing against it, yeah I guess you would say it's pushing against it.
> T: You said something about a reverse manner?
> S: Well just because I think of – if you put something on a spring, that something makes the spring go down, but I guess if you see it another way the spring is also holding that thing up from going, as uh, as far as it wants to go down.
> T: So do you see that as a different kind of push than the push you were giving with your hand on the book?
> S: (Pause) In one way yes, in one way no. I guess there seems to be more action in the spring than there is in the table, but it's still – my hand, I control my hand while the spring again is one of those things that it can't control its response to whatever is being placed upon it. But I guess it does have more of, it seems to have more of a, uh, impact on pushing back something than a table would.

The book on the spring, as a bridge between the first analogy and the original situation, seems to have had some impact on Ellen, even though she still remains to be fully convinced. Clearly part of what is needed here is a revision or reconstruction of Ellen's concept of "force." At issue is whether the concept's necessary features include volition and movement and a resident source of power in the exerter. The situation is shown in Fig. 10.3. Now the tutor sets out to build sub-bridges from B to C.

In Fig. 10.4 are shown the two sub-bridges between the book on the hand case (in which she is confident) and the book on the spring case. They are a hand pushing down on a spring, and a pile of books resting on a spring. After working though this series of analogies, she is brought to the point where it makes sense to her that the spring exerts a force on the book.

The final step in this teaching episode is documented below. T tells Ellen that the table bends when the book is placed on it. Ellen does not believe this and asks T to prove it. The instructor then introduces the explanatory model of spring-like bonds between atoms and molecules in the table.

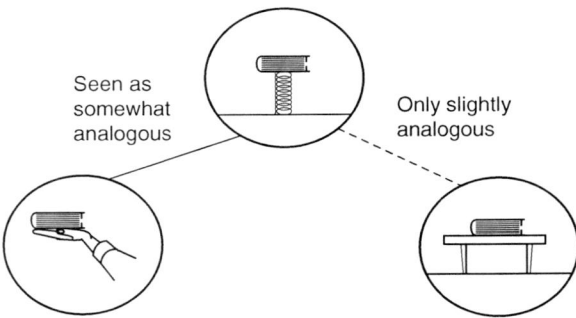

Fig. 10.3 Initial bridging analogy used with second student

10.3 Teaching Interviews

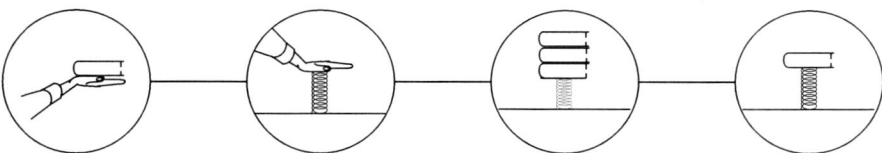

Fig. 10.4 Two sub-bridges between the book on the hand case and the book on the spring case

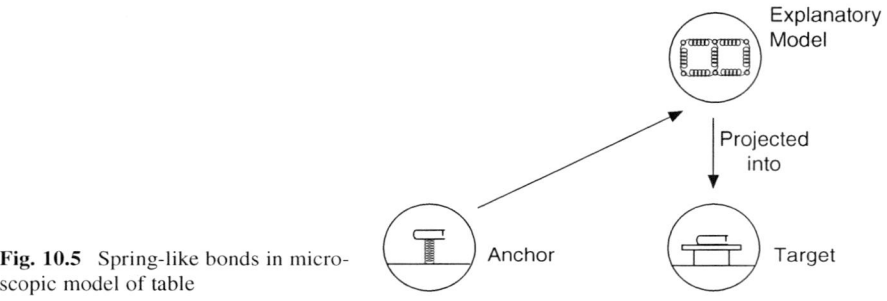

Fig. 10.5 Spring-like bonds in microscopic model of table

52 T: When you put this pile of books on it [the table], it's not bent very much but it is, it does bend a very slight amount....
S: But you're saying, even though with this amount of books on it this table is bending slightly?
T: Yes
S: How can you prove that? (pause) So you're saying that all things will bend? No matter, does it matter....

60 T: Well on the microscopic view if you wanted to look at it that way, um, a table, would you agree that the table is composed of molecules?
S: Sure.
T: And molecules, um, basically what they are is they're connected by bonds which are flexible, that are sort of like springs, they might be pretty stiff springs, but they're sort of like springs. And so this table is composed of (draws model in upper part of Fig. 10.5), each of these little circles is a molecule. You can think of it as being composed of a group of molecules which are attached by springs, each molecule has what's called a bond with other neighboring molecules which is something like a spring. It's not a literal spring, but it acts like a spring.
S: Mmm

67 T: ...a group of springs and I put this other group of springs on top of it, which is the book
S: Mmm hmm
T: and the two things kind of
S: Push against each other
T: Push against each other, right. Does that make sense at all?
S: Yeah that makes sense. So I can see why you would say the table would move.

73 T: So you're saying the molecules [picture] was helpful to you?

74 S: That was the most helpful, seeing the composition as being springs against springs, but, you know, the other ways, I would have just been agreeing with you for the sake of agreeing.
T: But the springs as molecules that did [help]?

S: Yeah, that did.
T: So if I were to ask you is the table pushing against the book what would you say now?
S: The molecules in the table are pushing against the book (laughs).
T: Okay now, what would you say if I were to ask you: "Is the table pushing against the book?"
80 S: The table pushing against the book? I could understand why you would say that. Molecule speaking.

In this last section, the tutor proposes that the spring and the table share the common property of deforming under a force. He also proposes a new explanatory model which draws on the anchoring idea of a spring as shown in Fig. 10.5. This model involves a picture of the table being made up of atoms and molecules connected by stiff springs and provides a deeper explanation for the bending property of the table. The explanation then finally seems to "make sense" to the student, after a fairly long prior period of disbelief on her part. In this case, however, the usual anchoring idea that springs provide elastic forces was not in place at the beginning of the interview and had to be constructed as a prerequisite. This was done via an alternative anchor and a series of bridging analogies as shown in Fig. 10.4.

10.3.4 Explanatory Models

The use of an explanatory model in this tutoring episode warrants further analysis. As discussed earlier in Chapter 6, Hesse (1967) and Harre (1961) identify two types of qualitative mental models:

1. A model which shares only its abstract form with the target (Hesse cites hydraulic models of economic systems as one example). Such an analogue may happen to behave like the target case and therefore provide a way of predicting what the target will do. Here we call this an expedient analogy.
2. A model that has become in Harre's terms a "candidate for reality," in which a set of material features, instead of only the abstract form, is also hypothesized to be the same in the model and the target situation (these features are often unobservable in the target at the time). The example used earlier was the elastic particle model for gasses, in which a gas is hypothesized not only to behave like billiard balls bouncing around, but to actually consist of something very much like tiny particles bouncing around.

We refer to the latter type of model as an explanatory model or mechanism. Thus an explanatory model is a predictive structure in which material elements of the model are seen as being in or operating in the target.

On the basis of statements like that in line 74 in the second case study, one can hypothesize that when molecules with spring-like bonds were introduced as an explanatory model, the anchoring example of a compressed spring was used as a basis for developing an image of a hidden structure or mechanism operating in the target. Because this protocol was audio recorded and not coded for hand motions, we lack enough data on imagery indicators to support the imagistic part of this hypothesis here more fully. Such indicators will be analyzed for other students to support similar conclusions in Chapter 18. However, the hypothesis provides a way to give what appears to be the simplest coherent explanation for the subject's statements in the present protocol.

At least two factors may make the learning of explanatory models difficult for a student: difficulty with the necessary spatial image manipulation skills (as in explaining the phases of the moon); or competition with a prior conception (e.g. seeing tables as rigid barriers rather than elastic sources of force). For the book on the table, the explanatory model is the image of the deformation of a springy substance causing a reaction force. This is easy to accept in the case of the spring, but in the case of the table, we hypothesize that this imageable mechanism must be projected by the student into the image of the table where the deformation is ordinarily unobservable.

Such explanatory models might seem more plausible or compelling to the student than an expedient analogy, since key elements of the model are seen as actually operating in the target. Thus the model involves concrete as well as abstract knowledge in that the model provides a structure or mechanism that could plausibly be imaged in the target. Such a plausible concrete mechanism may be a central aspect of conceptual change toward conceptual understanding.

10.3.5 *Abstract Transfer vs. Explanatory Model Construction*

This example may indicate a need for a more comprehensive theory of analogical learning than is common in education to date. A widely accepted view considers that an analogy is beneficial because it helps the student view the target in a more abstract way. In this view, by helping the student focus on the shared relational structure between the base and the target and downplaying the significance of the actual objects and object attributes, the analogy is thought to help lend relational structure to the previously poorly structured target situation (Gentner, 1983, 1989; Gick and Holyoak, 1983, 1980; Holland et al., 1986; Holyoak and Thagard, 1989). The learner is left with a mental representation of the target in which objects and object attributes are less salient, and abstract relational structure is more salient.

By contrast, in the successful intervention in this protocol, the analogies appeared to help *enrich* the students' conceptions of the target situations rather than (or at least in addition to) helping them view the situations more abstractly. This enrichment process is shown schematically in Fig. 10.5, in which the explanatory model, constructed by incorporating an anchoring intuition, is projected into the target situation. The student learns about a new concrete system as a mechanism that explains what is happening inside the target. We hypothesize that this enrichment of the target with new objects, object attributes, and casual relations (e.g. microscopic bonds, flexibility, and bending causing forces) may be a very important means for conceptual restructuring. In doing so, we separate the dimensions of concreteness and generality, believing that the concreteness of the imagined mechanism does not imply a lack of generality. For example, the idea of swarms of moving molecules in a gas is concretely imageable, but the model is very widely applicable.

10.3.6 Summary of Cases

We have only considered two subjects in this chapter, but the results of such case studies can be an important first indication of potential learning difficulties and of whether an interesting teaching method has been found. The protocols provide evidence for the students making some progress in changing their ideas at a fairly deep conceptual level. In both cases a preconception which is fairly deep-seated in many students was supplanted with other ideas. The main principles used in the approach were: (1) Socratic tutoring – in which questions posed to the students encouraged them to become actively involved in learning; (2) using key examples to activate useful intuitions possessed by the student; (3) building on and extending those intuitions by using analogical reasoning, and in particular, using the strategy of "bridging analogies," a strategy observed in the solutions of expert problem solvers; (4) incorporating anchoring intuitions into the construction of imageable explanatory models. Analyses of such transcripts should allow us to increase our understanding of the learning processes involved in dealing with deep-seated preconceptions.

10.4 Applications to Classroom Teaching

The tutoring strategy used in this study requires some modification in order to apply it to classroom instruction. Individual students differ in the strength of their beliefs in various preconceptions, and the classroom teacher cannot respond individually to each student. However, we have had some significant successes in generating discussions in high school classrooms by using the same basic strategy. Some of these discussions were quite animated. The conflicts between the strongly held views of different students were useful in that they seemed to be a powerful agent in promoting interesting debates. Minstrell (1982) also reported fairly good results in using a slightly less structured, analogy-rich approach.

10.4.1 Study of Classroom Lessons

The two case studies reported here and other tutoring studies led us to construct a group lesson for this topic based on the examples which worked well for the largest number of students. The resulting lesson is summarized by the concept diagram shown in Fig. 10.6 and is part of a curriculum (Camp et al., 1994). The

10.4 Applications to Classroom Teaching

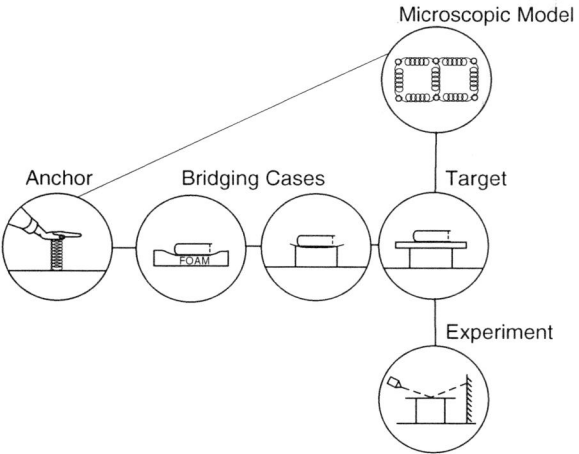

Fig. 10.6 Concept diagram for curriculum (Camp et al; 1994)

uppermost of the three levels in this diagram shows the explanatory model of spring-like bonds between atoms. The lowermost level shows an empirical demonstration in which a light beam reflecting off a mirror on a desk onto the wall is deflected downward when the teacher stands on the desk. This experiment is anomalous and provides dissonance for students who believe that desks are rigid objects that cannot deform to provide an elastic force. After establishing the existence of a normal force in the first lesson, the equality of forces in such cases was addressed using similar techniques in a second lesson. The central level contains a set of carefully chosen thought-examples or cases for discussion, including a target example, an anchoring example, and one or more bridging examples. Thus the three levels in the diagram illustrate a three pronged approach to dealing with a persistent preconception, involving work at three different levels of abstraction.

10.4.1.1 Measurable Improvement

In comparison with control groups, these lessons have shown large significant gain differences on the order of 1 standard deviation in size, as measured by pretests and posttests on problems which deal with students' preconceptions. Similar gains have been realized in five other topic areas in mechanics (friction, gravity, inertia, Newton's third law, and tension) using similar techniques including intuitive analogies and bridging strategies (Brown and Clement, 1989; Clement, 1993b). For example, in the latter study the average gain on three units for the control group was 28.2% whereas the average gain for the experimental group was 54.6%. Thus lessons using these techniques have been successful, and there is reason to be encouraged that students can, with help, deal with the problem of persistent preconceptions.

Qualitative observations from video tapes of these classes indicate that: (1) for target problems many students hold alternative conceptions (misconceptions) that are in conflict with accepted theory; (2) students appear to readily understand the anchoring cases; (3) however, many students indeed do not initially believe that the anchor and the target cases are analogous and resist changing their beliefs; (4) some of the bridging cases sparked an unusual amount of argument and constructive thinking in class discussions; for example, in the normal forces lesson the flexible board case usually promoted the greatest discussion, and a number of students switched to the physicist's view at this point; (5) the lessons led many students to change their positions about or degree of belief in the physicist's view; (6) some students changed their position toward that view during each major section of the lesson, e.g. after the anchor, bridge, model, and demonstration sections, leading us to hypothesize that each technique was helpful to some subset of students. (Brown (1987) reports evidence from tutoring studies which provides further support for this hypothesis); and (7) students were observed generating several types of interesting arguments during discussion, such as: generation of analogies and extreme cases of their own; explanations via a microscopic model; giving a concrete example of a principle; arguments by contradiction from lack of a causal effect; generation of new scientific questions related to the lesson; and even spontaneous generation of bridging analogies. This set of observations in (7) above gives us some reason to believe that even though the lessons were designed primarily with content understanding goals in mind, some process goals were also being achieved as an important outcome.

10.4.1.2 Effort Involved in Comprehending Analogies

However, the classroom discussion leading up to this learning took considerable time and effort – more than is usually allowed in classrooms. Not only did students not accept the aptness of proposed analogies initially, they appeared to need a chain of a number of analogies to make sense of the target problem. Thus, if this trend holds true for other areas, and our experience indicates that it does, methods which use nonformal reasoning to foster conceptual understanding will take more time and effort than is usually devoted to these topics. Furthermore, we view all the units studied above as prerequisites for coming to terms with one of the most persistent conceptual difficulties: most students think that constant motion requires a constant force to cause and sustain the motion. We do not believe that a simple bridging strategy can handle so large and deep a misconception quickly. Thus our experience is that it takes months to deal deeply with the full transition for force and motion. We have come to the position that this time is needed and well spent if it fosters in students an attitude that new explanations and ways of seeing the world can make sense in science, and if it preserves the students' use of model-based reasoning rather than memorized facts and algorithms as a basis for learning.

10.5 Conclusion

10.5.1 Persistent Misconceptions

Students are similar to experts in that they can get "stuck" in one particular model of a situation which resists change. Normal forces is a relatively elementary topic taught near the beginning of a mechanics course, yet as with S2's bending model, students can harbor misconceptions that carry some explanatory power and that are difficult to give up. Special techniques that foster dissonance, "unlearning," and restructuring, and even changes in the definition of individual concepts such as "force" may be needed to promote learning. When this is done, a series of modifications in the students model can lead to the desired overall conceptual change. A skilled teacher breaks the process down into a series of several small changes, but resistance may still be met at each stage.

10.5.2 Explanatory Models vs. Specific Analogous Cases

Returning to the concept diagram in Fig. 10.6, one can contrast the roles of explanatory models, specific analogous cases, and observation activities. The upper level contains more *general explanatory* models – in this case the visualizable mechanism of spring-like bonds between atoms. Explanatory models differ from specific analogous cases shown at the middle level in several ways. Rather than being a specific case, they represent a mechanism that is assumed to be present in many cases. Ideally, a well-understood model can be projected into any of the specific cases below in the diagram. Furthermore, they are not observed phenomena. During instruction they ordinarily must be constructed and not simply retrieved as a familiar example or abstracted from observed phenomena (one cannot observe atoms or bonds inside the table). Like other explanatory models in science they are imagined constructions (Hesse, 1966), (Brown and Clement 1989). Other interesting examples of explanatory model development appear in Driver (1973), White (1993), Steinberg (1992), Hafner and Stewart (1995), Rea-Ramirez (2008), Nunez et al. (2008a, 2008b), Zietsman and Clement (1997).

10.5.3 Two Roles for Anchors

Anchoring examples can play at least two roles in instruction. An analogy can be formed between two examples at the same level – that is, an anchor and a target (e.g. the spring and the table) – which can encourage a student to stretch the domain of application of a correct intuition and apply it to the target example. (Establishing this analogy may be aided by the intermediate bridging examples shown.) Secondly, an anchor can be used as a building block for an explanatory model. Here the two entities are at different levels of generality, and the anchor provides a starting point for a piece of the more general explanatory model being constructed. To our knowledge these two different roles for source analogues in prior knowledge have not been recognized previously in theories of analogy or of instruction. The second role will be examined further in Chapter 18.

10.5.4 Plausible Reasoning vs. Logical Proof Processes in Learning

As in the case of experts, in the cases of student learning the bridging strategy appears to work via knowledge representations that are qualitative physical intuition schemas and at a level that does not use formal notation. Analogies and bridging appear to be important plausible reasoning strategies for stretching the domain of applicability of an anchoring intuition to a new situation, that is, for making the intuition more general and powerful. Here these strategies are introduced by the teacher under careful guidance, but the student is eventually able to carry out the analogical evaluation and transfer operations with full participation. This means that although the teacher initiates and scaffolds these operations, the student is still "learning by doing" in a form of guided, but active, learning. It may very well be that such selected plausible reasoning processes are more powerful than logical proof processes for the development of qualitative ideas that make sense to students.

10.5.5 Role of Thought Experiments vs. Observation Activities in Instruction

In contrast to these thought experiments, actual demonstrations and laboratory experiences are shown at the lower level in Fig. 10.6. Although the demonstrations in these particular lessons tend to speak to the target question, demonstrations and laboratories in other lessons could also be used to support other elements such as an anchor or bridging case that needs added support. In our experience many physics teachers will try to find a single, quick demonstration that will remove the alternative conception and/or convey the physicist's conceptions. However, demonstrations that provide direct evidence for the physicist's qualitative models in these areas are hard to find. This makes sense because if one is trying to help students develop an explanatory model that is theoretical, the model itself will not be amenable to direct demonstration. Brown (1992) obtained evidence for conceptual change in tutoring interviews on static normal forces without using any demonstrations or laboratories. Evidence for change was also found for students using the above strategies minus demonstrations and laboratories in an instructional computer program described in

10.5 Conclusion

Murray et al. (1990). Thus it appears to be possible to affect students' alternative conceptions in some cases without relying on laboratories or demonstrations as a dominant method. Whereas authors like D. Kuhn (1988) have emphasized expert–novice differences in the way evidence is coordinated with theory as one side of the story, the emphasis in this book is on expert–novice similarities in the way people use nonempirical reasoning processes for learning and sense-making. It is granted that discrepant events can be important for providing dissonance with alternative conceptions. Our current hypothesis is that demonstrations and laboratories can and should play a powerful role in instruction but that they are only a part of what is needed. Discussions of rational thought-examples not only tap important anchoring conceptions, they may raise questions and conflicts that prepare students to see the significance of a demonstration and to think about it and discuss it actively rather than memorizing the result. Thus rational methods using analogy and other plausible reasoning processes that are neither proof-based nor directly empirical play a very important role in this approach. This provides an added incentive for the studies of plausible reasoning in this book. Other expert–novice similarities and their implications for instruction will be discussed in Chapters 18 and 21.

Part Two
Advanced Uses of Imagery and Investigation Methods in Science and Mathematics

Section IV
Transformations, Imagery, and Simulation in Experts and Students

If images are representations...that can be subjected to transformations such as rotation and scaling, then imagery could be used as an analog computer, to solve problems whose entities and relations are isomorphic to objects and spatial relations; and we might store a relatively uncommitted, literal record of the appearance of objects from which we can compute properties that we could not anticipate the need for knowing when we initially saw the object.

S. Pinker (1984, p. 55)

Chapter 11
Analogy, Extreme Cases, and Spatial Transformations in Mathematical Problem Solving by Experts

11.1 Introduction

The major issues motivating this chapter are the following. To what range of domains do the expert reasoning patterns of analogy use that have been identified so far in this book apply? Do they extend to mathematics? What nonformal reasoning strategies are found in expert problem solving in mathematics? Is it plausible that some of them involve imagery in spatial transformations? Mathematicians have for some time recognized the value of heuristics such as considering helpful analogous cases and extreme cases, breaking problems into analyzable parts, and performing simplifying spatial transformations (Wertheimer, 1959; Hadamard, 1945; Polya, 1954; Schoenfeld, 1985), but too little work has been done which actually documents their use in expert think-aloud protocols and which analyzes the subprocess used within each heuristic. This chapter uses such protocols to examine the use of these qualitative reasoning strategies in a mathematics problem. Presumably these processes allow talented scientists to attack problems outside the domain of familiar problems for which they posses established algorithmic procedures – problems of a kind they have never seen before. This short chapter can serve only to open the door on the above questions. I hope it may also serve to make an initial connection between the uses of nonformal reasoning in science and mathematics and to stimulate further work in this area.

11.2 Case Study of Analogical Reasoning in a Mathematics Problem

11.2.1 Method

A set of eight subjects were asked to solve the Torus or "Doughnut" problem shown in Fig. 4.7. All subjects were advanced graduate students or professors in technical fields. This chapter reports on results from the eight solutions to this problem and looks in detail at one of the solutions. Some behaviors parallel to those in the solutions of the spring problem discussed earlier have been identified, as well as some completely new behaviors.

A common analogy generated for this problem was to consider the case of the "straightened out" torus in the shape of a cylinder. Subjects conjectured that the volume of these two objects might be the same. The condensed transcript excerpt from subject S6 below gives one example of this approach.

001 S: O.K. So here's a doughnut. (Reads problem in Fig. 4.7) Now the question is how to get its volume. Uhh, the first thing that comes to mind is that it's probably pretty close to a worm, er,

002 S: I mean a cylinder. Where you know, if you laid out the doughnut on the ground, uh, if you cut it open and laid it out, it would basically be the area of the base times the length around the middle. So let's see, I'll put down here number 1 is my first approximation which in fact may turn out to be the exact thing. Uh, I'll just turn it into a cylinder.

003 I: Mmm

004 S: In other words my hypothesis here is this volume of doughnut – and so that would – Pi r2 squared would be the bottom of the cylinder and then, uh, you know, I think the relevant length of the cylinder would be not r1 but the distance uh, uh, to the middle there – namely r1 oops, r1 minus r2 – uh, right, that gets us to the middle, and then times 2 Pi....

005 I: When you thought about the cylinder, do you know how that arose? Did it just sort of flash in your head?

006 S: Well I mean I in fact, er, this little r2 was drawn here very nicely so I just imagine the knife cutting it open and you know, laying it out...

008 S: ...I mean to say I thought of cutting it at one edge and it sort of flopping down and then the uh, the doughnut becomes a cylinder.

009 I: OK

010 S: And so I guess uh, you know with er, I don't think I've made any real algebraic mistakes here. I mean I think this er, probably I mean I feel pretty confident about that answer because what happens when you open the doughnut is that the, the top of the doughnut kind of expands a little.

011 S: I mean er, the part of the doughnut which was the inside stretches out a bit but then on the other hand, the part of the doughnut which was the outer perimeter gets crunched a little bit. And so probably those two things cancel. And you know, if the world is made correctly. But now I guess...

012 I: Just put a confidence number on it at this point and then you can go on.

013 S: Ok. I would say that on that I'm probably like 80% confident.

The cylinder idea fits the definition of a spontaneous analogy as used here because it is a case which differs from the doughnut with respect to a feature (the shape of the doughnut) that is a fixed feature in the original problem. The observational definition of a spontaneous analogy used earlier was the following:

1. The subject, without provocation, considers another situation B where one or more features ordinarily assumed fixed in the original problem situation A are different.
2. The subject indicates that certain structural or functional relationships (as opposed to surface features alone) may be equivalent in A and B.
3. The related case B is described at approximately the same level of abstraction as A.

The act of violating a feature previously assumed to be fixed is the creative aspect of producing an analogy. The difficulty of such acts is presumably the underlying source of the finding in Wertheimer (1959) that many students do not think to modify the shape of a parallelogram in order to compute its area.

11.3 Results on the Use of Analogies for Eight Subjects

In this section, I will attempt to support three points: (1) analogies played an important role in many of the solutions; (2) the most common method for generating the cylinder analogy was via a transformation; (3) criticism and evaluation play an important role in using analogies. As shown in Table 11.1, all eight subjects wrote an equation for their answer that was correct in principle, with one subject making an algebra mistake.

Six of the subjects spontaneously considered the analogy of the cylinder and used it in their solution. The two other subjects used Pappus' theorem which states that if one has a surface of arbitrary shape and one rotates it around a line in the same plane as the surface (but not cutting the surface), the volume of the solid generated will be the area of the shape times the distance that its centroid covers in the circuit. The more interesting solutions were from subjects who did not think about Pappus' theorem, and of those six, four used the cylinder as the main route to their solution. In a fifth solution the cylinder appeared to be of equal importance to another approach, and in a sixth solution it played a confirming role. Thus, spontaneous analogies were observed to play an important role in most of these expert solutions. In saying that these strategies played an important role we mean that they were involved in a serious attempt to understand or solve the problem and were not just proposed by the subjects as an ornamental side comment or as a check on a firm answer.

11.3.1 Analogy Generation Methods

In Chapter 3, evidence for three methods of analogy generation in scientific problem solving was presented. These were generation via an association, generation via a transformation, and generation via a principle (Clement, 1988). It is interesting to ask whether the methods subjects use to generate analogies in mathematics might be similar to those used in science problems. A striking feature of S6's protocol above

Table 11.1 Solutions to doughnut problem (N = 8)

Correct	7
Correct in principle	8
Pappus' Theorem	2
Cylinder	
Played a role	6
Major role	5
Main route	4

Table 11.2 Generation methods for salient cylinder analogies (N = 6)

Evidence for generative transformation (Unbending)	4
Extreme case transformation ($R_1 \gg R_2$)	1
Unclear	1

is the explicit evidence for generating the analogy via a transformation rather than via an association. The most explicit criterion used to code for a generative transformation is the subject referring to changing a fixed feature of the problem. Here the subject makes statements like: "If you cut it open and laid it out" and …"I'll just turn it into a cylinder" (line 2), referring explicitly to changing the shape. This method contrasts to generation via an association, where the subject is reminded of a familiar situation via a direct association (for example, if the subject were reminded of another problem he had seen about an inner tube). This protocol, then, provides fairly explicit evidence for the process of generating an analogy via a transformation in a mathematics problem. As shown in Table 11.2, evidence for a generative transformation of this type was observed in four of the six cylinder analogies. In a fifth case the cylinder idea grew out of considering the extreme case of a very wide thin doughnut with r1 much greater than r2. In the sixth case the generation method was unclear.

11.3.2 Evaluating the Cylinder Conjecture

Some subjects, such as S6, critically evaluated the analogy relation they had constructed by questioning whether the volume of the cylinder they constructed was really the same as that of the torus and by seeking out alternative paths to the solution. For example, S6's transcript continues as follows:

013 S: Uhh, now what would happen if you did various things to the doughnut? Certainly you could argue that…this answer [the formula for a cylinder of length $2\pi(r_1 - r_2)$] is closer and closer to the correct one if uh, you know, if r1 is much, much greater than r2, then in that limiting case, you've got to get this. Because that's just…going to approach being a cylinder more and more. So whatever the correct answer is, it's got to have that [formula] as a limiting case if r1 is much greater than r2…

026 S: …I suppose the other way you could imagine doing it if you wanted to break it up would be to break it up into little wedges of doughnuts. So that if you were looking at it that way then you say OK, here's er, here's another

027 S: infinitesimal element which is a wedge like that, both faces of which have an area of uh, Pi r2 squared. Ha, ha. Of course the thing is what's the volume of a little wedge like that? Um, well, if it's small enough then it's just the thickness of it.

031 S: Volume of wedge would be Pi r2 squared and we'll call that now dZ and again you know, it would boil down to that same equation again, if you added them all up into

032 I: Yeah.

033 S: into equation 2, right? And dZ essentially would be the midpoint, distance there [the thickness of the middle of the wedge]. OK…I've, I mean I think my confidence level at this point would be – like 95%.

Earlier it was stated that the process of criticizing and evaluating an analogy is just as important as the process of generating it in solving science problems. This appears to be true in the case of mathematics problems as well. Subjects who think about an equivalent cylinder must choose a cylinder of the right length, and they often take pains to critically evaluate their choice of length. For example, S6 above chooses the central or "average" circumference of the torus, $2\pi(r1 - r2)$, as the length of the cylinder. But he then evaluates the plausibility of this choice in lines 10 and 11 by giving a qualitative compensation argument about the inside stretching and the outer part getting "crunched." He also evaluates his prediction further by using an extreme case in line 13.

11.4 Other Creative Nonformal Reasoning Processes

Other strategies observed in the doughnut problem solutions are shown in Table 11.3. For example, S6 cutting the "wedges" out of the doughnut above is an example of a partitioning process. Each of the processes is discussed in turn below.

11.4.1 Extreme Cases

Five of the subjects generated an extreme case in the problem and there were six extreme cases generated altogether. For example, several subjects thought about the extreme case where r1 is much greater than r2. Typically they reasoned that if r1 is very large, a small section would look locally very much like a cylinder since it would have very little curvature. Thus they felt that the formula derived from the case of the cylinder would be correct at least in that extreme case. This was the most common way of evaluating the cylinder analogy. Other subjects thought about the case where r2 goes to zero and checked whether the formula they had derived was correct in that situation.

11.4.2 Partitioning and Symmetry Arguments

As mentioned earlier, S6 partitioned the torus into wedges in order to help confirm his solution. Altogether there were 13 attempts to partition the problem generated

Table 11.3 Other strategies used in doughnut problem

	Number of subjects	Total cases
Extreme cases	5	6
Partitioning	5	13
Attempted reassembly of a partition	1	5
Embedding attempts	3	6

by five of the subjects. Subject S6 generated a second interesting partition by breaking up the doughnuts into smaller doughnuts as described below:

018 S: …let's see? Is there any other limiting case we can look at? (15 s pause) I suppose another way to you know, uh, increase my confidence on that is to say well suppose if I really believe – which I do – that this limiting case [r1 >> r2] is correct, then why not imagine the doughnut being made up of a lot of other little doughnuts you know, which are

019 S: tightly packed in there. In other words a whole series of thin doughnut rings that are all packed together in just the right way to give the slightly bigger, fatter doughnut. And then, you know, again, that would indicate that this [equation] is the correct answer. Uh – those are not really space filling though. There's

020 S: little interstices between those doughnuts…. In the final analysis I think that I feel very confident about that because you would – if you were to do the integral, you would break it up into doughnuts that have a square cross section…. And then you would just add all those up.

Figure 11.1 shows a cross section of the doughnut with tiny doughnuts which can be thought of as wires passing through the cross section. Although S6 does not complete the argument aloud here, we can use his imaginative way of partitioning the doughnut to show that the length of the equivalent cylinder should be the same as the length running through the center or midline of the doughnut. To do this we imagine the doughnut being filled with a multitude of tiny thin doughnuts or "wires." We consider a cross section of the doughnut and notice that the average circumference of the thin doughnuts in the cross section should be the same as the length of the conjectured cylinder. We can prove this to ourselves by drawing a vertical line down the center of

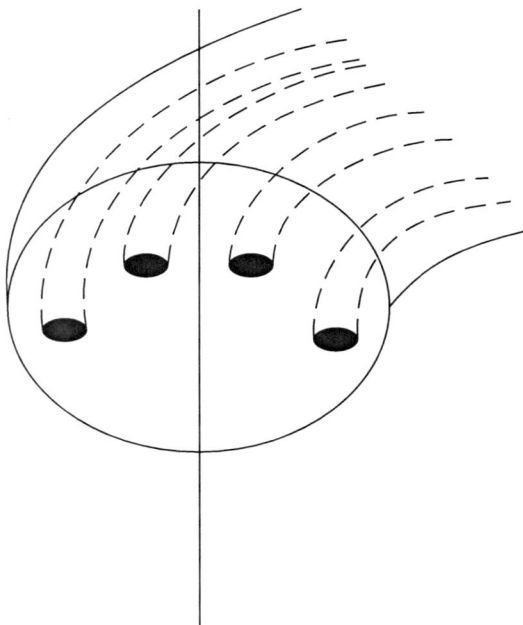

Fig. 11.1 Partition of torus into wires

11.4 Other Creative Nonformal Reasoning Processes

the cross section in Fig. 11.1. One then notices that for every wire on the left side of the line, there is a symmetrically placed wire on the right side of the line. The wire on the left will be of length $2\pi(r1 - r2) - d$, and the wire on the right will be of length $2\pi(r1 - r2) + d$. Thus each long wire on the left has a short counterpart on the right which cancels its extra contribution to the volume, and the average length of a strand is the same as the length of the conjectured cylinder. This argument is interesting because of its use of a creative partitioning strategy. It is also interesting because of the use of symmetry. One can cancel differences by creating a one to one matching between wires on the left and right of the cross section. This symmetry argument allows one to cancel an infinite number of contributions to the volume in one stroke even though each contribution has a different value.

11.4.3 Reassembly of a Partition

Another observed strategy is to partition an object in an attempt to rearrange the pieces into a more "congenial" (simpler or more familiar or more analyzable) object. Figure 11.2 shows a partition of the torus into what another subject, S2, called "apple rings." He convinced himself that the volume of each ring would be equivalent to a rectangular solid whose length is that of the mid-circumference of the annulus. This allows one to "restack" the slices in the shape of a cylinder. Five attempts to partition and reassemble the torus were observed in the solutions, but these were all generated by a single subject.

The classical example of creative partitioning and reconstruction of a problem is found in Wertheimer's (1959) discussion of the parallelogram, whose area can be found and understood by partitioning the parallelogram and reconstructing it

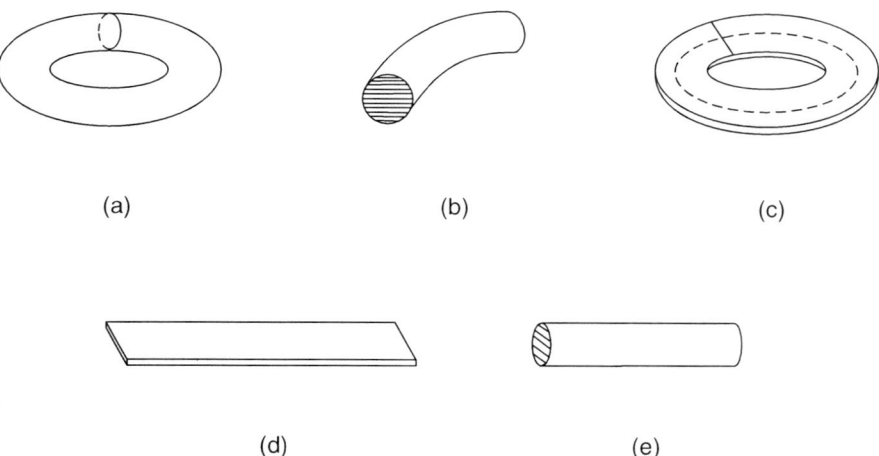

Fig. 11.2 Partition of torus into apple rings

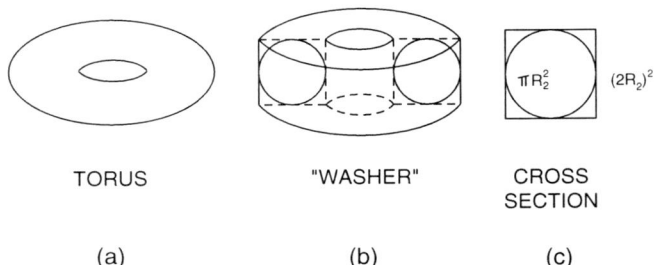

Fig. 11.3 Torus embedded in washer

into a rectangle. Some would not call the cutting and reassembly of a parallelogram into a rectangle an analogy when the two are known to be exactly equivalent with respect to area under this transformation. However with the torus, the partition and reassembly transformation is "rougher" in that there are small deformations in the resulting cylinder. Therefore subjects are dealing with an object that is similar to a cylinder, and arguing that under the right conditions we can treat it "as if" it were cylinder. What these "right conditions" are is not always easy to determine.

11.4.4 Embedding

Six attempts to embed the problem in a larger problem were observed in three of the solutions. For example, one subject embedded the torus in a "washer" shown in Fig. 11.3 which snugly wraps around the torus. The washer is a cylinder with a hole in it, and its volume is easy to calculate by subtraction. He then noticed that the ratio of the area of the torus' cross section to that of the washer could be calculated and that the volume of the torus and the volume of the washer should have the same ratio. Thus he determined the volume of the torus by embedding it in a larger object.

11.5 Discussion

As summarized in Tables 11.1–11.3, it is possible to document several kinds of creative reasoning strategies in expert solutions to mathematics problems; including analogy generation, extreme cases generation, partitioning, the reconstruction of the problem into a different shape, embedding the problem in a larger context, and the use of spatial transformations. Thus, creative nonformal reasoning played an important role in the problem solutions discussed.

11.5 Discussion

11.5.1 Imagistic Reasoning

The protocol also provides some initial evidence for the role of imagery in reasoning. First and most obviously there are ubiquitous references to spatial relations between objects that are primarily qualitative and often dynamic in nature, such as: (line 2) "You know, if you laid out the doughnut on the ground" and (line 11) "The part of the doughnut which was inside stretches out a little bit." The extreme cases most often include references to blowing up or shrinking a feature of the system that are most easily interpreted as a perceptual transformation. Many other passages include references to cutting the torus and reassembling it. All of this takes place with no real objects present in front of the subject. Passages of this kind suggest that the subjects may be: (1) imagining manipulating concrete or idealized objects; and (2) experiencing the anticipated outcomes of their manipulations via imagery.

Second, there are more explicit references to imagery. In Chapter 13, I will define an imagery report as occurring when the subject refers to imagining, picturing, "remembering a diagram for," hearing, or "feeling what it's like to manipulate" a situation. We refer to a dynamic imagery report if the reference is to imagining a situation which does not remain fixed but changes with time. In this book we are concerned only with spontaneous imagery reports where the interviewer does not ask the subject whether an image was used. Examples of dynamic imagery reports in the protocol are: (line 6) "I just imagine the knife cutting it open" and (line 26) "You could imagine…if you wanted to…break it up into little wedges of doughnuts." Thus it is possible to point to some initial evidence in these protocols which supports the hypothesis that reasoning via imagery is involved. I will discuss imagery indicators more thoroughly in subsequent chapters.

11.5.2 Conserving Transformations

S6's reference to cutting the doughnut into wedge-shaped pieces and computing their volumes documents the strategy of breaking a problem into parts – in this case the subject partitions the problem symmetrically into a number of equivalent parts. One could treat this as the simple application of a heuristic, but the trouble with the heuristic "break the problem into simpler parts" is that it does not tell you which parts to form. Such an act can require considerable creativity and ingenuity.

By speaking of an infinitesimal slice, S6 is in danger of breaching the request in the instructions to refrain from taking an integral. Indeed, creative cutting and partitioning of just this kind is an essential skill for applying the integral calculus to nontrivial situations. As in the case of analogies, the breaking into parts process is in effect an attempt to find a (volume-) *conserving transformation* which leaves one with one or more simpler problems. I will revisit the important concept of conserving transformations in Chapter 17 in the context of science problems. It is interesting to note that the wedges can be stacked alternately as shown in Fig. 11.4. In the limit, this can provide an elegant argument for the validity of the original analogy

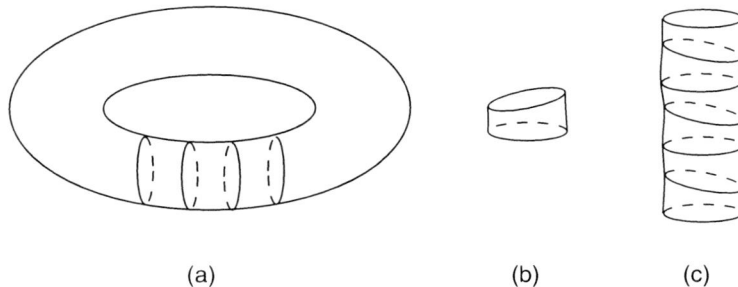

Fig. 11.4 Partition of torus into wedges

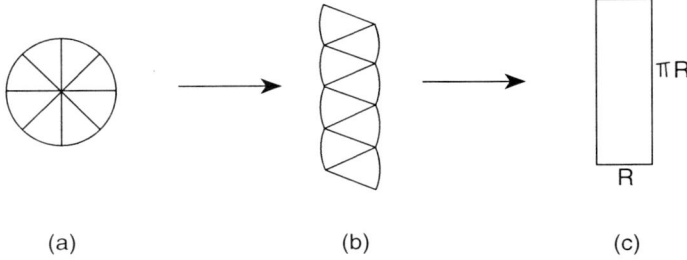

Fig. 11.5 Analogy for calculating area of a circle

to a cylinder of length 2Pi(r1 − r2). (The well-known analogous argument for calculating the area of a circle is shown in Fig. 11.5.) The discovery of such conserving spatial transformations and equivalencies can be a great source of satisfaction and appreciation for the interconnectedness of mathematical ideas. Physical–spatial partitions and transformations of this type played an essential role in enabling Archimedes to achieve the incredible feat of developing many fundamental ideas underlying the integral calculus over 2,000 years ago, in works such as his treatise, "On the Sphere and the Cylinder" (Polya, 1954; Dijksterhuis, 1987).

11.6 Conclusion

In these protocols we find many parallels to scientific reasoning processes documented earlier in the book. In particular, analogical reasoning and extreme cases played an important role in most of these solutions (Clement, 1983). It was found again that there is more than one process for generating analogies, and that the process of critically evaluating the analogy is very important. The fact that the analogy generation and evaluation processes identified were similar to those in the science protocols discussed earlier indicates that science and mathematics have some important nonformal reasoning strategies in common. In addition, the use of creative spatial partitions and transformations and other initial evidence for reasoning via imagery in these solutions provides a motive for gathering more specific evidence on imagery in the chapters to follow.

Chapter 12
Depictive Gestures and Other Case Study Evidence for Use of Imagery by Experts and Students

12.1 Introduction

Chapters 12 and 13 deal with issues concerning imagery and intuition in scientific thinking. The chapters attempt to identify new types of evidence from think-aloud protocols that can be used to identify places where scientists appear to be using various types of imagery in their thinking, including during mental simulations. Since think-aloud protocols have rarely been used for the purpose of studying imagery it is worth considering this in detail. In this chapter I will first examine the potential of depictive hand motions to provide such evidence. I discuss two case studies and examine evidence for the use and importance of imagery in each one, building toward the conclusion in Chapter 13 that the subjects are not only using static images, but are using perceptual/motor schemas to generate dynamic, imagistic, mental simulations to make predictions. In case study 1 of an expert solution I argue that some hand motions are direct reflection of the thinking process, rather than an indirect translation from verbal conclusions. In case study 2 of a student solution I examine evidence for the use of perceptual/motor imagery as a direct source of observed hand motions. I will conclude that certain protocol observations can be used to provide evidence for fruitful imagery use in higher-order thinking tasks. Chapter 13 provides more imagery indicators and evidence on the "presence and importance of imagery" question but also attempts to go beyond it by asking what generates the imagery. The answer proposed there is that perceptual/motor schemas can generate dynamic imagery in what amounts to a mental simulation. What people normally call physical intuitions are one example of such perceptual/motor schemas. This will lead to discussions in Chapter 13 of the active nature of knowledge, the role of implicit knowledge, and evidence that experts can use concrete physical intuitions in addition to abstract principles in higher-order thinking.

12.1.1 Hand Motions

An initial question addressed in the present chapter is whether the investigation of imagery and mental simulation processes can be aided by using data on depictive

Fig. 12.1 Hand motions used in thinking about the earth's direction of rotation

hand and body motions. In these instances, subjects make motions with their hands or body that depict objects, forces, locations, or movements of entities. Depictive motions are observed during important phases of problem solving where the subjects are trying to determine the qualitative aspects of movements and interactions in physical systems. For example, a research physicist shown in Fig. 12.1 stares at his own hand motions while solving a subproblem of determining the direction in which the earth rotates by thinking about a sunrise. And student S20 shown in Fig. 12.2 produces pushing motions while thinking about the added inertia possessed by an object when it has more mass, even in outer space. I will use the term "hand motions" rather than "gestures" because "gesture" carries the connotation of an act of communication. Since I do not wish to assume necessarily that these motions arise solely for purposes of communication, I use 'hand motions' as the more neutral term. This chapter proposes the hypothesis that hand motions can be an external manifestation of what is often a completely internal process of reasoning involving imagery of a visual and/or kinesthetic nature. If this hypothesis is supported, depictive hand motions are an important source of data concerning the form and content of internal imagery.

12.1.2 Imagery Questions and Hypotheses

Imagery hypotheses have been notoriously difficult to evaluate in the past, some authors even believing that they are undecidable (Anderson, 1977), while others believing that they can be supported or argued against (Kosslyn, 1980; Pylyshyn, 1981).

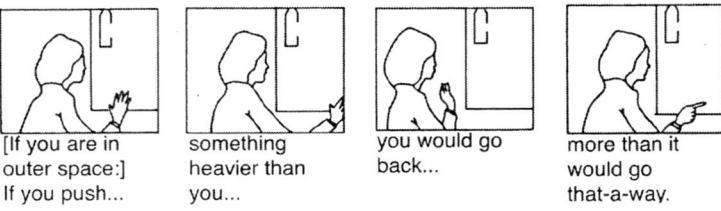

Fig. 12.2 Hand motions used in thinking about pushing an object in space

12.1 Introduction

Establishing a competitive paradigm that includes imagery as a representation is a very large enterprise, and the work of Pavio, Shepard, Kosslyn (and their students) and many others has achieved that. Based on their work, I believe it is reasonable to assume, as a starting point for this study, the existence of imagistic representations that have explanatory power for cognitive theory at a certain level of processing. The data to be presented here cannot eliminate the infinite variety of arbitrarily complex hypotheses that could be formulated to explain hand motions and imagery reports without using the concept of imagery. Instead, I will focus here on another question: "Assuming that imagery can play a role in thought, is there any evidence that it is used in higher level cognition?" I ask in this chapter whether protocols, and gestures in particular, have the potential to speak to this question. Three more specific sub-questions then for the next two chapters are: (1) Given the transcript of a think-aloud protocol that includes hand motions, can we say when it is likely that imagery was used? Can imagery play a significant role in such thought processes? (2) Is the imagery dynamic? (3) Is kinesthetic imagery used in this way? I will conclude that observations of depictive hand motions and other behaviors in transcripts can play a role in answering these questions. (Should one come to accept that such internal imagery is central to a higher-level solution process, beyond this there is a question of whether the external hand motions themselves actually aid the solution process. It may be that the movements sometimes, but not always, play a role in supporting an imagery process. As in the case of drawings, movements may not be *necessary* for thinking in adults, but they may serve as an aid to thinking. However we do not as yet have evidence which speaks adequately to this last point, so I will not address it in this book.) Instead we focus on whether the movements are evidence for imagery use.

To address the above questions I present case studies from two subjects: one expert physicist and one science student. So little is known about the relationships between hand motions, visual and kinesthetic imagery, and thinking, that we have to identify new observation categories as well as hypothesize basic models of cognitive processes in this area. The use of detailed case studies and transcript microanalysis is well suited to these initial tasks of hypothesis formulation and the construction of observation categories. There is a growing literature on the role of gestures in language and thought. I cannot give a comprehensive review of this literature here, but cite selected findings to motivate the present study.

12.1.3 *Previous Research on Hand Motions*

12.1.3.1 Internal Motor Representation

A small number of researchers have argued that motoric representations can play a role in thinking and linguistic meanings. Klatzky et al. (1989) found that priming subjects by having them enact certain hand shapes like "crumple" reduced the time it took to decide, for example, whether it is more appropriate to crumple a newspaper or crumple a window. They conclude that their "results suggest a cognitive/motoric representation of the hand with which actions on objects can be modeled, and the results can be interrogated" (p. 56). Johnson (1987) used patterns in the metaphorical structure of language expressions such as "she hit me with an accusation" to argue that motoric conceptions play a central role in the organization of core meanings underlying language.

12.1.3.2 Language and Gesture

A useful early review of the literature on gestures in language was assembled by Rime and Schiaratura (1991). They presented a synthesis of previous classifications of types of gestures, based on taxonomies by Efron and others. The major categories are: *ideational gestures*, including rhythmical markers that show emphasis and transition points in discourse; *evocative gestures* including deictic pointing and traditional stylized symbols, such as shaking one's finger at someone in blame; and *depictive gestures* referring to objects, shapes, spatial relations, and actions in the subject matter being described. These movements are not part of the traditional stylized gesture code, but are context specific. In this chapter *I will concentrate only on this last category of depictive hand motions.*

Rime and Schiaratura also described the following major theoretical positions on the role of gestures in speech: gestures as substitutes for speech in difficult to describe contexts Kendon (1972); gestures as tools which aid speech in communication (Freedman, 1972); and gestures as indicative of an independent internal source of meaning which can interact with verbal sources (McNeill, 1992). The last category corresponds to my own position, and involves two assertions: (1) gesture and speech often share an early computational stage of language generation and meaning; (2) gesture is often a direct rather than indirect expression of core meaning; that is, when hand movements *precede or co-occur* with speech they do not seem to be computed from the speech in a subsequent stage of processing. A competing view was provided by Hadar and Butterworth (1997), who believe that gesture is epiphenomenal to speech, and is either a derivative of, or designed to support, the speech production process.

12.1.3.3 Studies in Development, Problem Solving, and Instruction

Within studies of gesture in child development (e.g. Goldin-Meadow, 1999; Van Meel, 1984) a major finding is that ideas in young children are often expressed in gestures before they appear in language. A recent review of the role of gesture in educational contexts is provided by Roth (2001). There are few studies in educational contexts (but see Givry and Roth, 2006; Crowder and Newman, 1993; Clement and Steinberg, 2002); Alibali and Goldin-Meadow (1993) found that episodes where there are mismatches between gesture and language can indicate a transitional period where the subjects are particularly receptive to instruction or tend to advance through self instruction on a series of problems. Alibali, et al. (1999) found that strategies used to solve word problems varied systematically as a function of how those problems were represented by undergraduates in their gestures. Monaghan and Clement (1999) and

12.2 Constructing Observational and Theoretical Descriptors

Clement, Zietsman, and Monaghan, (2005) justified the use of gestures and three other categories of imagery indicators as evidence for imagistic reasoning processes.

12.1.3.4 Physiological Evidence for Motor Imagery

Other researchers have found physiological evidence for the involvement of the motor nervous system in tasks where subjects are requested to imagine an event. In a very early study, the inventor of the electromyogram found that subjects asked to remain immobile while imagining performing an action generated electrical EMG potentials in muscles very similar to those from corresponding real actions (Jacobsen, 1930). Decety and Ingvar (1990), Decety (1996), Jeannerod (2001) and Roland et al. (1980) found that blood flow patterns in the motor cortex were very similar for real and imagined actions. Other connections to previous literature will be developed in the discussion section.

12.1.4 Limitations of Previous Research

One limitation of the early research in this area is the tendency of studies to concentrate on relatively simple tasks. An unanswered question is whether hand motions are used in more sophisticated scientific reasoning tasks. Exceptions at the expert level are a conversation between two mathematicians reported in McNeill (1992), studies of scientists discussing graphically presented data in Trafton et al. (2004) and Trickett and Trafton (2002), and an early study of inventors retrospectively describing their process of invention in Krueger, (1981). If hand motions are found in scientific thinkers, can they combine with other evidence to indicate that concrete and dynamic forms of imagery are being used? A second limitation of previous research is the possibility that there may be new categories of hand motions not mentioned therein that can shed light on problem-solving processes. There is also a need to specify connections between observations and theoretical hypotheses with more precision.

12.2 Constructing Observational and Theoretical Descriptors

12.2.1 Proposed Set of Hypotheses

Some of the hypotheses that will be developed in this chapter are shown in Table 12.1. They are numbered roughly in order from those that are less difficult to those that are more difficult to support. Much of this chapter is concerned with the question of whether clusters of actual observations from protocols can be combined to jointly support these hypotheses. The list includes the hypothesis (1A) that word-like symbols are not the only type of internal representation being used during the solution; rather, the solution process can also use representations involving imagery.

Table 12.1 Hypotheses examined

(H-1A) The occurrence of *mental imagery*
(H-1B) The occurrence of *dynamic imagery* in which subjects imagine motions, changes, or interactions over time in a situation
(H-1C) The occurrence of *kinesthetic imagery*. In some cases the subject's perceptual/motor systems for controlling body movements and actions in space can play a role in generating imagery. This can involve (a) imagining moving an object without attending to accompanying forces; (b) imagining exerting forces on an object
(H-2) The processes accompanying hand motions were *important* for the solution

To say what is meant by "imagery,'" I begin with Ronald Finke's definition: "The mental invention or recreation of an experience that in at least some respects resembles the experience of actually perceiving an object or an event" (Finke, 1989. p. 2). Shepard (1984), Kosslyn (1980), and others explain the experience of imagery by assuming that subjects can generate a temporary representation capable of representing in at least a skeletal manner: (a) the shapes of objects; (b) spatial relations among them. This representation may use some of the brain's higher-level perceptual processing capacity which explains the similarity between our subjective experiences of imagery and of perception.

12.2.1.1 Dynamic and Kinesthetic Imagery

That processing capacity may also make available various image manipulation processes (e.g. orienting, transforming, and combining images). Shepard and Metzler's (1971) findings on mental rotation motivate an examination of the idea that dynamic imagery can function as a representation of object motions and other events. Based on their work as well as Freyd and Finke (1984) and our observations of hand motions and other dynamic imagery indicators to be discussed, I will hypothesize that subjects can imagine (c) object movements, changes, or interactions over time, not just as a way to identify two perspectives as the same object (as in Shepard's rotation experiments), but to represent an event in the world. Collins and Gentner (1987) referred to the prediction of future states of a target situation as a key feature of mental simulations and I will retain this as a minimal starting point for building a concept of "mental simulation." When I wish to hypothesize that the prediction of future states come from the use of mental imagery, I will refer to "imagistic simulations."

Recently, Barsalou (1999) has described a theory of perceptual symbols which represent schematic elements of perceptual experience and that can be integrated to produce dynamic simulations. This work will be discussed briefly in Chapter 13. Findings by Klatzky et al. and Decety and Ingvar cited above among others motivate my including imagery of forces and body motions (kinesthetic imagery) in the concept of imagery used here. Thus by imagery I mean the mental invention or recreation of an experience that in at least some respects resembles aspects of an actual perceptual or motor experience. Imagery may be useful in higher-order cognition because it constitutes a representation capable of representing, in at least a skeletal manner, aspects such as: (a) the shapes of objects; (b) spatial relations among them, and sometimes, (c) object movements, changes, or interactions over time. The use of imagery takes place in the absence of currently perceiving or acting on those aspects in the world. There is some evidence that images are generated by activity in certain layers of the brain's perceptual and motor systems (Kosslyn, 1994).

12.2.1.2 Paucity of Research on Imagery in Higher-Level Cognition

However, the level of most tasks used in previous research on imagery has been rather low, tending to involve, say, simple comparison tasks, rather than problem solving or explanation tasks. Rare exceptions are Krueger (1981), Craig, et al. (2002), Trickett and Trafton (2002), Hegarty, et al. (2003), Schwartz and Black (1996a, 1999), Clement (1994a, b). A major purpose of this and the next chapter is to examine whether imagery can be used fruitfully in higher order thinking contexts. The type of "higher-order" cognition this book concentrates on comprises solution processes for scientific prediction or explanation problems; for our purposes this includes the thinking of science students as well as experts. Although the thinking of naive subjects on these tasks is not as sophisticated as an expert's, it is at a much more complicated level than that involved in the simple recall or recognition tasks usually used to study imagery. It is also "content laden" compared to many tasks in previous research, dealing with knowledge that may be specific to certain situations as well as general spatial reasoning operations, and may therefore expose different forms of processing.

On the basis of expert protocol analysis, Clement (1994a, b) identified examples of (both kinesthetic and visual) imagery-related observations: personal action projections, depictive hand motions, and dynamic imagery reports. None are infallible indicators on their own, but together they were most plausibly explained there using a framework that includes flexible perceptual/motor schemas that generate and run imagistic simulations, via the extended application of a schema outside of its normal domain, implicit knowledge, and spatial reasoning. These themes are developed and expanded in the next four chapters.

12.2.2 Relations Between Observations and Hypotheses

Thus our main objectives here are qualitative existence proofs of (empirically based findings supporting the presence of) new phenomena worth observing, and of new mental processes suggested by these. In addition, there is the methodological question of whether protocol observations relevant to imagery from think-aloud protocols can be defined and collected at all. These goals are, methodologically speaking, prior to that of measuring frequencies of occurrences of certain behaviors in order to generalize to a certain population. We cannot count behaviors when we do not yet know which behaviors are relevant. The question here is more basic than: "How many occurrences?"– it is rather: "What sorts of occurrences are important and what do they indicate about the subject's thinking?" (See Clement (2000) for a discussion of how observation concepts must be constructed in addition to initial theories in the early stages of a physical science, and how, in cognitive science, generative case studies play a corresponding role.)

12.2.2.1 Triangulation and a Proposed Set of Observable Descriptors

In order to introduce some of the categories of observations identified, Table 12.2 shows a few of the many observable descriptors that can provide some support for Hypotheses 1A, B and C in Table 12.1. These observation categories and hypotheses were developed only after considerable number of cycles of data analysis, criticism, and revision. I present the categories here as an advanced organizer.

An expanded version of Table 12.2 with definitions appears in Table 12.3. Table 12.3 will be used in the analysis of the protocol segments. (Evidence for observations numbered 20 and higher will be discussed in Chapter 13.) This list of observable descriptors was constructed and refined over many years during repeated viewings and discussions of tapes in which we asked ourselves the questions listed in the chapter introduction.

Most of this chapter, then, discusses the possible connections of support between the observations like those in Table 12.2 and the hypotheses in Table 12.1. Hypotheses 1A, 1B, and 1C in Table 12.1 are difficult to evaluate, since there is no consensus on what counts as evidence for imagery-based processes and because it has been assumed that evidence for imagery is necessarily indirect. It may be that some types of depictive motions are as direct a source of evidence for the form of a subject's dynamic imagery as we are going to find. However, hand motions will still be challenged as a more subjective form of evidence than spoken words, since there is no standardized vocabulary for gestures in normal humans. To begin to bridge this gap, I have formulated descriptions of the movements at two levels, the more specific descriptions within the transcripts, and the more general observation patterns listed in Tables 8.2 and 8.3. In this way one hopes to contribute to the difficult problem of constructing an observation language for describing features in protocols that are relevant to evaluating and improving our hypotheses.

Table 12.2 Short list of some observations in transcripts related to imagery hypotheses

A. **Imagery indicators** (Support Hypothesis 1A)

> 3a *Imagery reports*: Subject states that s/he is imaging or imagining, "seeing" or "feeling"
> 1a *Depictive motions:* The subject makes nonstylized hand or body motions depicting objects, forces, locations, or movements of entities

B. **Dynamic imagery indicators:** (Support Hypothesis 1B)

> 2 *Motions depict dynamics*: Hand motions depict the form of a dynamic event, not simply a static picture
> 3b *Dynamic imagery report*: An imagery report where the subject indicates that they are imagining motions, changes, or interactions over time in a situation

C. **Kinesthetic imagery indicators:** (Support Hypothesis 1C)

> 3c *Kinesthetic imagery report*: Reports imagining own physical actions or muscular effort
> 16 *Personal Movement Projection or Analogy*: (a) Refers to movements of entities in target situation as if they were moved by a person, or (b) uses a personal analogy by referring to an analogous situation involving the body

12.2 Constructing Observational and Theoretical Descriptors

Table 12.3 Imagery indicators for transcripts: by category

I. **Imagery indicators** (Observations that, especially when they occur with others, can support the hypothesis of imagery use)

A. **General indicators**

3a *Imagery reports*: Subject states that s/he is imaging, imagining, "seeing" or "feeling" (or experiencing any other sensation)

Imagery enhancement

20 *Imagery enhancement report*: Adds markers to a situation or changes other perceptual features such as: (1) orientation or size; or (2) magnitude of problem variables – in conjunction with evidence for imagery. More direct evidence occurs when the subject speaks of modifying a problem situation in a way that makes it "easier to imagine".

23 *Orients body*: Body is oriented to reflect point of view (e.g. faces north when object in problem is facing north)

Hand motions

1a *Depictive motions*: The subject makes nonstylized hand or body motions depicting objects, forces, locations, or movements of entities

1b *Frozen action gesture*: The subject's hands are positioned as if manipulating or ready to manipulate an external object, but are stationary ("frozen")

1c *Depictive hand or pencil motions over a drawing*: These are not simply pointing to a word or picture, but indicate movements, locations, or shapes of objects or features that do not appear in the drawing

Motions not translated from verbal:

11 *Motions accompanied by ambiguous terms*: Uses ambiguous terms in conjunction with less ambiguous motions to refer to:
   (a) The shape or motion of an entity, e.g. "It looked like this," "It would go that-a-way"
   (b) The means used to exert a force on an object, e.g. "If you push this way"

12 *Motions with unrelated language*: Motions occur during language that does not describe them (e.g. S says "Well, let's see what we can find out here" during motions. Producing such oral language would be likely to interfere with other language that served as the source for a translation to gestures, and so such a translation is unlikely here)

13 *Particular motion contradicts a language error*: Verbal description of an event is inconsistent with concurrent hand motions representing the event, and these motions are consistent with later conclusions in the solution

Spatially coherent reasoning and point of view (these are not imagery indicators on their own, but support the assumption that hand motions can be one source of evidence for imagery)

17(a) *Descriptions of spatial reasoning fit motions*: The subject reports elements of reasoning via spatial relations between objects and this is coherent with hand motions

17(b) *Corresponding steps*: Steps in the reasoning being expressed by the subject correspond to steps in a series of motions

17(c) *Synchronized entrainment*: There is synchronized entrainment of movements and speech that describes the reasoning

18 *Matching points of view*: The point of view (particular angle of view) as expressed verbally by the subject during reasoning corresponds to the point of view represented by hand motions

Findings were not translated from verbal

24 *Indicates imagery as source of finding*: Indicates that "imagining" of some kind is the source for a finding

25 *Difficulty describing images*: Indicates that image or thinking in episode leading to finding is clear but is hard to describe

(continued)

Table 12.3 (continued)

B. **Dynamic imagery indicators:**
   2 *Motions depict dynamics*: The hand motions depict the form of a dynamic event, not simply a static picture
   3b *Dynamic imagery report*: An imagery report where subject indicates that they are imagining motions, changes, or interactions over time in a situation
   21 *Indicates time-extended process* (e.g. "So it's drifting along further and further and further") in conjunction with another imagery indicator

C. **Kinesthetic imagery indicators** (Imagery of muscular force)
   3c *Kinesthetic imagery report*: Reports imagining own physical actions or muscular effort
   14 *Uses force terms*: Imagery indicator occurs in conjunction with force context terms such as "pull," "push," "twist," "effort," "try to move it"
   15 *Personal force projection or analogy*: (a) Refers to forces exerted by entities in target situation as if they were exerted by a person or (b) Uses a personal analogy by referring to an analogous situation involving body forces
   16 *Personal movement projection or analogy*: (a) Refers to movements of entities in target situation as if they were moved by a person or (b) Uses a personal analogy by referring to an analogous situation involving the body
   (I refer to 15 and 16 collectively as Personal Action Projections or Personal Analogies)

II. **Evidence for importance of processing during episode containing an imagery indicator from Section I above:**

A. **Timing:** (When an episode containing one of the imagery indicators in section 1 above occurs in the following ways it supports the hypothesis that the imagery is occurring concurrently with reasoning about the problem)
   5 *Indicates thinking during episode*: Subject states an intention or need to think just prior to motioning or makes statement implying they are thinking during episode
   6 *Episode precedes finding*: Episode occurs just prior to finding of prediction or increase in confidence
   7 *Indicator follows question*: Relevant motions occur immediately after a specific question is asked, so that movements appear to be associated with the process that answers the question

B. **Increased attention or effort:** (When an episode containing one of the imagery indicators in section 1 occurs in conjunction with the following it supports the hypothesis that the subject pays increased attention to processing during imagery):
   10 *Minimizes visual stimulation*: (a) Closes eyes or
      (b) stares at ceiling or other minimally stimulating area
   4 *Gap in speech*: Subject becomes silent and there is a noticeable gap in the train of speech while running through the motions
   9 *Episode repeated*: Episode is repeated as subject indicates an attempt to confirm an answer to the same question
   1b *Frozen action gesture*: The subject's hands are positioned as if manipulating or ready to manipulate an external object, but are stationary ("frozen")
   22 *Effortful episode*: Indicates episode was effortful
   (The following indicators are discussed in Chapter 13)

C. **Imagery enhancement reports:** (These indicators support the hypothesis that imagery played an important role in reasoning)
   20 *Imagery enhancement*: Adds markers to a situation or changes other perceptual features such as: (1) orientation or size or (2) magnitude of problem variables – in conjunction with evidence for imagery. More direct evidence occurs when the subject speaks of modifying a problem situation in a way that makes it "easier to imagine."
   23 *Orients body*: Body is oriented to reflect point of view (e.g. faces north when object in problem is facing north)

(continued)

12.3 Case Studies

Table 12.3 (continued)

III. Related indicators for simulation

19 *Anticipates new states*: Indicates new anticipation of states of a system or continuous change of state. This is the minimal indicator for a simulation, not an imagery indicator, but, but when it appears along with imagery indicators we take it as evidence for an imagistic simulation

IV. Status unclear

(The following observations are interesting when they occur but they were not used as imagery indicators in this book as a conservative measure. The last four may be defendable as indicators in other studies.)

8 *Stares at own hand motions* (Eventually this might be used in support of a theory about how hand motions can facilitate cognition by off-loading some of the processing as an external memory or model.)

17d *Describes plausible method of spatial reasoning*: The subject reports reasoning via spatial relations between objects. A necessary but not sufficient requirement for this to be true is that the described objects (or a scale model), if actually assembled and manipulated, could be used to attempt a solution

26 *Generates a diagram or drawing*

27 *Focuses on a drawing while making inferences about spatial transformations, movements, or spatial or physical relationships* not depicted in the drawing

28 *Intuition report*: Spontaneously reports using an "intuition" or uses terms that indicate they are proceeding primarily on the basis of a nonformal "feeling" or "sense" of what will happen to a system

Observations 20-25 will be discussed in Chapter 13.

It will be noted that an observation can support two or more hypotheses that are part of a larger framework. As in ecological systems and many other areas of science, the system we are talking about is complex enough to preclude attempts to describe it using a one to one or many to one relationship between observations and theoretical hypotheses. It is necessary to use many–many relationships (Easley, 1974).

12.3 Case Studies

12.3.1 An Expert Protocol

The first protocol excerpt to be examined is from an expert – in this case a professor of physics actively engaged in physics research. While working on a problem about the displacement of a ship at the equator, he sets a subgoal for himself: "Now I have to decide which way the earth is turning." While answering this question he exhibits several complex hand movements described in the protocol on the following page (see also Fig. 12.1) as he simulates the motion of the earth with respect to the sun. Our purpose here is to ask whether the hand motions and other observations can be used to gather evidence on the presence and form of imagery processes. Another interesting feature is the fact that the subject actually misstates the direction of the earth's rotation as he represents the motion correctly with his hands in line 3b. This suggests that his internal representation of the motion is quite different from the language he uses to refer to it. For convenience, Table 12.3 is rearranged in Table 12.4 in numerical order in Appendix 2 of this Chapter.

Protocol of S15

Code numbers for observed features from Table 12.4, Appendix 2	Protocol	Movement Observations
(Ob-1b, 5)	3 (a) Ah, OK, now I have to decide which way the earth is going and whether, let's see (M1a), the earth turns from	(M1a) Keeping the palms of his right and left hands open and facing each other as if grasping a 5" sphere from the top and bottom, he freezes hands in this position for 2 s, then (See Fig. 12.1)
(Ob-1a,2,7,8)	(b) (M1b) east to west…	(M1b) *twists them counterclockwise… while staring down at his hands* (axis of rotation is vertical)
(Ob-1,2,8,13,17a,b,c)	(c) The(M2) sunrise, OK…	(M2) While keeping right hand in grasping position as if still holding a sphere, waves left hand in air immediately to the left of right hand.
(Ob-1,2,6)	(d) So that the ship, is going to the east…(M3)	(M3) holds left hand stationary with open palm up as if supporting a ball, *moves his right hand around the far side of left* hand with fingers pointed to his left while staring up at the wall
(Ob-19)	4 (a)…is going in the same (M3 again) direction as the earth's motion and the plus sign is for east.(M4)	(M4) Picks up pen, raises it toward paper, stops, puts pen down, turns to stare at the wall.
(Ob-1,1b,2,4,7,9,10b,17a)	(b) Is that right? (M5)	(M5) Stops talking. *Repeats original twisting motions* with palms of hands facing each other, *while staring at the wall*, then keeps hands frozen in position for 3 s
(Ob-1,2,6,8)	(c)…and it's in the (M6) – yes that's right, OK	(M6) Points index finger of right hand forward alone and bends his wrist so that finger sweeps from right to left, while staring at his hand.

12.3.2 Analysis of S15's Protocol

12.3.2.1 Basic Solution Strategy

The first of several depictive motions made by this subject appears in line 3b. From his motions and comments we infer that his method is to imagine the earth rotating from a vantage point in outer space, and simultaneously to imagine a person standing on the earth facing east. If the sun would appear to rise to that person, the direction of rotation is correct. The problem then becomes: which way should I turn this ball so that the sun will appear to rise to a person standing on the side of that ball? After making an initial determination of the direction of rotation while staring at his own hand movements, he says to himself, "Is that right?" stops talking, and proceeds to silently perform an elaborated version of the original set of hand movements while fixing his gaze on the wall in line 4b. He then says, "Yes that's right." The way S15 repeats the motions silently in a second episode is interesting because it suggests strongly that he is repeating some mental process while he is making the second set of hand movements that lets him check the direction of the earth's motion, and that these motions are not for purposes of communication.

12.3.3 Evidence Supporting the Use of Imagery in the Solution

The following observations from this transcript support the idea that the subject is using imagery during his reasoning. We can triangulate from multiple observations, when available, to provide stronger support for imagery use than can be provided by any one type of observation.

> Ob. 1a *Depictive hand motions*. The subject displays depictive hand motions during the problem solution. These motions depict locations and movements of entities described in the problem.
> Ob. 13 A *particular motion contradicts a language error in line 3b*. A verbal statement of the subject is inconsistent with concurrent hand motions representing the event, and these motions are consistent with later conclusions in the solution.

We first give arguments that it is reasonable to take depictive hand motions as one source of evidence for an internal imagery process. (A more detailed argument for this statement is given in Appendix 1 and Fig. 12.6 there.) The last observation above provides evidence that the motions were not translated from verbal statements. The fact that the subject misstates the earth's direction of rotation suggests that a linguistic representation is not the only one operating during his correct solution. Also the fact that the motions occur just after a question is asked as well as just before an answer to the question is given suggests that the motions were made concurrently with the subject's thought process. His misstatement is consistent with the hypothesis that he is using a cognitive process involving imagery which is independent of, and in this case eventually acts to correct, his language-based processes. In this interpretation his speech may be actually a less-direct indicator of his processing here than are his hand motions – the speech may be the result of a faulty translation from the correct imagery in his original processing. (See McNeill, 1992 for similar observations in simpler recall tasks).

Finally, the analog nature of the depictive hand motions and the fact that they depict continuous, perceptual properties of entities involved in the solution (e.g. shape, trajectory of motion), suggests in turn that the subject is using imagistic, perceptual/motor, internal representations during his solution process which are able to generate the hand motions. Together, these constitute arguments that it is reasonable to take depictive hand motions as one source of evidence for an internal imagery process. A more detailed argument for this conclusion is given in Appendix 1 to this chapter.

12.3.4 Argument Structure

Assuming that one can take depictive hand motions as one source of evidence, Fig. 12.3 then shows a larger initial argument structure or triangulation network that can be used to argue that imagery is involved in S15's solution. The arrows in this figure mean "supports and is explained by" (not "implies"). Here the statements about imagery are divided into two major hypotheses: first that the subject uses imagery during the solution, and second that processing during the time of the imagery episode was important to the solution. Taken together, these hypotheses in turn suggest that imagery played a significant role in the solution process, and that this is the simplest explanation for this subject's behavior. I have already discussed the first hypothesis in Fig. 12.3 and will now discuss the second.

12.3.4.1 Evidence for Importance of Processing During Depictive Motions

Even if it is accepted that the subject is experiencing dynamic imagery, it could still be argued that the imagery did not play a role in the actual solution process. To begin to answer this objection, several additional kinds of observations are shown on the right side of Fig. 12.3. They support the following hypothesis:

Hypothesis (H-2): The hand motions accompanied processes that were important to his solution; this hypothesis is supported by the following observations on timing:

> Ob. 7 Movements occur immediately after a question.
> Ob. 6 Movements occur immediately before he/she reaches a finding.
> Ob. 5 Subject's statements imply the subject is thinking about the problem during motioning.

In these three observations, the timing of the movements argues that they were closely tied to the solution process. The subject's statement that he "has to decide which way the Earth is going" in line 3a implies that he is thinking about the problem during the motioning, as does the fact that it occurs immediately prior to the motioning. The way that he reaches findings correctly just after motioning suggests that he makes the motions simultaneously with his mental solution process rather than afterwards.

In addition, hypothesis (H-2) above is supported by observations indicating that the subject is paying increased attention to his processing during the motions:
Increased attention:

> Ob. 1b "Frozen action" gesture. The subject's hands are positioned as if manipulating or ready to manipulate an external object, but are stationary ("frozen").

12.3 Case Studies

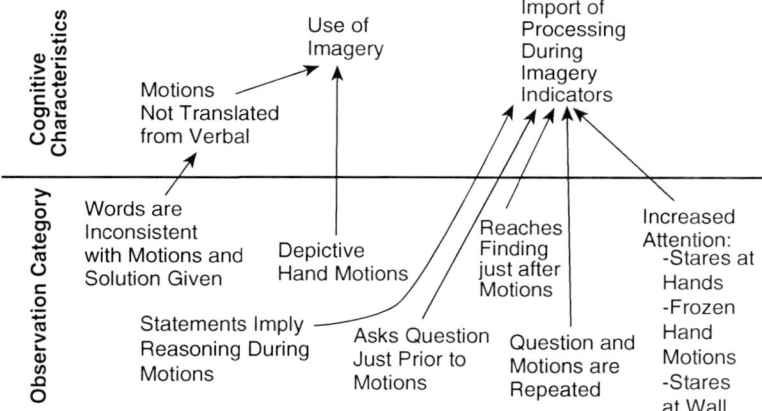

Fig. 12.3 Observations relevant to use of imagery and its importance

Ob. 9 Movements are repeated (here after repeating the same question) as if to check the answer by repeating the process.
Ob. 10B Stares at wall, ceiling or other minimally stimulating area.

These final observations supplement the previous ones in supporting our view that processing during the hand motion episodes was important to the subject's solution. When this finding is combined with hypothesis (H-1), the presence of imagery in these episodes, the most parsimonious view is that imagery played an important role in the solution process.

Figure 12.4 is an expanded version of Fig. 12.3 connecting the behavior categories to protocol data. Transcript and hand motion data at the bottom are tied to observation categories and from there to cognitive characteristics at the top. Thus we can point to a larger network of support for our imagery hypotheses in this case, even though each piece of evidence on its own provides only a limited amount of support. This gives an initial indication that it is feasible to gather, from multiple observations, evidence for the importance of imagery in a solution process – including evidence from hand motions. In particular, there is evidence that depictive motions can be a partial reflection of significant problem solving processes, and that these processes can involve imagery.

Still, to provide more adequate substantiation for these hypotheses requires the analysis of other protocols and other indicators. Our aim in this chapter is simply to begin the process of identifying potential imagery indicators from think-aloud protocols. Other very important imagery indicators such as spontaneous imagery reports where the subject describes "imagining" or "picturing" a situation, and imagery enhancement episodes where the subject actually adds "markers" to an image or takes a new point of view in order to make it easier to "see" an image are discussed in Chapter 13.

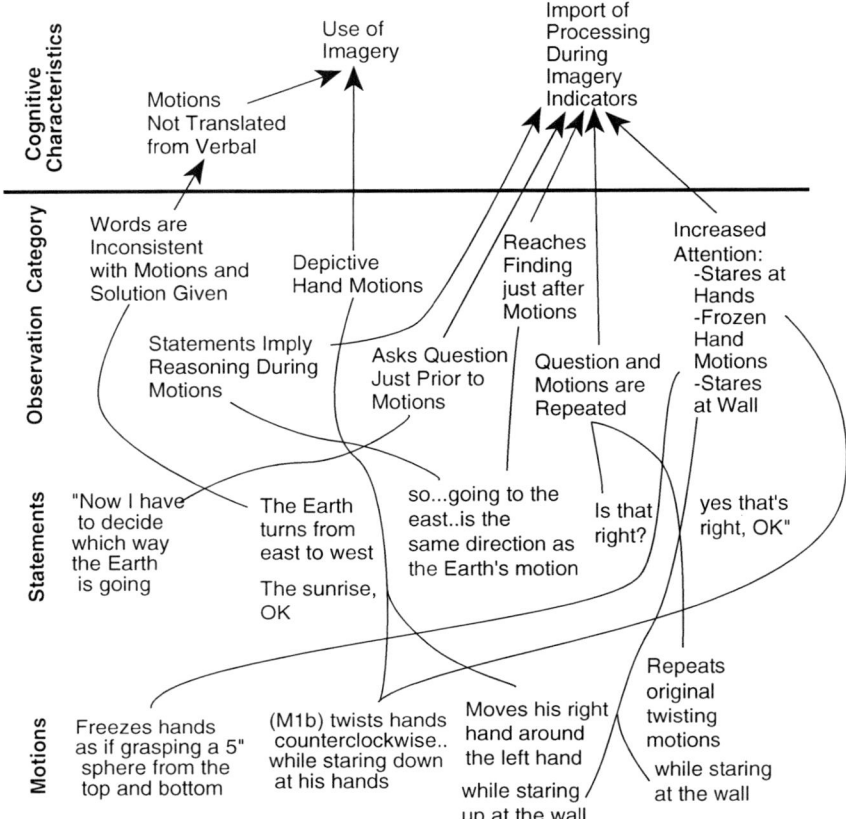

Fig. 12.4 Hypotheses supported by imagery indicators for S13

12.3.4.2 The Methodology of Collective Abduction and Multiple Support

It is worth considering the general methodology underlying Fig. 12.4. Collecting evidence on imagery processes is inherently difficult. When multiple observations triangulate together on the same hypothesis, they give one a stronger basis of support. Single observations can have explanations other than the hypotheses given above the horizontal line in Fig. 12.4, as symbolized by pairs of lines reaching upward from each observation symbol in Fig. 12.5. But when a number of the observations have a single hypothesis in common, they can point to the hypothesis as *the most plausible explanation* for that cluster of observations, shown schematically as hypothesis E in Fig. 12.5.

Thus the relation between observations and hypotheses used here is not one of unique implication but rather of collective abduction and support. Pierce (1958) and Hanson (1958) used the term "abduction" to describe the process of formulating a hypothesis which, if it were true, would account for a set of phenomena. Our aim

12.3 Case Studies

Fig. 12.5 Hypothesis (e) with multiple sources of support

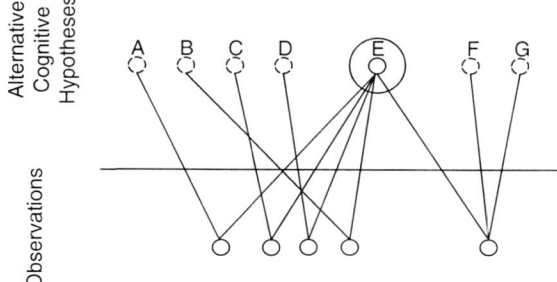

is to generate, winnow, and converge on hypotheses which provide the most plausible explanations for the greatest number of observations in these protocols. The statements on methodology in the above paragraph are pertinent to most of the analyses done throughout this book.

Categories in Table 12.3 were generated by a team of two analysts, applied to several protocols as observations (as in the protocol of S15 above) and linked to implications about imagery (as in Fig. 12.4). In early stages, these linkages were often then criticized by one or the other analyst as invalid, leading to revisions in the observation definitions and in their placement in Table 12.3. This cycle was then repeated many times until agreement was reached. (I am indebted to William Barowy for his early role in this work.)

Figure 12.4 provides a way to explain why the process of sorting out the observation categories and relations of support was quite challenging. The larger task we faced can be described by imagining six episode descriptions instead of one at the bottom of the Fig. 12.4 as the database, plus about 40 candidates for observation categories, and about ten candidates for cognitive characteristics. The categories and hypotheses effectively formed a maze of possible choices and relationships that could only be improved by repeated cycles of criticism and revision. For example, it took time to see the value in separating hypotheses about the presence and use of imagery from hypotheses about the import of processing during episodes containing imagery indicators. These hypotheses have reached a stage where they can be articulated publicly for further criticism and revision.

12.3.4.3 Dynamic Imagery and Mental Simulation

We can go on to ask whether there is any evidence here for the use of *dynamic* imagery. I will assume that dynamic imagery possesses characteristics proposed by Cooper and Shepard (1973), namely, it operates over a period of time and passes through a large number of intermediate states that should correspond to intermediate states in the perceived system. Thus dynamic imagery in this case means an internal flow of imagined perceptions and/or actions over time that causes the subject to experience some aspects of the flow of perceptions or motor actions that would exist if the subject were actually viewing and/or causing such events.

Using this concept, we can ask whether there is any support for hypothesis 1B in Table 12.1 in the case of S15:

(1B) Dynamic imagery is used in which subjects imagine motions, changes, or interactions over time in a situation.

If, from prior considerations, one assumes that the subject is using *some type* of imagery, the fact that, as the subject makes predictions about a dynamic system (the earth's motion relative to the sun), his hand motions depict the form of a series of future states of the system suggests that he is using *dynamic* imagery during the problem solution. This contrasts to the use of hand motions to paint a static "picture in the air." (Another observation not present here that would support hypotheses 1B is a *dynamic imagery report* where the subjects speak of imagining motions, changes, or interactions over time in a situation. These will be discussed in the next chapter.)

As stated earlier, I will build on Collins and Gentner's (1987) focus on the prediction of future states of a target situation as a key feature of mental simulations. If dynamic imagery is included as part of our model of mental simulation, then we have a richer explicit meaning for the term simulation than is often used in the literature. I will refer to a mental simulation that involves the use of imagery as an *imagistic simulation*.

12.3.4.4 Do Hand Motions Aid Thinking?

If one assumes that the hand motions occur during important thought processes, there is the additional question of whether the external hand motions themselves actually aid the thought process. Although we do not have any strong evidence on this issue here, it is possible that movements may sometimes, but not always, play a role in providing partial support for an imagery process. In this case the fact that S15 stares at his own hands suggests the possibility that the movements can play a role in supporting an imagery process, similar to the way in which a rough sketch might provide external support for a more detailed imagery process. As in the case of drawings, the movements may not be *necessary* for thinking here, but some could still serve as an aid to thinking. However, because there is little evidence on this issue from our protocols, it is not a focus of this book. Rather, our immediate focus is on the potential of depictive motions to serve as indicators of imagery and imagistic simulation processes, along with other indicators.

The case of S15 gives an initial indication that it is feasible to gather evidence for the idea that depictive motions can be a partial reflection of imagery during significant, problem-solving processes. A summary of the kinds of evidence used to support this position is given in Fig. 12.6 in Appendix 1 to this chapter. We will examine more evidence for the use of dynamic imagery in the next case study and in Chapter 13.

12.3.5 *A Student Protocol*

The second protocol to be examined is that of the freshman engineering student S20 described in Chapter 9, who appears to use *kinesthetic* imagery. The problem concerns the behavior of a metal cart being launched by an elastic band. In the first part of the interview S20 launches the cart across the top of a table using the elastic band (as shown in Fig. 9.1a) and watches it roll to a stop. He correctly predicts that its maximum speed will be lower if pieces of steel are placed in the cart to increase its mass. He is then asked what will happen when two adjacent carts are launched with an elastic band of the same strength, but the upper cart in the problem drawing has more mass. The physicist predicts that the more massive cart's speed should be lower, since although the *weights* of the objects (the force of gravity acting on them) are both zero in deep space, their *masses* [resistance

Code numbers for observed features	Protocol for S20
(Ob-14, 15a)	31 S: Uh, I'd say that – it would still go slower (points to upper cart and slides finger to right while staring at drawing). If you pulled this out with the weight on it, it would go slower than if you pulled this out (points to lower cart and slides finger to right while staring at drawing) without the weight.
(Ob-1,5,2,14,15b)	32 S: Without the weight you'd go faster and with the weight you'd go slower, because I can still think of, uh, you pulling (holds right hand up and moves it toward himself) something very heavy and pulling something very light. (Repeats hand motion)
	33 I: In space?
(Ob-1,2,6) (Ob-11a,14,15,16, 18)	34 S: All right, in space. I've never been in space – Yeah, in space, too, I guess, in space. If you push (moves right hand away from chest) something heavier than you, you would go back (moves head and right hand back) uh, more than it would go that-a-way (points forward). (See Fig. 12.2)
	35 I: Mmm.
(Ob-1,2,14,15,16, 18,19)	36 S: I just think this. I figure if you pushed (moves hand forward) the [large] rocket, you'd go back (moves hand back) more than the rocket would.
	38 S: So if, uh, so if you pulled this (slides finger to right of upper cart while staring at drawing) with a heavier [weight in it], it would go slower....
	46 S: (a) So (looks back at drawing) if you had the rubber band here, (slides finger to right while staring at drawing) it would still pull the lighter rocket [cart] faster than it would pull the heavier rocket [cart]. The heavier rocket would stay slower.

to acceleration] are different, being unchanged from their masses on earth.) During the interview, S20 apparently prefers to think of the two carts being launched from elastic bands as "small (auxiliary) rockets" being launched from the side of a large rocket ship. In this protocol I will focus on possible types of evidence for the use kinesthetic imagery. Care has been taken to note all hand motions in this section of the transcript; other indicators from Table 12.4 in Appendix 2 of this chapter are shown on the left.

12.3.6 Analysis of S20's Protocol

S20's comments all seem to relate to the question of whether an object's greater mass makes it harder to accelerate in space, and he is initially not at all sure of the answer to this question. However, he does appear to arrive at a strong, correct answer to the cart question, based on nonformal arguments. He does this apparently by considering a series of analogous cases in the form of thought experiments, each accompanied by depictive motions or body movements. The themes of two of these thought experiments are "pulling" the cart with his hand, and "hitting" a rocket with his hand. These are analogies in the sense that, like the carts problem, they are cases where the amount of matter in an object may make a difference when one is attempting to accelerate it in space. (See Fig. 12.2 for exact tracings of video images from S20's second thought experiment theme.)

12.3.6.1 Use of Imagery

Evidence for the use of imagery here can be described as follows. First, depictive hand motions occur in each of the thought experiments above. Pulling and pushing motions accompany his respective thought experiments about pulling and pushing objects. The thought experiments are concrete situations that appear to be invented by the subject. Specific support for the hypothesis that he is using imagery comes from the following:

- In line 34, he uses ambiguous terms in conjunction with motions to refer to the motion of an object; this provides some evidence against the alternative hypothesis that the motions are translations from a linguistic representation.
- In lines 31, 38, and 46 we see the display of hand motions over drawings that depict the action of an object in the drawing, as the subject stares at the drawing. In this situation it is highly likely that an internal perceptual representation is already in use that corresponds to the drawing. Since the hand motion is used to represent the action of an object and is aligned to the drawing, the idea that the motion is generated by an image of the action within the internal representation of the drawing is highly plausible.
- The first two themes are *personal analogies* in the sense that he puts himself in the place of one of the objects in the problem situation. For example, in lines 31 and 32, he replaces the elastic band with his own hand: "I can still think of you pulling (holds

right hand up and moves it toward himself) something very heavy and pulling something very light." Since there is evidence from the literature that adults can easily form images of body motions (see chapter introduction) the imagery hypothesis is further supported by the fact that he uses personal analogies in lines 31, 32, and 34, by referring to forces exerted by objects as if they were exerted by a person.

These observations provide evidence for the involvement of imagery in his thinking.

12.3.6.2 Dynamic Imagery

We can also use the last two observations to support the further hypothesis that his imagery is dynamic. The subject uses hand motions to depict forces in lines 32, and 34, and to depict movements in lines 13 and 34. All of these are motions depicting the form of a dynamic event rather than "painting a static picture in the air" with his hands. And the personal action projections suggest imagery of forces, not just a static picture. This suggests that he is thinking about the forces via dynamic imagery of pushing and pulling actions with his own hands.

12.3.6.3 Kinesthetic Imagery

The personal analogies involving body forces and movements also support a final hypothesis (1C) that the subject is making a direct appeal to intuitions that involve kinesthetic imagery as a special form of dynamic imagery. His movements and statements suggest that he thinks in terms of muscular actions that represent forces exerted by objects in the original problem. I hypothesize that one of the reasons he goes to the trouble to create thought experiments is to tap into this intuitive, perceptual/motor knowledge about inertia. This last type of hypothesis is discussed more extensively in Chapters 13 and 18.

In summary, the simplest hypothesis that explains the observations above is that the subject's visual images of moving objects and kinesthetic images of pushing and pulling on objects are reflected directly in his hand motions. This is further supported by the coherence between the actions depicted by his motions and those described by him during other imagery indicators such as personal action projections.

12.3.6.4 Importance of Reasoning Processes Occurring Concurrently with Imagery Indicators

The importance of the reasoning processes occurring at the time of the above observations is supported by the following observations:

> Ob. 5 In lines 32 and 36, that he looks away from the interviewer and indicates that he is thinking during the motioning.
> Ob. 6 His motions occur just before subject reaches a finding.

12.3.7 Summary of S20 Analysis

If these interpretations are correct, this protocol provides an example of visual and kinesthetic imagery being used to improve one's understanding of a physical event. The imagined actions appear to be part of a runnable model for understanding the effect of mass on the acceleration of a cart in space. This allowed him to make good progress on overcoming a common misconception in physics by generating and running his own invented set of related analogies comprehended via imagistic simulations, a rather remarkable achievement for a freshman student.

12.4 Discussion

12.4.1 Types of Processes Associated with Motions

If we now accept that the two protocols involve mental simulation processes that include dynamic imagery, it appears that there are at least two different processes producing movements. We can contrast S20's protocol about the cart in space with the "direction of the earth's rotation" protocol from S15. In the latter there are movements which reflect the comprehension of spatial relations between moving entities – such as the way in which the earth's rotation causes the sun to appear to rise. In these episodes, it appears that the subject is simulating movements without thinking about the forces that produce them – simulating the motion of a system by imagining operating on a model of it with his hands. On the other hand there are episodes where the subject appears to focus on images of muscular forces, such as the one where S20 imagines pushing on a heavy object in space. The observation types which would be used to distinguish the second process from the first are numbers 11b, 14, and 15 in Table 12.4 in Appendix 2. Observations 14 and 15a and b are visible in S20's protocol. A third possible mode occurs when depictive hand motions simply represent a *static* object by "drawing it in the air".

We are now in a position to raise more difficult questions. Assuming that it involves imagery, is S15's earth turning episode an example of kinesthetic imagery? The content of S15's reasoning appears to be merely kinematic and not to involve attending to forces per se. He speaks of attending to movements but not muscular forces. Yet his hand motions seem very deliberate and convey a way to control object movements carefully, suggesting that the manipulation of the imagined globe with his hands could be a relevant part of his reasoning method. He needs to be able to control the globe so that it can turn in either direction around a particular axis. In addition to forces, the kinesthetic system transmits information about muscle and joint position changes, and this would be a useful source of information in moving an object carefully. Therefore I consider it to be possible that S15 is using

12.4 Discussion

kinesthetic feedback as part of his imagery. When this is true I will include such manipulative cases where there is no focus on forces in our concept of "kinesthetic imagery." This manipulative type of imagery can be contrasted to non-kinesthetic, non-manipulative (but still dynamic) types such as remembering what the trajectory of a bird was like. When it is necessary to refer more specifically to cases involving a focus on forces like S20's "pushing the cart," I will refer to "muscular force imagery" as one type of kinesthetic imagery. I summarize this view in hypothesis (1C): In some cases the subject's perceptual/motor systems for controlling body movements and actions in space can play a role in generating imagery. This can involve (a) imagining moving an object without attending to accompanying forces; (b) imagining exerting forces on an object.

12.4.2 Can Depictive Hand Motions be a Direct Product of Imagery?

In cases of muscular force imagery, it is easiest to argue that there should be a significant isomorphism to hand motions. One can state this hypothesis in the following way: some depictive gestures are direct, non-suppressed outputs of thinking about muscular effort. These can occur when there are no objects to touch. For example, I hypothesize that hand motions for pushing and pulling in S20's case are the result of motoric thinking about muscular actions. Imagery involving a completely internal form of this kind of thinking would have the motor output suppressed. I will assume that hand motions can provide direct (although possibly incomplete) evidence about the form and content of this type of imagery. When the imagery is sufficient to anticipate future states of a target situation, the hand motions provide evidence about the form and content of an imagistic simulation.

In generalizing this point beyond the case of hand motions that reflect imagery of muscular forces, we need to be more careful, since hand motions apparently vary in the degree of detail with which they represent images of situations. Theoretically, depictive hand motions (where no real objects are touched) can represent:

1. Prior experience of a hand manipulating an actual object.
2. The manipulation of an object with a focus on movements rather than forces (as we have hypothesized for S15, the first subject, simulating the rotation of the earth in line 3b).
3. The independent motion of an object (as in someone indicating the trajectory path of a bird with their hand).
4. The force of one external object on another (as in depicting an elastic band pulling on a cart).
5. The relative position or location of objects (as in S15's gesture showing the position of the sun relative to the earth in line 3c).
6. The shape of an object (e.g. "drawing in the air" or S15's stationary gesture in line 3a where he is "holding the earth").

We refer to all six types above as *depictive* motions. Cases of imagery at the top of this list can be indicated by hand motions very directly with almost complete isomorphism to one's experience, in contrast those at the bottom, where the isomorphism is only partial. Thus we say that depictive hand motions are a *partial* reflection of the subject's processing. Nevertheless, in all of these cases hand motions may be the most direct type of evidence we have concerning the form of internal imagery. Types 2–5 above will be of most interest in the cases in this study.

12.4.3 Summary of Relations Between Observations and Hypotheses

One can group the observations discussed in this chapter according to the hypotheses they can support as shown in Table 12.3. The table provides a list for think-aloud protocols of observation categories that can be used as evidence for the hypothesis that imagery can be used in an important way in problem solving.

More examples of concurrent indicators of this kind will be presented in Chapter 13. Thus the most important conclusion of this chapter for the purposes of this book is that by using the tools developed here, *it appears to be feasible to gather evidence for the presence and importance of dynamic imagery in expert thinking*. The indicators listed in Table 12.3 give us a "tool box" for doing this, and depictive motions are one of the most important indicators. The more specific hypothesis that *depictive hand motions can be derived from and reflect the presence and form of visual or motor imagery* has initial support in the protocols in this chapter, according to the argument structure shown in Fig. 12.6 in Appendix 1. However it needs more support. I will gather some here by making connections to related literature and then provide more evidence on this question from protocols in many of the following chapters in this book.

12.5 Relationship of These Findings to Others in the Literature

In this chapter we have been exploring the feasibility of using video data on depictive hand motions as a comparatively direct indicator of the form of thought processes such as mental imagery. Because the present work has its origins in an early study of problem solving and explanation processes as opposed to linguistics, the observations and hypotheses were formulated prior to a consideration of the linguistics literature on gestures, such as Rime's, Alibali's and McNeill's. Convergences between work reported here and theirs can thus be taken as coming from independent sources. Some examples of these are given below.

12.5 Relationship of These Findings to Others in the Literature

12.5.1 The Existence of Kinesthetic Imagery

Kosslyn (1980) conjectured that "imagery is a way of anticipating what would happen if a person, or an object, were to move in a particular way." Thus, Kosslyn theorized that simulations involving dynamic imagery are not just a by-product of thinking but can be an integral part. As mentioned in the introduction, Jacobsen (1930) found changes in EMG potentials in appropriate muscles during imagined actions, while Decety and Ingvar (1990) and Roland et al. (1980) found increases in blood flow to appropriate motor control areas of the brain during similar conditions. These findings support the view that subjects can engage in motor imagery. This lends support to the hypothesis of a continuous spectrum of levels of internalization in motoric thinking, ranging from motor imagery with no movement, to "frozen action" depictive hand positions, to depictive hand motions, to real actions on real materials, suggesting that the underlying thought process used in each of these modes may be similar. This supports hypothesis 1C, that subjects can use kinesthetic imagery.

12.5.2 Depictive Motions Are Not Simply Translated from Sentences

Going beyond thinking with motoric content to other kinds of content, Rime (1982) reported the surprising finding that speakers exhibited only slightly fewer hand motions when they were separated so that they were out of sight of the listener. In a study of four blind subjects, Iverson and Goldin-Meadow (1998) found that all of the blind speakers gestured when addressing an experimenter they knew to be blind. They believe that these findings leave open the possibility that the gestures which co-occur with speech may themselves reflect the thinking that underlies speaking. This provides indirect support for the hypothesis concerning hand motions as a direct, spontaneous output of thought processes.

12.5.3 Movements Are a Partial Reflection of Core Meaning or Reasoning

McNeill (1985) arrives at the following conclusions, arguing: (1) that gestures are not a translation of the sentence into a different medium; (2) that speech and gesture share an early computational stage; (3) that the shared computational stage is not a verbal plan from which the sentence is generated; and (4) that "compared to the concurrent spoken linguistic string, gestures are more direct manifestations of the speaker's ongoing thinking process." (p. 367). He emphasizes as evidence the fact that people sometimes misspeak with a correct gesture, but very rarely misgesture with correct speech. Speech is also repaired more often than gestures

(McNeill, 1992). This evidence corresponds to observation 13. I believe that it supports: hypotheses (H-3b) and (H-3c) in Fig. 12.6 in Appendix 1: Movements are not translations from verbal sentences and movements are partial reflections of reasoning. He also believes that the "process of utterance has both an imagistic side and a linguistic side. The image arises first..." (McNeill, 1992, p. 30). The main evidence for this is that many gestures have a preparation phase that puts the hands in position to execute the main stroke phase of the gesture. Preparation movements often precede speech that is synchronized with and corresponds to the meaning of the subsequent stroke movement. His conclusion requires other corroborating evidence, but that the observation of a preceding gesture preparation phase supports the idea that gesture and speech share an early stage of processing, and the idea that the gesture is not simply a translation from the sentence.

Returning to S15's "Earth turn" protocol, the frozen hand motions observed in line 3a are reminiscent of McNeill's observation of a communicatory preparation phase that puts the hands in position to execute the main stroke phase of a gesture. S15's frozen hand movements may also indicate a kind of *cognitive* preparation: that is, the setting up of a mental model involving static imagery, that provides a framework for subsequent dynamic imagery or simulation.

Van Meel (1984) observed that gestures tend to be coordinated and synchronized with speech content in adults, but also observed 4- to 6-year old children making gestures *before* the beginning of their verbal answer to questions. In a study of 60 adult gestures, Krauss (1998) found that all gestures were initiated either prior to or simultaneously with the onset of articulation of the lexical affiliate. These findings support the view that processes producing motions can occur during an early stage of processing. In fact, it suggests that nonverbal processing may in some cases occur prior to and therefore independently from verbal processing.

Going beyond timing studies, Alibali et al. (1999) found that strategies used to solve word problems varied systematically as a function of how those problems were represented by undergraduates in their gestures. She concludes that gestures can be used to infer important information about problem-solving strategies and problem representation. In an experiment where subjects described navigation information in maps, Lozano and Tversky (2005) found that gesture types corresponded to the type of information subjects later remembered.

The studies in this subsection support the hypothesis that movements can be a partial reflection of reasoning, and they are consistent with the possibility that movements can be the direct result of a non-verbal thought process such as imagery.

12.5.4 Gestures Can Reflect Imagery

The possibility that gestures can reflect imagery use in general is supported by the following findings. Feyereisen and Havard (1999) found that speakers produced more beat (rhythmic emphasis) gestures when speaking about abstract topics and more representational gestures when speaking about topics that were judged to involve visual or motor imagery. Krauss (1998) found that Gesturing during "spatial content phrases" was nearly five times more frequent than it was for the remaining nonspatial phrases. Krauss states that this data is consistent with his conjecture that lexical gestures reflect spatio-dynamic features of concepts, and other studies have concluded that representational gesture appears to be associated with visuospatial processes (Iverson and Goldin-Meadow, 1997; Hostetter and Alibali, 2004; Alibali, 2005.) Similar conclusions are beginning to appear concerning dynamic imagery (mental animation) (Trafton et al., 2004). Gesture has been used to help identify when subjects are first using a new imagistic model (Schwartz and Black, 1996a). In a study where subjects inferred motion from static diagrams, Hegarty et al., (2005) found participants gestured on most problems and most gestures expressed information about the motion that was not contained in their words. Collectively, the findings in the above sections lend initial support to the hypothesis that hand motions can be derived from and reflect visual or motor imagery.

12.6 Conclusion

12.6.1 Sources of Information About Imagery and Simulation

Previous research on imagery has now taken us to the point where the existence of mental imagery in thinking has been accepted as at least an arguable theory. However, the level of the tasks used in that research has been rather low. On the basis of protocol data we have developed a fairly large initial set of indicators in Table 12.3 which can be used to provide evidence for imagery in scientific thinking. In particular the two case studies discussed in this chapter indicate that depictive hand motions can be used as an important source of evidence for the hypotheses in Table 12.1.

In the case of the first subject, we found evidence that S15's hand motions were a partial reflection of the use of dynamic imagery for thinking about the earth's relationship to a sunrise, via the imagined manipulation of an object with the subject's arms and hands. In the second protocol, the analysis suggested that S20 imagined novel events involving pushing objects in space to make predictions about them. Here there is strong evidence that part of the core meaning in the solution process was the kinesthetically imagined forces on an object exerted by the subject's arms and hands, as depicted in his hand motions.

We have presented arguments from both previous research and the present protocols that;

1. Although actual hand movements are not *necessary* for thought in adults, they are often not simply extensions or translations of verbal language. This is supported

by studies showing that gestures can precede speech both developmentally and in specific sentences, and that gestures sometimes convey meanings that differ from speech meanings.
2. Depictive gestures can be a natural result and external reflection of the reasoning processes of the subject. In this case they become a natural "window" (although by no means a fully transparent one) via which we can gain information about imagery processes. (This is supported by previous findings, such as Alibali's, showing that gestures vary systematically with types of problem representations and strategies used, as discussed in the introduction to this chapter.)
3. This allows us to use hand motions as one kind of evidence to support the first two hypotheses in Table 12.1: (A) imagery can be used fruitfully in higher-order thinking contexts such as scientific problem solving; and (B) subjects can use *dynamic imagery* in which they imagine motions, changes, or interactions over time in a science problem.
4. In particular, the second protocol involving muscular force imagery here supports the conclusion that hand motions can reflect a subject's use of *kinesthetic* imagery. (The possibility of the existence of kinesthetic imagery is supported by neurological studies showing plausible brain locations for such imagery and studies of priming effects of movements for question answering about movements (discussed in the introduction)).
5. I have also proposed a set of other imagery-related observation categories in Tables 12.3 and 12.4 can be used together with depictive motions to provide evidence for dynamic imagery use in scientific thinking. Additionally, in later chapters, when these appear along with the prediction of multiple states of a system, I will treat this as evidence for an imagistic simulation process.
6. Other indicators in Tables 12.3 and 12.4 allow one to support hypotheses about the importance of these imagery processes to the solution. Many of these are summarized in Fig. 12.3.

12.6.2 Limitations

One would expect others to add more imagery indicators to our initial list. In fact it is eminently reasonable to assume that when a subject describes spatially embedded relationships or transformations not depicted in a diagram while staring at or making a picture or diagram, that he or she is using imagery. I believe that this indicator will serve well in the future, but I have retained a conservative stance here and have not used this indicator in this book except in conjunction with others.

Further support for the above conclusions requires the examination of more protocols. Chapters 13–18 will provide more cases with evidence on the "presence and importance of imagery" question. There additional imagery indicators will be identified. A final problem is that the hypotheses presented so far describe important cognitive characteristics of simulations but do not begin to specify a process model for how simulations work. For example, I have not addressed questions about whether the source of S20's knowledge of inertia resides in his images or somewhere else; I have only hypothesized that dynamic imagery is somehow involved in his thinking. What is the source of that knowledge? Chapter 13 begins the task of developing more detailed models of processes underlying mental simulations that involve the coordinated use of both imagery and physical intuition schemas.

12.7 Appendix 1 – Detailed Justification for Using Evidence of Imagery from Hand Motions in S15's Protocol

This appendix examines more carefully the question of whether depictive motions can be taken as evidence for imagery. Unlike mental speech, imagery has had no recognized natural behavior correlate that can be used as an indicator. However part of the conclusion in this chapter is that depictive gestures may be viewed as a candidate for such a correlate. This is a difficult relationship to pin down however because of the variety of possible cognitive theories and the complex relationships between them and observations. The relations between a higher-level cognition hypothesis such as imagery use and observations from transcripts are necessarily going to be high inference ones. Under these circumstances our best strategy is to look for a variety of observations in order to triangulate on a hypothesis concerning the origin of hand motions that is the simplest explanation for several different behaviors.

This appendix attempts to show how such a strategy could be implemented in a particular protocol. It gives an argument for why depictive motions can be taken as evidence for imagery in the protocol of S15 using the following strategy. We first ask whether the observations from this protocol can provide any evidence for the following three hypotheses (shown in Fig. 12.6). We then use these partial findings along with other observations to support the hypothesis that appears to most parsimoniously explain all of the observations: that the motions reflect imagery used during reasoning in the protocol. The preliminary hypotheses are:

- (H-3A) Movements do not always occur as an addendum after the subject thinks about an issue, but can occur concurrently with part of the solution process.
- (H-3B) The movements are not translations of verbal sentences.
- (H-3C) Movements can be a partial reflection of the solution process, rather than an indirect translation of it.

12.7.1 Motions Are Concurrent with Solution Process

In movement Episode 1, the following three observations taken together provide some support for hypothesis (H-3A) above:

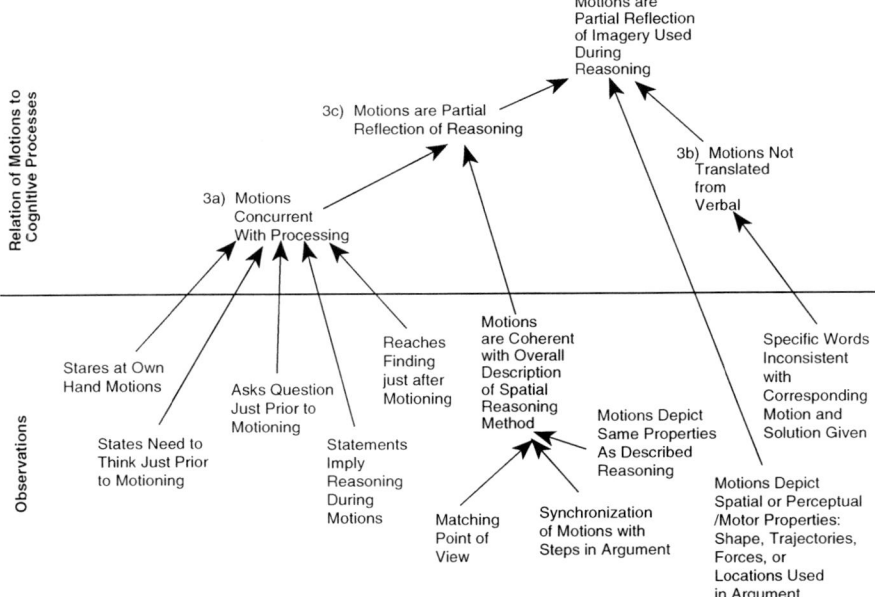

Fig. 12.6 Argument that depictive motions are evidence for imagery during reasoning in S15's protocol

- 5- Ob. He states an intention or need to think just prior to motioning; other statements imply that he is thinking during the motions.
- 7- Ob. The movements occur immediately after a question.
- 8- Ob. He stares at his own hand motions.
- 6- Ob. The movements occur immediately before a finding is reached.

These observations all suggest that his processing occurs concurrently with the motions. They argue against the idea that the motions are translated from some other representation after thinking is over. There is also evidence for hypothesis (H-3A) above when S15 *repeats* the motions in line 4b. The last two observations above also apply to this episode. The fact that he repeats the motions immediately after re-asking the same question and confirming his finding in this manner reinforces the inference that the motions are concurrent with his solution process.[1]

---

[1] It is possible to collect evidence relevant to hypothesis (H-3a) because of the extended reasoning required for our tasks, in contrast to the tasks used by McNeill (1992), who tends to use recall tasks. This difference in tasks may also explain the contrast between our finding that hand movements can occur without speech and McNeill's tendency to assume that speech always accompanies gestures.

12.7.2 *Motions Can Be a Direct Product of Solution Process*

Evidence for the first hypothesis that the motions were concurrent with solution processes provides support for the further hypothesis that:

(H-3c): The movements can be a partial reflection of his reasoning process.

This support relation is shown in Fig. 12.6. Hypothesis (H-3c) above is also supported by the observation that the motions were coherent with his overall reasoning method. This comes from three more specific observations: the motions were synchronized with steps in the argument; the motions depict spatial or visual properties such as shape, trajectories, and location that were used in the argument; and the described argument's point of view matches that of the motions (i.e. both include a point of view from above the solar system, rather than, say, from the sun.)

12.7.3 *Motions Not Translated from Verbal Sentences*

Another key feature of this protocol is the fact that S15 actually misstates the earth's direction of rotation at the beginning of the protocol in line 3a (he says east to west instead of west to east), and simultaneously represents the motion correctly with his hands. Only after he has gone through two sets of motions does he make a correct verbal statement in line 4a. These observations support hypothesis (H-3b) above that the motions were not translated from a verbal sentence. This discredits a major competitor to the hypothesis, discussed in the next paragraph, that the motions are a partial reflection of imagery used during reasoning.

12.7.4 *Evidence for Imagery*

Finally, the view that his motions are a partial reflection of his processing, along with the continuous and analog character of the motions, provides a platform of support for hypothesis (H1A) that S15's processing involves the use of imagery.

If one believes it is likely that the hand motions observed are a partial reflection of his thinking processes, then this lends support to the hypothesis that significant parts of the thought processes are imagery based. Arguing that the gestures are not translations from verbal sentences (external or internal) discounts a major competing hypothesis to this point of view. Since the hand motions are continuous and analog in nature, and depict perceptual properties such as shapes, trajectories, and/ or locations, the most parsimonious hypothesis concerning their origin is that of an internal imagistic representation that was used during the reasoning. These support relations are summarized in Fig. 12.6.

In addition to the indicators shown in Fig. 12.6 for S15, another indicator from S20's protocol that could support the argument that depictive motions are evidence for imagery is the display of hand motions over drawings. These depict the action of an object in the drawing, as the subject stares at the drawing. In this situation it is highly likely that a perceptual representation is already in use that corresponds to the drawing. Since the hand motion is used to represent the action of an object and is aligned to the drawing, it supports the idea that the motion is generated by an image of the action within the internal representation of the drawing. Also, the use of motions in conjunction with ambiguous propositions (by S20) was used to support the view that the motions were not translated from verbal statements, and his described projection of personal forces into the problem in ways that were coherent with the hand motions was used to support the presence of kinesthetic imagery.

The above considerations provide examples of how one can support the position that depictive hand motions can provide one source of evidence for imagery, and in fact are a partial reflection of the form of that imagery. This conclusion will be supported further in the next chapter where hand motions will be seen to co-occur with other imagery indicators such as spontaneous imagery reports, kinesthetic imagery reports, intentional episodes of "imagery enhancement," and personal action projections.

Appendix 2

Table 12.4 Observation categories in numerical order

1a *Depictive motions:* The subject makes nonstylized hand or body motions depicting objects, forces, locations, or movements of entities

1b *Frozen action gesture:* The subject's hands are positioned as if manipulating or ready to manipulate an external object, but are stationary ("frozen")

1c *Depictive hand or Pencil Motions over a Drawing:* These are not simply pointing to a word or picture, but indicate movements, relations, locations, or shapes of objects or features that do not appear explicitly in the static drawing

2 *Motions Depict Dynamics:* The hand motions depict the form of a dynamic event, not simply a static picture

3a *Imagery Reports:* Subject states that s/he is imaging, imagining, "seeing" or "feeling" (or experiencing any other sensation)

3b *Dynamic Imagery Report:* An imagery report where subject indicates that they are imagining motions, changes, or interactions over time in a situation

3c *Kinesthetic Imagery Report:* Reports imagining own physical actions or muscular effort

3d *References to Perceptions:* These are similar to imagery reports but not as direct. The subject refers explicitly to the sensation of perception while describing visual or other perceptual aspects of scene during thinking by using phrases such as "the car probably looks like it is going that way"

4 *Gap in Speech:* Subject becomes silent and there is a noticeable gap in the train of speech while running through the motions

5 *Indicates Concurrent Thinking:* Subject states an intention or need to think just prior to motioning or makes statement implying they are thinking during episodes

(continued)

Table 12.4 (continued)

6 *Episode Precedes Finding:* Episode occurs just prior to finding of prediction or increase in confidence
7 *Indicator Follows Question:* Relevant motions occur immediately after a specific question is asked, so that movements appear to be associated with the process that answers the question
8 *Stares at Own Hand Motions* (status unclear as imagery indicator)
9 *Episode Repeated:* Episode is repeated as subject indicates an attempt to confirm an answer to the same question
10 *Minimizes Visual Stimulation:* (a) Closes Eyes or (b) Fixes Gaze or (c) Stares at Ceiling Or Other Minimally Stimulating Area
11 *Motions Accompanied by Ambiguous Terms:* Uses ambiguous terms in conjunction with less ambiguous motions to refer to:
   (a) The shape or motion or action of an entity, e.g. "It looked like this", "It would go that-a-way"
   (b) The means used to exert a force on an object, e.g. "If you push this way"
12 *Motions with Unrelated Language:* Motions occur during language that does not describe them (e.g. S says "Well, let's see what we can find out here" during motions. Producing such oral language would be likely to interfere with other language that served as the source for a translation to gestures)
13 *A Particular Motion Contradicts a Language Error:* Verbal description of an event is inconsistent with concurrent hand motions representing the event, and these motions are consistent with later conclusions in the solution
14 *Uses Force Terms*: Imagery indicator occurs in conjunction with force context terms such as "pull," "push," "twist," "effort," "try to move it"
15 *Personal Force Projection Or Analogy:* (a) Refers to forces exerted by entities in target situation as if they were exerted by a person or (b) Uses a personal analogy by referring to an analogous situation involving body forces
16 *Personal Movement Projection Or Analogy:* (a) Refers to movements of entities in target situation as if they were moved by a person or (b) Uses a personal analogy by referring to an analogous situation involving the body
17(a) *Descriptions of Spatial Reasoning Fit Motions:* The subject reports elements of reasoning via spatial relations between objects and this is coherent with hand motions
17(b) *Corresponding Steps*: Steps in the reasoning being expressed by the subject correspond to steps in a series of motions
17(c) *Synchronized Entrainment*: There is synchronized entrainment of movements and speech that describes the reasoning
17(d) *Describes Plausible Method of Spatial Reasoning*: The subject reports reasoning via spatial relations between objects. A necessary but not sufficient requirement for this to be true is that the described objects (or a scale model), if actually assembled and manipulated, could be used to attempt a solution. (status unclear as imagery indicator on its own)
18 *Matching Points Of View*: The point of view (particular angle of view) as expressed verbally by the subject during reasoning corresponds to the point of view represented by hand motions
19 *Anticipates New States*: Indicates new anticipation of continuous change of state or sequence of states of a system. This is the minimal indicator for a simulation, not an imagery indicator, but when it appears along with imagery indicators, we take it as evidence for a imagistic simulation
20 *Imagery Enhancement*: Adds markers to a situation or changes other perceptual features such as: (1) orientation or size or (2) magnitude of problem variables – in conjunction with evidence for imagery. More direct evidence occurs when the subject speaks of modifying a problem situation in a way that makes it "easier to imagine."

(continued)

Table 12.4 (continued)

21 *Indicates Time-extended Process*: (e.g. "So it's drifting along further and further and further") in conjunction with another imagery indicator
22 *Effortful Episode*: Indicates episode was effortful
23 *Orients Body*: Body is oriented to reflect point of view (e.g. faces north when object in problem is facing north)
24 *Indicates Imagery as Source of Finding*: Indicates that "imagining" of some kind is the source for a finding
25 *Difficulty Describing Image*: Indicates that image or thinking in episode leading to finding is clear but is hard to describe
26 *Generates a Diagram or Drawing*
27 *Focuses on a drawing while making inferences about spatial transformations, movements, or spatial or physical relationships* not depicted in the drawing
28 *Intuition Report*: spontaneously reports using an "intuition" or uses terms that indicate they are proceeding primarily on the basis of a nonformal "feeling" or "sense" of what will happen to a system

Not all of the above are used as imagery indicators in the present study; see "Status Unclear" category at end of Table 12.3.

The last four observation types are defendable as internal imagery indicators and are used in some studies but were not used in this book as a conservative measure.

Chapter 13
Physical Intuition, Imagistic Simulation, and Implicit Knowledge*

13.1 Introduction: Issues in the Area of Imagery, Simulation, and Physical Intuition

This chapter discusses evidence from think-aloud case studies that indicates that part of the knowledge used by expert problem solvers consists of concrete intuitions rather than abstract verbal principles or equations, and that these intuition schemas can be used to generate imagistic simulations. An intuition, as used here, does not refer to a mysterious reasoning process, but refers very specifically to a qualitative, concrete element of knowledge about the world that is self-evaluated and stands without the need for further explanation or justification.

Chapter 12 put forward a set of observable descriptors that can be used to provide evidence for when imagery is occurring in think-aloud protocols. I will continue to use these descriptors as evidence that imagery can be involved in an important way in higher-order thinking. The term imagery was used for a mental process that involves part of the perceptual/motor systems and produces an experience that resembles the experience of actually perceiving or acting on an object or an event. Indicators providing evidence for dynamic imagery and kinesthetic imagery were also identified. Processes where dynamic imagery appeared to be involved in generating predictions of changes or movements were termed imagistic simulations. Here I will examine more evidence for important instances of imagery use by experts and also address the question of how such images are generated. The central idea is that the dynamic imagery in a simulation can be generated (driven) by a schema. In order to account for cases where subjects appear to be "running a simulation" of an event, a model will be presented in which a permanent schema generates a temporary dynamic image of a situation. I will also examine physical intuitions as one very important type of schema that can generate dynamic imagery. Recalling the protocol of the last subject in the previous chapter, S20, will motivate the view taken here. His images of pushing and pulling on objects suggest that perceptual/motor physical intuition schemas can generate dynamic imagery to

---

*Some portions of this chapter are based on findings reported in: Clement, J. (1994). Use of physical intuition and imagistic simulation in expert problem solving. In Tirosh, D. (Ed.), *Implicit and explicit knowledge*. Norwood, NJ: Ablex.

produce findings in a mental simulation of an event. The problem is to specify what one might mean by "physical intuition schema" and its relationship to dynamic imagery. Examining physical intuition schemas before other types in the research agenda has the advantages that: (1) they are a relatively simple, elemental type of schema; (2) they may produce more hand motions than some other types of schemas, giving us additional data on the nature of schemas and imagery; (3) they exemplify how both the perceptual and motor systems can be involved in imagistic simulation; (4) they may exemplify how relatively concrete, embodied forms of thinking can still be powerful in expert cognition.

13.1.1 Abstract vs. Concrete Thinking in Experts

The documentation to be presented of the use of concrete physical intuitions by experts can be contrasted with the more usual characterization of the distinguishing features of expert knowledge as predominantly abstract. For example, Chi et al. (1981) state that experts in physics use: "abstract physics principles to approach and solve a problem representation (p. 121). Novices, on the other hand, 'base their representation and approaches on the problem's literal features…" (p. 121). This characterization of experts appears to conflict with reports of scientists such as Einstein's: "The words or the language…..do not seem to play any role in my mechanism of thought. The…elements in thought are certain signs and more or less clear images…of visual and some of muscular type" (quoted in Hadamard, 1945, p. 142–143). One purpose of this chapter is to investigate whether concrete, nonabstract knowledge in the form of elemental physical intuitions can play an important role in expert thinking. Work in this area is also motivated by previous studies on (1) the important cognitive roles played by actions involved in scientific experimental practice (Tweney, 1986; Gooding, 1990) and (2) imagery in science (Shepard, 1984; Nersessian and Greeno, 1990; Miller, 1984; Qin and Simon, 1990).

A major theoretical question is how to sort out the relationships between imagery, intuitions, and mental simulation. In particular, I will ask whether the use of an elemental physical intuition concerning a single causal relationship can involve imagery in a mental simulation. Prior work exists on causal inferences involving propositional networks of causes with many links (e.g. de Kleer and Brown, 1983; Forbus, 1984). However there has been little work on the nature of the underlying elemental simulations in humans involving a single causal relationship.

Thus, this chapter attempts to formulate a framework for thinking about how physical intuitions and imagery can combine to create a new prediction from a mental simulation.

13.2 Initial Examples of Physical Intuition

A first example of an episode from a solution to the spring problem that appears to involve physical intuition follows. This excerpt comes from the solution of the Nobel laureate in physics mentioned in Chapter 2:

13.2 Initial Examples of Physical Intuition

> 027 S1: You don't have to know any formulas to see that...God almighty! Of course it [the wider spring] goes way down. You know. How could it do otherwise?
> That's a seat-of-the-pants feeling I would trust beyond any of it.... I would bet a thousand to one.

This example counters the idea that intuitions are used only by those who lack formal reasoning capabilities. However, most of the ten subjects studied for the spring problem did not have so strong an intuition about the target problem itself, but had intuitions about related problems. For example, in Chapter 2 we saw subject S3 consider the analogous case of weights on the ends of long and short horizontal rods:

> 010 S3: My intuition about that is that if you took the same wire that was fastened on the left here [short horizontal rod] and doubled the length...that...it would bend considerably farther.

He then attempts to transfer this intuition to the case of the spring, using it as the basis of his prediction that the wider spring will stretch farther. This last example suggests that physical intuitions can be responsible for what we have called anchoring conceptions that provided a base for an analogy, as discussed in Chapters 2–6.

13.2.1 Intuition Reports

Both excerpts above contain an example of what I call an *intuition report*, where the subject spontaneously reports using an "intuition" or uses terms that indicate they are proceeding primarily on the basis of a nonformal "feeling" or "sense" of what will happen to a system. However, we cannot attach too much importance to a subject's use of the term, since, for one thing, its meaning in natural language is somewhat broad and vague. Therefore this observation should be only one of several possible indicators used to provide evidence for intuitions. One aspect of the vagueness of the word "intuition" is that the word can refer to both knowledge structures and nonformal reasoning processes. I will avoid the latter use here, so that I will not use the term intuition for processes such as induction by enumeration, analogical reasoning, or heuristic strategies for problem solving. Instead, I will focus on elemental knowledge structures as basic units of knowledge.

The concept of intuition as used here shares many features with the concept of a schema and indeed I will develop a theoretical view that describes most intuitions as rooted in a schema. Perhaps the most unique feature of an intuition that sets it apart here is the property of self-evaluated plausibility. This separates it from rote knowledge or other knowledge whose plausibility depends primarily on an external authority.

13.2.2 Defining Features and Observable Behaviors Associated with Intuition

On the basis of transcripts like those above, one can identify a cluster of phenomena that suggests the existence of *intuitions* as a type of cognitive structure. In the solutions to problems that have been discussed so far in this book, subjects make many inferences and arguments. However, there are always some assumptions at the base of these inferences and arguments that are not justified or explained, such as the idea that a longer rod will bend more that a shorter rod. These instances suggest the presence of a knowledge structure that is activateable, permanent, self-evident, self-evaluated, concrete, and more general than the memory of a specific incident. I will treat these as the central properties of an intuition. Although the discussion of more instances is needed for support and will take place over the next several chapters, I discuss this list of properties here as an advanced organizer. Cognitive properties for intuitions are underlined below, along with some of the observable properties that can indicate them. diSessa (1983) refers to certain kinds of similar knowledge structures as "phenomenological primitives." I will use the term *elemental intuition* here in a way that shares several of the features of a phenomenological primitive, including the following ones.

Knowledge Structures. An intuition is a knowledge structure (schema) that resides in long-term memory and that can be activated to provide an interpretation of or an expectation about a system. This can produce predictions about the behavior of the system, even when that particular system has not been observed before.

Explanation, Justification Unnecessary. Subjects do not feel a need to further justify, derive, or explain intuitions. They are self-evident. This does not mean that an explanation or finer analysis is impossible, but that none is needed because the behavior is self-evident to the subject.

Modest Generality. An intuition is more general than the memory of a specific incident. As diSessa points out however, the degree of generality is often not nearly so large as that of the concepts used in Newtonian mechanics.

In addition to the features identified by diSessa, I also point to some other characteristics below. Later I will identify other observation patterns associated with intuitions and will attempt to outline a theory of how intuitions can produce new predictions via mental simulations.

Self-evaluated. Intuitions have some self-evaluated plausibility. In a strong intuition I would refer to this as a self-evaluated conviction. Strength of belief in a intuition depends largely on internal criteria rather than being dependent on the evaluation of an authority.

Oriented to Concrete Entities. Intuitions contain knowledge about concrete entities and manipulations of or relationships between entities, rather than about a symbolic result that must be interpreted.

One is tempted to consider adding "self taught" to the list of defining features for intuitions. I have not done this because it would imply that intuitions are unteachable. In keeping with the examples given in Chapter 10 of instruction that works with students' intuitions, the concept of intuition used here is designed to allow for intuitive knowledge structures that are changeable, learnable, and to some extent teachable by guiding the student through practical experiences (or memories of them).

13.2.3 Physical Intuitions

With regard to the last concreteness feature above, we can refer to *physical intuitions* when they contain knowledge about a concrete physical phenomenon or system and manipulations of or relationships between entities in the system. Virtually all of the examples of intuition in this book will be physical intuitions, but there are surely others, such as intuitions about living things and interpersonal relations.

13.3 Imagery Reports and Imagistic Simulation

13.3.1 Moving from the Findings in Chapter 12 to Models of Imagistic Simulation

The above features do not provide us yet with anything like a process model of physical intuition; rather they suggest a cluster of characteristics of physical intuitions to be explained by the theory of intuition to be developed. But first, I need to add two important items to the list of observable imagery indicators developed in Chapter 12: imagery reports, and dynamic imagery reports. Then I can develop a parallel set of intuition indicators, and finally, some initial models of elemental simulation processes that use both imagery and intuitions.

13.3.1.1 Adding to the Collection of Imagery Indicators: Imagery Reports

I emphasized that the imagery indicators developed in Chapter 12 were *partial* indicators in the sense that no one indicator on its own is conclusive, but that two or more appearing together are more convincing. In the cases to be examined below subjects spontaneously use terms like "imagining," "picturing," "hearing," a situation or "feeling what it's like to manipulate" a situation." I refer to such statements as *imagery reports*. These refer to several sensory modes, including kinesthetic imagery. In contrast to most of the literature on imagery, I am concerned here with *spontaneous* imagery reports where the interviewer does not ask the subject whether an image was used. For example, S2, thinking about the related problem of comparing short and long springs says:

041 S2: I'm imagining (Raises Hands 10 inches apart in front of chest, Closes fingers as if holding something between hands, moves hands together slowly in 5–6 small movements while speaking until they almost touch just before the word 'origin') that one applies a force closer and closer to the origin [top] of the spring, and…it hardly stretches at all.

Although an imagery report is not conclusive evidence that a subject is using an image, I will take it to be one source of evidence. The evidence in this case is bolstered by another indicator in the form of depictive hand motions. Similarly, I define a *dynamic imagery report* as an imagery report in which the subject refers to imagining motions, changes, or interactions over time in a situation.

13.3.2 Schema-driven Imagistic Simulation Processes

An important observation from the above transcript excerpts and those in Chapter 12 is that *imagery* indicators can co-occur with subjects' *predictions* about a system. Perhaps imagery and its cognitive characteristics tell us all we need to know about the nature and origin of predictions from mental simulations. However, I do not believe that the capacity for imagery alone can explain such predictions. Images are temporary cognitive representations, and do not arise from nowhere – there must be a source that generates the image. In the simplest case this could be thought of as an episodic memory or "stored snapshot." But many of the cases considered here were unfamiliar to the subjects and required generating the image from a less specific source. Another source of difficulty here is that most discussions of imagery involve visual imagery alone, whereas physical intuition often appears to involve imagining *actions* taken on objects as well. I have still not outlined an explanation for how a new image that leads to a new prediction can form.

Clement (1994a, 1994b, 2003, 2006) accounted for episodes like that of S2 above by hypothesizing that subjects are running through a *schema-driven imagistic simulation* where a schema generates a dynamic image to produce expectations about behavior of a system. In the broad use of the term, mental simulation is a process by which one can anticipate changes in a system over time. The fact that humans seem to be able to do this for some system configurations they have never seen before, and that they are not always able to describe a logical sequence of reasoning operations they use to accomplish this, makes simulation a subject of intense curiosity and investigation on the part of cognitive scientists. The concept of mental simulation to be used here takes as its starting point Collins and Gentner's (1987) concept of a "running a generative mental model." For them, a generative mental model is a cognitive entity that leads to new inferences by allowing one to predict new future states of a target situation. I will retain the idea of anticipating a change in a system or a sequence of system states as part of the concept of simulation and will begin the investigation of these ideas using the concepts shown in Fig. 13.1. This figure shows the basic mental phases of a schema-driven *imagistic*

13.3 Imagery Reports and Imagistic Simulation

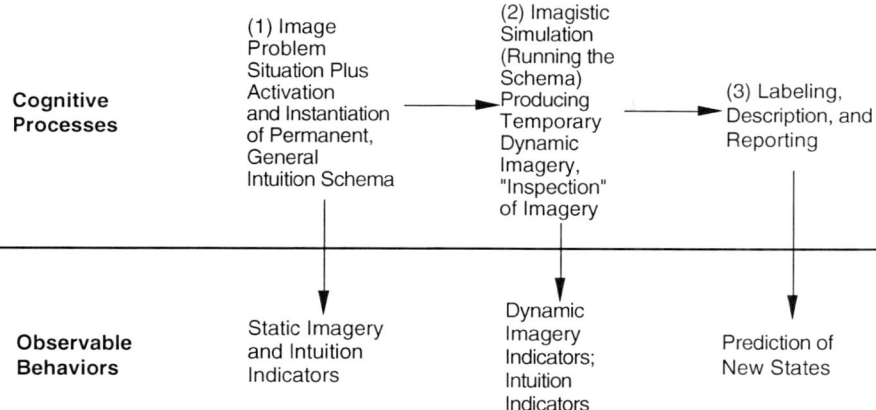

Fig. 13.1 Basic phases in an imagistic simulation and the observable behaviors they may generate

simulation above the horizontal line, and some of the observable behaviors these can generate below the horizontal line. There the *schema* is capable of generating a mental simulation in order to make predictions about a system.

The two subjects in Chapter 12, S15 and S20, appeared to use dynamic, perceptual/motor imagery in mental simulations as part of their solution processes. This hypothesis goes beyond the statement of Collins and Gentner (1987) that mental simulation involves the prediction of some of the future states of a system over time. This hypothesis says that *imagery* produced by the operation and activity over time of the perceptual and/or motor systems is part of the thought process that produces the prediction. I have referred to these cases as *imagistic simulations*.

The following excerpt provides evidence for the simultaneous use of imagery and a physical intuition schema. At this point the subject has decided that a twisting deformation in the wire is one of the consequences of stretching the spring. (Twisting of the wire and the resulting torsion do in fact play a predominant role in determining the behavior of a spring.) He is trying to decide what effect widening the spring will have on the twisting deformation by imagining himself twisting straight horizontal rods:

137 S2: If I have a longer (raises hands apart over table arriving at stationary position with the word "rod") rod and I put a twist on it (moves right hand as if twisting something, as shown in Fig. 13.2) it seems to me – again physical intuition – that it will twist more. Uhh, I'm – I think I trust that intuition.

138 I: Can you stop thinking ahead and just think back on that; what that intuition is like?

139 S2: I'm (raises hands in same position as before and holds them there continuously until the next motion below)…imagining holding something that has a certain twistiness to it and twisting it.

141 S2: Like a bar of metal or something like that. Uhh, and it just seems to me as though it [a longer bar] would twist more.

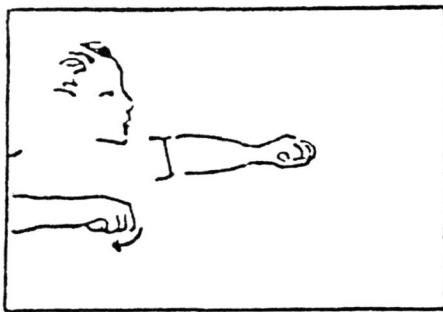

Fig. 13.2 Hand movement during imagistic simulation for S2. Drawing is an exact tracing from photograph of video image

The subject eventually uses this result as a central assumption in order to make inferences about the spring problem. The last two episodes have included simultaneous observations of imagery reports and depictive hand motions. This adds to the findings assembled in Chapter 12 arguing that depictive hand motions can be taken as one form of evidence for the use of imagery.

The result here appears to come from a dynamic image of acting on the rod, but what is the origin of the image? I hypothesize that the origin is a perceptual/motor schema for twisting various materials that the subject has used for simple actions in daily life, and that the schema is able to operate here, "acting out" its control program and anticipations even though no actual materials are present. Instead, here it assimilates an image of a situation (a thin metal bar-like object) and "runs through" an act of twisting on the image, generating certain expectations about the effort involved. The subject presumably goes through *two* simulations with a short and a long rod here, after which he is able to compare them. In the transcript above one can observe a number of the phenomena under discussion: co-occurrence of an imagery report (line 139) and a stated prediction (137, 139), intuition reports (137, 139), self-evaluation of the intuition (137, 141), and depictive hand motions (137, 139). In addition the imagery report in line 139 is really a *dynamic imagery report*, where the subject describes elements in an imagery report as changing or interacting over time. Furthermore this is a personal-action analogy to the original spring problem in which the subject is applying forces instead of a weight. The co-occurrence of intuition reports and dynamic imagery reports motivates proposing the hypothesis that a physical intuition schema and imagery are used in tandem here. It is hypothesized that the subject is going through an *elemental, schema-driven, imagistic simulation* process wherein a *schema for performing the action of twisting objects* assimilates the *static image of a particular object* and produces expectations about its behavior by generating a *dynamic image*. In this case the simulation is grounded in a physical intuition, so we can also refer to it as *intuitively grounded*. It is elemental in the sense that this episode involves only a single intuition schema rather than an assembly or chain of schemas. "Running" the intuition has a simple natural meaning here of having the motor schema for twisting "go through its paces" internally with the actual muscular output suppressed.

The diagram above the horizontal line in Fig. 13.1 gives an overview of this process. In this case it is assumed that the physical intuition about how an object behaves is an expectation embodied in a permanent and somewhat general schema that can assimilate a variety of objects. By "assimilates" I mean "locking onto and coordinating dynamically with" a system or image of a system. The process consists of three steps: (1) activation and instantiation of the physical intuition schema; (2) imagistic simulation (running the schema so that it generates a dynamic imagery of an event); (3) inspection of the dynamic image for the variables in question and description of the result in language. It is assumed that all of the cognitive processes shown in Fig. 13.1 before the description process in step (3) can be nonverbal in character. The description process enables the subject to describe his or her findings in words after it occurs via imagery.

13.3.2.1 Perceptual/Motor Nature of Many Physical Intuitions

The diagram in Fig. 13.3 shows a blow up of the simulation process labeled (2) in Fig. 13.1. In both Figs. 13.1 and 13.3 the terms below the double line denote observable behavior patterns in the transcripts that can provide evidence for the hypothesized cognitive structures and processes shown above the line. For S2 imagining twisting a rod in the transcript above, the imagistic simulation is the process of applying a perceptual/motor schema that would also be entirely capable of controlling a real action; in the world. However, instead of acting on a real object here, the schema is applied to an image – a particular image of a 1 ft long bar of metal. During the "running" process (symbolized by the two horizontal arrows showing activity over time) the schema assimilates ("locks on to") the image, "performs an (imagined) action on it" by generating a dynamic image over a period of time on the order of a few seconds (as indicated by the downward pointing dotted arrows), thereby generating an expectation about its behavior. It is an important part of this view that the process requires two major components: a more general schema that is a permanent resident in memory (this helps explain where the knowledge being used comes from); and a temporary image of a specific example (this helps explain how one is able to think about twisting a novel new example). Some of the relevant categories of evidence for an intuitively grounded imagistic simulation are shown below the horizontal line in Fig. 13.3.

S2's "twisting rod" transcript above is exceptional in having more than the usual number of indicators since each of the observables in Fig. 13.3 can be seen there. This includes the last three observations in the list below of observable indicators that provide evidence for use of physical intuitions.

- *Predictions*: Subjects make_predictions about the behavior of a system, even when that particular system has not been observed before.
- *Explanation, justification unnecessary*. Subjects often refer to a physical intuition as a starting point and do not express a need to further justify, derive, or explain it. They are self-evident.

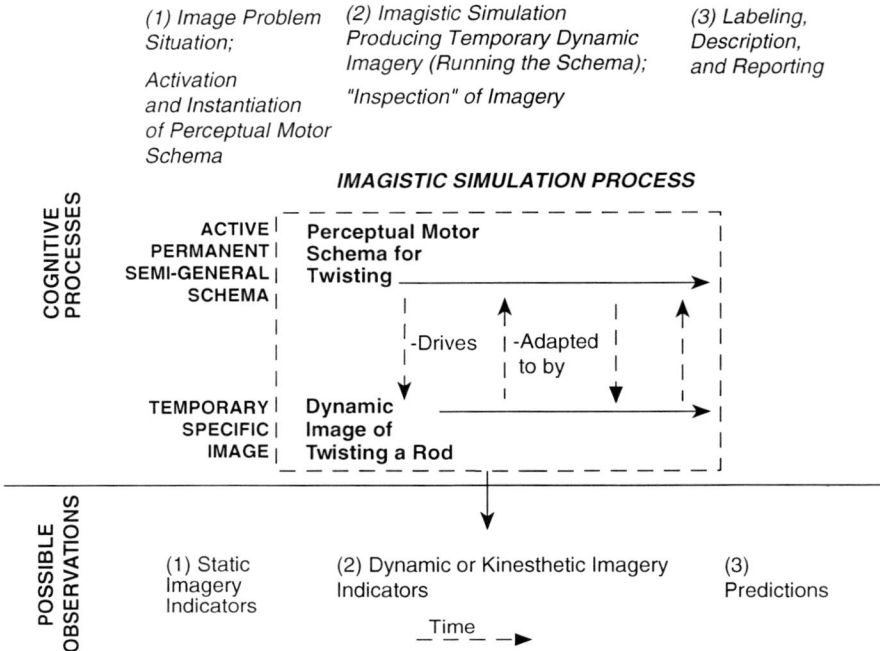

Fig. 13.3 Imagistic simulation process showing hypothesized mental processes and observable behaviors they may produce

- *Modest generality.* Predictions or explanations made by subjects from an intuition are more general than the memory of a specific incident. However, the generality is not as broad as that of the concepts used in Newtonian mechanics.
- *Self-evaluated.* Strength of belief in a physical intuition is described as being determined largely via internal criteria rather than being dependent on the evaluation of an authority.
- *Oriented to concrete objects.* Subjects speak of an intuition as knowledge about objects and manipulations of or relationships between objects, rather than as a symbolic result that must be interpreted.

Defining characteristics are defining indicators in the sense that if negative evidence is found on any one of them, it discounts the possibility that an intuition is being used. We have the most complete evidence for an important, intuitively grounded imagistic simulation if the subject exhibits evidence for: (1) new changes or states for a system; (2) imagery; (3) intuition use; (4) the importance of processing during the latter three indicators. As was the case with our imagery indicators, we use the principle of triangulation here. No one indicator is compelling on its own, but when several appear in the same episode, there is more evidence for triangulating on a hypothesized mental process as the simplest possible explanation.

13.3.2.2 Terminology to Be Used

I will hypothesize that most of the imagistic simulations discussed in this book are intuitively grounded in the sense discussed above. Since "intuitively grounded, schema-driven imagistic simulation" is a somewhat lengthy phrase, I will continue to simply use "imagistic simulation" as a shorter name for this idea, and add the phrase "intuitively grounded" when I wish to emphasize that the schema is an intuition. From this point on, when I use the term "imagistic simulation" I have a process like that shown in Fig. 13.3 in mind.

13.3.3 Precedents in the Literature on Perceptual/Motor Schemas

A perceptual/motor schema is hypothesized to contain at least three major subprocesses, as shown in Fig. 13.4: a subprocess for assimilating (instantiating) an object in the environment based on preconditions for application; a subprocess for implementing and tuning or adjusting the action so that it is appropriate for that particular object; and a third subprocess for generating expectations about the results of the action – here, an image of how much the rod will turn. Although the elements in the diagram are necessarily labeled with words here, the schema is thought of as a nonverbal interpretation and action control structure which is active over a period of time and which keeps track of and acts on external objects (or images of objects) over that period.

Expectations are built into a schema in the form of an initial rough template or shape for the action, a readiness for feedback signals from the objects during the action to help control it, and standards providing information about whether the action was completed and successful. If a schema can generate a flow of bodily actions on

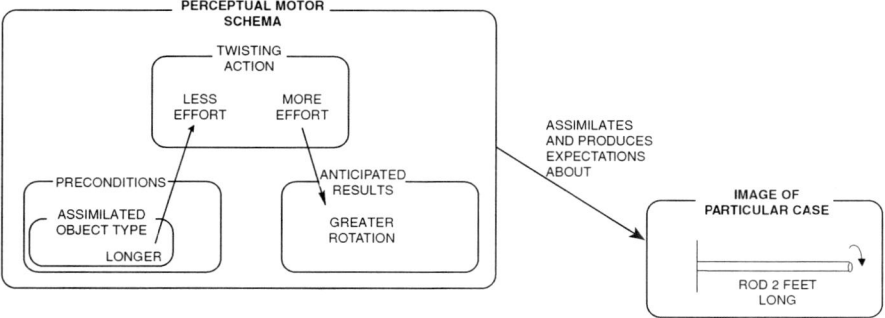

Fig. 13.4 Perceptual/motor schema driving an imagistic simulation for twisting objects. Schema incorporates implicit expectation relations of a longer object requiring less effort to twist (for same resulting rotation) and more effort leading to greater rotation (for the same object length)

real objects and accompanying perceptual expectations of the results, it is not so hard to believe that it can "run through" that same flow of actions and expectations vicariously in a dynamic image, assuming there is a way to "turn off" the final output to the muscles. This is what I mean by a schema that drives a dynamic image above the horizontal line in Fig. 13.3.

It is likely that perceptual/motor schemas that represent physical intuitions are not the only kind of schemas capable of generating imagistic simulations, but in the examples discussed here they appear to serve as one very important and analyzable type. The perspective that a motor schema can have generality through a pattern of actions and expectations developed over time, with parameters adjusted to a particular situation in a process of tuning has precedents in Piaget (1976), Neisser (1976), and Schmidt (1982). Piaget (1976) proposed that very early sensory motor schemas (in his terms, *schemes*) can assimilate and accommodate to more and more situations to expand their domain of application and can combine to form more complex cognitive structures in the child, all prior to the development of language. This view is compatible with Neisser's (1976), who describes schemas as accepting information, directing movement, and, importantly, providing nonverbal anticipations. In his view even visual perceptual schemas control actions (e.g. exploratory eye movements that track objects and seek new information). As he puts it, schemas involve feed forward units and efferents as well as receptors and afferents. A schema coordinates processing over time at many levels: sensory, perceptual, cognitive, action, and muscle control. Thus, it is capable of participating in a *continuous proprioceptive interaction with the environment over time* when active.

Subsequent work by authors such as Schmidt (1982) and Arbib (1981) further developed the concept of schemas or "programs" in motor control theory. Schmidt proposed a theory of motor schemas in his discussion of generalized motor control programs that provide a plan for coordinating parallel overlapping sequences of muscle actions over time. He describes such motor programs as general in the sense that a single program can produce a large variety of responses depending on the values of certain input parameters. He also refers to expectations that anticipate the results of the action. In addition, as discussed in Chapter 12, physiological research suggests that the motor control system can be involved in imaginal activity, even when no actual movements occur (Jacobsen, 1930; Decety and Ingvar, 1990; Jeannerod 2001; Romero, et al. 2000). Generalized motor schemas are therefore prime candidates for a type of knowledge structure that has pre-symbolic components and yet has some degree of generality and flexibility. They may therefore provide a foothold for beginning to develop a theory of imagistic simulation processes. Schwartz and Black (1999) studied subjects who have difficulty predicting the degree of tilt required to make water start pouring from a glass. They found that when subjects were asked to hold and tilt an empty glass with a line marking the height of water while closing their eyes, they performed significantly better than the previous subjects. This suggests that the involvement of the motor system may be important for some tasks that involve imagery.

Transcript observations in the present data that support the view that perceptual/motor schemas are being used include those listed under Kinesthetic Imagery

13.3 Imagery Reports and Imagistic Simulation

Indicators in section IC of Table 12.3: personal action projections (describing a system action in terms of a human action), use of force terms, and kinesthetic imagery reports; as well as hand motions that depict actions performed by the subject. In episodes like that of the twisting rod above, it is possible that the subject is able to focus on an implicit variable relationship embedded in the tuning function of a perceptual/motor schema and describe it explicitly for the first time. In Fig. 13.4, the perceptual/motor schema for twisting objects, has embedded in its tuning parameters the expectation of a longer object requiring less effort to twist (for same resulting rotation). By conducting two imagistic simulations of the long and short twisting rods and "inspecting" the force needed for each, the conversion from implicit to explicit knowledge occurs.

One may object that if this "inspection" is conducted by a homunculus containing the "minds' eye" that it has little explanatory value and could lead to an infinite regress of homunculi. But here one can adopt the view of Shepard (1984) that the "minds' eye" is just the upper layers of our perceptual system being driven in a top-down manner rather than a bottom-up manner. In this view the cognitive system consists of layers of schemas and subschemas, and the perceptual system contains many layers which interact with the bottom layer of the cognitive system. Layers that represent objects and spatial relations between objects in a scene during perception would be in the top perceptual layer. Then one can hypothesize that when there is significant processing in that layer, in the absence of seeing the real objects, the subjective experience of imagery occurs. This could happen if it was activated in a top-down manner by higher cognitive layers. Similarly, kinesthetic imagery would involve top-down activation of upper layers in the motor control system, but with suppression of the lowest layers which produce and sense actual movements. During a real interaction with objects in the world, a perceptual/motor schema is conceptualized as monitoring or controlling a select hierarchy of processes in both perceptual and motor systems. "Inspection" during a nonreal imagistic simulation then refers to the act of focusing attention on a particular feature of the experience during the truncated "running" of the perceptual/motor schema which involves both the motor and visual systems. During the time the image is maintained, focusing attention can activate and/or instantiate other schemas that interpret the image. Thus a certain amount of parallel processing is assumed. Because the buck stops with the schemas interpreting and responding to the image layer, just as it would stop there during the perception of a real event, there is no infinite regress of homunculi in this view.

Coming from a very different tradition and area of focus in studying categorization, Barsalou (1999) has described a theory of perceptual symbols which represent schematic elements of perceptual experience and that can be integrated to produce dynamic simulations and combined in productive clusters. A number of his concepts share many features with the concepts of imagistic simulation introduced in Clement (1994a, b). Because the studies arrive at similar conclusions from a different theoretical direction and different sources of support, this is an interesting example of convergence. Barsalou emphasizes the application of these ideas very widely to categorization and language, areas that will not be

13.3.4 Relations Between Observations and Hypotheses

The left half of Fig. 13.5 helps one keep track of the growing population of observable indicators and hypotheses in this section and the question of the relationship between them. Whereas in Chapter 12 we asked this about *imagery* indicators, now that we have a more elaborate model of the source of the imagery, we will ask about the broader set of *imagistic simulation* indicators shown in Fig. 13.5. The figure shows imagery, simulation, and intuition indicators and the hypotheses they support. Since, according to Fig. 13.3, the imagistic simulations in question involve an intuition schema generating an image, we may see evidence for both intuition use and imagery in the same episode. Figure 13.5 also shows several simulation indicators – these provide evidence that the subject is forming new anticipations of multiple states of the system in question. Thus this diagram can be thought of as an extension of Table 12.3, including not only imagery indicators but also intuition and simulation indicators.

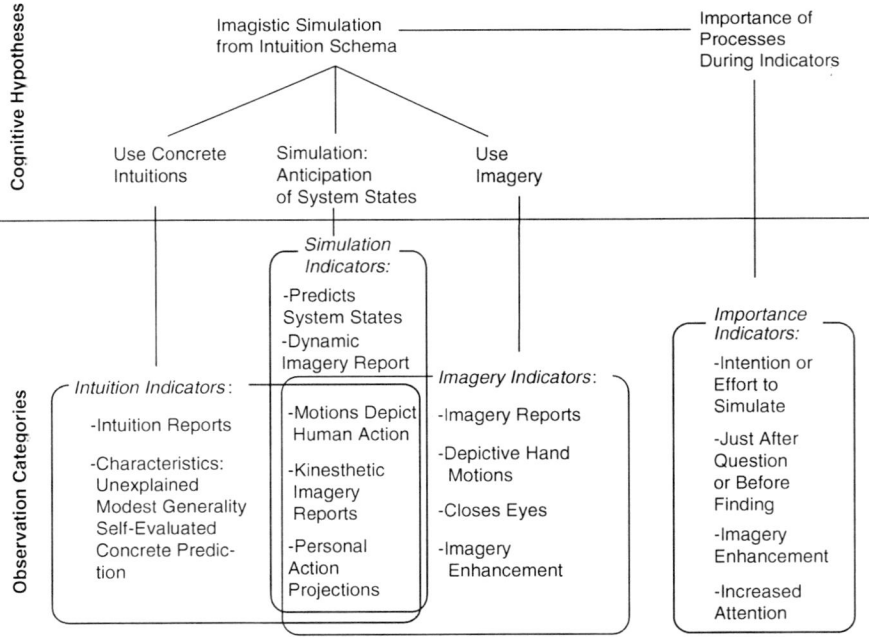

Fig. 13.5 Possible observable indicators for intuitively grounded imagistic simulation

As in any complex area of science such as cognition, there is going to be a many-to-many rather than a one-to-one correspondence between indicators and cognitive processes. Some indicators, such as kinesthetic imagery reports, may be indicators for more than one process, as indicated by the overlapping boundaries. Upward pointing lines in this diagram indicate relations of support. The strongest evidence supporting the hypothesis that a physical intuition schema was used would be the co-occurrence of a number of the observable behaviors shown. However, because subjects are not used to thinking aloud in this degree of detail, in practice we are only likely to see a small number at best in any one episode.

13.3.5 *Importance of Concrete Intuitions and Imagistic Simulation*

We are now in a position to revisit the question raised in Chapter 12 of whether these processes are important to the solution or merely side effects. I also wish to extend the target of the discussion in Chapter 12 from the importance of processes occurring *during an imagery indicator* to the importance of processes occurring *during an entire imagistic simulation*. In Fig. 13.5 indicators of the *presence* of imagery and physical intuition processes are shown on the left, and indicators of the *importance* of processes occurring during the indicators are shown on the right.

The following protocol from a research physicist, S5, indicates that imagistic simulation is sometimes quite effortful, and I will infer from this and other findings that the subject considers it an important, rather than inconsequential process. This episode comes after a follow up question from the experimenter about how one might determine that twisting is involved in the spring. At first he is uncertain about how to "imagine" it:

022 S5: [S]uppose I had a big spring and I could make little paint dots on it all along its length....and saying...would I see a torsional displacement of the paint dots. And what would it look like? And I have a hard time imagining that because you know, the torsional displacements that come to mind are very small.

024 S5: (Makes drawing in Fig. 13.6a of spring with paint dots on outside of wire.)

036 S5: So...the other parts are going to twist such that...little dots on the surface will tend to move up....

038 S5: The mass is going down and so now – these portions of the spring–Hmmmm

040 S5: ...I'm just getting a hard time envisioning what's going on 3-dimensional space, and so I'm having a hard time seeing which way this is going to rotate.

041 S5: Well I want to imagine that the portion here up to the cross section(draws cross section circle on wire near top of spring)...is fixed. So I'm pulling down on the weight or the weight's pulling itself down, and that's causing these coils to elongate. I'm trying to decide how it's gonna twist this portion of the wire....

But eventually he is able to make a prediction in Fig. 13.6b by thinking about a "frictionless cross section" labeled in the figure:

> 042 S5: If you imagine the extremes, if you pull it up and down, this little line...on the outside of the spring you know, would...rotate down till it's at the bottom.
>
> 046 S5: I guess I'm – I'm quite satisfied with that.
>
> 072 E: Were you thinking about an equation there?
>
> 073 S5: Oh, no. This is all er, I think very experimental. What I think I have – this image of this line of paint dots on a spring and you know I'm pulling on the weight. I'm going pull and release, pull and release and so I'm constantly putting it through its paces. And asking you know, how would I see the dots move.
>
> *(Protocol from Clement, 1994b, pp. 217–219.)*

13.3.5.1 Imagery Enhancement

I will refer to his references to imagining "little paint dots" on the spring as examples of an "imagery enhancement report" where the subject adds markers to a situation or changes other perceptual features such as: (1) orientation or size; or (2) magnitude of problem variables – in conjunction with evidence for imagery. More direct evidence occurs when the subject speaks of modifying a problem situation in a way that makes it "easier to imagine" or "easier to see". The attempted enhancement is successful if it increases the subject's confidence in the results. Other examples of imagery enhancement are the generation of the extremely narrow spring in Chapter 14 episode 6, or the frictionless cross section (as labeled in the drawing in Fig. 13.6b) that helps the subject to see the direction of movement.

Since S5 makes a drawing of the spring in line 24, does that mean that it replaces and makes redundant any internal imagery that occurs? It cannot replace the

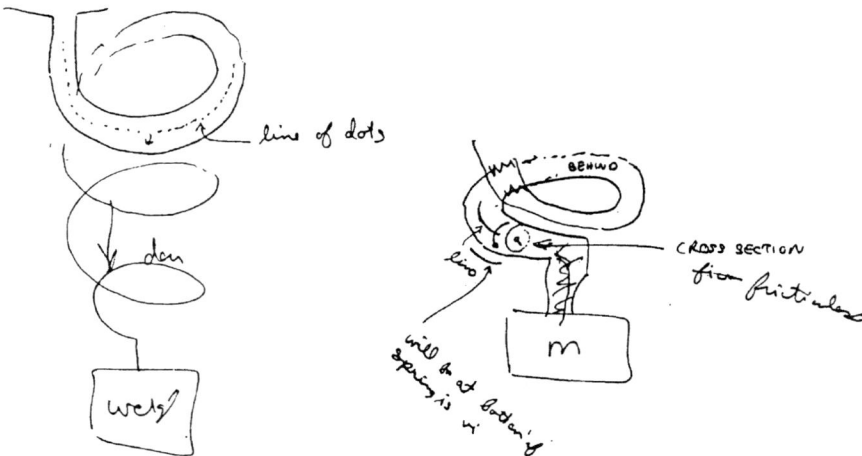

Fig. 13.6 (a, b) S5's drawings for spring problem

13.4 Implicit Knowledge

imagery, since he speaks of imagining movements, and the drawings do not move. This makes it is reasonable to assume that the drawing reflects and supports static features of the subject's initial image, but cannot replace any dynamic imagery.

13.3.5.2 Effort Can Be Required for Simulation

The observations for this protocol then fit most of the patterns (except hand motions) shown below the horizontal lines in Fig. 13.5, and support the hypothesis that the subject is running through imagistic simulations. He describes the simulation as a process extended over time (73), and makes a "personal action projection" by substituting his own pulling for the force of the weight (41, 42, 73). The fact that he: says "I want to imagine" the situation (41), says that it is difficult for him (22, 40), makes repeated and extended efforts to do it anyway, and uses imagery enhancement techniques such as the "painted dots" (22, 36), is evidence that: (a) he intends and tries to set up the imagistic simulation as an extended process very different from "remembering a fact"; (b) the act of imagining is important to him as a technique; (c) the process involves considerable effort. These findings, along with the fact that he "asks questions of" the simulation (22, 73), and that the simulation is the main technique used to give his answer, argue that the intuitions and imagery involved in the simulation are not simply unimportant side effects of some other process, but are effortful processes that are central to his thinking here. Thus he must work hard to adapt and apply old schemas to a newly imaged, unfamiliar situation in order to construct new knowledge. This argues that the grounded imagistic simulation process was important rather than inconsequential to the subject's reasoning.

13.4 Implicit Knowledge

This section discusses the question of whether intuitions can be implicit as well as explicit. This will have an important bearing later on the question of how a person can learn from running an imagistic simulation. Perhaps one source is some kind of implicit knowledge "buried" in the intuition in some way.

When an intuitive expectation is conscious during a real or imagined event and is represented verbally or mathematically (in addition to other possible representations), this will be termed an explicit intuition. Thus, more than one type of knowledge representation may underlie an explicit intuition. Alternatively, it is plausible that a perceptual/motor intuition could be implicit in the sense that a parameter could be taken into account in controlling an action without being consciously differentiated perceptually or articulated verbally. For example, when one is saying the word "with," one is usually not aware of the sequence of action components performed with the tongue and mouth even though one may be aware that they have said the word. I will use the term "unconscious knowledge" to mean *never having been consciously aware of the knowledge*. (I do not intend it to carry Freudian overtones in this context. Unfortunately, other possible choices for terms here, like "preconscious" or "nonconscious," have misleading overtones as well.) This is the stronger sense of the term "implicit" used here.

It may be possible for some elements of previously unconscious knowledge to become a conscious intuition. For example, I can imagine saying the word "with" without actually saying it, and identify at least two separate component motor movements involved, while answering a question like, "how many times does the tongue touch the teeth in saying this word?". Here imagery generated by implicit, unconscious knowledge is "inspected" and used to produce explicit knowledge.

13.4.1 Distinguishing Different Levels of Implicit Knowledge

Here I use quotations from Clement (1994b, p. 227) to construct a table: "Rather than a strict dichotomy between implicit and explicit knowledge, it is useful to propose three levels on a dimension running from more implicit to more explicit knowledge, shown in Table 13.1. The levels describe types of knowledge one might use in completing a task or comprehending a scene.

The terms implicit and explicit have often been used more simply as a dichotomy which distinguishes level 1 from level 3 only – unconscious from verbally explicable knowledge – ignoring the possibility of conscious but undescribed knowledge at level 2. I will use the term "implicit knowledge" to refer to knowledge of type 1 or type 2 above (Clement, 1994b, pp. 227–230).

Table 13.1 Implicit to explicit knowledge dimension

(More implicit)

1. *Unconscious knowledge*: I am unaware of an element of knowledge I use, even as I use it, such as the sequence of individual limb placements during crawling. ("I may be aware of a larger unit of knowledge than the knowledge element in question, such as the whole act of crawling and the idea that I am crawling. But this awareness is holistic and there is not a differentiated awareness of the component knowledge elements…in question" [Piaget, 1976]). Similarly, I am usually not aware of the sequence of actions performed with my tongue and mouth as I say the word "with" even though I may be aware that I have said the word. The knowledge elements are representationally undifferentiated – not separated out from the whole experience…

2. *Conscious but nonverbal knowledge*: The entity in question is differentiated from the context (e.g. imagistically) but no verbal or mathematical description in terms of discrete symbols has been constructed. For example, a child may be able to watch someone make a sequence of novel arm movements and imitate them afterwards without any need for verbal description. Or I may suddenly notice contrasting types of cloud formations in the sky without describing them verbally. By "differentiated" I mean that the knowledge element is separately represented (e.g. as an image) and recognized as distinct from its context so that it can be attended to and reflected upon.

3. *Conscious and verbally descriptive*: The knowledge element is conscious, differentiated from its context, and verbally or mathematically labeled or described in terms of discrete symbols.

(More explicit)

13.4 Implicit Knowledge

One could if necessary divide the first category of unconscious knowledge into two subcategories: processes that are forever inaccessible (such as consciousness of movements of individual parts of the heart) and processes that are accessible or inspectable via imagery or observation or training. In this book we will be interested in those unconscious processes that are inspectable via imagery.

13.4.1.1 Examples of Undescribed Knowledge

For example, in the twisting-rod episode, I have hypothesized that S2 runs through an intuitively grounded imagistic simulation in order to (a) activate an intuition; (b) focus on the variables of rod length and twisting force: (c) run the simulation and experience a conscious expectation; and (d) give a verbal description of the effect. This is a process for making implicit knowledge more explicit. The fact that the subject refers to his "imagining" and "physical intuition" as the source of his prediction is evidence that it is at least implicit in the sense of type 2, undescribed knowledge. This makes plausible the interpretation that the source of knowledge is a nonverbal imagistic simulation and that it is verbalized in a subsequent step.

This and another indicator for nonverbal sources of knowledge were listed in Table 12.3 Part IA:

Findings were not Translated from Verbal

24 Indicates that "imagining" of some kind is the source for a finding.

25 *Difficulty Describing Images*: Indicates that image or thinking in episode leading to finding is clear but is hard to describe.

In a second example S2 has a direct intuition about narrow and wide springs at the very beginning of his solution:

05 S2: I'm going to try to visualize it to imagine what would happen…my guess would be that it would stretch more…a kind of a kinesthetic sense that somehow a bigger spring is looser.

Here again he refers to a nonverbal source for his finding. Why the subject made an effort to attempt a visualization or simulation as a source for findings can be explained by the hypothesis that he is able in this way to access implicit knowledge that has not been described before. These two examples provide some evidence against several competing hypotheses. In the first part of the quotation above, S2 indicates that he anticipates that an extended imagery process (running through a simulation) may be helpful in producing a finding. This argues against the idea that explicit, verbally described knowledge was retrieved in a direct manner – that it was not the retrieval of a specific fact. Also, there is no evidence in either example for the competing hypothesis that the situation is a complex inference that would lead to announcing a verbal result inferred from other propositions in memory. Here the subject refers to only a single action. A viable explanation is that his findings come from nonverbal elements of an intuition schema expressed in an imagistic simulation and are eventually translated into words.

13.4.1.2 Other Possible Sources of Evidence for Implicit Knowledge

Thus under the definition of "implicit" used here, evidence that linguistic encodings are not the source of the knowledge being gained can also support the hypothesis that the subject is tapping implicit knowledge in the broad sense of undescribed knowledge. Table 12.3 Part IA identifies a number of other observations which could also provide evidence for implicit (nonverbal) knowledge.

Motions Not Translated from Verbal

Ob-11– Uses ambiguous terms in conjunction with less ambiguous motions to refer to the motion or interactions of an object

Ob-12 – Motions occur during language that does not describe them (e.g. S says "Well, let's see what we can find out here" during motions. Producing such oral language would be likely to interfere with other language that served as the source for a translation to gestures, and so such a translation is unlikely here)

Ob-13 – Verbal description of an event is inconsistent with concurrent hand motions representing the event, and these motions are consistent with later conclusions in the solution

When hand motions depict events deemed important to the subject's findings, the indicators above can provide evidence that the source of the findings is nonverbal and implicit in the broad sense. Observation 11 above was noted for S20 in the last protocol in Chapter 12, where he appeared to generate imagistic simulations of pushing an object in space:

(Ob-1,2,6)34 S: All right, in space. I've never been in space – Yeah, in space, too, I guess, in space. If you push (moves right hand away from chest, Fig. 12.2) something heavier than you, you would go back (moves head and right hand back) uh, more than it would go that-a-way (points forward).

(Ob-11a, 14,15,16, 18)

Since the original problem here did not involve people pushing or pulling, this excerpt provides another example of a personal action projection as an indicator of intuition. The trouble the subject goes to in order to generate the analogy of pushing with his own body argues that this intuition was part of his solution attempt, not just a side effect. The sense of the ambiguous statement "it would go that-a way" is indicated by his hand motion rather than a verbal description, suggesting that a language-like representation was not the source of this idea. Of course, a critic could always advocate that some "deeper" propositional description was the source of his intuition. But translations to language should be trivial from such a representation in cases like this, and when indicators like the ones above are present, a simpler hypothesis is that the subject's imagery is the source of both the hand motions and the conclusion.

Thus, where we have evidence that the source of knowledge is nonlinguistic, that is evidence for the presence of implicit knowledge in the broader sense of the term as used here. In order to record or report their knowledge, subjects will ordinarily then convert it into an explicit statement.

The five observable indicators for nonverbal knowledge discussed in this section above could also be added to Fig. 13.5 below the horizontal line but have been left out for simplification purposes. They would point to a finding above the line of "nonverbal source of knowledge" and would provide evidence against a major competing hypothesis to that of imagery use.

13.4.2 Evidence for Unconscious Knowledge

For certain examples there is also some weaker evidence that the knowledge used was implicit in the stronger sense, that of being previously unconscious. The subject S2 in line 05 above describes his findings about the spring as new information rather than something he was familiar with but never described. (An even stronger form of this would occur if the subject expresses surprise at the result.) Also, he refers to it as coming directly from "visualizing" or "imagining it" and a "kinesthetic sense," so this new information does not appear to be the result of a chain of inferences. Thus a plausible explanation is that he is using perceptual/motor knowledge that was previously unconscious in the sense that he has never heeded it before. This may also be true for the twisting rod episode, where S2's statements imply that he is realizing for the first time that it will "twist more" rather than remembering it as a familiar image.

Perceptual/motor schemas have been described as coordinating processing at many levels from sensory up to cognitive, and back down to muscle control. This gives one another way to describe the "upper" or cognitive levels of a schema with more explicit knowledge, and the lower levels with more implicit knowledge, represented as being above and below the horizontal lines in Fig. 13.7 (adapted from Clement, 1994b). There a triangle represents an active perceptual/motor schema whose upper part includes only very skeletal, depictive representations of the preconditions and expectations for an event. The lower part includes the substructures supporting the perceptual/motor knowledge involved in carrying out an action that is performed more or less automatically by subschemas and that is not ordinarily under conscious control. This distributed and parallel knowledge comprises the "lower" part of a schema whose other part is in conscious attention. (I am not referring here to other schemas weakly activated at a subliminal level by "horizontal association" to a schema in conscious attention.)

When the schema is used in a planning process, only the small upper part of the schema's processing need be in conscious attention for very brief moments. By making an effort to focus on and inspect features in an imagistic simulation how-

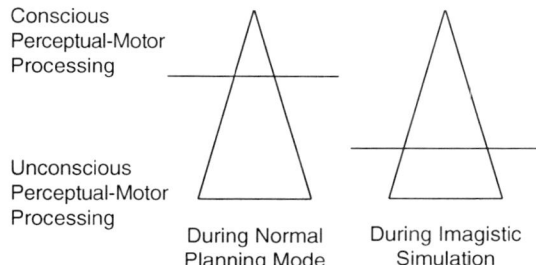

Fig. 13.7 Different proportions of conscious and unconscious processing during normal planning vs. imagistic simulation

ever, more of the schema's processing can be brought into conscious attention, revealing knowledge that was previously unconscious.

Studies show that infants 18-months old or less can anticipate whether an object is light or heavy in preparing to pick it up (Mounoud and Bower, 1974).These force level estimates are presumably embedded (embodied) in control (tuning) processes of a perceptual/motor action schema that are parallel and distributed in nature rather than being encoded as discrete, language-like symbols. Similarly, when S2 reports his intuition about twisting different rods, he may be consciously attending to and articulating it for the first time, even though he may have used the knowledge previously to control actions. This would account for the subject's experience of an "intuition" or "kinesthetic sense" that is not explicated further as the source of knowledge in this case.

It is reasonable to assume that perceptual/motor schemas can have tuning parameters which are unconscious (Schmidt, 1982). In general, however, discriminating between the unconscious and conscious but nonverbal forms of implicit knowledge on the basis of transcript data is difficult, and I do not consider the examples discussed here to be definitive on this finer point. We need to develop better methods for determining when knowledge is unconscious. Looking at cases where a motoric response differs from verbal predictions is one approach (Piaget, 1976; Krist et al., 1993).

Historically, researchers such as Gelerntner (1959), Larkin and Simon (1987), and Lindsay (1988) proposed the hypothesis that there is a special mode of nondeductive spatial inferences made possible by diagrammatic representations. Such inferences are thought to be derived from knowledge of spatial relationships and constraints which becomes usable when a static diagram is "examined" and described. The perspective I am proposing in this section can be viewed as related to theirs but in the domain of imagery for dynamic actions. That is, there are implicit relationships and constraints that are built into action schemas which can become explicit when the schema is applied to an image of a particular situation in a dynamic imagistic simulation with a particular question in mind. The view that intuitions can take the form of knowledge in a perceptual/motor schema provides one possible reason for why subjects can take them to be "givens" that are not in need of further explanation or justification. They are something the subject just "knows how" to do that requires no justification. And some aspects of that knowledge may be implicit in that they have never been described or attended to before, much less explained.

13.5 Knowledge Can Be Dynamic

13.5.1 Different Uses of the Term "Simulation"

de Kleer and Brown (1983) used the terms "envisioning" and "simulation" to refer to reasoning via chains of causal inferences. In their approach, knowledge elements take the form "A causes B" or "an increase in A causes an increase (decrease) in B" and these are treated as passive data structures rather than processes. Other rules

13.5 Knowledge Can Be Dynamic

are then used to combine these elements in order to make chains of inferences in an information processing model of causal reasoning.

13.5.1.1 Passive Strings of Symbols

However one is tempted in such a perspective to take the computer metaphor too seriously and to assume that each such knowledge element in a subject's working memory is like a static string of symbols. In that view while reasoning and learning are seen as active processes, knowledge is seen as a network of symbols consisting of static and passive "data" elements. The printed word is the implicit, underlying metaphor for knowledge in these approaches. It is true that not all features of a cognitive model need to be taken as literal features of the real system under study. However, it is important to note that very few authors have actually disavowed static symbols as central features of these models. Here I suggest the possibility of a much richer and more dynamic representation as an alternative basis for knowledge.

13.5.2 Knowledge Experienced in Imagistic Simulations Is Not Static

de Kleer and Brown (1983), as well as Forbus and Falkenhainer (1990) use the phrase "running a simulation" to refer to the process of making *inferences from collections* of such static causal knowledge elements. However, in this book I will use the term to describe a more dynamic process of running through the action or movement involved in a *single* causal knowledge element. That is, instead of just modeling *reasoning* as a dynamic process acting on knowledge elements as static symbols, I will also model *each knowledge element* as a dynamic process.

13.5.2.1 An Alternative View

A more dynamic view of individual knowledge structures is suggested by protocols such as S2's statements immediately after he has constructed the two dimensional zigzag spring analogy in Fig. 6.4b:

> 23 S2: And when we do this what bends...is the bendable bars...and that would behave like a spring. I can imagine that it would; I can kinesthetically pull on it and it would stretch, and you let it go and it bounces up and down (waves hand up and down, palm down, fingers extended). It does all the things.

In cases like this the subject gives evidence for: making an imagery report, making a prediction, and indicating the use of an extended-in-time process either verbally or via hand motions. This provides evidence that he is *experiencing an event over time* in order to produce a prediction for a single causal relationship as he "runs through" the event mentally in real time (or some speeded up or slowed down version of real time). These features are also present in S2's imagistic simulation of the twisting rod depicted in Fig. 13.2. In addition, I have already discussed the case of S5 who uses considerable time and effort attempting to determine the single causal relationship between stretching and the direction of twisting deformation in the spring wire. His protocol suggests that establishing a single knowledge element of the form "A causes B" can require a significant effort, and that imagistic simulation can play a major role in such an effort.

This raises the question of why subjects were willing to run through effortful imagistic simulations. That they did so provides one more argument that imagistic simulations allowed them to apply knowledge that was not stored as a linguistic description. For if it were, then why announce the intention to form an image of the situation and make the effort to run through a simulation of it (see S2 line 5 above)? Since the task is to produce a verbal prediction, why not just report it? The presence of dynamic imagery reports, hand motions, imagery enhancement techniques, and the effort put into imagistic simulations support the view that imagistic simulations of a single causal relation are very different from a static symbol structure. The fact that these refer to only a single causal relation discounts the alternative view that the effort involved comes from making some kind of chain of reasoning inferences. A real motive for using the term "simulation" comes from the subjects' reports of experiencing the effects of actions occurring over time (S5, line 73, S2 line 23 above). They suggest that the subjects are experiencing some aspects of the rich flow of perceptions and/or motor actions over time that would be present if they were actually perceiving such events. This is a different meaning for "simulation" than a symbol manipulation procedure which steps its way through a series of inference rules operating on a set of word-like tokens. The intent here is not to deny the power of linguistic abstractions and mathematics in human thinking, or the idea that once an intuition is labeled or described in language, that such a linguistic description can be stored with the intuition schema. But the present findings support the view of simulation as a thought process that can take place outside of a linguistic system, before it is translated into one. This suggests that causal knowledge elements in working memory are not simply static, language-like symbols. An alternative hypotheses is that part of one's knowledge in working memory can be based in a dynamic, distributed flow of imagined perceptions and actions or in a state of readiness for such a flow. In this view the knowledge experienced in an imagistic simulation does not just consist of an end state of static tokens, but is a process or activity which takes time to experience and have meaning for the subject. To be sure, the subject can then go on to describe such a finding in language. This is what I have called converting implicit, undescribed knowledge into explicit knowledge. The view of knowledge as dynamic in the context of imagistic simulations is consistent with positions that have developed from very different types of data in areas such as "representational momentum" (Freyd and Finke, 1984; Freyd, 1987.) It is also compatible with the classic observation of Shepard and Metzler (1971) that subjects take longer to imagine large rotations of a figure than small rotations.

Another strength of the present view is its potential for providing a more adequate view of meaning and consciousness than the view of knowledge as passive data structures. When a perceptual/motor schema is activated and interacting over time with events in the world, this interaction can be seen as a primitive and primary locus of conscious experience and meaning. In this view, a closely related and powerful but less direct locus of meaning occurs when the schema is activated and interacting with less real but still meaningful internal images of such events.

13.6 Conclusion: The Role of Concrete Physical Intuitions and Simulations in Embodied Thinking by Experts

13.6.1 Summary of an Initial Framework for Modeling Physical Intuition and Mental Simulation via Perceptual/Motor Schemas and Imagery

Chapters 12 and 13 have attempted to show that think-aloud protocols can contribute to several research goals. First, although the role of imagery in lower-level tasks is becoming more accepted, there is a need to determine whether it is also used in higher-level thinking, and to develop relevant concepts of observation for detecting its probable presence during think-aloud protocols. Second, one can use protocols to begin to study the nature of mental simulations. Although rules for making inferences from simulations of larger networks of causal relations have been studied, we lack models which analyze the nature of mental simulations underlying a single causal relationship. A final larger goal of this chapter was to construct a plausible initial framework which proposes basic relationships between physical intuitions, perceptual/motor schemas, visual imagery, kinesthetic imagery, and simulations.

13.6.2 Imagery

In this chapter I have tried to extend the analysis of imagery from Chapter 12. At the level of observations, three very important imagery indicators have been added: imagery reports, dynamic imagery reports, and imagery enhancement reports. The phenomenon of imagery enhancement is also particularly compelling as evidence for the *import* of imagery as a thinking tool. These indicators were included in Tables 12.3 and 12.4. Cases were cited where an imagistic simulation appeared to be at the very center of a new inference. Thus there is further evidence in this chapter for the use and importance of imagery in expert cognition. Cases were also cited where observations of imagery reports and hand motions occurred simultaneously, supporting the conclusion from Chapter 12 that depictive hand motions can be taken as one form of evidence for the use of imagery.

13.6.3 Intuitions and Imagistic Simulation

At the theoretical level, a plausible answer to the knotty problem of the source of new and even novel dynamic imagery has been provided: that dynamic imagery can be generated by a perceptual/motor schema which "operates" on an image as if it were a real object. This view treats images as temporary constructions and schemas as more permanent residents of long-term memory. The case studies described in this chapter provide some initial evidence for framing the following hypotheses: (a) In addition to abstract principles, experts can use schema driven simulations involving imagery (here termed imagistic simulations). (b) The schemas analyzed in this chapter were concrete, self-evident, physical intuitions of modest generality. These played an important role in problem solutions that was more than simply a "start-up" role – imagery and intuitions were part of the central argument in these solutions, not just side effects. The kinds of evidence that can be used to support this first pair of hypotheses are summarized in Fig. 13.5. (c) A grounded imagistic simulation can be used to make implicit knowledge more explicit.

These findings conflict with the typical assumption, as in Chi et al. (1981), that the most distinguishing feature of expert knowledge is its use of abstract principles. In the expert protocols discussed here, much of the knowledge used was present at the concrete, perceptual/motor level via imagistic simulations and was "embodied" in this sense. However, in the case of the Chi study, their focus on abstract knowledge may be partly due to the fact that their study used problems that were easy or routine for the experts, whereas in the present study the experts were engaged in nonroutine model construction. In the present study, the experts often did not have an adequate set of ready made abstract principles with which to understand the system. Thus the context was closer to that on the frontier of science, where principles are being constructed rather than only used. In addition these findings conflict with a view of knowledge as a collection of static symbols, since perceptual/motor intuitions are experienced over time as depictive action images and need not be interpreted.

13.6.4 How Is New Knowledge Generated from an Elemental Simulation?

At a somewhat more detailed level, as illustrated in Fig. 13.3, the proposed model for elemental simulations says that intuitions are hypothesized to be expectations embedded in a schema and that such a schema produces imagistic simulations. This separation of a schema activated from permanent memory and an image in temporary memory (generated by a top-down and/or bottom-up process) allows us to

13.6 Conclusion: The Role of Concrete Physical Intuitions

separate what is old, familiar, permanent, and more general (the schema) from what can be new, novel, temporary, and more specific (the image of particular objects and movements) in an elemental simulation. The "running" of an elemental imagistic simulation involves the following:

1. Such simulation involves a permanent perceptual/motor schema (such as a schema for twisting). When acting on real objects in the world in its normal, nonimaging role, such a schema controls perceptual tracking and/or motor coordination processes over a period of time.
2. As such, at least at the level of organization being discussed, parts of the schema in working memory are analog and active or depictive in character, rather than being a descriptive set of static symbols.
3. The schema can have expectations which "set up" a readiness to perceive certain expected events in a top-down manner.
4. Some of these expectations can be implicit (undescribed or in some cases possibly unconscious).
5. In addition to real situations, the schema can assimilate an image of a situation.
6. It is possible for the expectations to operate in the absence of the real situation and "drive" a temporary dynamic imagery representing an event.
7. The subject can focus on and "inspect" certain features in the resulting image.
8. The subject can then describe these verbally. In this manner, implicit knowledge in a schema can become explicitly described knowledge.

In this model a new prediction can be derived from an imagistic simulation that does not depend on inferences from chains of word-like symbols. Two of the ways that this can happen are the following:

1. The subject can apply an existing schema to an image of an unfamiliar situation outside of its normal domain of application, and the schema can assimilate and adjust its expectations about the situation via tuning mechanisms. This takes advantage of the natural flexibility of perceptual/motor schemas, which can be fairly adaptive (e.g. I can imagine using a crate as a chair for sitting; or a wrench for hammering). It is also possible that the case being considered can be familiar, if the subject brings a new question to the situation so that an aspect of "running" the schema that has not been described (or in some cases even thought about) before is attended to and articulated.
2. Implicit knowledge in an intuition schema can be tapped as previously undescribed expectations in the schema generate images which can be interrogated and described, thereby being converted to explicit knowledge.

These two sources will be viewed as "elemental" sources of knowledge from simulations as a foundation for more complex sources later, such as thought experiments.

The idea that people can gain new knowledge from running a new experiment in their head is a strange one – one that I will discuss in Chapter 15 under the heading of "the fundamental paradox of thought experiments." After surveying a broader range of thought experiments there, the list of sources will be extended to address the paradox more fully. For example, in a more complex simulation several schemas might interact with the same image.

The activity of a perceptual/motor schema that is interacting with the world can be seen as a primitive and primary locus of conscious experience and meaning. The present theory proposes that a similar locus of conscious meaning occurs when the schema is activated and interacting with internal images of such events.

13.6.5 Using Perceptual/Motor Schemas as an Initial Foothold for Understanding the Use of Intuitions and Imagistic Simulation

Another issue that is beyond the scope of this chapter is the origin of intuitions. The cases discussed support the theory that some physical intuitions originate in perceptual/motor schemas. It is reasonable to assume that most physical intuitions are grounded in personal experience with physical phenomena. Like other knowledge, however, intuitions could be biased during construction by other prior knowledge or rational tendencies which would mean that direct experience does not play the only role in their construction. Also, some intuitions may be based on more rationalistic ideas such as symmetry. Virtually all of the examples of intuition in this chapter were modeled in terms of perceptual/motor schemas with motoric components. But it is not clear that all intuitions, including notions of symmetry, can be explained in this way. Perhaps they can be, if the idea of a conserving transformation is taken as fundamental for symmetry. On the other hand, perceptual operators may be involved in some cases that do not depend on the motor system. A full discussion of the cognition of symmetry is a topic that is beyond the scope of this book and an important area for future research.

Thus, I do not wish to imply that schemas with motoric components are necessarily the only source of intuitions or of imagery. However, situations involving the motor system are a good starting point for developing new cognitive models in this area, partly because of the need to include images of actions, and partly because we can draw on existing theory of presymbolic flexible processing. Some of the interesting properties of generalized motor schemas that have been discussed are listed in Table 13.2.

I have added the property of implicit knowledge at the end of the list, and have discussed the idea that it can be examined and articulated during imagistic simula-

Table 13.2 Interesting properties of generalized motor schemas

- Provides a plan for coordinating actions in parallel over time
- Anticipates consequences of actions
- Presymbolic
- General – Responds to input or tuning parameters
- Flexible
- Implicit knowledge

tions. In addition, motor theory provides paradigmatic examples of, and the beginnings of a theory for, schema flexibility during application to new situations. By studying this very interesting property we may gain a foothold for beginning to explain how schemas are used in the construction of flexible scientific models. So although I am not saying that all schemas used by subjects were simply motor schemas, I do think that a number of important ones had motor components. Looking at motor schemas provides a very interesting starting point for developing theories of dynamic, flexible knowledge structures.

13.6.6 Imagery, Intuitions, and Anchoring

The findings in this chapter give us a new way to look at the concept of "anchoring" developed in the first section on analogies in this book and extended in Chapters 8–10 on student thinking. Physical intuitions, as a self-evaluated, primitive source of knowledge and conviction, may be a natural source of anchors for analogical reasoning. We have seen a number of expert cases that fit this hypothesis:

- S2 above anchors his understanding of the spring in his physical intuition about the analogous case of a twisting rod.
- S2 in Chapter 4 anchored his understanding of the wheel in his physical intuitions about the analogous case of a lever.
- S1 discussed as the second case early in this chapter and in Chapter 2 used his physical intuition about bending long and short rods as an anchor for understanding the case of the stretched springs.

These examples provide initial grounding for the hypothesis that such runnable intuition schemas, capable of generating imagistic simulations with a degree of conviction, can serve as analogical anchors for constructing imageable models in science. Thus the accessing of runnable prior knowledge schemas in general, and physical intuitions in particular, may provide a powerful reason for experts to use analogies in science. Lakoff and Nunez (2000) have theorized that mathematical ideas are grounded in elemental perceptual/motor ideas and their study provides an interesting precedent to the present one in this regard. A comparison to their study appears in Chapter 20, section 20.5.3.

The findings in Chapters 8–10 document the fact that *students* can also use concrete intuitions. The literature on alternative conceptions in science education provides many other examples, and there is also evidence for useful intuitive student preconceptions that can ground qualitative scientific models (Clement et al., 1989). Although the separation between experts and naive subjects is significant, the finding that experts use concrete physical intuition schemas makes it less sharp. In Chapters 9–10 and 18, I argue that significant educational benefits can be derived from making full use of positive intuitions in students, illustrating how students can benefit from analogies that appeal to anchoring conceptions that are intuitive. The ideas developed in the present chapter suggest the hypothesis that these benefits may have accrued because the anchoring intuitions were runnable perceptual/motor schemas that could be exploited and transferred in the process of developing student understanding in the form of visualizable models. This idea will be developed in detail in Chapter 18, as will the idea that transferring imagery and runnability to an explanatory model may contribute to its flexibility and usefulness.

Section V
Advanced Uses of Imagery in Analogies, Thought Experiments, and Model Construction

Models, about which I shall have nothing further to say in this paper, are what provide the group [of scientists] with preferred analogies, or, when deeply held, with an ontology. At one extreme they are heuristic: the electric circuit may fruitfully be regarded as a steady-state hydrodynamic system.... At the other, they are the objects of metaphysical commitment: the heat of a body is the kinetic energy of its constituent particles.

Thomas Kuhn (1977a, pp. 297–298)

Where the senses fail us, reason must step in.

Galileo Galilei (1954, p. 60)

Chapter 14
The Use of Analogies, Imagery, and Thought Experiments in Both Qualitative and Mathematical Model Construction

14.1 Introduction to Chapters 14–16

Scientists often say that the reasoning methods they use in their research *reports* are more formal than the primary processes such as analogy and special case analysis by which they *discover* their findings. Polya coined the term "plausible reasoning" to describe the nonformal processes that mathematicians use on the way to a new discovery. This chapter presents a longer and more complete collection of plausible scientific reasoning operations on the spring problem than have been dealt with in previous chapters, and it culminates in a quantitative solution. In doing so, this chapter bridges from the domain of qualitative modeling into the realm of mathematical modeling. Arguments collected from the solutions of several subjects are combined together to give a more complete and coherent solution. The analysis of this composite solution will be the topic of the next three chapters. Other purposes of Chapters 14–16 include the following:

- To identify five major stages and a fairly large number of subprocesses in the model construction enterprise
- To examine the role of imagistic simulation and thought experiments in advanced model construction (Chapter 15)
- To use the composite protocol as a stimulus for outlining a coherent theory of model construction processes. This will include an examination of separate description, explanation and mathematization cycles as well as abductive generation processes (Chapter 16)
- To bring some closure to many of the questions raised in the transcripts about the spring problem, such as how to tell whether the spring is going through bending or twisting during stretching, whether there is cumulative bending, whether the stretch increases in a linear way with changes in coil diameter, etc. A major theme is that conjecture and counter conjecture – criticism and revision – are key characteristics of productive scientific thinking

Another purpose here is to examine the use of nonformal reasoning strategies and to ask whether they can simply be applied in random order as a list of heuristics, or whether they can be structured in some well-defined order or hierarchical procedure. I will compare these findings with the "heuristics semirandomly applied" view that Polya used in describing advanced problem solving in mathematics.

14.1.1 Stages in Model Construction Leading up to Quantitative Modeling During the Solution

This chapter also attempts to begin bridging the gap between qualitative and quantitative reasoning methods, as the spring's behavior is examined at higher and higher levels of precision. Traditionally, mathematical processes are thought of as a very different kind of reasoning than qualitative scientific methods. It is interesting to ask whether this is really true – whether at some point in the solution there is a marked shift to another set of processes, or whether there is a gradual transition indicating a large overlap between processes in qualitative and mathematical modeling in science. The solution illustrates a number of stages in model construction, beginning with very rough qualitative models and culminating with a precise level of quantitative modeling, as shown in Table 14.1. Since no single subject exhibited all the steps in this transition, in this chapter I have combined reasoning operations from several subjects into a hypothetical composite solution. This is also motivated by the practical difficulty of capturing advanced creative behavior in science – one needs a fairly hard problem to elicit creative work, but one may only have an hour of any one subject's time to work on it. Successful innovations on hard problems do not occur as fast as every few minutes. So another purpose of the composite protocol is to provide a richer variety of moves than one could find in any single protocol.

Actual quotations from individual subjects' reasoning episodes have been retained in quotes, but unlike previous chapters, the adjacency and exact sequence of these operations for any one individual have not. Lakatos (1978), in his most famous book, *Proofs and Refutations*, presented a reconstructed history of the mathematics of polyhedra in the form of a hypothetical dialog where characters

Table 14.1 Stages in model construction leading up to quantitative modeling during the solution

Qualitative modeling
  I. Efforts to develop an initial *qualitative description or prediction* for the targeted relationship
  II. *Searching for and evaluating initial, qualitative explanatory model elements*
  III. *Seeking an imageable, causal mechanism with elements that are fully aligned and connected spatiotemporally*

Mathematical modeling
  IV. *Seeking a geometric level of precision* in the spatial and physical relationships in the model
  V. *Developing a quantitative model* on the foundation of the improved qualitative and geometric models

14.1 Introduction to Chapters 14-16

present the arguments of historical figures. He therefore referred to it as a "rational reconstruction" of the history, and this allowed him to emphasize the major conceptual issues and make them more transparent. The composite protocol in this chapter is a "rational reconstruction" in the spirit of Lakatos (1978), but instead of historical arguments, it uses quotations from several subjects' think-aloud protocols and (primarily for the quantitative section) the author's written record of his own attempts to generate a more complete model for the spring.

The initial analysis of this composite protocol in this chapter will be a simple listing of the different types of reasoning used in the solution. The fuller analysis of the organization or orchestration of these reasoning strategies is more complex. This will be given in Chapter 16. It involves at least six different types of iterative cycles; many of these can call each other as subroutines; some are recursive in that they can call themselves; and a cycle can sometimes be interrupted by an "insight" that causes a sudden switch from the present process to one that has been put on hold. When first looking at a particular expert's raw transcript, many of these patterns are not easy to discriminate; they can be confused with or hidden "underneath" another process. This presents a huge challenge to the analyst. Add to this the ungrammatical nature of natural speech, and the complexity of the blind alleys that a subject goes down before making productive progress, and one has a formidable problem in analysis as well as a problem in communicating with one's readers. This is another reason for assembling a simplified, composite protocol from several sources. (Chapters 6 and 7 provide a more precisely naturalistic case study restricted to transcript data from a single subject.)

The core of the composite protocol is the development of a viable qualitative explanatory model of a mechanism for how the spring works. The protocol contains more ideas and fewer detours, fuzzy concepts, repetitions, and backtrackings than did the transcript from any single subject. However, it is quite realistic in the sense that each segment is centered on an actual transcript episode of expert reasoning, and it exposes many more of the "hidden" steps of qualitative reasoning, cross-checking, impasses, conjecture, and refutation than would be included in the polished version of a scientific paper.

The solution is written in the voice of a "dialogue within one person" or dialectic inquiry. Arguments which most clearly represent the various reasoning methods that were used to develop models of the spring system are linked together to make a coherent story of problem solving and discovery. Data sources for this dialogue are: (1) transcribed protocols from the ten subjects interviewed on the spring problem, as reported in Chapter 3; (2) the protocol of one additional expert subject interviewed after the original study; (3) handwritten records from the author's own work, over a period of many months, on modeling the spring problem and its extension into quantitative modeling (used only after episode 13 in preparing the development of a mathematical model). The author has training in physics but, like the other subjects, is not a mechanical engineer and did not have expertise specific to the spring problem*.

---

*Thus the essential breakthroughs to develop the foundational explanatory model were done by the external subjects, while the subsequent geometric/quantitative model was developed by the author.

14.1.2 Issues in the Field

As opposed to focusing on data-driven inductive processes for generating certain scientific laws, here I will argue that hypotheses can be formed by a less data-driven abductive process, leading to explanatory models rather than mathematical descriptions of observation patterns. Some previous approaches do recognize the presence of qualitative scientific models (Larkin, 1983; Tweney, 1985), but our knowledge about the connections between qualitative and quantitative models is rather weak. This is partly because there is an inadequate description of what it means to have a visualizable qualitative model as a foundation for a mathematical model. There is also too little appreciation of the need for intermediate stages of a model between the purely qualitative and the fully developed mathematical model. Several intermediate stages will be discussed in this chapter.

Forbus and Falkenhainer (1990) provided an important precedent in describing the design of an algorithm for integrating qualitative and mathematical models by specifying rules for going from qualitative models to certain numerical ones in a "math model library." Once equations have been selected, causal relations in the qualitative model can be used to select an order of computation for the equations. Some similarities will be seen in the analysis given here, but there are differences as well. The initial models are assembled from and grounded in qualitative runnable intuition schemas, as described in Chapter 13 (see also Schrager, 1990), and the initial model must be extended and elaborated in another stage to align science schemas imagistically, not just algebraically, within the model. The transcript then indicates the use of "imagery alignment analogies" to help in aligning geometric schemas visually with the model, leading to the development of a more precise geometric/spatial (Euclidean) model before the algebraic model can be developed. The grounding in runnable perceptual/motor intuition schemas, the intermediate steps involving visual representations, and transformations for imagistic alignment separate this from the processes used in the Forbus and Falkenhainer model.

Another purpose to be realized in Chapter 15 is to continue to identify episodes that help us illuminate the role of *imagistic simulation* in scientific thinking. I will highlight several such episodes and attempt to identify several types *of thought experiments* that use imagistic simulation and several examples of *imagery enhancement* techniques used in their solutions. To this end I have included additional data on hand motions in the present chapter in transcript excerpts where they occur. Previous discussions of thought experiments in science have highlighted their role as a way to test for inconsistencies in a scientific theory (Kuhn, 1977d); as a way to tap a priori knowledge (Koyre, 1968); as a way to access hands-on knowledge built up through experience in the laboratory (Gooding, 1992); or as a way to access prior knowledge of experiences in the everyday world through mental simulation (Nersessian, 1991). However, these are history of science studies without access to minute by minute processing. By contrast, the present study examines the specific roles that thought experiments (defined in Section 15.1.1 as imagining and giving a prediction for the behavior of an untested, concrete system) can play in model construction on the basis of evidence from think-aloud protocols.

14.1.3 Ways to Read this Chapter

Interpretive comments appearing in italics after each episode are brief preliminary hypotheses for the type of cognitive processes that can account for that section of transcript, such as "generates an extreme case." More detailed analyses appear in later chapters that use examples from an episode to focus on a particular cognitive process. In particular, claims about imagery and simulation use, and thought experiments are supported in more detail in Chapters 15 and 18 via connections to the imagery indicators developed in Chapter 12. Discussion of larger patterns, such as the solution stages shown in Table 14.1, are given in the discussion section at the end of the chapter. These solution stages are also reflected in the major section headings in the monologue below. *Verbatim transcript sections from subject interviews are indicated by quotation marks.* The first half of the protocol (episodes 1–13, 15a,b, and 17a) is based on excerpts from these interview subjects (referred to here as "external subjects"). In fact, all of the main ideas presented in episodes 1–13 are from external subjects; this covers the lion's share of the qualitative model construction sections. Since these subjects were not asked to give a quantitative solution, the second and more mathematical half of the protocol is based largely on the author's records of his own work on the problem (the "internal subject").

In the end, in a beautiful way, the symmetries in the spring system allow it to be seen to be equivalent to a much simpler system. It is hoped that by the end of this chapter the reader will not only be convinced *logically* of the truth of this equivalence but also that the reader will find that elements of the spring can be "seen *as*" this much simpler system by being able to visualize it as a hidden mechanism operating in the spring. That is, readers should be able to ground their understanding of the system in intuitions which they trust, thereby increasing their sense of conviction, understanding, and sense of having a satisfying explanation of how the spring works. Thus, one productive way to read this chapter is to engage the subproblems, do the suggested mental simulations, and then evaluate them as to whether they facilitate one's own understanding. I have also begun many episodes of the monologue with a question. Those readers interested in a challenge or who want to investigate their own reasoning processes, can attempt to answer these questions before reading further.

14.2 Composite Protocol Monologue for the Spring Problem

The monologue begins after the scientist has read the spring problem in Fig. 2.1.

14.2.1 I. Efforts to Develop an Initial Qualitative Description or Prediction for the Targeted Relationship

(Part 1) OK, to begin with, there is a well-known formula associated with springs: F = kx, where F is the force that you apply to a spring, x is the distance that the free end of the spring moves, and k is a constant of proportionality that depends on the properties of the spring. That is Hooke's Law. So my first question is, will that help me? (Part 2) That formula does not contain a variable for the width of the spring coils, however, "so the equation that says that the length [of the stretch] is related to the spring constant is not going to help me." If the width had any effect on the stretch it would be reflected in a change in the constant k, and since I do not know how k is calculated from the geometry of a spring, I am still no further along than when I started the problem. I need to think of some other approach.

[*Attempts to **apply established principle** above and fails.*]
(Part 3) "I guess...I'm going to try to visualize it to imagine what would happen – my guess would be that it would stretch more – this is not really visualization, it's kind of a kinesthetics sense that somehow a bigger spring is looser – Umm, that's high uncertainty."

[*This prediction appears to come from a **direct physical intuition** which in this case appears to be **kinesthetic**. Most subjects had a weak but direct intuition that the wide spring would stretch more.*]
(Part 4) "Here's a good idea. It occurs to me that a single coil (makes circular motion with index finger of l.h.) of a spring...is the same as a whole spring." It would be simpler just to think about a single coil.

[***Partitions** the spring into unit segments that repeat and then focuses on one unit.*]
(Part 5) "If we had a case where the second one went – had huge diameters compared to the first, it would appear to sag a lot more. It just feels like it would be a lot more spongy." That seems to help because the wider it is, the looser it feels. In fact if I had the right material like coat hanger wire, I could make a spring coil, say, as wide as a dining room table and it would be so flexible that it would be very easy to pull down compared to a coil that was 3 in. wide made of the same coat hanger wire. I find that much more convincing and my confidence level is becoming fairly high. So the larger spring will stretch further, unless I am overlooking some mistake I have made.

[*Generates **extreme case to enhance the use of a physical intuition in a simulation** to predict the direction of an effect; **also makes the simulation more concrete by using familiar material** to increase confidence*].
(Part 6) I could also imagine a very narrow spring.
"So the way to really eke out my intuitions would be to take the coiled spring in 1 down to an extremely tightly coiled spring. It's almost no distance from side to side of the spring. And obviously in that case it can't stretch very far." "if you...imagine shrinking the coils to a very small diameter, the wire would be practically straight and you could barely stretch it at all. There'd be no 'give' to it." So narrower springs should stretch less.

[*Another extreme case used to enhance use of a physical intuition in imagistic simulation. The subject's use of the term "eke out my intuition" supports the interpretation of "enhancing the use of physical intuition."*]

Parts of the following episodes will include some of the data already presented in Chapter 6, but I will analyze it in a different way.

14.2 Composite Protocol Monologue for the Spring Problem   243

(7a) "Now what if I recoiled the spring and made the spring twice as long – instead of twice as wide? Uhhhh – it seems to me pretty clear that the spring that's twice as long is going to stretch more. Now that's a – again, a kinesthetic intuition, but now I'm thinking – I'm imagining (raises hands 10 inches apart in front of chest, closes fingers as if holding something between hands, moves hands together slowly in 5–6 small movements while speaking until they almost touch just before the word 'origin') that one applies a force closer and closer to the origin of the spring, and again, as clo – as you get closer to the origin of the spring, it hardly stretches at all. Therefore, the further away you are along the spring, the more it stretches. So a spring that's twice as long, I'm now quite sure, stretches more.
Now if this is the same as a spring that's twice as wide, then that should stretch more – Uhhh, but is it the same as a spring that's twice as wide?" I'm not sure, therefore I guess I can not use this analogy right now.

[*Subject **generates an analogy** to a double length spring; then imagines a specially constructed **extreme case** and inspects the amount of effort required, in order to gain a **higher level of confidence in the behavior of the base of the analogy**. However, the subject cannot confirm the validity of the analogy relation to the original target problem, so the analogy is dropped.*]

(7b) "What else could I use that stretches, instead of a spring…would that be helpful? Like rubber bands…let's say…that doesn't suggest anything to me…."
"The problem might actually be simpler if you talked about…what would happen to a piece of foam rubber if you had foam rubber of the same amount that had small little cells, and suddenly you decided you'd deal with…cells in it that would be much bigger…If you have foam rubber with gigantic air cells in it it's going to be very squishy. So I guess I would feel fairly confident that it [the wider spring] is going to stretch more."

[*Generates **analogous cases**. These cases appear to be generated by association. In the case of the foam rubber analogy, the subject is able to make a prediction and infer a parallel result in the spring problem.*]

(8a) Is there another analogy that might help?
"I'm going to unroll these things [*the two springs*] and see if that helps my intuition any. Um, if I essentially, uh, uncoil or project the spring into a wire…the wire will actually go from here to here (draws horizontal line.) That's if I actually unroll the wire." That may be an interesting idea. "It occurs to me that a spring is nothing but a rod wound up, uh, and therefore maybe I could answer the question for a rod (Fig. 2.3)." "I have a strong intuition, a physical imagistic intuition that this [*rod a*] will bend a lot more than that [*rod b*] will…."

"I'm imagining what it's like to bend the rod (Places hands together in front as if gripping a horizontal rod and moves them very slightly)."

[*Generates **an analogous case** B (the rod), to help think about the original problem, A. In this case the analogy is generated by **transforming** (unwinding) the spring. The analogous case is conceptualized via a **physical intuition** about rods.*]

(8b) Is there any way to confirm that? "In fact, the intuition is confirmed by taking it to the limiting case. It becomes intuitively obvious to me that as one moves the weight closer and closer to the fulcrum [attachment point] that the thing will not bend at all, and that therefore, uh, there must be a continuing bend as one goes farther out." If that analogy is valid it indicates that the wider spring would stretch more.

[*Generates an extreme case whose apparent role is to enhance the original physical intuition for the dependence of bending on length of the rod*].

(8c) I now have six different arguments above all arguing that the wide spring stretches more and so I am starting to have considerable confidence in my prediction. But I am still not satisfied at all with my understanding – I have no explanation of how the spring works, that is, I have not really examined how the spring wire itself is deforming locally to produce stretching in the whole spring.

14.2.2 II. Searching for and Evaluating Initial, Qualitative, Explanatory Model Elements

(9) Here is one explanation. The rod analogy suggests that "the springiness of the spring – the real spring – is…a matter of its bendability." If the spring wire is bending, that would explain why it stretches.

Promotes bending idea from being a mere analogy of a bending rod to an **explanatory model** *of deformations that may actually occur in the spring wire.*

"But then it occurs to me that there's something clearly wrong with that metaphor, because…it [the spring wire] would (raises hands together in front of face) droop (moves r.h. to the right in a downward curve) like that, its slope (retraces curved path in air with l.h.) would steadily increase as you went away from the point of attachment whereas in a [*helical*] spring, the slope of the spiral is constant." And if that is correct it makes me wonder whether the rod is a good analogy.

"I'm imagining a diagram in my mind. This is the diagram in my mind [draws] – We have the spring swooping around like this if it's really bending downward, gradually bending, downward, there should be a curve – this is quite clear but I'm not quite sure how to express it in words." On the other hand, "the slope of the [real] spring is uniform. That is to say, if it's stretched to a certain degree and you go around the spring, uh, you're always sliding down a 20° slope say. But if it were bending – if you imagine a solid bar bent – the slope is constantly increasing as you go toward the lower end."

"You get a spring which stretches more and more at the bottom. The loops are wider apart there."

"But that isn't the case. You can imagine a spring just as well as I can, and you know that they're uniform all the way around. And of course, anyway, there's no difference between the top and the bottom. It's a symmetric situation." This contradiction creates an anomaly for the bending model that makes me worry about it.

[*Criticizes the validity of bending as an explanatory model for the spring as well as the analogy relation between the spring and the rod by (a) imagining the implications of* **running the bending model** *and (b) using prior knowledge about the appearance of springs to set up a* **conflict** *with those implications.*]

"I'm still bothered by the anomaly; I'm not satisfied with my understanding of the situation…. Uhh, I – I'm 90% confident [of the prediction wider stretches more]...because of the analogy with the straight rod – because I think there's something right about the analogy, even if there's something wrong about it. And because my physical intuition says the same thing...I have a certain...convergence of kinesthetics imagination and this model and the er [extreme] smaller coil model and for all those reasons, I'm...I'm inclined to feel fairly confident of the answer.... Without, however, feeling that I really understand what's going on...and I'm not very happy about that.... This bugs me, not to conceptualize it adequately."

(10a) Maybe the stretch has to do with shear forces rather than bending. In that model "(Draws a spring with elements descending in a helical staircase so that each element is slightly displaced vertically with respect to the next) the mechanism of force communication is this sort of shearing-like property (holds both hands side by side, palms open, oriented vertically and facing each other with thumbs up)... this picture of transverse – of communicated forces from one little block to a neighboring block is...the way in which force gets communicated up the coil (makes a wide tracing of a coil upwards with his right hand)."

[*Subject **generates** a microscopic **explanatory model** of vertical shear forces and displacements in the wire.*]

"But now the next thing I want to try to understand is, ok...given that mechanism, explain why then that the very loosely wound coil (makes a wide tracing of a large coil with his right hand) and the tightly wound coil have different elasticities....

The net displacement is proportional to how much relative motion (holds palms flat, face down, so index fingers are touching and slides one hand upward and one down) the communication of force...calls for...the winding (makes a tracing of a coil upwards with his right hand) it around is sort of a...(pause) I don't entirely believe myself....

[According to the shear model] I could take this thing, [one piece of wire] and I could wind it tightly, or I could wind it loosely, and I have the sense that if my model's correct, the rate that I wind it is gonna be...(pause). Oh, ok. Now I'm extremely unhappy.... The spring...is made out of...steel which is very stiff(motions as if hanging an object from his left hand).

The stiffness of the spring is just a function of its length – the length of the material – and not a function of how tightly I wind it because the forces I'm talking about are brought into play just between different pieces (holds palms flat, face down, so index fingers are touching and slides hands up and down in opposite directions)...that's what's bothering me – the contradiction...that I could wind this thing as tight as a spring...or as a very loose spring...and it would still have the same displacement for the same force, given that the length was the same." The shearing model predicts that stretch in a loosely coiled wire and tightly coiled version of the same wire should be the same. But this is counter to my intuition since I imagine that the loosely coiled wire would stretch much more. So I am going to give up on the shear model.

[*Here the comparison of a single piece of wire coiled loosely or tightly yields a counterargument for the shear model. This is not the same as the original problem because the wire length in this case is the same. This is a **Gedanken experiment** (a formal definition of Gedanken experiment is given in Section 15.1.1.) that pits* *results of running the shear model against strong intuitions about the behavior of loosely and tightly coiled material to produce a conflict.*]

(10b) Going back to the bending idea, "I'm visualizing a single coil of a spring.... I'll be damned if I see why it [the coil] should be any different from that case [the rod]."

[*Stubbornly reconsiders the bending rod analogy on the basis of a feeling that it is valuable*].

"What could the circularity [of the coil in contrast to the rod] do? Why should it matter? How would it change the way the force is transmitted from increment to increment of the spring? Aha! Now let me think about; Aha! Now this is interesting. I wonder, what if I start with a rod and bend it once, and then I bend it again."

[***Transforms*** *rod into new **analogous case***].

"What if I produce a series of successive approximations to the circle by producing a series of polygons! Maybe that would clarify because maybe that's constructing a continuous bridge, or sort of a continuous bridge, between the two cases [the rod and the coil]. Clearly there can't be a hell of a lot of difference between the circle and say, a hexagon." (Draws hexagon in Fig. 6.6a)

[*Polygons serve as **bridging analogies** ((see Chapter 4) between the circular coil and rod cases in order to help him evaluate the rod analogy.*]

(11a) Let us take a look at the hexagon.

"Surely springwise, that would behave pretty much like a circle does. Now that's interesting. Just looking at this it occurs to me that when force is applied here [arrow on hexagon in Fig. 6.6a], you not only get a bend on this segment, but because there's a pivot here (point x), you get a torsion [strain from twisting] effect. Aha! Maybe the behavior of the spring has something to do with twist (makes twisting motion with right hand) forces as well as bend forces. That's a real interesting idea. That might be the key difference between this (rod in Fig. 2.3) which involves no torsion forces, and this [hexagon]."

[*Insight episode: Discovers a new critical variable – the torsion effect in the hexagon. (See Appendix to Chapter 6 for primer on torsion and torque.) The hexagonal [polygonal] coil idea is generated as a bridging analogy in order to evaluate the previous bending rod analogy. But the new simulation appears to trigger a new schema – the idea that torsion may be involved. Thus a strategy used for analogy evaluation (bridging) is **interrupted by the recognition of an idea useful for the higher goal of model construction**.*

"Let me accentuate the torsion force by making a square (draws square in of Fig. 6.6b) where there's a right angle. I like that, a right angle. That unmixes the bend from the torsion."

[*Chooses square as **simpler case** to analyze, presumably in order to facilitate running the following **compound simulation** involving more than one idea or schema.*]

"Now I have two forces introducing a stretch."

[*Begins to analyze square.*]

"I have the force that bends this segment a (in Fig. 6.6b) and in addition I have a torsion (makes twisting motion with right hand) force which twists [side b] at vertex, um, x." It is as if side "a" were a wrench acting to twist side "b."

(11b) "Now let's assume that torsion and bend (makes bending motion with hands together) don't interact...does this (points to square) gain in slope – toward the bottom? ... So if you ignore the (traces in air over one side in drawing with pen) increase in slope ... which starts over again at each joint. Indeed, we have a structure here which does not have this increasing slope as you get to the bottom. It's only if one looks at the fine structure; the rod between the Y and the X, that one sees the flop (moves left hand horizontally in a downward curve) effect."

[*Square coil is a structure in which the cumulative increasing slope difficulty discussed in (9) does not occur.*]

(12a) What does the square model predict for the original question?

"Now making the sides longer certainly would make the [square] spring stretch more.... The longer the segment (holds hands up in front as if holding something between them) the more (makes bending motion with right hand) the bendability."

[*Prediction from* **kinesthetic physical intuition**.]

(12b) "Now the same thing would happen to the torsion I think, because if I have a longer (raises hands apart over table arriving at stationary position with the word 'rod') rod and I put a twist on it (moves right hand as if twisting something), it seems to me – again, physical intuition – that it will twist more. Uhh – I'm – I think I trust that intuition.... I'm (raises hands in same position as before and holds them there continuously until the next motion below)...imagining holding something that has a certain twistyness to it, a – and twisting it. Like a b-bar of metal or something like that. Uhh, and it just seems to me as though it would twist more."

"Again, now I am confirming (moves right hand slowly toward left hand) that by using this method of limits. As (moves right hand even more toward left hand until they almost touch at the word 'closer') I bring my hand up closer and closer to the original place where I hold it, I realize very clearly that it will get harder and harder to twist. So that confirms my intuition so I am quite confident of that. So, doubling the length of the sides, it will clearly stretch more. Both for reasons of torsion and for reasons of the segment."

[*Prediction for twisting rod analogy from* **kinesthetic physical intuition** *confirmed by using an* **extreme case** *to enhance the intuition. He transfers this result immediately to infer that the wider spring made of square coils will stretch further.*

*Engineers do in fact believe most of the restoring force in a spring is due to torsion. At this point the spring is seen as analogous to a single coil, which in turn is analogous to a square coil (as in Fig. 6.8d). The effects in the square coil are seen as resulting from those in a bending rod and a twisting rod. The schemas and imagery for bending and twisting are now "***protomodels***" for what is happening in the spring – they are the* **beginning elements of an explanatory model**.

At the more formal level of explicit relationships between named variables, this can be represented by the diagram notation (Driver, 1973; Clement, 1979) in Fig. 14.1. In this diagram an arrow going from the top of one bar to the top of the other indicates that "an increase in X causes an increase in Y" whereas an arrow going from the top of one bar to the bottom of the other would represent the conception that "an increase in X causes an decrease in Y." In this diagram an increase in coil width causes both an increase in bending and an increase in twisting, for example.

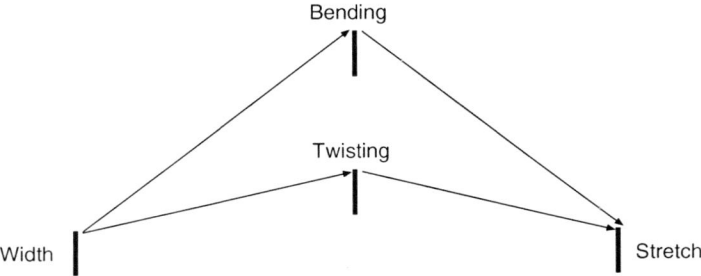

Fig. 14.1 Ordinal, direction-of-change relations between variables in the spring

*These in turn would cause an increase in stretch. These are **ordinal, direction-of-change relations** and are less precise than, say, proportional functions, but they are a step toward that **level of precision**.]*

(12c) "Before this torsion insight, my confidence in the answer was 95% but my confidence in my understanding of the situation was way way down, I felt that I did not really understand what was happening. Now my confidence in the answer is near 100% and my confidence in my understanding is like 80%."

[*The subject distinguishes between **confidence in the answer** and **confidence in understanding**, and indicates that the torsion analysis has increased his subjective feeling of understanding.*]

(13a) Is there any other way to see whether the twisting has to occur in a single loop of the spring?

"There's an implication of this which is sort of a way of testing it. You shouldn't be able to make a spring that behaves like the spring – it occurs to me – if the spring had no torsion, if the substance had no torsion, that is to say if it was a very twisty, twisty substance."

"Easy to twist, something made up of a chain of rings let us say like ball bearings, which twist against one another in perfect ease. You just couldn't make a spring out of it, it wouldn't matter if it was coiled or not, coiling would be irrelevant, really irrelevant...." I might have something like a series of cylinders in a chain which form the coil of the spring, a single coil, shown in Fig. 14.2. And the connection between each cylinder is a frictionless bearing which allows each cylinder to rotate freely along the common axis with the one next to it, but they have normal resistance to bending. Would that act like a spring?

[*This is an evaluative **Gedanken experiment** constructed in order to test the importance of torsion as a factor. If the new case **did** behave in the same way it would argue that torsion was **not** a necessary cause of restoring force in the spring.*]

I can imagine that such a spring would fall down to an almost straight wire! If the coil is held at A, and the cylinders were free to turn at A, then C could fall in an arc to a point directly below A and then E could fall in an arc to a point below C, and so on. And further straightening would occur as well. So this coil could just fall

Fig. 14.2 Evaluative Gedanken experiment: torsionless spring coil with frictionless bearings between joints would collapse

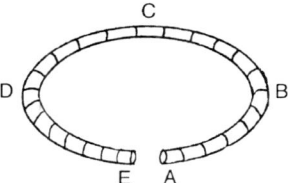

down into a crooked line. And that means the spring loses virtually all of its stiffness when it cannot sustain torsion. This supports the argument that torsion is involved and necessary for the spring to have a restoring force that resists stretching.

So in jumping to the bending-rod model early on, "that was my mistake at first off I thought coiling would be irrelevant – now I see that towards my insight, coiling isn't at all irrelevant because it's the coiling that introduces the torsion forces. It's the coiling that allows the mechanical device to take advantage of torsion, the straight spring doesn't take advantage of the torsion properties of the material, that's what's really nifty and clever about a coil spring. That's really interesting."

14.2.2.1 Gedanken Experiments and Imagery Enhancement

At this point in the protocol a piece of "simulated empirical data" was added to the problem that more or less confirms what has been discovered so far – that torsion is an important factor in the spring's mechanism – and that poses the question of how torsion could be more important than bending. Subjects who had sufficient time left in the interview read the following statement: "Measurements are taken on small segments of the spring below and it is found that the primary deformation in the segment is a twisting or torsion effect. How can stretching a spring twist the wire without bending it much at all?" This question initiates Part B of the Spring problem. Only material from part A of the Solution was included in the previous chapters; the present and subsequent chapters also include material from part B. Note that most of the qualitative investigation, including the discovery of torsion and certain thought experiments favoring it over bending, were conducted before this input, so that the new information is primarily confirmatory. Also, the input mentions only the presence of torsion. It gives no information on how torsion is created nor does it give geometric or quantitative factors, so that subsequent advances in these areas are unaffected.
(13b) Now how can I check on whether twisting could be the major source of stiffness or restoring force in the spring as opposed to bending?

"I have an idea about a spring made of something that can't bend. And if you showed that it still behaved like a spring you would be showing that the bend isn't the most important part, or isn't particularly relevant at all maybe somehow. I have those two images…. Now how could I imagine such a structure? Now this is an interesting idea. I'm thinking of something that's made of a band. Since it's cross section is like that (draws Fig. 14.3)

Fig. 14.3 Evaluative Gedanken experiment: spring made of vertically oriented band eliminates bending but still stretches

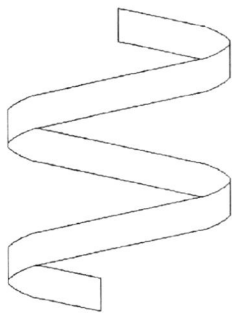

...it can't bend in the up-down (indicates up/down directions with hands) direction like that because it's too tall. But it can easily twist (motions as if twisting an object.... That could behave pretty much as an ordinary spring would.... I have an image of the wood shavings that peel off one of those charcoal pencils (makes motion like pulling paper off pencil).... Actually it's not wood, its paper...it makes a kind of (motions as if stretching a spring in front of self) spring with just this structure. And it's very springy (repeats motions)." Therefore the stiffness of the spring does not seem to depend on bending.

[*Evaluating the plausibility of the bending mechanism by eliminating bending in order to see whether the spring stretches considerably less. Note the imagery indicators in the form of imagery reports and depictive hand motions. This Gedanken experiment in the form of a bend-proof vertical band spring makes plausible the idea that vertical bending is not necessary as a mechanism operating in the spring!*]

(13c) "If we imagine a piece of paper...and imagine you hold (pretends to hold piece straight in front of self) it straight and you pull down (makes downward motions with one hand) on one end of it like that. There's practically no force at all resisting you pulling down on one end of it. Now if you imagine (motions with finger tip) it coiled into a spring – this is (holds hands out as if holding strip of paper) a thin strip of paper – you could imagine that it does have a certain (moves hand up and down) springiness to it. You can take a circle (draws spiral on paper) of paper and you can cut it like this.... And if you (moves hand upward from coil on paper) lift that up, you do get something that sort of dangles down and is (moves hand up and down) springy.... But, if you took an equivalent length of a straight paper it would just (makes downward motion with hand) flop.

And the argument there would be the same, that it has to be the torsion that makes it (moves hand up and down) springy because it sure isn't the bendiness that makes it springy because a piece of paper a yard long, like this would be, that was just straight – it would just flop."

14.2 Composite Protocol Monologue for the Spring Problem

[*Another attempt to design a **Gedanken experiment** that argues against bending as the primary mechanism for stretching in the spring. The subject's hand motions, imagery reports, and the projection into the problem of his own actions as the source of force provide evidence for the use of **imagery** and **imagistic simulation**.*]

(14) Suppose now that we take the spring and we take a single coil and stretch it all the way out into a straight wire hanging vertically. Can I see a twist in that straight wire?

[*Using an **extreme case to check on the existence of an effect**; in this case this also gives the **direction of the effect**.*]

Now to keep track of what is happening to the wire I will need some markers and a way to 'watch' for twisting. So I am going to imagine the coil made of a flat horizontal ribbon and little horizontal match sticks attached to the ribbon at the ends, as in Fig. 14.4a.

[***Imagery enhancement***: *Intentionally modifying the situation to make it easier to visualize a particular feature or variable. In this case that feature is the new hidden mechanism itself: twisting.*]

As I stretch it I will constrain the ends so that the match sticks end up still being parallel and horizontal in Fig. 14.4c. And by imagining stretching it I can see that the single stretched-out loop may produce some twist in the ribbon.

However, I still find it very difficult to imagine the exact number of turns of twist in the ribbon; I wonder if I could modify the experiment so that it would be easier to see? I'm finding that a half loop seems easier to "see" than a whole loop. Also a fairly wide flexible ribbon made of something that would stretch, say 3" wide with a half loop 12" in diameter is easier; in addition I can concentrate on the edges if I imagine a transparent ribbon with two dark black elastic strings forming the edges all along it as in Fig. 14.5a. As I imagine holding the ends horizontal and move one

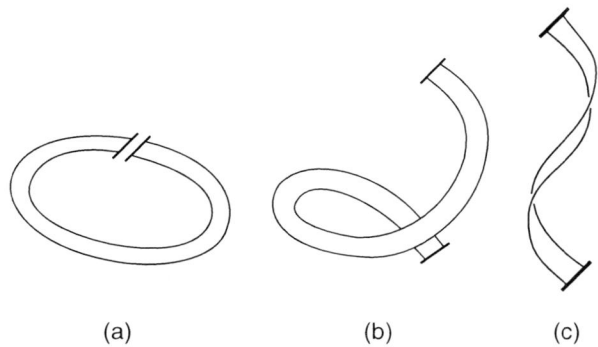

(a) (b) (c)

Fig. 14.4 Evaluative Gedanken experiment: horizontally oriented flat ribbon exposes twisting when stretched

Fig. 14.5 Evaluative Gedanken experiment: simplified ribbon for imagery enhancement

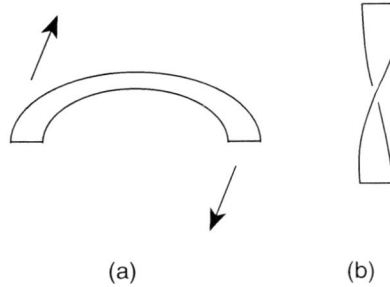

(a)   (b)

end down almost directly under the other end I can see that there will be a 180° twist in it, as in Fig. 14.5b. In fact the reverse action is easier: if I start with a vertical ribbon of this size with a half twist (of the right orientation) in it, I can imagine how by raising my lower hand so that it is beside the other hand, the ribbon will form a half loop. So I am beginning to feel quite confident that there is indeed a twisting effect in the circular spring as well as in the square spring and that it produces 360° of twist when completely stretched into a wire. This predicts that the maximum twist that can be introduced into a spring wire is one full 360° twist times the number of coils.

[*Further attempts at **imagery enhancement**. The choice of a half loop may be for simplification, but the 3" × 12" size choice with "dark strings on the edges" appears to be aimed at maximizing the clarity with which images can be inspected and interpreted, and this may be true of the "half loop" choice as well. This is compatible with the early finding of Kosslyn (1980) that there is an optimal range of sizes of objects in the visual field that people can imagine with ease.*]

(15a) "I'm going back to imagining a [normal] spring... I was thinking if you stretched it all the way out so that the kinks got all out of it – then you could say...'look where all that distance came from. It sure didn't come from the bending of it.' No, but, no, that's just quite unconvincing. That didn't seem to pan out conceptually...." But, "Now that's funny. The spring is a spiral...and if you straighten the spiral in the vertical direction, the spring doesn't have a curvature. All its curvature is sort of horizontal – or near horizontal, as the coils curl around." When the spring wire is almost stretched straight by pulling down hard on it, it looks like a slightly wavy wire. This makes me think that stretching is "unbending" rather than bending the spring. That is, stretching is *removing* curvature from the spring!

That may make sense because the original spring is curved, at least in the horizontal direction to make the loops of the spring. And if you stretch the spring out as much as possible into a wire, that curve has disappeared. Therefore there must have been some unbending going on to get rid of the curviness of the spring. The spring started out bent into a curve and if you stretch it into a wire it ends up "unbent."

[*Insight episode:* A new possible **mechanism**, **unbending**, **emerges from a thought experiment** *involving an extreme case. This is supported by a more indirect argument that a* **transformation** *(unbending) was required to get from a curved state to an uncurved state.*]

(15b) Therefore it appears that spring stretching involves at least two kinds of deformation: both twisting the wire and unbending it, which would both be sources of restoring force acting to restore the spring to its original shape. But in imagining this kinesthetically I see that the force required to do the unbending part is much higher toward the end of this operation of stretching the spring into a wire, not at the beginning. That is, when it is stretched nearer to the shape of a straight wire, it will take a lot of force to get any of the remaining curvature out. I see another thing happening in that the width of the coils (the horizontal diameter of the spring) has been reduced at that point. "The coils...get narrower, and that's where the length comes from...the diameter shrinks" at this stage. Returning to the case of modest stretching in a normal spring, my imagining it tells me that very little unbending is taking place. The circular coils are still mostly there. So it is probably the case that twisting is the dominant effect in normal use, where a spring is not stretched too much.

[*Conjecture about the* **shape of a function from imagistic simulation** *involving* **visual and kinesthetic imagery**. **Simulation used to estimate the relative size of** *two forces involved in stretching.*]

(15c) Imagining stretching still further raises the issue of whether the extreme case of the completely stretched spring as a straight wire will actually have elongated under tension into a longer wire than the original one. I imagine that for a not very stretched spring, this elongation effect would provide a negligible contribution to the stretch of the spring, because forces are being applied in a vertical direction and very little of those will act along the lines of the wire which is almost horizontal. In addition I have the kinesthetic sense that a straight wire is much easier to twist than it is to elongate (using a pliers in each case so that gripping is not a problem). So that effect can probably be ignored here and I will make the simplifying assumption that we are dealing with a wire that does not elongate.

[*Use of* **kinesthetic imagery** *to determine* **which effects are insignificant**.]

I now recognize three possible mechanisms for restoring force in the spring: torsion, tension, and 'unbending' of the curvature in the coils. The tension factor is clearly insignificant. My early intuition that bending was involved in the spring was correct, but not at all in the way I imagined. I had it oriented in the wrong way. The bending is in a different plane (horizontal) than I had imagined (vertical), my guess is that it is a much smaller factor than twisting in normal use, and it is unbending rather than bending!

(15d) By the way this talk of horizontal vs. vertical aspects is suggesting another argument against the idea of bending in the vertical plane. It makes me think about what would happen if I tipped the spring on its side and stretched it horizontally between my left and right hands, keeping my right hand stationary and moving my left hand away. By symmetry, there should be no reason for the coils to be closer together on the left or the right – implying no preferential direction for bending. Why should that change depending on which hand I move? This argument adds to my qualms about the original bending idea.

[*This is a **symmetry argument** for uniform stretching in the spring and for why it cannot bend in one direction throughout.*]

(16) Now I want to see if I can pass from a calculation of torque and twisting for the hexagon to one for the spring coil as the limiting case of a polygon with very many sides. I use Fig. 14.6 to think about the torque exerted by force F on segment w. This is equivalent to an identical force F′ exerted on a lever arm L that is exerting a torque on segment w. "What happens to the torsion in the limit...as one keeps making the segments smaller...yeah and the angle of the bends smaller – Ahhh – I'm just trying to imagine that if you make the segments smaller there's a shorter segment and the fulcrum it [lever L] makes – the distance it is from the axis of twist – gets smaller too, so there's less leverage; but then everything is getting smaller." Assuming an equal and opposite torque on the other end of W, I can calculate the twist in w and sum all of the twists to get a total twist in the coil. And if I write the expression for this and pass to the limit as the number of sides goes to infinity, I get zero for the total twist! (This calculation is left as an exercise for the interested reader.) Something is very wrong here; I do not know what it is; so I am going to regroup and go back to reexamine the original analogy between the square coil and the circular spring.

[*Subject makes a first attempt to **quantify the** torsion **model** and fails, because he has chosen the wrong lever arm, one that goes to zero length in the limit as the number of sides increases, failing at a qualitative level to **align** the torque schema correctly with the image of the model in the target. A sophisticated mathematical model has been built on a correct qualitative idea, but with an **incorrect imagistic alignment** to the target case. But this is useful in that it signals something is wrong with the analysis.*]

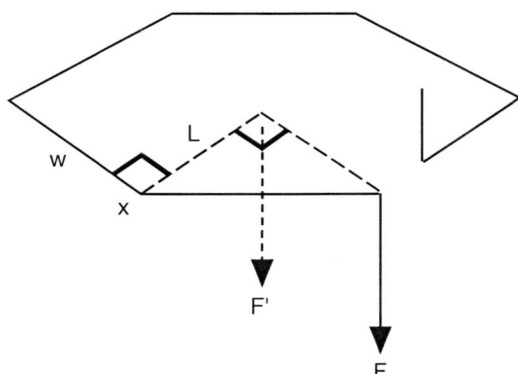

Fig. 14.6 Applying torque schema to hexagonal coil

14.2.3 III. Seeking a More Fully Imageable and Causally Connected (Integrated) Model: Attempts to Align and Elaborate the Model So as to Have Elements That Are Fully Connected Spatiotemporally

(17a) It has been shown that there is a torsion effect operating in the square. But is there really torsion in a circular spring? Is the square coil a valid analogy to the circular coil in this respect? Is there a way to **see** the source of the torsion effect or twisting effect in the circular spring? I have shown that there is clearly a source of twisting operating in the square coil but have not yet shown where it is in the circular coil. "The answer in principle is apparent from the square design. If I take a bar here and then a bar at right angles to it, call it A and B, it's easy to see how I could gain length downward so that this [side] A twists, and let's say...B is rigid; completely rigid, So if you push that [B] down, you (twisting motion w. r. hand over drawing of square coil) gain your distance downward by the twist in [side] A.

The actual round spring is just a case of that happening infinitesimally uh, all the way around the spring and all your distance down is gained by the kind (twisting motion with r. hand) of twist effect.... Let's...consider this (marks segment w in circular coil in Fig. 14.7a) an infinitesimal place where twists can occur.... And the pull that you could think of as twisting that is...uh, 90° around [at d] (makes twisting motion with r. hand)....as putting a (makes twisting motion with l. hand over drawn increment w) torsion on that increment of the spring."

[*Attempting to imagine the mechanism* operating in an actual circular spring coil. Transforms the system by "softening" part of it at w to form a close analogous case which makes the possibility of twisting effects obvious. Softening appears to

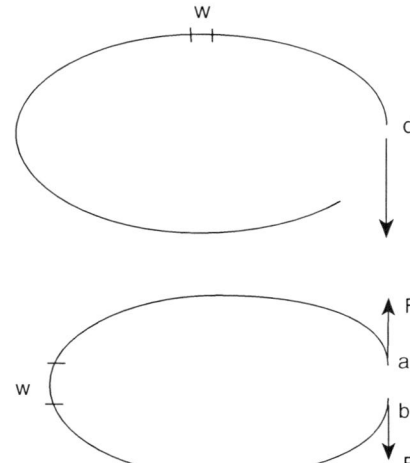

Fig. 14.7a Envisioning torque and twisting in circular coil with force applied at 90°

Fig. 14.7b Envisioning torque and twisting in circular coil with force applied at 180°

*enhance the ability to simulate a twisting effect. I will call this an **imagistic alignment analogy** because it allows one to align the torsion idea visually with the target and facilitates the process of improving the torsion model. The subject is for the first time able to align the point of application of force relative to an element in order to **actually envision how forces can create twisting** in a circular coil.*]

(17b) Pulling down at c would, I am now quite sure, clearly twist that piece of rubber at w. I can accentuate the twisting effect by imagining a spring made of a single coil as in Fig. 14.7b and imagining w were made of a soft material like rubber. Mounting wires are attached at each end of the coil. In this situation there is even more leverage for twisting segment w.

This gives an initial indication of how twisting is possible in the circular spring wire. But it does not show whether it is the *dominant* effect because there could still be a lot of bending between w and a and that would account for most of the displacement at a.

(18) Now I am looking back at Fig. 14.7b and I am worried because it looks so asymmetrical. It seems that there would be more twisting at w than at point a, where there is no twisting effect at all that I can see, and I am wondering, if I make it more of a symmetrical spring like most springs are in real life as in Fig. 14.8a, whether it would simplify by making the forces the same everywhere – that is, if forces are applied from the center, where they are equidistant from all parts of the loop? So in Fig. 14.8a the forces are applied to the end of these arms, and for simplicity I will think of the arms as being rigid and unbendable. And that actually helps me see the twisting effect because now this arm, m, looks like a handle or wrench for twisting the wire at a and similarly for n at point b.

[*Construction of a **symmetrical case** as a **simplifying modification**.*]

(19a) But now what about the cause of the twisting? How in Fig. 14.8a does the force up affect point c in the coil; how is that force transmitted? I am thinking of a

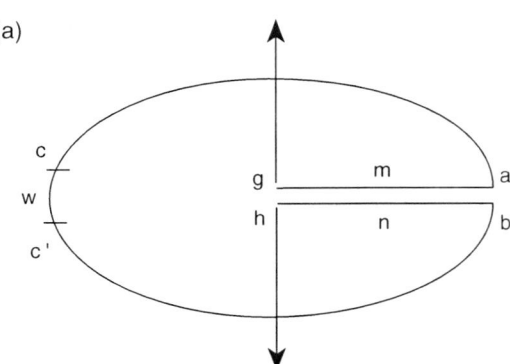

Fig. 14.8 (a) Forces on spring with "handles" exert torque on w; (b) equivalent lever arms for torque on w

14.2 Composite Protocol Monologue for the Spring Problem

situation where we have a twistable segment of the coil w, and I am thinking of the segment ac as being rigid. I am not going to allow it to bend or twist by making it out of extremely rigid material; I am just going to make a solid, curved, steel bar out of it. And similarly I will make c to a to g, and c′ to b to h rigid. I have now turned this wire into rigid handles for twisting segment w. And it is clear to me that those rigid handles in Fig. 14.8a are equivalent to the situation in Fig. 14.8b where I simply have two straight rigid levers operating on w in order to produce a torque [intuitively a "twisting force"] in w. That definitely helps me see how the forces applied at g and h can produce torsion in the wire at w.

[*Analogy via conserving transformation* to the simplest symmetrical case. ***Refining and simplifying the explanation*** *by introducing the concept of "handle" gac in Fig. 14.8a – as the device that transforms force to twist. This gives the length of the torque arm, gc. Making segment ac rigid may also serve as a **bridging case**. It lies between the case of the straight handles in Fig. 14.8b, which are well understood since it is very close to the form of a standard torque schema diagram, and the earlier view of ac made of flexible wire in Fig. 14.8a, where the transmission of torque is poorly understood.*]

(19b) And it looks like I can apply that argument to any short segment, w, of the coil even a segment near point a.

[***Generalizing*** *the specific finding from the analogy about w to all parts of the coil. The same preconditions for the argument are true anywhere on the coil.*]

(19c) Now the question is whether this argument will apply to a spring of more than one coil. So in Fig. 14.9a, I have 2 coils and I want to ask whether the segment w shown there experiences a torque. And I can use the same argument by thinking of sections hc and gc as rigid. Those sections could be replaced in terms of their effect on w by two rigid bars gw and hw as shown in Fig. 14.9b. And those in turn could be replaced by radial bars perpendicular to the forces as in Fig. 14.9c, and these would supply torque.

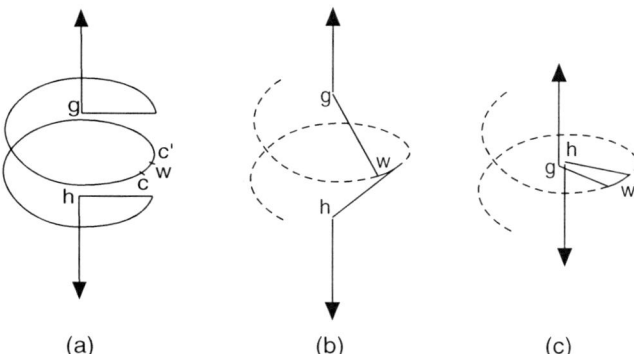

Fig. 14.9 (a) Extending the model to multiple coils; (b) analogous case proposed as equivalent to (a); (c) analogous case proposed as equivalent to (b)

258   14 The Use of Analogies, Imagery, and Thought Experiments

[*The investigator is trying to gradually* **expand the domain of situations to which he can apply the** "*rigid handles*" **analogy, thereby elevating it from a specific, analogous case to a more general model** *for springs.*]

(19d) Now I finally feel that I can see how forces at the end of the spring are translated into "twisting forces" (torques) on each element of the spring. And since as w twists in Fig. 14.8a, m and n will move apart from each other some, I can see that the resulting twisting deformation in each element is going to spread the coil around it apart a certain amount and add to the stretch of the spring by moving m and n further apart. So I feel I have a model here for how the spring "works" in terms of seeing how force causes twisting, and seeing how twisting causes stretching. This model is also much more compatible with the idea that the coils are equally spaced apart everywhere than was the bending model.

[*Investigator now has a* **fully imageable and causally integrated qualitative model** *with elements that are* **fully connected spatiotemporally**.]

I can now see a possible reason for why my calculation in episode 16 failed. Although the idea of torque and twisting occurring is correct, the force being exerted to put torque on a segment via the lever arm should be acting from the center of the coil, not the adjacent segment. I will try to develop this approach carefully in what follows.

14.2.4   *IV. Increasing the Geometric Level of Precision of the Spatial and Physical Relationships Projected from the Model into the Target Until They Are Ready to Support Quantitative Predictions*

(20) Now I can ask whether the torque on w in Fig. 14.9a is equal to the torque in Fig. 14.8a and 14.8b? And following the same argument as in episodes 19b and c above, at each step the same amount of torque is produced, and I have convinced myself that even for a long spring of many coils, the torque on any segment w is equal to the torque in Fig. 14.8b.

[*Retraces the argument in 19b and c at a level of precision where the effective lever arm segments are geometrically equal in length (and therefore torques are equal).*]

(21a) Now I need to define a variable for stretch per unit coil. After stretching, a and b will move apart as shown in Fig. 14.10 where rod m is thought of as being directly over rod n in the same vertical plane. And this produces a gap between the ends a and b of the wire, and that is the amount of the stretch S from that coil, assuming the end rods are held horizontal.

[**Defining the variable** "*stretch per coil*" *with quantitative precision.*]

(21b) And because of the equal torques everywhere the stretch in each coil should be equal. Therefore my evaluation of my level of understanding is increasing since

Fig. 14.10 Forces on coil produce a stretch s

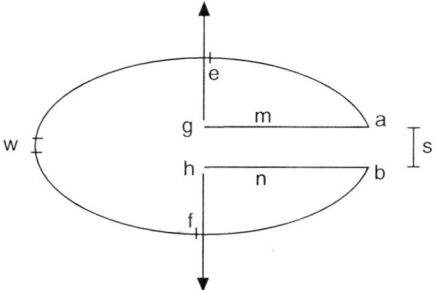

this model explains the feature that the coils are equally spaced apart everywhere and that implies that the slope in the stretched spring is the same everywhere.

(22a) But how do I know that the end bars stay horizontal during stretching? One argument is to imagine a horizontal matchstick glued perpendicular to the exact midpoint of the spring wire in segment w in Fig. 14.8a. By symmetry, during normal stretching it should remain horizontal. (This intuition is aided by tipping the coil 90° and stretching it in a horizontal direction.) This should be true for the midpoint of each half of the spring in Fig. 14.8a as well. I can repeat the argument in successive halvings for intermediate points in smaller and smaller segments. This convinces me that the stick will be horizontal at any segment I would care to choose within the helix of the spring.

[*A **symmetry argument** supports a new feature detail; **repeating the argument** recursively **generalizes** it to other parts of the spring.*]

(22b) And I see another argument for why the bars in Fig. 14.10 should stay horizontal during stretching. If I divide the coil into quarters as shown there, with an upward force at g, the deformation of all of the small individual twisting elements in the quarter coil wE will be to make end a of the bar tilt upward relative to g (considering gaE as temporarily rigid). Whereas the deformation of the matching elements in quarter EA (when EA is relaxed to be nonrigid) will act to make end g tilt upward by the same angle. Therefore the bar should remain horizontal. A similar argument can be made for bar n.

(22c) However this reminds me that it is still counterintuitive to me that the wire in Fig. 14.10 is twisting at all, because if rods m and n are still parallel, how can the wire be twisted? If they are acting like "wrenches" on the wire, it seems they should turn like a wrench.

[***Criticizing the mechanism*** *via a possible reason for no twisting, from features in Fig. 14.10.*]

But in my previous argument about the flat ribbon coil being stretched into a straight ribbon in Fig. 14.4c, I recall that even though the match sticks at the ends remain horizontal and in parallel in the straight vertical ribbon, there will still be a full twist in that ribbon. This helps reassure me that there should be twisting in the wire in Fig. 14.10 as well.

14.2.5 V. Developing a Quantitative Model on the Foundation of the New Qualitative and Geometric Models

In this section a quantitative function for how stretch is related to radius is developed by applying established principles such as trigonometric relations and the torque version of Hooke's law (twisting being directly proportional to torque). However these arguments utilize and rely heavily on the imageable analogies and qualitative models for analyzing deformation in the spring that were developed earlier. Readers wishing to skip these quantitative considerations can jump to the discussion at the end of the monologue.

(23) I am leaning more toward Fig. 14.11 as a good model of the spring. Because even though section ac in Fig. 14.8a may twist in a real coil, the net torsion at c on w will still be determined by the force down and up at h and g. And the distance from g to any segment w′ in the coil is this same radial distance and therefore it looks like a segment taken at any point in the coil would experience the same torsion. I am still hoping, then, that I can string the w's together in a straight line and use Fig. 14.11 to calculate my answer for torque in the spring.

(24) I wonder if these torsion effects I have examined predict a direct proportion relationship between the width of the coil and the amount of stretch you get for a given force. Let me think about the angular twist in segment w in Fig. 14.9c. I will imagine g to be fixed. Then the force at h will produce a torque on w in proportion to the radius of the coil and according to standard principles of elastic deformation the twist produced will be proportional to the torque.

[*Applying the quantitative aspect of standard torque and elasticity schemas. The supporting causal direction-of-change relationships so far are symbolized by the horizontal chain in the middle of Fig. 14.12, plus the first relationship for the radius affecting torque. (Other equations shown at the bottom of the figure will be added later.)*]

(25) But if one doubles the radius, it is as if we doubled the size of everything in Fig. 14.9c. Therefore the amount of wire to be twisted, w, will also double. I have already established that the longer segment w is, the easier it is to twist the near end of w through a certain angle θ. I am imagining doubling w as two rods of length w attached end to end. The two rods will twist the same amount under the same torque, and when we add them together, we get a doubling of the twist. Therefore doubling the length gives you twice the twist and the "twistability" is proportional to the length of w. We now have a second linear dependency on the radius of the spring r and it looks like the stretch is at least proportional to r squared.

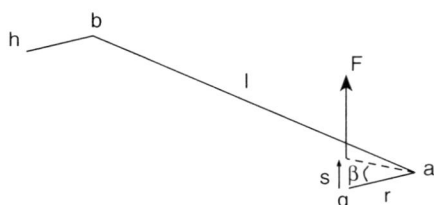

Fig. 14.11 Simple torsion case equivalent to Fig. 14.8a

14.2 Composite Protocol Monologue for the Spring Problem

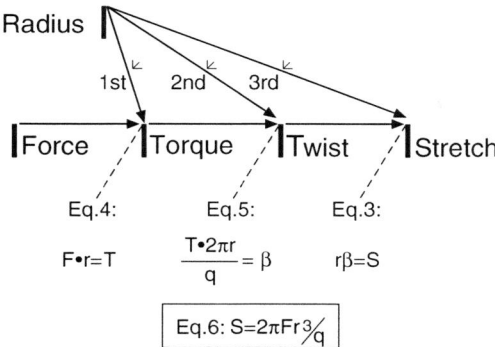

Fig. 14.12 Qualitative causal relations and quantitative relations in spring model

[*Causal and proportional reasoning;* **determining mathematical linearity based on an imagistic simulation**. *This simulation represents the spatial arrangement of the two rods in series and apparently indicates* **by symmetry** *that the torque will be "felt" equally strongly by both rods, and* **by spatial reasoning** *that the twists of the two rods will add together directly.*]

(26) And I see another dependency: as I imagine generating a certain twist angle θ in w, I see that the stretch contribution L in Fig. 14.8b will be the length swept out by the end of the rod h as it moves down caused by the twisting of w. And for small angles, the vertical drop of the end of the rod h is going to be proportional to the radius r of the sweep. [Applying a geometric schema; Determining the stretch that results from a given twist in w.]

Therefore the bigger the radius is, the more stretch is swept out by h, and that is another factor proportional to r which contributes to the stretch, shown as relation 3 in Fig. 14.12. So according to this model the stretch L should depend on r cubed! That would mean the double width spring would stretch eight times as far!

(27) So it does appear that Fig. 14.8b is a good model and that each segment w of the spring behaves as if it were a straight rod with two lever arms applying a torque on each end that are one radius long, and if you add up the little twists produced by that torsion in each w, you should get the total twisting effect.

And since it should not matter how you line those w's up, you could line them up in a line, as in Fig. 14.11, and just apply the forces perpendicular to the ends. And that should give you the same amount of twist. For the same arguments given for the coil above, if one doubled the size of everything in that figure, the distance (for small displacements) the handle at one end would go up with the cube (eight times). That is if you held the other end fixed and used the same force, and for small angles only. That makes me feel somewhat more comfortable with the R cubed relationship.

[*This analogy appears related to the "partitioning and reassembly" heuristic discussed in Chapter 11 for volume problems, except that here we have* **"dynamic partitioning and reassembly"** *of elements and their motions.*]

(28) But to write an equation I am still unsure about how that total twisting deformation in the wire would translate into a certain amount of vertical stretching of the spring. The next challenge is to see if I can determine how much stretch S will be caused by a certain twist in the wire in Fig. 14.10. I prefer here to reverse the causal direction for ease of thinking temporarily to determine the twist caused by a certain stretch. One might think that a stretch of s produces angle gwh in Fig. 14.10 very directly as the amount of twisting in w. However there are questions – perhaps awb is instead the appropriate angle of twist, if one thinks of the twisting in w being caused by the separation of wire in the coil at the opposite side from w. I have to find a way to imagine this more carefully.

I think the best way to try to calculate it is by imagining causing the twisting in segment w in Fig. 14.13a in a very physical way one step at a time. And let me start by tilting the half-loop from a to w up around w, and the half-loop from b to w down as if everything in this coil were absolutely unbendable and untwistable except section w which is made of some kind of elastic material. My idea is that this "opening of the clamshell" produces a certain twist angle θ in w and that I can then unrigidify aw and bw, and given the symmetry of the spring, with equal torques being applied everywhere, the twist stored up in w would now distribute itself throughout the whole coil with only the bars m and n remaining rigid. And I would still have that same net twist angle θ, but in the whole coil.

[*Extending previous qualitative analogy to a quantitative argument. This analogy allows the alignment of a mathematical schema for angular displacement to the qualitative models already developed*. Notice how important the qualitative models are to the analysis. These include the single coil, the force causing

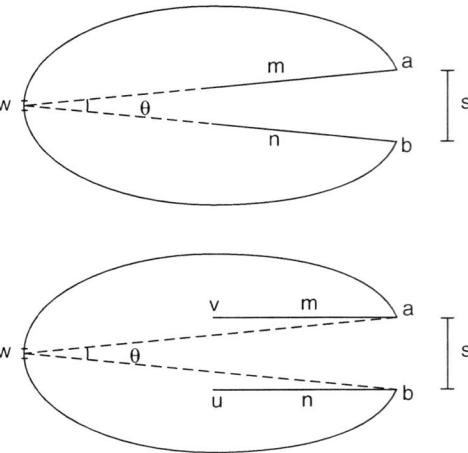

Fig. 14.13 Determining the relationship between twisting and stretching: (a) coil opens like clamshell; (b) "handles" are returned to parallel positions

torsion causing twisting causing stretch model, as well as the analogy of the "rigid handles," and the strategy of partitioning into small segments "w."]

(29) But wait – I have previously argued that the endbars would remain horizontal. I am going to have to tilt the bars in Fig. 14.13a so that they become horizontal as shown in Fig. 14.13b. By imagining holding a and b fixed in Fig. 14.13a and tilting the upper bar to the horizontal with my hands, I can see that sweeping m up is going to transmit more twist to the wire aw – and in the same direction that w was twisted by moving point a up in Fig. 14.13a.

[**Breaking a larger change into an equivalent sequence** *of two smaller changes so they can be aligned to and assimilated by a standard mathematical schema. Using dynamic imagery to get the correct sign of the addition]. Again the subject appears to be "partitioning the dynamics" of the system – that is, breaking trajectories down into pieces that can be simulated accurately by using or adapting available schemas.*]

I see that sweeping the rods to make them horizontal in Fig. 14.13b adds an additional 1/2 θ apiece, so that the net twist angle β in the coil should be 2 θ.

$$\beta = 2\theta \tag{1}$$

and that is satisfying because it resolves the anomaly raised in episode 22c about how there can be any twist in the wire when the "handles" remain parallel.

So in a sense, stretching does make the handles sweep through an angle as they twist the wire; but instead of making the top handle angle upward, the upward force on it has the effect of simply canceling out exactly the downward tilt of the handle that should be caused by the coils spreading apart! There is more symmetry built into this everyday object than I realized!

[*Analysis has exposed more ordinarily hidden mechanisms at work which underlie observable behavior. These cohere with some previously identified features. From this point on, the solution passes into the realm of mathematical equation manipulations, bringing in a new layer of algebraic, trigonometric, and other processes, the cognitive analysis of which is beyond the scope of this chapter. However, I wish to carry the solution to its conclusion.*]

(30) And now I want to relate the total twist angle β in a single coil to the stretch S, as depicted in Fig. 14.13b, taking D as the diameter of the coil,

which in a small angle approximation is:

$$S = D\,\theta \tag{2}$$

[*Writing expressions for geometric constraint relations; use of a standardized approximation scheme.*]

Since $\beta = 2\theta$ and D = 2r, this may be written:

$$S = r\beta \tag{3}$$

This equation tells me how much stretch one sees in the coil for a given twist in the wire. That is very interesting, because it looks like the same answer one would get in the case of the "uncoiled spring" in Fig. 14.11. That is, the distance swept out as g moves up, the arc length S, is equal to the radius times the twist angle β. The equation is the same as Equation 3 above, therefore it looks like the stretch of the spring can be computed after all by thinking of the distance S' swept out by point g in Fig. 14.11 where the rod ab, which is the same length l as the wire in the spring, is twisted through an angle β! And the restoring forces should be the same if all my assumptions are correct.

(31) How can we examine the issue of an R cubed relationship? For the first two equations, we use the twisting rod analogy in Fig. 14.11. The torque on the rod is simply the force F times the length of the lever arm, r (Equation 4). The total twist angle β in the rod is the torque times the length of the rod ($2\pi r$), divided by a constant for the torsional stiffness constant q of the material (Equation 5).

$$t = Fr \tag{4}$$

$$\beta = T(2\pi r)/q \tag{5}$$

Finally, the total stretch of the coil is given by Equation 3 above. Combining these three Equations 3, 4, and 5, the total stretch S of each coil is given by the equation:

$$S = 2\pi Fr^3/q \tag{6}$$

where q is a constant representing the stiffness of the material. Thus the cubic relationship can be derived from the twisting rod analogy, once we are convinced that that analogy is valid.

[*We can create a representation of the subject's conceptions, by adding some detailed mathematical relations to Fig. 14.12 that show the causal relationships in the length of wire in a circular coil, and they have reached a mathematical level of detail. The figure illustrates the sense in which a mathematical representation can be built on the foundation of a qualitative-ordinal representation.*]

(32) Can I tell whether the equation makes sense? Certainly I can imagine that a spring of very small radius made of coat hanger wire would be hard to stretch, and that fits the prediction of the equation as r gets very small. And the same would be true if the stiffness were very large. So the equation is in the right ballpark according to those tests.

[*Uses **thought experiments** in the form of **extreme cases** to test the equation.*]

(33) It would be interesting to know whether unbending plays an important role in addition to torsion, as argued in episode 15a. My sense from imagining stretching a single coil of a spring is that the curvature of the wire does not change much at

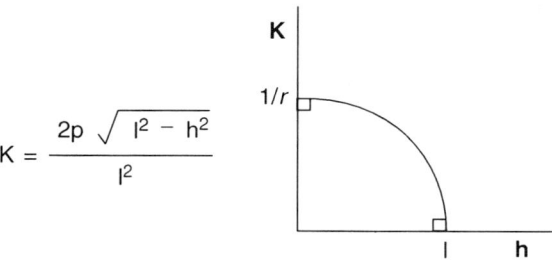

Fig. 14.14 Curvature of a spring coil as a function of stretch

the beginning of the stretch but that it changes a lot during the last third of the stretch as one moves toward stretching it into a straight wire.

[*Estimating the magnitude of an effect from an imagistic simulation*]

Perhaps I could calculate the curvature in the coil as a function of stretch. And now I have looked up a formula for the curvature K of a helix:
(See Fig. 14.14)
where l is the length of wire and h is the distance between coils (stretch). The graph of curvature vs. distance stretched in Fig. 14.14 is a circle and shows a very small slope, that is a very small amount of "unbending" per unit stretch, when the wire is near the shape of a ring, and a large slope or amount of unbending per unit stretch when the wire is nearly straight. Therefore most of the unbending occurs during the last part of the stretching process. Since the coils in normal springs stay very close to the shape of a ring, this says that unbending should play a minor role in comparison with torsion for normal stretching. Ultimately to be sure of this I would have to know the coefficients for (un)bending and torsion and calculate the amounts of force produced by each.

(34) Finally! I have generated a theoretical model for springs which predicts that the primary mechanism producing a restoring force is twisting in the spring wire. Further, it predicts correctly that stretch varies with the cube of coil diameter, and the amount of stretch is predicted by S in the analogy of the "unwound spring" (using the same length of wire) shown in Fig. 14.11. The symmetry and simplicity of the final model is remarkable. Equation 6 is now understandable and even compelling to me.

14.3 Stages in the Solution

14.3.1 Some Possible Precision Levels for Relationship R Between X and Y

The arguments exhibited in this chapter illustrate the power of nonformal reasoning strategies in analyzing unfamiliar systems. The entire solution can be seen as an effort in five stages (shown in Table 14.1) to understand, in more and more detail, the mechanisms involved in converting force to stretch in the spring. The overall pattern of the solution can be described as follows: after initial attempts at direct qualitative predictions, the protocol gradually develops an explanatory model for the spring, beginning from the rough analogy of a bending rod. This analogy is criticized repeatedly as a model and modified until a model of twisting deformations in each small element of the spring is developed. Further creative transformations are performed in order to identify constraints at a quantitative level of precision, allowing physical principles to be applied and leading finally to a mathematical function relating coil width and stretch.

However, rather than simply having a dichotomy of a qualitative level and a quantitative level, there appear to be several possible levels of precision for describing the causal relationships involved, as shown in Table 14.2. The description of each relation moves up this hierarchy, and eventually reaches the most precise level of quantitative functions. In order to accomplish this, the solution follows a pattern of development shown in Table 14.1. An index for the episodes in the solution is given in Table 14.3.

14.3.1.1 Summary of Each Stage

We can summarize episodes in the solution which fit the five stages in Table 14.1 as follows.

14.3.1.2 Qualitative Modeling

I. Efforts to Develop an Initial Qualitative Description or Prediction for the Targeted Relationship (Episodes 1–8c). Early in the solution a rough, direct simulation of the original targeted relationship between coil diameter and stretch led to a weak prediction that larger diameter causes greater stretching. Extreme cases of very narrow and wide coils were then used to enhance and increase confidence in this prediction. This prediction was at a fairly low level of precision, at the "direction of the effect" level, in Table 14.2.

Table 14.2 Some possible precision levels for relationship R between X and Y

1. Existence of effect: X affects Y?
2. Ordinal direction of effect: If X increases, does Y increase?
3. Complete (topological) mechanical connectedness: can envision complete mechanical chain from X to Y?
4. More detailed, pre-quantitative (e.g. geometric) features and equivalencies; e.g. geometric schema such as equal angles in spatial alignment with model image for how X is connected to Y
5a. General category of mathematical function: Is Y proportional to X? Additive? etc.
5b. Detailed mathematical function or procedure: Is Y a particular function of X?

Table 14.3 Index for the episodes in the composite solution

Episodes

I. Efforts to develop an initial qualitative description or prediction for the targeted relationship
1. F = kx
2. Critique F = kx
3. Direct visualization of spring
4. Single coil
5. Very wide coil
6. Very narrow coil
7a. 2x-long coil
7b. Rubber bands, foam analogies
8a. Unroll to wire, rod
8b. Moving closer to fulcrum extreme case for rod
8c. Evaluate confidence

II. Searching for and Evaluating Initial, Qualitative, Explanatory Model Elements
9. Model of spring elements bending, generates conflict
10a. Shear model; discounted by wide vs. narrow coil Gedanken
10b. Single coil, polygons bridge
11a. Hexagonal coil, torsion insight, square coil
11b. Uniform slope in square
12a. Predicts wider spring stretches more for bend
12b. And for torsion too
12c. Confidence in the answer vs. confidence in understanding
13a. Torsionless coil
   Addition of confirmatory simulated data indicating detection of torsion in spring wire
13b. Bend-proof vertical band spring Gedanken
13c. Horizontal paper experiment
14. Stretched ribbon
15a. Unbending mechanism
15b. Width of spring is reduced with stretching; unbending negligible
15c. Tension mechanism
15d. Symmetry argument for lack of increased distance between coils
16. Troubling zero torque result for limit calculation using adjacent segments

III. Seeking a fully [imageable and] connected (integrated) causal model: attempts to elaborate and align the model so as to have elements that are fully connected spatiotemporally
17a. Soften segment and imagine torsion in quarter coil of circular spring
17b. Imagine torsion in half coil
18. Insure force operates at center of spring
19a. Transform target to simpler case; soft with rigid handles
19b. Generalize argument to all elements in coil
19c. Generalize to more than one coil
19d. How twisting causes stretch

IV. Increasing the geometric level of precision of the spatial and physical relationships projected from the model into the target until they are ready to support quantitative predictions
20. Torques equal everywhere generalize to multiple coils
21a. Define variable of stretch per coil quantitatively
21b. Same torque everywhere
22a. Matchstick symmetry arguments show each element stays horizontal
22b. Quarter coils yield counteracting tilts for endbars
22c. Critique twisting level 1; ribbon argument

V. Developing a quantitative model on the foundation of the new qualitative models
23. Leaning to Fig. 14.11
24. Apply physical/mathematical torque schema
25. "Twistability" joining two rods experiment
26. Sweep arc length: apply mathematical schema, R cubed relationship
27. Add twists to get total twist; equivalent to straight rod twisting; partition and reassembly
28. Stretch from twist; clam transform; apply arc length math schema; assume twist distributes through whole coil
29. Endbar correction; partitioning the motion; reinterpret horizontal endbars as resulting from canceling sweeps
    Begin equation manipulations
30. Geometric constraint equation
31. Calculate R cubed quantitative function
32. Extreme cases to check equation
33. Curvature formula for unbending
34. Final model equivalent to twisting straight wire of same length

II. Searching for and Evaluating Initial, Qualitative, Explanatory Model Elements (Episodes 9–15). The goal shifts to become deeper: constructing a model of the hidden deformation operating in the spring wire to explain stretching. At first the model of bending is proposed. After this is criticized, a model that includes *twisting* in the wire, is introduced in episode 11. More possible mechanisms are generated and evaluated for their possible role in explaining stretching. After several thought experiments, twisting emerges as the most likely model. This is a major achievement. New causal mechanism elements and variables have been discovered in the system. The subject has overcome an initial Einstellung effect which tied the subject to an inferior model. Support for the twisting model has been recruited from several independent arguments. Other models have been eliminated by running thought experiments. Evidence for the involvement of imagistic simulations in thought experiments that feed this process will be examined in Chapter 15. A premature attempt to proceed to a mathematical level of modeling at the end of this stage fails, so the subject returns to qualitative modeling to improve the foundation in the next stage.

III. Seeking an Imageable, Causal Mechanism with Elements that are Fully Connected Spatiotemporally. Attempts to add enough detail to the model to have elements that are fully connected spatio-temporally (episodes 17–19).

As impressive as the breakthroughs in stage two were, at the end of that stage the subject believes that twisting probably plays some sort of causal role in the helical spring, but exactly how this happens to increments of the spring is not known. Thus the subject's model consists of certain *mechanism characteristics or model elements, but these are not visually aligned with the spring or fully connected together*. In order to make further progress and advance to a new level of precision in modeling, it is necessary to take twisting seriously as a mechanism that is occurring in elements of the wire, and to understand what is causing that twisting and how it deforms the circular coil. The third stage in Table 14.1 commences when the subject appears to project into an element of the circular spring coil an image of how force causes twisting and how twisting causes stretching (cf. "seeing as"). This is extended until its occurrence in the entire spring can be envisioned. The final result of this additional qualitative modeling process is shown in Fig. 14.15, where the anchoring conception of twisting and torsion is used to build up a fully connected explanatory model of twisting in small elements of the circular coil. (It will be noted that earlier diagrams have placed the target on the left and the base of the analogy on the right consistent with an earlier emphasis on generation and the order of appearance, whereas the base appears on the left in this diagram. This shift will in fact be followed throughout the rest of the book to emphasize showing the information flow from the source on the left to a model or target on the right.)

14.3.1.3 Mathematical Modeling

IV. Seeking a Geometric Level of Precision in the Spatial and Physical Relationships in the Model. These are projected from the model into the target (e.g. equal segments, angles, areas, etc.) until they are ready to support quantitative predictions (episodes 20–22). A higher level of precision is then asked for in episode 20 as

14.3 Stages in the Solution

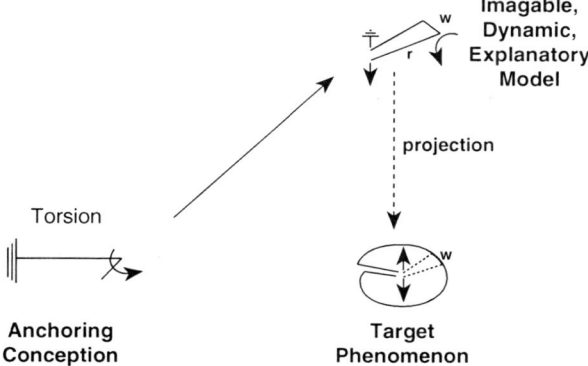

Fig. 14.15 Source schema for torsion is aligned with spring via an incremental model of torsion in the spring

the precision of the twisting model is pushed toward a level that can support mathematical description. Analogous cases are used to argue that the torque is uniform everywhere. Another question is: What are more exact geometric constraints on the spring as it stretches? Here increasingly precise details are worked out about the effect of stretching and twisting on the shape of the spring until the model is ready to support quantitative inferences. No new causes are identified, so this stage seems primarily one of increasing precision to a fully geometric level.

V. Developing a Quantitative Model Built on the Foundation of the New Qualitative and Geometric Models (Episodes 23–34). After all of the above issues are resolved it is finally possible to construct detailed quantitative functions in episodes 23–34 by applying established principles such as trigonometric relations and the torque version of Hooke's law (twisting being directly proportional to torque). These arguments utilize and rely heavily on the imageable models for analyzing deformation in the spring that were developed at a qualitative level. These enable a transition to algebraic notations for representing the geometric and physical relationships involved, and these allow the use of standard algebraic problem-solving techniques to solve for the function relating coil width and stretch. Again, no new causes are identified, so this stage seems primarily one of increasing precision to a quantitative level.

14.3.2 Transforms to "Close" Analogies in Later Stages of Solution

This chapter described the use of analogy and extreme case transformations in the later stages of quantitative modeling, to find equivalent systems that can be conveniently analyzed using standard mathematical schemata. Chapter 11 catalogues a variety of such transformations for calculating volumes of odd-shaped forms. In the present chapter various transformations are aimed at calculating angles and torques conveniently. These appear to be expedient analogies realized via conserving transformations, such as "softening" a part of the coil so that all of the twisting is concentrated there; unlike the original qualitative analogies to the rod and square coil, the transformations for quantitative calculations must be conserving at a much finer level of precision – e.g. the magnitude of radius arms, force, torque, and angles must be conserved. The subject is no longer concerned about constructing or adding basic mechanisms for explaining how the spring can stretch and therefore does not need divergent new anchoring analogies for developing a qualitative explanatory model. Rather the concern is for increased precision in describing geometric and quantitative details about the twisting mechanism. Therefore these "close" expedient analogies generated by conserving transformations become very valuable, even though they are not so divergent, when used in the later, mathematical stages of this solution.

14.3.3 Summary

In contrast to a simple dichotomy of qualitative vs. quantitative work in this solution, Table 14.1 identifies five different stages that start from rough, qualitative ideas, and progress gradually toward much more precise quantitative models. A key step in this process is the formation of an explanatory model of hidden deformations in the target. I expand on this idea below.

14.4 Building a Theoretical Distinction: Explanatory Models vs. Expedient Analogies

In this section I use the new imagery constructs developed in Chapters 12 and 13 to begin to develop a theory of the relationship between analogies and models on the basis of the protocol in this chapter. In Chapter 6, I cited the pioneering work of Campbell (1920) who articulated the role of analogy in science using the example of the elastic particle model of gases. Because the gas is made of entities that are too small to observe directly, one has to invent a hidden explanatory model that fits and explains the observable behaviors of the gas that one can see.

14.4.1 Expedient Analogies

I introduced a distinction between expedient analogies and explanatory models as two different kinds of scientific model in Chapter 6, a distinction anticipated by T. Kuhn in his comment quoted just before the beginning of this chapter. If one examines the final qualitative model developed in this chapter – the twisting model at the

14.4 Building a Theoretical Distinction: Explanatory Models vs. Expedient Analogies

top of Fig. 14.15 developed at the culmination of stage 3 – one can see how it contrasts to initial "expedient analogies" in the investigation, such as the foam rubber analogy in episode 7b. The two pieces of foam rubber with large and small holes is simply an analogous case, not a system operating within the target, whereas the image of the element, w, of the spring coil twisting is an explanatory model proper, since it can be projected into the target as something that could be actually happening in the target case. An expedient analogy like the foam on the other hand, shares abstract relationships with the target, and can be used for predictions without making any assertions about or commitment to its role in the causal mechanism of the spring.

14.4.2 Source Analogues

In contrast, Hesse (1966) pointed out that analogies involved in generating explanatory mechanisms are of a special kind; they have "material similarity" to the target phenomenon in that they are imagined to be actually occurring in the target. For example, a swarm of tiny elastic particles is thought of as a "candidate for reality"– a system that could be operating within a gas (Harre, 1961). (The "candidate" metaphor extends nicely to theories being "elected" or confirmed for an unlimited term but always subject to parliamentary repeal.)

Thus, not just any analogy will do for an explanatory model; analogies used in explanatory model construction are going to have concrete elements that are part of the final model and be materially connected to the system in this sense. However, in my view, it is a mistake to take this important and valid point too far – to treat the analogy and the final explanatory model as the same entity– where the model "is" the analogy. Here I will build on Harre's (1961) view that the analogy to colliding balls is a source analogue that should be thought of as a separate entity from the final model. I will take the view that the analogy to bouncing balls is only the first step in developing an explanatory model, and that the initial analogy should be thought of as a ***starting point*** for building the final explanatory model.

It is possible that when the elastic particle idea was first formulated, it took a form something like: "perhaps the gas is made up of something like small rubber balls bouncing off of each other." At the point of first insight, this idea appears as a mere analogy to something the gas might be like, if one is lucky. It is easy to lose track of this tentative perspective once we have been trained in the science. Initially it is simply a rough analogy to a more familiar system. However, as the model grows in complexity and refinement, it may draw on several other analogies or schemas, and certain differences between it and the toy balls become apparent. In this view the final model is not equivalent to the initial analogy; rather the subject uses the analogy as a source analogue (source of ideas) that becomes elaborated and modified as it is incorporated into the model.

The idea that explanatory models are mechanisms thought to have some material similarity to the hidden structure of the target is consistent with the idea that many preferred scientific models are imageable entities. In this view material similarity results from the projection of image elements from a source analogue into the image of the explanatory model being developed, and I will explore this possibility in the chapters that follow. Analogies used to seed explanatory model construction may contribute ideas for the addition of concrete material elements to the model and these must be plausible components that could actually be in the system. I would include the requirement that the model must "fit" spatially, temporally, physically, and behaviorally into the targeted system in a coherent way.

14.4.3 Triangular, Not Dual, Relation in Model Construction

An analogy can be viewed as involving two main elements: the target case and the analogous case. However, the above considerations mean that it is desirable to take a *three*-element view of the relation between a target, an analogous case, and an explanatory model, as shown in Fig. 14.16 (Clement, 1989b). A source analogue or anchor (billiard ball reflection) is used as a starting point for building up an explanatory model (elastic particles and reflection). The model can then be projected into the target and used to explain the target phenomenon. Such an explanatory model is neither deduced from axioms nor induced as a pattern from repeated experiences. Rather it is a construction pieced together from various sources, designed in such a way as to provide an explanation for the target phenomenon. Most previous theories have not distinguished clearly between the source (anchoring) conception and the final explanatory model (exceptions are Harre (1961), Nersessian (2002), and Clement (1989b)).

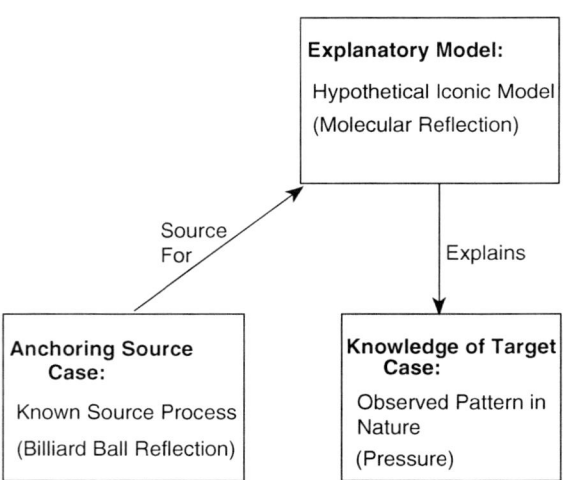

Fig. 14.16 *Three*-element view of the relation between a target, an analogous case, and an explanatory model

14.4 Building a Theoretical Distinction: Explanatory Models vs. Expedient Analogies

In Chapter 6, I argued that since subject S2 (1) talks about having an explanation for the lack of observable bending in the spring (2) refers to "increments of the spring" (line 117 in Chapter 6), and (3) considers a model with deformations in microscopic elements that are unobservable to him, I can infer that he eventually does work with an explanatory model in his solution. In the protocol in this chapter, that model is developed considerably further. The triangle in Fig. 14.15 expresses the view that the explanatory model of torque producing twisting in the coil is a separate (and more complicated) entity from the anchoring source analogue of the twisting rod.

14.4.4 Source Analogues are Projected into the Composite Model and Must Be Imagistically Aligned

Once the twisting hypothesis has been generated, the solver must determine whether twisting is a plausible mechanism that could be operating in the circular coil without producing contradictions. Informally speaking, Fig. 14.7a is a breakthrough in this regard, since it allows one to "see where the twisting comes from" in the circular coil, as do Fig. 14.4 and 14.5. Here the torque idea is beginning to be aligned correctly within the model so that one can "see" how forces cause twisting in the wire. Figure 14.9a then "stretches" this newly acquired "seeing as" ability to a more difficult case: here the solver imagines sections such as gc to be "rigid handles" equivalent to those in Fig. 14.9c, in order to "see" the source of twisting. The drawing in Fig. 14.7a actually adds no new concepts to the spring mechanically; but it changes the imagistic alignment of the concepts, making it possible to project torque and twisting appropriately into the image of the spring coil. This type of "stretching" or extension of the domain of application of one's anchoring intuitions (here the intuition of twisting) appears to be one of the principle functions of nonformal plausible reasoning.

Therefore model projection is seen as a very important and sometimes difficult activity. A major theme here is that even after the key anchoring analogy (to a twisting rod) is found, considerable development needed to occur in order to construct a viable explanatory model (of how torque and twisting operate in the spring). Even though an anchor had been identified, exactly how and where to apply it was not at all clear initially. In this case the role of the analogy is as an *initial starting point*, whose major impact is to facilitate the larger enterprise of model construction. *Thus explanatory model construction involves more than finding a key source analogue in a familiar system, although that can be a starting point.* This is the meaning of the triangular, three-part relationship in Fig. 14.15. The model is usually more complicated or refined than a simple analogy.

In addition, an explanatory model is often constructed from multiple analogies and should therefore be thought of as a separate entity from any one analogy. *In sum, I hypothesize that the explanatory models being considered here incorporate assemblies of image-producing prior knowledge schemas, and that this assembly is modified as the composite model is refined.* Evidence for image-producing schemas will be examined in the next chapter, but I have stated these hypotheses here as an advanced organizer and to provide a context for their analysis.

14.4.4.1 Special Evaluation Criteria

To highlight another difference between ordinary analogies and explanatory models, I hypothesize that there are criteria for evaluating an explanatory model that do not apply to expedient analogies, such as the requirement that an explanatory model be able to "fit" spatially into or onto the target system in a plausible way. For example, the size and number of molecules imagined to be in a gram of metal must not exceed the space allotted. The model of molecular structure must "fit" into the spatial context of the target, whereas with an expedient analogy like the foam rubber analogy this need not be the case. Extending this criterion by asking for a model that is fully connected spatiotemporally, when attainable, can also take one well beyond the initial analogy. (Other additional desirable features for explanatory models are described in Chapter 20). Developing detailed subprocesses of investigation processes that operate differently in using expedient analogies and explanatory models will be one of the challenges faced in Chapter 16.

14.5 Conclusion

14.5.1 Plausible Reasoning and Stages of Investigation

Evidence for a variety of medium-sized plausible reasoning processes was observed in this solution, including: simulations; analogies; extreme cases; formation and running of an explanatory model; and thought experiments. The category of thought experiments appears to be quite broad and these may appear in conjunction with many of the other reasoning processes. I will examine the role of thought experiments more carefully in the next chapter.

Table 14.1 showing five stages or modes of investigation is the result of attempting to find categories of episode types at a larger-sized time scale. The five-mode categorization provides some large-scale distinctions between initial descriptions, initial explanatory modeling, modeling via a connected mechanism, geometric modeling, and quantitative modeling. The stages organize the protocol episodes in a plausible but idealized order. The empirical content of this table is reflected more in these five categories of episode types or modes of investigation, rather than in their ordering, since the protocol is a composite one. However, the present ordering does not seem far out of line with most of the work of individual subjects. The extent to which it is possible in the future to map out orderings or partial orderings in individual protocols will depend on how structured experts' control structures are and how "volatile" other processes are in their tendency to interrupt a currently running subprocess.

14.5.2 Parallels and Differences Between Qualitative and Mathematical Modeling

The investigation presented in this chapter demonstrates the possibility of a gradual transition from qualitative to quantitative modeling, with overlapping processes used in each case. There are several aspects to this transition. First, the investigation shows that *a significant part of the mathematical analysis of a new system can actually be qualitative in nature.* Referring to the levels of precision shown in Table 14.2, one can see that much of the solution was occupied with the qualitative relationships in levels 1–3. Apparently the qualitative work was done to find a way to see the system in pieces that are analogous to paradigmatic cases handled by established principles in mathematical form. This constitutes an enormous amount of work as a prerequisite to the simple imperative to "apply" equations. The major role of the mathematical model in this case is to add precision to the relations developed in the qualitative causal model. However, mathematical modeling also affected the qualitative model in at least one way, since the calculation in episode 16 led to the discovery that the alignment of the qualitative model was inadequate. So the mathematics was a way of extending the inferences that could be made from running the qualitative model to a level of precision that exposed a serious mismatch. Thus the trajectory of the investigation was not a simple linear pass through the five stages with no looping.

Second, many of the same reasoning processes identified in qualitative stages discussed earlier, such as analogy, extreme case analysis, and thought experiments, were still applied during the mathematical stages of the solution. In some cases even the same particular analogous case was used at both levels (e.g. the analogies of "freezing" parts of the spring to be rigid to facilitate imagining or calculating the deformation in other parts.) In addition it appears that, model criticism and revision cycles were still applied, even when the investigation extended into the domain of mathematics. These reasoning methods appear to be applicable during both qualitative and mathematical stages of model construction. They are therefore candidates for being considered reasoning methods of scientific analysis at the highest level of generality.

However, there are several important aspects of the protocol that have not been accounted for. Analogies used in the solution appear to be far, intradomain, rough analogies at the beginning of the investigation, and close transformations of the system toward the end. Some of the stages appear to emphasize the identification of new causal factors and explanations, whereas others are focused on adding to descriptions of the system without developing new causes. How these features arise poses interesting questions for the more detailed model to be developed in Chapter 16. To enable that to occur, however, it is necessary to first examine the central role of imagistic simulation in Chapter 15. Finally, I proposed a triangular, three-part relation between a target, a source analogue case, and an explanatory model. Modeling the nature of the relations between these entities will be another issue of focus in the rest of the book.

Chapter 15
Thought Experiments and Imagistic Simulation in Plausible Reasoning

This chapter investigates the origins of conviction in thought experiments. To do this I build on the set of sources of knowledge identified in Chapter 13 for imagistic simulations, since imagistic simulations are seen as playing a key role in thought experiments. This leads to the development of an initial theory or process model for thought experimenting, addressing the paradox implied by the sensation of "doing an empirical experiment in one's head." I then extend the theory to discuss how thought experiments are used within more complex reasoning modes such as analogy and model evaluation. These reasoning modes will be used as components in the larger theory of scientific investigation processes to be presented in Chapter 16.

15.1 Nature of Thought Experiments

15.1.1 Fundamental Paradox of Thought Experiments and Two Definitions

When S2 makes his torsion discovery in the spring problem, he appears to immediately "run" and make predictions from several thought experiments involving torsion in order to examine the consequences for stretching behavior. What is intriguing here is how subjects are able to do this kind of experiment effectively without examining real materials. They appear to be able to perform thought experiments in their heads which somehow examine consequences of a deformation like twisting to see if unexpected or contradictory results arise. This leads to the following question.

> "How can findings that carry conviction result from a new experiment conducted entirely within the head?"

I will refer to this as the *Fundamental Paradox of Thought Experiments*. This is one of the most intriguing paradoxes in cognitive science.

One of the difficulties that makes this question hard to answer is the lack of agreement on the meaning of terms like "thought experiment." ("Gedanken experiment" is another related term that has been used in scientific contexts.) Unfortunately,

Table 15.1 Hierarchy of possible criteria for defining the term "thought experiment," from very broad use at the top to narrow use at the bottom

Label	Criteria
Imagining a situation	Free imagining of a concrete situation- not necessarily predictive
Predicting states of a system	Subject predicts future states of a system
Untested thought experiment (in the broadest sense used here)	(My preferred criterion for broad sense). Predicting states of a system where subject has never observed the results
Evaluative Gedanken experiment (in the narrowest sense used here)	(My preferred criterion for narrow sense). Subject designs a case to evaluate explanatory or mathematical model or concept

there is not a consensus on the definition of "thought experiment" in the literature, and it is used for a very broad spectrum of entities. Table 15.1 shows a number of increasingly restrictive criteria that could be used define the term thought experiment. Each item in the left-hand column is thought of as a subset of the item above it. Some authors have used the term very broadly for any situation where a subject predicts future states of a system, and everything appearing below that in the table. I will attempt to increase the precision of the term by restricting and subdividing it as follows.

> An *(untested) Thought Experiment* (in the broad sense) is the act of generating or considering an untested, concrete system (exemplar, case, "experiment") and attempting to predict aspects of its behavior. Those aspects of behavior must be new and untested in the sense that the subject has not observed them before or been informed about them. "Concrete system" here means one potentially perceivable via the senses or via instruments; i.e. the experiment is one that would yield empirical observations if it could actually be performed.
>
> An *Evaluative Gedanken Experiment* (in the narrow sense) is more sophisticated than a simple thought experiment. It is the act of making a prediction for an untested, concrete system designed or selected by the subject to help *evaluate* a scientific concept, model, or theory. For example, an evaluative Gedanken experiment could be used to evaluate an explanatory model and/or its mathematical elaborations.

The first and broader definition is appropriate for expressing and discussing the fundamental paradox introduced at the beginning of this chapter. The second one is appropriate to some of the more sophisticated thought experiments in the history of science, and experiments of this kind will be analyzed later in this chapter.

15.1.2 Nersessian

Nersessian (1991) proposed, based on her careful reading of certain historical records of investigations in science, that simulation may play a role in thought experiments and scientific theory formulation. She refers to the "paradox of empirical force" in thought experiments. If one takes this to mean the paradox of being able to generate findings that feel virtually as strong as an empirical observation and that are done completely within one's head, then that version of the paradox focuses on thought experiments where extremely high convictions are attained. My own phrasing of the paradox above considers a broader range of thought experiments because I think they can play an important and impressive role, even when they lead to more modest levels of conviction.

Gendler (1998) argued that the power of certain TE's cannot be assigned to an underlying formal argument, and Shepard (2006) proposed that they may draw on imagined transformations depending on innate knowledge of 3 dimensional space and symmetry. Nersessian (1991) hypothesized that thought experiments could involve simulations, thought of as "depending on spatial representations and perception-like mental capacities." The grounds for this were primarily Maxwell's references to novel moving systems (e.g. gears and idler wheels throughout space) that he could not build and therefore presumably must have somehow imagined. However, it is difficult to provide more support for this hypothesis or to expand it in more detail based on the historical record because that record does not provide evidence on minute by minute processing.

15.1.3 Focus of This Chapter

Clement (1988, 2002, 2006) provided initial documentation of certain thought experiments in experts from think-aloud protocol data and noted that they were generated cases rather than cases retrieved from memory. In what follows, I will attempt to address the above-mentioned paradox by extending this documentation and providing a more detailed accounting of *how* imagistic simulation could explain the origins of conviction. This question is most interesting for cases where a prediction appears not to be reached via a memory of a direct observation – that is, for untested thought experiments in the broad sense defined above. These cases are particularly interesting because the origin of the "experimental observation" remains enigmatic. However, the definition does not rule out a partial role for remembered actions that had been done before on real objects, as long as the experimenter is also examining some previously unexamined aspect in the prediction.

Note that the definition does not specify whether thought experiments are implemented via imagistic simulations; this is a question I would like to answer on the basis of evidence from the protocols. The definition itself leaves open the possibility that a thought experiment could be implemented via logical deduction, for example. To anticipate though, a key hypothesis generated by the protocol analysis will be that schema-driven imagistic simulation is a mechanism that can account for the benefits of many thought experiments. These imagistic simulations in turn derive their predictive power from new combinations or new applications of existing knowledge schemas, including physical intuitions, which generate a dynamic image in a way that incorporates spatiotemporal constraints. This will require developing further the theory of how imagistic simulations work and what

Table 15.2 Some complex plausible reasoning strategies that utilize elemental imagistic simulations

Compound simulations
Analogy
Running a qualitative explanatory model
Evaluative Gedanken experiments

their various powers and functions are. Untested thought experiments provide a good arena to do this in because new knowledge is being generated by the simulation, not just a replay of old knowledge.

Later I will examine evaluative Gedanken experiments as a special case. I will consider evaluative Gedanken experiments to be part of a group of more complex reasoning strategies listed in Table 15.2. I will also examine evidence later that each of the lower three categories in Table 15.2 can utilize one or more untested thought experiments (carried out via imagistic simulation) as a subprocess. To my knowledge the interconnection between imagistic simulation and thought experiments in experts has not previously been examined on the basis of think-aloud protocols.

15.1.4 What are Some Major Functions of and Benefits from Untested Thought Experiments?

Table 15.3 shows four types of functions that can be hypothesized for untested thought experiments from observations in the present protocols. I begin by giving an example of each type.

15.1.5 Primary Function

1. *Primary predictive function.* An example of function A is the subject in episode 12b thinking about whether a long rod would twist more easily than a short rod, with accompanying imagery reports and depictive hand motions. After making this comparison he was fairly confident that the long rod would twist more easily. This is an example of a direct prediction from an untested thought experiment. The novelty of the specific experiment makes it unlikely that it has been previously tested with real objects.

    By definition, all untested thought experiments described in this chapter involve such untested predictions, and so the prediction process just described is viewed as primary.

15.1 Nature of Thought Experiments

Table 15.3 Major functions or benefits from untested thought experiments

A. Primary predictive function
B. Secondary evaluative function: new dissonance relation between experiment and an explanatory model
C. Secondary evaluative function: new coherence relation between experiment and an explanatory model
D. Secondary generative function: generation of a new variable or factor

15.1.6 Secondary Functions

In addition there are three secondary benefits that can sometimes accrue immediately from a thought experiment; the experiment may generate a new evaluative relation of coherence or dissonance or it may activate a prior knowledge schema for the first time.

2. *Evaluative function: new dissonance relation.* Evaluations using thought experiments can also lead to a negative result in the form of dissonance (conflict) with prior findings. An example of this is the case of a spring made of a vertically oriented band of material, such as the paper unwound from a grease pencil or the metal unwound from a coffee can (episode 13b in Chapter 14). This sophisticated thought experiment was designed to evaluate bending as the model for what is happening in the spring wire. When the subject imagined stretching such a bend-proof spring, he sensed that it would still be quite stretchable and springy. This created dissonance with the model of bending as a necessary causal factor.
3. *Evaluative function: new coherence relation.* In episode 11b, when the subject analyzed the square coil by imagining it being stretched, he discovered that it did not posses the overall property of increasing slope that he had been so worried about in the simple bending model of the spring. This created a new coherence relation between a recalled property of springs (uniform slope) and the model of the square coil involving twisting plus bending. Again, the novelty of the experiment makes it unlikely that it has been previously tested with real objects.
4. *Generative function: identifying a new variable or causal factor.* The most important example documented here is the recognition of torsion as a new causal variable while the subject imagined stretching the hexagonal coil (episode 11a).

These examples suggest that there is not one, but a wide range of possible beneficial functions of untested thought experiments, as summarized in Table 15.3.

15.1.7 Can Schema-based Imagistic Simulation be Involved in Untested Thought Experiments with These Different Functions, and If So, What is Its Role?

In this section, I cite evidence for the involvement of imagistic simulation as a process in each of the above examples of untested thought experiments and others. I also develop hypotheses for how it makes a central contribution.

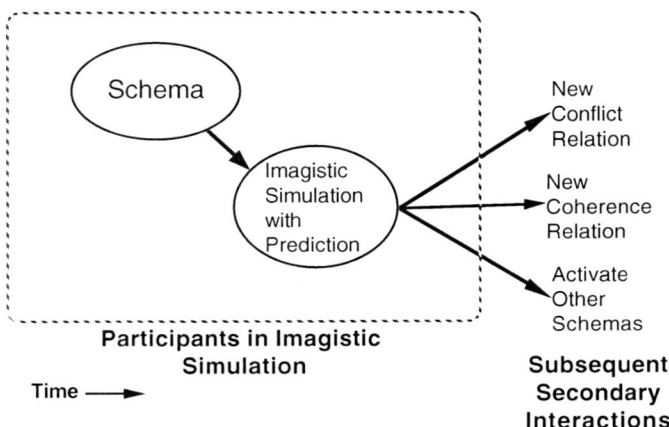

Fig. 15.1 Basic benefits of imagistic simulation in an untested thought experiment

A. *Predictive Function.* In Chapter 13 it was hypothesized that the core mechanism that allows a subject to make a prediction is his or her use of a schema to simulate the case directly, as shown on the left side of Fig. 15.1. In episode 12b the subject states: "if I have a longer rod (moves hands apart), and I put a twist on it (moves hands as if twisting a rod), it seems to me – again, physical intuition – that it will twist more." The personal action projection (spontaneously redescribing a system action in terms of a human action), hand motions, dynamic imagery report, and intuition report provide evidence that he engages here in a schema-driven imagistic simulation, as shown in Figs. 13.3 and 13.4. This was described as a process involving perceptual/motor anticipations that can produce dynamic imagery for a series of states of a system over time. In this case the simulation appears to be an efficient process the subject can use in generating a prediction. This constitutes the beginning of an explanation for how untested thought experiments can work. While the broad term "untested thought experiment" refers to the idea that the subject is considering a somewhat unfamiliar system and making a prediction about it, the term "imagistic simulation" refers to a specific process that can be used to carry out the thought experiment.

Can the ability to carry out a thought experiment be explained by imagery alone without positing a role for schemas? There is some evidence for the theory that schemas are separate entities from the more specific images they assimilate and operate on. The same twisting schema appears to be able to run on different images here (long, short, and very short rods, as well as the square and hexagonal coils). And conversely, one can find evidence in other cases of a single image being assimilated by different schemas, as in the case of both the bending and twisting schemas applied to the single image of the square coil; or applied to the single image of a straight rod. These considerations motivate the idea of having a two-element theory for an imagistic simulation with one or more schemas operating on a specific image. In an exploratory clinical study, accounting for multiple instances like these helps constrain the theory by means of the need for it to explain different episodes.

15.1 Nature of Thought Experiments

B. *Dissonance.* In type B, running a simulation may also generate a new conflict relation with an existing schema, as shown on the right-hand side of Fig. 15.1. In the protocols examined dissonance appeared to be generated primarily in two ways:

1. *Spontaneous dissonance.* For example, when the subject envisioned bending in the helical spring, he realized it would cause an increasing slope as one traveled down the helix of the spring. This conflicted with his image of what springs look like. Evidence for imagery is provided by the hand motions in episode 9:

> But then it occurs to me that there's something clearly wrong with that metaphor, because...it would (raises hands together in front of face) droop (moves r.h. to the right in a downward curve) like that, its slope (now retraces curved path in air with l.h.) would steadily increase

This appears to be a case of spontaneous dissonance. I hypothesize that the subject runs the model on a case, and a feature emerges in the simulation that conflicts with the subject's prior knowledge. Figure 15.2a shows this process as the formation of a jagged line representing dissonance between of feature of the simulation and a prior knowledge schema.

2. *Evaluative Gedanken experiments.* In an evaluative Gedanken experiment the subject designs a special case, makes a prediction for it using an independent schema, and compares the result to that predicted by the current model for the same case. In the case of the "vertical band spring" in episode 13b the subject provides evidence for conducting an imagistic simulation with a number of indicators, including imagery reports and hand motions:

> How could I imagine such a structure?.... I'm thinking of something that's made of a band... we're trying to imagine configurations that wouldn't bend. Since it's cross section is like that (see Fig. 14.3)... it can't bend in the up-down (indicates up/down directions with hands) direction like that because it's too tall. But it can easily twist (motions as if twisting an object).

He appears to be evaluating the plausibility and necessity of the bending mechanism by eliminating bending in order to see whether the spring stretches considerably less. Here an evaluation of the bending model is generated in a more deliberate way. Rather than casting around for a random analogy, the subject has done something more complicated by designing a specific experiment that isolates a key causal variable. Because the mechanism is more complicated here it is not shown in Fig. 15.2. I will discuss such cases below in the section on Gedanken experiments. In summary, in these two episodes imagistic simulation is apparently involved in generating dissonance from a thought experiment.

C. *New coherence relation.* The simulation of the square coil in episode 11b appears to produce the emergent property of lack of cumulative bending which solves the problem of increasing slope that was present in the simple bending model, leading to a new coherence relation as represented by the double line in Fig. 15.2b. The hand motions in this segment provide one piece of evidence for imagery use here. (This example shows that the newly generated coherence relation need not always be simply a confirmation of the answer to the target problem, i.e.

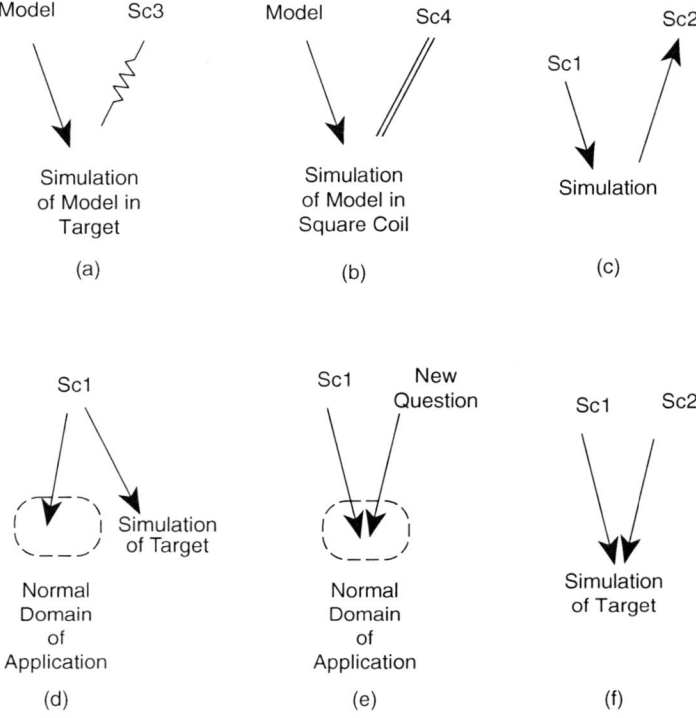

Fig. 15.2 Relationships between schemas and the running of imagistic simulations: **a.** Feature is detected in the simulation that conflicts with the subject's prior knowledge; **b.** Simulation triggers a new coherence relation between of feature of the simulation and a prior knowledge schema; **c.** Simulation leads to the identification of a new variable or causal factor; **d.** Schema generates a simulation for a target case outside of its normal domain of application; **e.** Schema being applied to a case within its normal domain of application, but with a new question being asked; **f.** Compound simulation where two or more schemas are coordinated in the same simulation

wide springs stretch further. In this case the coherence relation connects to a different property – whether slope changes at different points in the spring – a now recognized constraint in the target.)

As a second example, in episode 12, bending and twisting actions are projected into the image of the same stretched square coil to predict that the wider coil will stretch more. This creates a second coherence relation with previous findings. The hand motions, personal projections, and intuition reports in episode 12 also provide evidence that the subject is using intuitively grounded imagistic simulations here. There are also hand motions throughout other episodes of analysis of the square coil both immediately before and after segments 11b and 12 suggesting that the subject is imagining bending and twisting movements that cannot be shown in his drawing.

D. *Identifying a new variable or causal factor.* Chapters 6–7 gave an initial perspective on how models are generated and evaluated in construction cycles. Here we wish to look at how individual variables or causal factors can be discovered. In the protocols examined here this appeared to happen primarily in two ways:

1. *Schema activation.* Imagistic simulation can be involved when a thought experiment leads to the identification of a new variable or causal factor. In the important case of the hexagonal coil triggering the torsion insight in episode 11A cited above, there was evidence of this from his hand motions. (This is consistent with the fact that the subject was staring at a drawing while he made statements projecting forces and deformations into the system – where the objects in the system but not the movements were shown in the drawing.) This is shown as type c in Fig. 15.2.

This thought experiment leads to a scientific insight in a more unplanned and divergent way in the form of a sudden recognition. One can hypothesize that certain spatial and dynamic features of the simulation of the hexagonal coil in this case triggered the activation of a new schema Sc2 (torsion) unexpectedly in this case. The suddenness of this activation may be part of the feeling of having an "insight" that causes the subject's Aha! reaction at this point. If this interpretation is correct, it means that running an imagistic simulation can serve as an important mechanism of schema activation in addition to the verbal association mechanism that is cited traditionally.

2. *Emergence of novel image feature.* A somewhat different source is suggested by the example in episode 9 discussed above of the recognition of increasing slope in a spring when the bending model is applied. Here one can hypothesize that the image of Fig. 6.3 emerges from the original simulation itself, rather than from activating a schema.

The novelty of this image suggests that it has very likely never been seen or imagined by the subject before. If one assumes that the process of analyzing and describing it as a spring with "increasing slope" comes after the formation of the images, then the formation of a new and novel image via the extended application of a schema is the origin of the newly identified factor or effect. This interpretation is supported when he later comments on the same idea, saying: "this is quite clear but I'm not quite sure how to express it in words." A similar process may occur in episodes 20a where "unbending" is identified as a new causal factor in the image of the drastically stretched spring. It is also quite possible that new factors might emerge in compound simulations and indeed this may be involved in the former example above where bending is being applied to multiple increments of the helical spring and the effects are added together cumulatively. Representing these possibilities requires a more complex model, depicted in Fig. 15.3, to be discussed in the next section.

15.1.8 Summary

In addition to predictions, other possible functions and benefits of thought experiments observed in the protocols are listed in Table 15.3. Returning to the question: "How can a subject understand and predict new behavior in an untested thought experiment without making new observations or implications from a formal theory?" My first and most basic response is: subjects appear to use schema-driven imagistic simulations to run the experiments, as depicted in Fig. 15.1. I elaborate on this view in the next section.

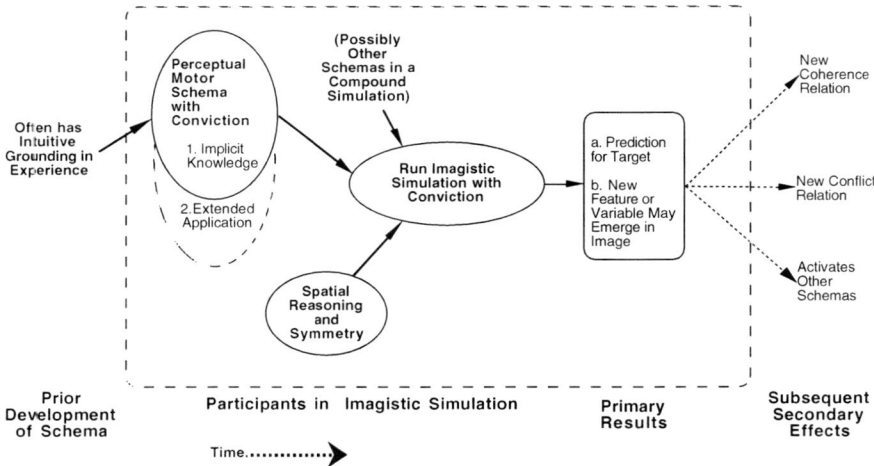

Fig. 15.3 Imagistic simulation process with possible benefits on the right and four possible origins of conviction on the left

15.2 Addressing the Thought Experiment Paradox: How Can an Untested Thought Experiment Generate Findings with Conviction?

15.2.1 Introduction

Several of the previous examples show that confidence in the outcome of an untested thought experiment can indeed be very high. The strong version of the question in the above heading is: how can an untried thought experiment done completely in the head appear to have comparable power to one done in the lab and to be equally compelling? The milder version of the question is: How can one derive any knowledge – even tentative knowledge – from an experiment done in the head that has no real new empirical input?

I will speak to these using the expanded model of an imagistic simulation in Fig. 15.3. It proposes that the properties of runnable schemas in conjunction with spatial reasoning can allow for considerable flexibility and confidence, and at times even conviction with empirical force, in solving unfamiliar problems via imagistic simulation. *A runnable schema is defined as one that is capable of generating an imagistic simulation.* In Chapter 13, I pointed to two ways that a new prediction could be derived from an imagistic simulation that would not need to depend on inferences from chains of word-like symbols: flexible adaptation in the *extended application* of a schema to a new situation, and *tapping implicit knowledge* in a schema. These properties are shown in Fig. 15.3 as two inputs from the left. They are complemented by other processes such as spatial reasoning – i.e. generating inferences

15.2 Addressing the Thought Experiment Paradox

about spatial relations or actions that embody spatiotemporal constraints on any system of objects, such as the constraint that solid objects may not occupy the same space, or that the face of an object turning on a vertical axis will disappear and reappear. In contrast to domain specific schemas, these operations are domain general, and can include imagined actions involving object manipulation or navigation frames.

The above sources are shown to the left of the prediction/simulation process in Figure 15.3, with the secondary outcomes of the simulation shown to the right. Separating these two stages of a thought experiment helps greatly to clarify the theory. In the next few sections, I will speak to the thought experiment paradox by focusing on the left-hand side of Fig. 15.3 to provide the foundation for the answer given here.

15.2.2 Sources of Conviction: Perceptual/Motor Schemas

I will hypothesize that there are not one, but several sources of conviction that operate through imagistic simulation, shown in Fig. 15.3 and Table 15.4.

15.2.2.1 Adaptability from Extended Application of Schema

I begin with an example of the extended application of a schema. An example of this source is episode 9 quoted earlier of projecting bending into the helical spring, leading to a prediction that stretching would arise, assuming bending were present in the spring wire: "it would (raises hands together in front of face) droop (moves r.h. to the right in a downward curve) like that, its slope (now retraces curved path in air with l.h.) would steadily increase as you went away from the point of attachment." I hypothesize that the prediction arises from the extended application of a dynamic schema (bending) to an image of an object outside its normal domain of application (the helical spring). It is outside because the subject is unlikely to have thought about bending an object that is already curved in another plane and therefore it should not be immediately clear to him how this would work.

Table 15.4 Sources of conviction in an isolated untested thought experiment performed via imagistic simulation

- Self-evaluated conviction of intuition schema in its normal domain
  Intuitive grounding; Prior coherence with rational and empirical (experiential) knowledge
- Schema flexibility in imagistic simulation
  Adaptive, extended application of schema outside its normal domain
  Tapping implicit knowledge in the schema
- Spatial reasoning
  Symmetry
- Compound simulation

(Note: This prediction about the *results* of bending is correct, even though the subject in the same breath correctly suspected that the assumption of this kind of bending being present is false.)

Here I follow common practice in thinking of schemas as knowledge structures that can generate certain predictions or explanations within their domain of applicability. They are also thought of as having at least modest generality; the schema applies to more than one situation. However, I am interested in untested experiments, thought of as being outside the normal domain of application of a schema, or responding to a question that has not been asked before and that refers to relationships that have not been heeded by the subject in this context before.

This source is also illustrated by the figure labeled d in Fig. 15.2. which shows a schema generating a simulation for a case outside of its normal domain of application. In general, running any schema on a "strange" or novel case would be an extended application.

The subject is confident that if the spring wire is bending downward in a vertical direction that the spring would stretch. I assume that the subject's confidence in such cases comes from a flexible, confident schema that can be adapted and extended to apply to an unfamiliar but imageable case being run and to the question being asked.

Flexibility here means (1) that the schema can assimilate elements of an unfamiliar target and (2) that the expectations from the schema can generate imagery of events that can lead to a possibly novel occurrence such as asymmetric stretching in the spring.

15.2.2.2 Experience

It seems likely that intuition as self-evaluated conviction in a schema ultimately comes largely from experience for most of the examples discussed here. One could say "generalized from experience" but we should be careful to understand that: (1) perceptual/motor schemas may develop from a process that is very different from formal induction; (2) Some intuitions may conceivably be largely innate and nonempirical (this is why I resisted requiring them to be generalized from observations in the definition); and (3) by experience I mean the mind interacting with the world, not "one directional empirical input."

A second very important example is the more imagistically aligned application of the ideas of torque and twisting to the helical spring in episode 17. The subject complains that even after considerable analysis he is still not able to "see" how twisting would arise in the spring wire. To become convinced that stretch is going to introduce twisting in every element of the helical spring constitutes a culmination in formulating the first full qualitative model of how the spring works. I interpret this as coming to be able to extend the torque and twisting schemas beyond their normal domains of application to apply to a shape as odd as the helical spring.

15.2.2.3 Intuition, Implicit Knowledge and Compound Simulation

My examples of extended schema application here have focused on perceptual/motor examples of physical intuition where parameters could be "tuned" to adapt a schema for an imagistic simulation. As discussed in Chapter 12, there is an established literature in motor control theory that describes this kind of flexibility in motor schemas within normal domains of application. Most of the cases here are not normal, but novel, so I am hypothesizing an extended type of tuning capability. It is remarkable that mental simulations with some accuracy can be performed on such cases. Schema flexibility in an extended mode is posed as the explanation for this ability, using the examples involving motor schemas as paradigmatic examples. Some of the major primitive sources of conviction (Clement 1994a) that I will discuss are:

- Extended application of a flexible schema to a new context outside its ordinary domain of application in an elemental imagistic simulation.
- Inspection of implicit relations not attended to previously within an elemental imagistic simulation generated by a single schema applied to either a familiar or a new context. This may occur because a new question about the situation has been asked or a new feature has been noticed.
- Chaining or coordination of multiple schema-driven simulations in a compound imagistic simulation (as in the square coil).

In the above cases new knowledge can also emerge from using particular spatial constraints of the immediate problem context and spatial reasoning operations, including imagined actions involving object manipulation or navigation frames. Table 15.5 shows a hierarchy of theoretical terms to be used in this theory where each category includes all of the categories below it, but not above it.

Intuition. The examples of extended application above happen to also be examples of the use of intuitions. As shown in the taxonomy in Table 15.5 and defined in Chapter 13, this means that the schema is self-evaluated with at least some conviction. As shown on the leftmost part of Fig. 15.3, conviction within the schema is thought of as stemming from coherence relations between the schema and previous experience (whether empirical or nonempirical). A schema has conviction when it has developed from multiple experiences or innate tendencies, has few or no dissonance relations with other cognitive entities and has ties of support to others. The conviction will apply most strongly to cases like ones the subject has experienced before, in the schema's normal domain of application. It would not be

Table 15.5 Taxonomy of theoretical terms for simulation

(Mental) Simulation in the broadest sense (prediction of future states or changes in a system)
Imagistic simulation (simulation with use of imagery)
Schema-driven imagistic simulation (SDIS) (imagistic simulation generated by a runnable schema with perceptual and/or motor components and modest generality)
Intuitively grounded or intuition-based simulation (self-evaluated SDIS, with at least some conviction present)

surprising if confidence was not as high in the extend case as in a familiar case, but it might still be high if the extension feels feasible. So some original level of conviction in the schema would appear to be necessary (but not necessarily sufficient) for running an untested thought experiment with confidence.[1]

Implicit knowledge. The case of the twisting rods quoted in the preceding sections and in episode 12b illustrated how knowledge that was implicit (not previously described or heeded) in the schema could be made explicit via such a case. Presumably, the subject had twisted objects before, but had never consciously considered or paid attention to, the factors that make one object easier to deform by twisting than another. I hypothesize that in the case of the twisting rod this led to a high confidence simulation because the schema used was rich enough in analog perceptual/motor information to be "mined" for additional information beyond its usual purpose. This implicit knowledge was drawn out in running an imagistic simulation on a particular imaged case designed to answer the new question. Drawing e in Fig. 15.2 shows this as a schema being applied to a case within its normal domain of application, but with a new question being used to interrogate the simulation that has not been asked before. Such a question may tap implicit knowledge in the schema via the simulation.

15.2.3 Sources of Conviction: Spatial Reasoning, Symmetry, and Compound Simulation

15.2.3.1 Spatial Reasoning

Finally, each of the processes above can occur in conjunction with spatial reasoning operations which embody very general spatial and physical constraints on the cases being simulated. As an elementary example, adults "just know" that two solid blocks cannot occupy the same space or that a chair cannot fit into a breadbox. And a subject should be able to imagine that in stretching a spring, the coils will not collide and interfere with each other, whereas in compressing a spring, this could be a problem. It seems reasonable that adults would be able to reason to this conclusion in response to the question "Can the ends of a spring be compressed toward each other to a distance as small as one likes?" even if they had never actually compressed a spring to its compressional limit. In this view these examples reflect very general inferencing capabilities about spatial relationships that are not domain or schema specific. This includes the ability to compare size, shape, and distances

---

[1] The word "conviction" carries the desired connotation of self-evaluated rather than evaluated from authority. However, a terminological problem here is that "conviction" also connotes "belief" in a proposition, whereas some of the schemas being discussed here are simply action schemas that may not be driven by propositions. "Sureness of operation" may be a more appropriate description in some cases than "conviction," although "with conviction" is somewhat broader. But for the sake of simplicity I will simply use the term conviction here.

15.2 Addressing the Thought Experiment Paradox

as well as to imagine the results of basic manipulations for arranging objects in space: rotating, moving, nesting, aligning, etc. I follow Kosslyn (1980), Finke (1990) and Lindsay (1988) in assuming that if an image is not too complex, most adults can form an image of such simple manipulations and anticipate the outcomes in terms of the new spatial relationships they set up.

Such spatio-physical reasoning operations may use a fairly primitive knowledge system for how objects can be arranged in space that may develop quite early on from experiences with playing with blocks or other objects, and perhaps utilize elements that are to some extent "built in" to the developed perceptual system. (The brain's system for locomotive navigation is another branch of the basic spatial reasoning system.) I will distinguish these very general and commonly used spatial reasoning capabilities from the anticipations generated by perceptual/motor schemas that are more domain specific. Whereas conceptions of twisting, torque, acceleration, and rolling are more specialized schemas, I will assume that basic spatial reasoning operations for rearranging objects are used ubiquitously and are in some sense "built in" to the imagery system.

As shown in Fig. 15.3 then, such spatial reasoning processes can be used in conjunction with a specific schema that is generating an imagistic simulation. For example, I assume that the spatial reasoning operators of translation, rotation, zooming, etc. are always available to help in the assimilation of an untested case to a schema. A very simple metaphor here is that just as we may plan to rotate a coffee cup in real life in order for a grasping schema to execute grasping its handle, similar perceptual transformations may be used with other imagined objects so that other schemas can assimilate and operate with them. Unfortunately these processes for rotating and arranging objects may be so fast and automatized that they are rarely reported in a think-aloud protocol. I therefore feel that I have a limited amount of data on these operators and for that reason basic spatial reasoning operations cannot be analyzed deeply in the present theory. However, a theory that hypothesizes the presence of imagery in these episodes should also hypothesize the availability of these basic spatial reasoning operators if the system is to have normal inference making power.

A more sophisticated use of spatial reasoning occurs in episode 14 and Figs. 14.4 and 14.5 where a circular ribbon is extended to show the resulting twisting effect. Here the challenge to the imagery system is essentially to keep track of the edges of the ribbon during the deformation. This ability might also be considered a geometric skill involving specialized mathematical schemas, and this cannot be ruled out entirely. However, the informality of the subject's language argues that this inference is likely to have involved the heavy use of general spatial reasoning capabilities, among the factors shown in Fig. 15.3.

In some cases the spatial reasoning system that we have developed for planning everyday activities in dealing with objects may be very efficient for doing certain kinds of reasoning with the mental objects in mental models. When these same spatial reasoning skills are used, say, to quickly eliminate some possibilities for how a foreign molecule might fit into a host crystal because of its shape, the scientist may experience this as being close to "getting that part of the solution for free" with very little effort.

15.2.3.2 Symmetry

Some thought experiments appear to rely on symmetry as a source of conviction. For example, the horizontal spring case in episode 15d is used to argue against the possibility of asymmetric stretching shown in Fig. 6.3. This argument says that if one pulls on both ends of a uniform horizontal spring to stretch it with the same force on each end, there is no possible reason for why one side of the spring should stretch any differently than the other side; therefore since this is equivalent to the vertical situation in Fig. 6.3, the outcome shown there is impossible. (This equivalence depends on knowing that forces applied at each end of the spring are equal and opposite and the given assumption that the mass of the spring wire is negligible.) The symmetry principle used for the horizontal case is very general and has to do with one's view of and faith in nature as operating according to consistent principles: in equivalent causal circumstances one should see identical outcomes (with "equivalent" meaning situations related by conserving transformations). Another way to say this is: if two simulations are performed at different times, orientations, or locations but in the same causal circumstances, then the outcomes should be the same (where changing the time, orientation, or location is seen as a conserving transformation). Symmetry cases like the horizontal spring can be thought of as involving the detection of an equivalence between comparative simulations (in each half of the spring). This source of conviction for a thought experiment is of philosophical interest because some would argue that it is a nonempirical source. Because of the connection to spatiotemporal transformations in this analysis I have lumped symmetry together with spatial reasoning in Fig. 15.3, but they could have been shown separately. A similar symmetry argument is used in episode 28 to conclude that the spring wire twists uniformly throughout the spring.

15.2.3.3 Compound Simulation

I also need to hypothesize that new predictions can emerge from the coordination of multiple schemas in a compound simulations such as the square coil case involving both twisting and bending schemas. Diagram f in Fig. 15.2 depicts a compound simulation where two or more schemas are coordinated in the same simulation. I infer that the subject running multiple instances of the bending and twisting schemas in the orthogonal sides of the square coil in episode 11 led to the emergent prediction of no cumulative bending in the square spring in Fig. 6.6b.

It also seems necessary to assume that spatial reasoning is used in this simulation. The situation is so novel that it is likely that he needed to use very general knowledge of spatial and physical relationships to imagine the way movements in adjacent sides would add or cancel out during stretching, and to conclude that there would be accumulated stretching, but no accumulation of increasing slope or bending in such a coil. This appears to be an impressive result of spatial reasoning in a compound simulation – one that accomplishes quickly and efficiently what would

15.2 Addressing the Thought Experiment Paradox   293

take an inordinate amount of mathematical work if it were done formally. The process of generating a new image from the interaction of one or more schemas plus spatial transformations has also been studied by Finke (1990).

Another kind of compound simulation occurs when multiple instantiations of the same schema are run simultaneously or in sequence. An example is thinking about multiple elements of the spring twisting with their effects adding together cumulatively in episode 25. One can hypothesize that when more than one schema is applied or when a schema is applied to several linked objects in an image that there are new relationships added along with each new element of the simulation, and that one of the primary functions of the imagery system is to attempt to keep track of these relations, including how the effects of more than one schema action accumulate. In this view these relationships and effects can be "interrogated" and "inspected" if needed, as long as the capacity (or ability to represent detail) of the imagery system is not exceeded. (I put "inspected" in quotes here, because it cannot be carried out by a literal "mind's eye," but can be thought of as an attentional function which compares aspects of the image at a certain level of depth in the layers of the perceptual processing system – aspects corresponding to a query of some kind.) I follow Kosslyn (1980) in referring to this process as image interrogation and inspection, as discussed in Chapter 13. (Kosslyn, however, discusses only static images; here the subjects appear to have done this with images of events and actions.) Ippolito and Tweney (1995) discuss how abilities related to these may also guide the interactions of scientists with real experiments – by allowing them to make interpretations of observations that have a top down component and in some cases by preparing them to make difficult observations by enlarging the range of their perceptions.

Thus the concept of compound simulation here includes the ability to chain imagined events or causal actions together. It therefore can be seen as an imagistic process involved in imagining chains or networks of actions (up to the limits of the memory systems involved) in everyday problem solving. However, it is also theorized to enable imagining a limited number of actions that are coordinated and performed simultaneously.

15.2.4 Summary

The sources summarized in Table 15.4 and Fig. 15.3 thus provide an initial answer to the thought experiment paradox by describing the sources of conviction in untested thought experiments (which, according to the definition, involve a direct prediction). The overall hypothesis can be stated briefly as follows: confident predictions in untested or even novel thought experiments can come from several sources. The first set of sources were foreshadowed in Chapter 13, where it was concluded that the inherent flexibility of perceptual/motor schemas is perhaps a paradigmatic example of the origins of new knowledge in thought experiments. The initial sources cited there were the self-evaluated conviction of an intuition schema, its schematic, runnable nature that allows it to generate imagistic simulations, and the flexibility of such schemas to adapt in extending outside their range of normal application to unfamiliar cases.

Along with the evolved coherence of a schema and its ability to make implicit knowledge explicit under inspection in such simulations, this first set of sources provides several ways to produce new confident conclusions from new adaptations and extensions of confident existing prior knowledge structures (Clement, 1994b, 2006).

Other sources of conviction include the use of spatial reasoning to incorporate spatial constraints, and the use of symmetry. Some cases also require the subject to use new combinations of schemas in compound simulations with the additional use of spatial reasoning to coordinate these and combine effects. We saw these factors at work in a variety of cases. Thus there appear to be a number of fundamental sources of confidence in untested thought experiments, and they can yield a strong conviction for some novel cases, even when no new real experiment has been performed.

15.3 Imagery Enhancement Phenomena Support the Proposed Answer to the Paradox

15.3.1 *Limitations on Simulation Ability*

It is clear that the resulting confidence in a simulation can vary widely from very low to very high. For example, the transcript contains examples of direct simulations where the subject seems uncertain of the outcome (e.g. Episode 14 – subject uncertain of number of twists produced in fully stretched coil; S6 uncertain of the direction of twisting in Figure 13.3). One can use the basic theory summarized in Table 15.4 to infer some of the factors that could influence confidence and these are shown in Table 15.6.

These factors could produce a spectrum of difficulty levels and outcomes with respect to confidence in simulations. The factors are implications of the theory of imagistic simulation and sources of conviction shown in Fig. 15.3 and Table 15.4. In particular, factors 4–7 in Table 15.6 speak to questions about the difficulty of running imagistic simulations in conjunction with spatial reasoning in a cognitive system where imagery ability is limited in scope. This suggests that with regard to

Table 15.6 Some factors influencing ability to run an imagistic simulation with conviction

- The availability of a relevant schema or relevant spatial reasoning operator
- The degree of experience behind or conviction in the schema being used
- The "nearness" of the new case and question being examined to previous experience
- The complexity of the imagery being attempted
- The magnitude of the effect being imagined
- Other features that would influence the "detectability" of the imagined effect (such as size and orientation of the overall case)
- The level of detail being asked by the question and whether imagery can support that level of detail

15.3 Imagery Enhancement Phenomena

Table 15.7 Heuristics for designing cases for imagistic simulation

In designing cases for thought experiments:

- Find a case close enough to something you already know to make a confident prediction
- Find cases that are as simple as possible in order to image them easily
- Imagine an object whose particular features (such as size and orientation) are such that the experimental variables would be easy to detect if it were a real object:
  - Use an extreme case
  - Imagine "markers" on the object to make it easier to track changes in key features or variables
- Imagine an object of a size and orientation such that the variables in question could be manipulated with one's hands

---

the final thought experiment, level of conviction is a function of a triple rather than a single entity. It depends on the schema being used, the particular target case and the question being asked about it. The remarkable finding however, is that even a very novel thought experiment can sometimes lead to a confident or even compelling result for a subject.

The factors in Table 15.6 in turn suggest a number of heuristics that could be valuable for someone using imagistic simulations, shown in Table 15.7. These heuristics are essentially designed to make possible the generation of perceptible sensory or kinesthetic images from a runnable schema or from the use of spatial reasoning.

15.3.2 Imagery Enhancement Focused on Enhancing the Application of a Schema in a Simulation

In this section, I develop a theory of extreme case analysis and other types of what I call "simulation enhancement" or "imagery enhancement" strategies. Because they appear to be a way to increase conviction, these enhancement cases provide important further evidence on the sources of conviction in thought experiments. They can therefore be used to further evaluate the theory just developed concerning those sources.

Very short rod case. The extreme case introduced earlier from S2 (shown in Fig. 15.4) of twisting a very short rod provides a first example of imagery enhancement. There are three steps in this episode:

The subject first compares a longer to a shorter rod and makes a prediction that the longer one will twist more under the same force; he indicates a medium level of confidence in this conclusion. He then generates an extreme case by moving his hands closer together. This case leads to the same prediction that a shorter rod would be harder to deform by twisting, but with a much higher level of confidence.

> 137 if I have a longer (raises hands apart over table arriving at stationary position with the word "rod") rod and I put a twist on it (moves right hand as if twisting something), it seems to me – again, physical intuition – that it will twist more."

Fig. 15.4 Extreme case for imagery enhancement generated by S2

Uhh, – I'm – I think I trust that intuition.

138 I: Can you stop thinking ahead and just think back on that-; what that intuition is like?

139 S: ....I'm (raises hands in same position as before and holds them there continuously until the next motion below)... imagining holding something that has a certain twistiness to it, a-and twisting it.

141 S: Like a b-bar of metal or something like that. Uhh, and it just seems to me as though it would twist more.

Again, now I'm confirming (moves right hand slowly toward left hand) that by using this method of limits. As (moves right hand slowly toward left hand until they almost touch at the word "closer") I bring my hand up closer and closer to the original place where I hold it, I realize very clearly that it will get harder and harder to twist.

143 S: So that confirms my intuition so I'm quite confident of that."

I will first state an overall hypothesis that explains the effect of the extreme case in increasing confidence and then consider its support in transcript evidence compared to other rival hypotheses. I hypothesize that the first prediction in (1) above comes from an imagistic simulation produced by a practical intuition schema for twisting objects. Such a schema would probably have been used rarely, such as for putting a twist in a ribbon or some other object. One's attention in using this schema would be on whether the object had been put into the desired shape. The question being asked about the force required to deform different-sized objects is novel enough to assume that it would be a new question for the subject, about something he had never attended to before in using this schema. The simulation allows him to examine implicit knowledge in the schema and convert it into explicit knowledge. This would actually involve a pair of simulations where the kinesthetic sensations are compared for a long and short rod. Although he has never paid attention to this before, he can run through the simulation and focus for the first time on the issue of which length of rod would be harder to twist, by interrogating and inspecting that aspect of the imagery resulting from running the schema on these particular cases. This leads him to the first prediction, but the process is new and the difference in the comparison is not large, therefore he reports it with only medium-level confidence.

He then generates the extreme case of the very short rod and in step (3) makes the comparison again to the longer rod during another imagistic simulation. This leads to the same prediction with a much higher degree of confidence. This comes from increasing the differences between the two images being compared and making that difference more detectable under inspection of the images – here the kinesthetic

15.3 Imagery Enhancement Phenomena

difference in the torque or twisting force applied to a "normal" rod and a very short rod in order to put a certain amount of twist in it. Thus, I hypothesize that this is a case of "simulation enhancement" and that the role of this extreme case is *to enhance the subject's ability to generate and compare imagistic simulations with high confidence*. In this case the main source of conviction in the simulations is the tapping of implicit knowledge and its conversion into explicit knowledge.

The subject's saying, along with hand motions, that: "As I bring my hand up (moves right hand slowly toward left hand) closer and closer to the original place where I hold it, I realize very clearly that it will get harder and harder to twist" provides interesting evidence that the subject is actually imagining a special case and his actions upon it, inspecting or focusing on the amount of effort required, and gaining a higher level of confidence from the extreme case comparison.

Very Short Spring Case. Another example was provided by the extreme case of the very short spring generated in episode 7a, where the subject says:

> 039 Now what if I recoiled the spring and made the spring twice as long-instead of twice as wide?...Uhhhh...it seems to me pretty clear that the spring that's twice as long is going to stretch more....Now that's a- again, a kinesthetic intuition,... but now I'm thinking.... I'm imagining Raises Hands 10 inches apart in front of chest, Closes fingers as if holding something between hands, moves hands together slowly in 5–6 small movements while speaking until they almost touch just before the word "origin," shown in Fig. 15.4 that one applies a force closer and closer to the origin of the spring, and again, as clo – as you get closer to the origin of the spring, it hardly stretches at all. Therefore, the further away you are along the spring, the more it stretches....So a spring that's twice as long, I'm now quite sure, stretches more.

Here the extreme case is generated in order to make a few predictions for the base of an attempted analogy to a double length spring. Again, the subject saying with accompanying depictive hand motions: "I'm imagining that one applies a force closer and closer to the origin of the spring, and it hardly stretches at all." provides interesting evidence that the subject is actually imagining a specially constructed extreme case and inspecting the amount of effort required, in order to gain a higher level of confidence.

15.3.3 Analysis of Transcripts

I will now consider in more detail hypotheses that are supported by the transcripts in both episodes above.

15.3.3.1 An extreme case can raise confidence in an earlier prediction via an untested thought experiment

In both episodes the subject starts from a nonextreme case comparison, makes a prediction with medium confidence, generates an extreme case, and then makes a prediction with high conviction. In other words, the extreme case serves to raise the subject's confidence in an earlier prediction for the same problem. The novelty of the extreme cases argues that they are untested thought experiments. So far these findings are primarily observational ones, but the next three findings concern cognitive mechanisms.

15.3.3.2 Intuition-based Imagistic Simulations were Present

Using the guide in Table 12.3, and focusing on both the original and the extreme cases together, there are several indicators giving evidence for intuition-based imagistic simulations here. Imagery reports occur in both cases; in both cases these are also dynamic imagery reports and kinesthetic imagery, reports; the depictive hand motions are also evidence for imagery, and the fact that they portray a human action suggest the involvement of a self-evaluated intuition schema; finally, this is also supported by personal action projections present in both cases where the subject is imagining the amount of force applied by his hands, since the original problem does not refer to any human actions of this kind.

These observations provide evidence that intuition based imagistic simulations were present. The two examples also add to the arguments in Chapter 13 that the phenomenon of imagery or simulation enhancement provides evidence for the importance of the imagistic simulations. This is supported by the following observations: predictions are reported immediately after the imagery indicators; and if one assumes the involvement of imagistic simulation in the thinking about both the original and extreme cases, the fact that the subject goes to the trouble, after an initial prediction, to generate a second "improved" case and run it, argues for the importance to the subject of imagistic simulation in the solution.

15.3.4 Sources of Conviction in Imagery Enhancement

The extreme case episode of the very short twisting rod poses an interesting challenge for theory because it simply seems to repeat the same reasoning as the previous twisting episode, but yields a much higher level of confidence. In this section, I argue that the conversion of implicit to explicit knowledge is the most plausible hypothesis for the source of conviction. Weld (1990) proposed that one mechanism for the effectiveness of an extreme case is to allow access to the second of two data points (pairs) for the values of two related variables. If one assumes a monotonic relationship one can predict an increasing or decreasing function from knowing two data points. But how can considering the extra extreme cases above add so much confidence since the subject has already just consulted his knowledge on this issue and already has the equivalent of at least two "data points"? It is difficult to see how this small change in the value of one variable in the extreme case could generate a

15.3 Imagery Enhancement Phenomena

new deduction about the variables to produce considerably greater conviction. And his saying for the very short rod: "I realize very clearly that it will get harder" indicates there is something special about the extreme case that makes it count more than simply adding a third data point from which to induce a pattern. It is more plausible to interpret this process as "imagery enhancement" (or "simulation enhancement") – that the role of this extreme case is to enhance the subject's ability to run or compare imagistic simulations with high confidence, and that this comes from increasing the difference between the two images being compared and making that difference more detectable under inspection of the images. In this case the main source of conviction in the simulations appears to be the tapping of implicit knowledge embedded in a motor schema and its conversion into explicit knowledge. The extreme case makes differences in implicit expectations more "perceivable" in this case.

With regard to the hypothesis that the extreme case activates a new "data point" – a new specific fact or episodic memory – the novelty of the extreme cases argues that the origin of this new confidence is the imagistic simulation process repeated on the extreme case rather than the activation of another fact or episodic memory. And both extreme case sequences have a quasi-continuous character where the subject talks about a variable changing as he moves his active hand closer to an end point. This argues that we are seeing another trial of the same thought experiment using the same perceptual/motor schema in the extreme case, not invoking a new and more abstract propositional rule or episodic memory. In fact in the case of twisting an extremely short rod, the subject maintains the very same hand positions continuously as if holding a rod in front of him during both the original long/short comparison and the extreme case comparison. Thus, the language and actions in the transcripts of the extreme cases appear to indicate that the subject is simply "rerunning" the simulation in his head during the extreme case by running through the action and examining the outcome, not activating a new memory or propositional rule. This lends support to the imagery enhancement hypothesis.

A second problem that questions the adequacy of describing this as "accessing a stored data point symbolically" is its difficulty in explaining the hand motions and imagery reports. Why did the subject exhibit these? The fact that he did so suggests the view that he was applying knowledge that was not stored as a linguistic description. And if it were already explicitly described, then why speak of forming an image of the situation and making the effort to run through a simulation of it? Why not just report it?

Thus the imagistic simulation concepts developed so far appear to offer the best explanation of the effectiveness of the extreme case at the end of the transcript above as an example of "imagery enhancement," or more specifically in this case "image comparison enhancement." The fact that this simple change of degree in one of the targeted variables increases the subject's confidence significantly is difficult to explain at this level via models using a general symbolic rule. On the other hand, the imagery-based explanation can explain the presence of a weak and a strong prediction for very similar cases by referring to the degree of contrast between images being compared.

15.3.5 Implications of These Extreme Case Examples for a Theory of Thought Experiments

The phenomenon of simulation enhancement provides an important source of support for the general theory of sources of conviction in thought experiments proposed earlier, since that theory, summarized in Tables 15.4 and 15.6, includes elements that are important for explaining detailed transcripts on how extreme cases work as untested thought experiments.

Element 3b in Table 15.4, Tapping implicit knowledge, was used in explaining the effectiveness of extreme cases like the very short twisting rod, by saying that because the extreme case increases the contrast between the images being compared, confidence in the result rises. If, as I have argued, one can eliminate the alternative hypothesis discussed above of the extreme case activating some separate explicit memory, then tapping implicit knowledge via imagery enhancement emerges as the most plausible theory and provides some initial support for the general theory of sources of conviction.

I hypothesized that this knowledge was accessed weakly by the original comparison of longer and shorter rods, but it was only accessed strongly in the comparison using the extreme case. In daily life we receive sensory signals, some of which are too weak to notice or detect with confidence. The above explanation makes sense if what is being sensed internally is something like a strong vs. weak "perception" of an event or comparison. This is hard to explain in a model where knowledge is stored explicitly and symbolically because it should either be there or not there, not vary on a "strength of signal" continuum. It can be explained in this case by saying that perceptual/motor knowledge implicit in the schema was converted to explicit knowledge in an imagistic simulation comparison, and that the contrast involved was easier to inspect for the extreme case. This explanation draws on two of the items listed in Table 15.6 as factors that could influence the ability to run an imagistic simulation with conviction. The lack of "nearness of the question being examined to previous experience" helps explain the initial weak predictions prior to the generation of the extreme case. And the effectiveness of the extreme case is explained by its increasing "the magnitude of the effect (in this case the contrast) being imagined."

The transcripts are also difficult to explain in terms of a new symbolic rule being deduced from other rules. When the extreme case is run, the answer or hypothesized relationship has already been verbalized explicitly; yet the subject chooses to return to an imagistic simulation and reenact it as if consulting the imagery produced was the main source of information and confidence. There should be no reason to make the extreme case comparison if the first comparison was deduced from other rules and propositions. And deductive rules should operate in just the same way on the normal and extreme cases – there should be no added value in using the extreme case finding since its qualitative form is exactly the same. The subject seemed to be still striving in the extreme case for a clearer image comparison that would raise confidence. Thus this is a case where "stating a rule does not a conviction make" and the simulation of the extreme case itself appears to be very important to the subject's convictions. This is consistent with the idea that implicit knowledge is being tapped and converted into explicit knowledge. This would seem to be a case of *generating* a symbolic rule from deeper analogue sources than recalling a rule or inferring it from prior rules. This is reminiscent of the classic example of being asked to count the number of windows in one's dwelling from memory, where most people report imagining being in rooms of their dwelling to do this, and are successful. A natural way to explain this is that one can generate an image, which under inspection using a new question, contains implicit knowledge that can be made explicit.

One might conjecture that these extreme cases move the subject *away from* rather than *closer to* what is familiar to them and therefore should be *less* effective for confident prediction. Yet using the idea of extended schema application (3a in Table 15.4), one can explain that the subject's schema is able to extend adaptively to make a prediction for the extreme case in each instance, and when this is done successfully it sharpens the contrast between images being sought.

In summary, the effectiveness of the above extreme cases has been explained as enhancing the drawing out of implicit knowledge in an extended application of a perceptual/motor schema. This explanation uses some of the elements shown in the theory of sources of conviction in untested thought experiments developed earlier in Fig. 15.3. Since that theory provides the basis for the most plausible explanation found for the extreme case's effect on the subject's confidence, this is also a source of support for those elements of the theory. In the next section, I will examine other types of enhancement that appear to be aimed at increasing the effectiveness of other sources of conviction shown on the left-hand side of Fig. 15.3, thereby lending further support to the theory.

15.3.6 Imagery Enhancement Focused on Enhancing Spatial Reasoning or Symmetry or Compound Simulations

Enhancing spatial reasoning by adding markers. In Chapter 13, I introduced the concept of "imagery enhancement" with the example from S6 of mentally putting "paint dots" on the spring wire so that he could try to imagine their direction of movement during stretching. I defined an imagery enhancement report at an observational level as the subject: adding markers to a situation or changing other features that do not affect causal relationships such as: (1) orientation or size or (2) magnitude of problem variables – in conjunction with evidence for imagery. More direct evidence occurs when the subject speaks of modifying a problem situation in a way that makes it "easier to imagine." These are indicators of successful enhancement if they increase the subject's confidence in the results. I hypothesized that adding the paint dots was an attempt to improve the case being simulated so that it was easier to study the image produced in the simulation. In the protocol the subject: indicates that the imagery being attempted is difficult; imagines dots on the wire; reports imagining pulling on the spring and "asking you know, how would I see the dots move."

> 022 S5: ...suppose I had a big spring and I could make little paint dots on it all along its length...and saying... would I see a torsional displacement of the paint dots. And what would it look like? And I have a hard time imagining that because you know, the torsional displacement that come to mind are very small.
>
> 024 S5: (Makes drawing in Fig. 13.6a of spring with paint dots on outside of wire)

Although he indicates the imagery is still difficult, he does in the end reach a point where he is "quite satisfied," and when asked whether he was thinking about an equation, he says:

> 073 S5: Oh, no. This is all er, I think very experimental. What I think I have – this image of this line of paint dots on a spring and you know I'm pulling on the weight. I'm going pull and release, pull and release and so I'm constantly putting it through its paces. And asking you know, how would I see the dots move? (Complete transcript appears in Chapter 13, section 13.3.5.)

The paint dots episode suggests another heuristic of the form: imagine "markers" on the objects to make it easier to track changes in key features or variables" in Table 15.7. I put this case of enhancement in a closely related but slightly different category from the extreme cases discussed above: I hypothesize that the role of the markers is to enhance the imagibility of a result with high confidence. Whereas the role of the previous extreme cases is to enhance the subject's ability to run a schema with high confidence.

15.3.7 *Enhancing Spatial Reasoning Via Image Size and Orientation*

Earlier in this chapter, I cited another use of spatial reasoning occurring in episode 14 and Fig. 14.4 where the spring wire is replaced by a flat circular ribbon extended drastically to enhance imagery and "make visible" the resulting twisting effect. Since this appears to involve less physics and more thinking about how an object will deform in space, I hypothesized that spatial reasoning was the major source of conviction in this case. A big challenge here is to keep track of the edges of the ribbon during the deformation. Subsequently, the subject improved the simulation by adjusting the ribbon to be a *half* coil of transparent ribbon with black edges of a specific size and shape, and running the simulation backwards (Fig. 14.5). This involved changing the "detectability" by adding visual "markers" (the black edges) and reducing the complexity of the imagery being attempted (by going to a half loop). Presumably this enhanced ability to track and follow movements of parts of

15.3 Imagery Enhancement Phenomena

the image. (The advantage of running the simulation "backwards" may have to do with imaging the more difficult state (twisted edges) first before performing the transformation.) In sum, I hypothesize that these changes enhanced the spatial reasoning process in Fig. 15.3 of imagining new spatial relationships as the ribbon is deformed. The success of this explanation in accounting for the transcript supports the spatial reasoning element of the theory of sources of conviction in Fig. 15.3.

This case is reminiscent of Kosslyn's (1980) finding that subjects reported that for an everyday object such as an apple, it was more difficult to image an extremely tiny or extremely large apple than a normal apple. There appear to be optimal ranges for imagery that mimic to some extent the ranges of normal perception, and part of the skill of using thought experiments may be to design objects of an optimal size and shape for imagistic simulation. In this way microscopic effects may become imageable. This finding supports the validity of some of the possible heuristics listed in Table 15.7.

This case also illustrates that there can be more than one round of enhancement. This suggests viewing enhancement as part of a "case evaluation and modification cycle" or "enhancement cycle" for improving and tuning cases for imagibility. This is similar to the evaluation (GEM) cycles identified earlier in this book for improving analogies and models.

15.3.8 Symmetry Enhancement

Enhancement may be possible in the case of symmetry arguments as well, but since there is only one example here this is difficult to speculate about. The move in episode 15d from a vertical spring to considering a horizontal spring with forces applied at both ends can be construed as a way to enhance the intuitive symmetry of the situation to establish a clear result. This result must then be projected back to the vertical spring using the problem assumption of negligible mass for the coils of spring wire. Whereas the earlier extreme case examples enhanced the detection of a *difference* in comparative simulations, symmetry enhancement cases like the horizontal spring appear to enhance the detection of an *equivalence* in comparative simulations (by arguing that the slopes and coil separations should be identical in both halves of the horizontal spring).

(Note: Is symmetry one aspect of spatial reasoning? A question for theory here is to decide whether to consider symmetry arguments to be a form of spatial reasoning. This proposal seems reasonable since the term symmetry has a strong perceptual connotation. An argument against the proposal is that the kind of symmetry arguments discussed here and in other areas of physics involve the symmetry of *actions*, not just static perceptual patterns (e.g. the actions of pulling on both ends of the spring above). On the other hand, the meaning of spatial reasoning in this theory is also heavily imbued with action, since it includes processes for imagining rearrangements of objects and the new spatial relations generated by those actions. Therefore the decision might hinge on factors such as whether we want to include ideas about the symmetry of forces under spatial reasoning. I have not taken that path in the present theory, preferring to think about forces as represented by more domain-specific schemas that are less general than spatial reasoning operations. So I leave symmetry processes in a separate category in Fig. 15.3. But this choice might be modified in the future, depending on factors such as whether new findings on modular brain systems for spatial reasoning are shown to include ideas about forces.)

15.3.9 Compound (or "Linearity") Enhancement

The ability to run a compound simulation involving more than one schema operating on the same case suffers from the problem that there may be unanticipated interactions between two processes operating in nature at the same time. The subject reflected this worry when immediately after his key torsion insight in the hexagonal coil case he changed to thinking about the square coil in episode 11A, saying:

> Let me accentuate the torsion force by making a square (Fig. 6.6b) where there's a right angle. I like that, a right angle. That unmixes the bend from the torsion.

One way to view his move here is as a way to enhance a compound simulation, that is, he feels that it will be easier to run the compound simulation accurately for the square case than for the hexagonal case. Physicists might call this an enhancement to seek linearity, or independence of causal factors in, the system.

15.3.10 The Effectiveness of Enhancement Can Be Explained Using the Present Theory of Conviction in Thought Experiments

In the four cases discussed in this section, the subject appears to attempt to enhance an imagistic simulation by making a small modification to the problem situation. (e.g. paint dots or shape of wire). As in the section on extreme cases we can pose the alternative hypothesis that these superficial modifications might have activated a new memory in the subject and accessed additional knowledge rather than enhancing an imagistic simulation or spatial reasoning process. But again the novelty of the enhancement cases of the flat metal ribbon and the paint dots on the spring argue against these being sources of specific facts or episodic memories. And in the extreme case and paint dots episodes it appears that in the enhanced case the subject is conducting the same simulation he has run before in a way that suggests that the purpose of the paint dots is to improve the imagery.

I can now review the processes involved in increasing conviction via enhancement in untested thought experiments as shown in Table 15.8. The finding that subjects

15.4 Imagistic Simulation in More Complex Reasoning Modes

Table 15.8 Imagery enhancement heuristics for improving cases for imagistic simulation

A. Enhancing the application of a schema in a simulation
   (e.g. by generating an extreme case that helps tap implicit knowledge in the schema by increasing the contrast in comparisons between simulations)
B. Enhancing spatial reasoning
   1. Use of "markers"
   2. Making features more detectable with image size and orientation, etc.
   3. Using a simpler case
C. Enhancing equivalence in comparisons via symmetry enhancement
D. Minimizing variable interactions in compound simulation via a "linearity enhancement"

make the effort to maximize imagibility/runnability through enhancement strategies attests to the value of imagibility to them. This *supports the hypothesized importance of imagistic simulations in the thought experiment process.* The marked increased confidence in prediction from extreme cases was explained by their role in enhancing the process of making implicit analog knowledge in a perceptual/motor schema explicit during imagistic simulations. This is difficult to explain in other ways such as the use of a general rule structure operating on static symbols.

Other examples I have discussed show how subjects who go to the effort of using different kinds of enhancement strategies can increase the effectiveness of most of the sources of conviction shown on the left-hand side of Fig. 15.3, including imagistic simulation, implicit knowledge, compound simulation, spatial reasoning, and symmetry. Thus *the phenomena of simulation enhancement provide important additional support for the theory of untested thought experiments and sources of convictions* presented in that figure. When these sources are strong, they can explain the experience of strong conviction from a simulation of an unfamiliar case in a thought experiment. The theory presented in the above sections is a more detailed and empirically grounded answer to the thought experiment paradox than I have encountered previously. However, many details still remain to be worked out and there is a need for further research. I have dealt with this question so far primarily for the case of isolated thought experiments. More complex and sophisticated forms of reasoning such as "custom designed" evaluative Gedanken experiments will be discussed starting in the next and in later sections. However, the factors discussed in the present section will be viewed as the most fundamental sources of conviction in thought experiments.

15.4 How Are Imagistic Simulation and Thought Experiments Used Within More Complex Reasoning Modes?

15.4.1 Four Important Types of Plausible Reasoning

Now that I have presented evidence for the role of imagistic simulation in untested thought experiments and sources of conviction, I will discuss the way that these processes can be used within the four more complex plausible reasoning modes in

Table 15.2: compound simulations, running analogies, explanatory models, and Gedanken experiments. In this section, I wish to address the following questions:

- In which of these more complex reasoning modes are untested thought experiments observed?
- Can imagistic simulation be used in each of these modes?
- How are predictions made in these modes?

In order to address the above questions I will outline a very basic process model for each type of plausible reasoning mentioned above. First, I will review data showing that each can involve untested thought experiments. Secondly, I will propose that all of them can involve imagistic simulation as a core subprocess. I will then use this analysis of imagistic simulation within modes of plausible reasoning to comment further on questions like the thought experiment paradox question. Table 15.9 shows how the examples of thought experiments given so far fit into these categories

Table 15.9 Generative plausible reasoning processes that can utilize imagistic simulation in an untested thought experiment

Thought experiment used within what mode of reasoning?	Examples	Imagery evidence from subject interview?	Typical stage in qualitative explanation process	Common function
Elemental imagistic simulation in an Isolated thought experiment	– Stretch original two springs in problem – Very narrow spring – Drastically stretched spring	Yes Yes	Early on its own and throughout within other processes below	Predict behavior of a system
Analogy	– Foam rubber – Hexagonal coil stretching[a] – Twisting rod – Very short twisting rod	 Yes Yes Yes	Early or middle	Generate predictions for target or generate ideas for constructing model
Running a qualitative explanatory model	– Twisting in stretched square coil[b] – Bending in spring[c] – Paint dots on spring – Twist in quarter coil of spring	Yes Yes Yes Yes	Middle	Provide explanatory model of mechanism
Evaluative Gedanken experiment	– Flat circular ribbon vs. Straight horizontal ribbon – Vertical band spring – Tight and loosely coiled piece of wire – Torsionless coil[c]	Yes Yes Yes	Late	Evaluate model

[a] Also a compound simulation.
[b] Also a compound simulation. Although the square coil begins its life as an analogy, the subject eventually analyzes and develops a model for its operation as a target of its own.
[c] Also a compound simulation since it involves multiple elements.

15.4 Imagistic Simulation in More Complex Reasoning Modes

by showing whether the experiment was simply a direct running of an isolated case in row 1 or a subprocess to a more complex mode of reasoning in the other three rows. A fourth mode of complex reasoning that can operate in conjunction with others, compound simulation, is indicated by superscripts. A "yes" in column 2 indicates that there are one or more imagery indicators present in that episode and that this came from an external subject's think-aloud interview.

There are close calls here in deciding whether to call cases like the twisting rod and the hexagonal coil analogies or explanatory models. These are borderline cases – see Chapters 6 and 7 for a further discussion. In some cases it is necessary to distinguish between the intent for which the case is created, for example, the hexagonal coil as a bridging analogy, and a later role that the case eventually plays. (The hexagonal coil eventually provokes activation of torsion as a provocative analogy.) Since every example in the table incorporates a thought experiment, the table shows that for each of four modes of reasoning, untested thought experiments can be a useful subprocess within that mode. The extension of this finding to mathematical Gedanken experiments and models will be discussed in a later section.

Figure 15.5 shows the most basic hypothesized relationships between these processes. This diagram is not intended to depict the order in which processes

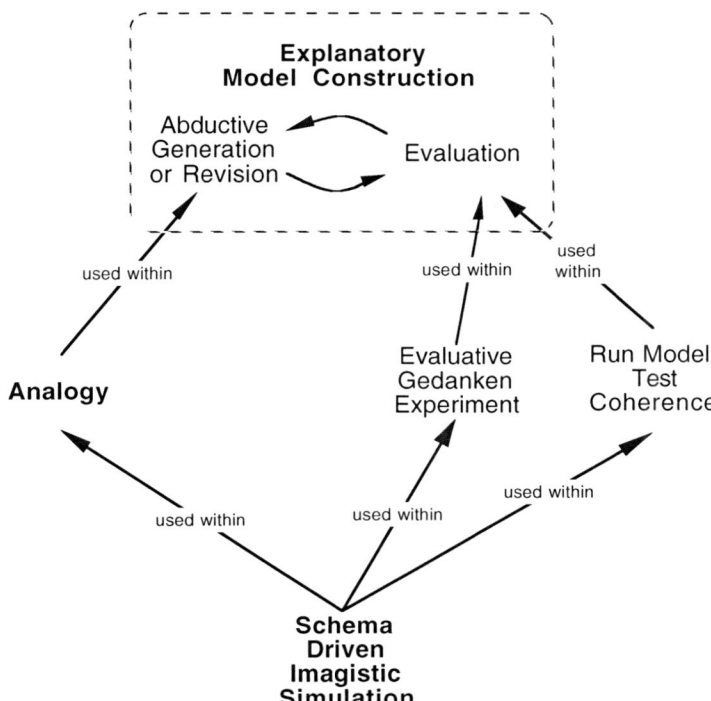

Fig. 15.5 Some major nonempirical reasoning subprocesses used in qualitative explanatory model construction

occur, but simply to show dependency relations between processes and subprocesses involved in explanatory model construction. Three sections of the book so far have been organized around the three important processes shown in bold type on the left but other reasoning processes are now being added. Imagistic simulation is shown as a fundamental subprocess for all of the others. Explanatory model construction on the other hand utilizes all of the other processes. This figure signals a move to a new level of description, since the first part of this chapter focuses on how isolated simulations work in thought experiments, whereas the present section focuses on how such simulations can be utilized and combined within larger reasoning patterns, as illustrated in Figure 15.6.

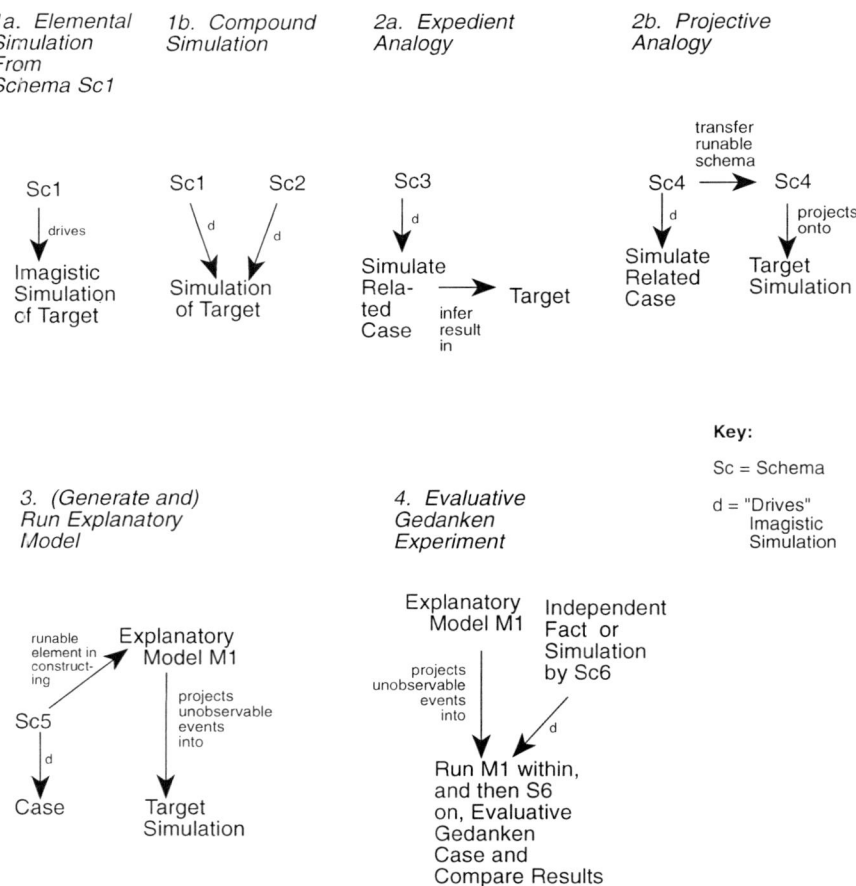

Fig. 15.6 Schemas generate imagistic simulations in four types of nonformal reasoning. For new target cases, all can involve imagistic simulation(s) driven by runnable prior knowledge schemas as sources of knowledge

15.4 Imagistic Simulation in More Complex Reasoning Modes

There are two levels shown in each part of the figure: below, an imagistic representation that captures features of a particular exemplar and its behavior; and above, a schema-based representation that is more general and embodies conceptions of a physical process (such as bending) and that is capable of driving an imagistic simulation for a particular exemplar. The figure does not show complete processes of knowledge development or order of activation but includes only information on various sources of new knowledge during each reasoning process, as follows:

(Part 1) *Elemental imagistic simulation in an isolated thought experiment*. We have already analyzed several of these, as shown in Table 15.9. They were explained in this chapter as involving the direct application of a single perceptual/motor schema to a case in an imagistic simulation. For example, in the spring solution in Chapter 14 (episode 3) the subject imagined stretching the original two springs in the problem and had a direct physical intuition that "a bigger [wider] spring is looser, and would stretch more," with accompanying depictive hand motions, based on what he called a kinesthetic sense. Thus his transcript contains imagery indicators. Earlier this subject had said that the problem was unfamiliar and that "he might have real problems conceptualizing it," so I do not treat this "kinesthetic sense" as the routine application of a schema in its normal domain. I describe this process as using an existing schema to generate an elemental simulation to answer a question that has not been answered before. Part 1a in Fig. 15.6 shows a schema being applied directly to a question that is new and outside its normal domain of application but that is within a range that the schema can adapt to. This gives us our lowest level source of new knowledge, as a schema is extended or "stretched" to apply to a new situation in a thought experiment.

(Part 1b) *Compound simulations*. In 1b two (or more) schemas are applied together to the same case. This occurred for example with the case of the square coil, where the subject imagined bending and twisting in the elements occurring simultaneously. One form of reasoning falling into this category is the familiar chaining of operators found in classical problem solving theory where a chain of actions transforms one state into others. However, here this is thought of as a sequence of image producing simulations rather than manipulations of symbolic tokens. A less traditional form is the coordination of two schemas operating in parallel on the same image. This would occur in the square coil if one attempts to visualize whether bending interacts with twisting to cancel or enhance it. A simpler example is imagining the joint action of two forces on the same object. Combinations of schemas operating in compound simulations for a complex case can generate unexpected interactions that produce new knowledge.

(Part 2) *Analogies*. Part 2 of Fig. 15.6 shows an analogical inference occurring from a well-understood analogous case to a target situation in two possible ways:

(a) *Expedient analogies*. In an expedient analogy such as the "two pieces of foam rubber with large and small holes" in Fig. 3.1 if, the novelty of the case argues for its untested nature. Figure 15.6 shows the theory that only a relation such as the end result "the one with larger gaps deforms more" is transferred in such cases.

(b) *Projective analogies.* However, I hypothesize that a projective analogy (such as the wheel seen as a pulley analogy in Chapters 4 and 17) does more than allow one to add a discrete symbolic relation in the target; a schema-driven imagistic simulation is also projected onto the target from the analogous case.

Kinesthetic imagery reports provide evidence for such imagery in statements such as S7's in the wheel problem:

> "106 S: And you're over here pulling like this. That feels like you're on the outside of a pulley pulling up."

The subject uses the same drawing to talk about the wheel and the pulley analogy. Here the analogous case is imagined to "act in the same way as" the novel, untested, target case in its operation, rather than merely adding a new discrete symbolic relation to the description of the target. One can hypothesize that a simulation involving dynamic spatial relations is projected from the analogous case to the target here and that the full process for a projective analogy, would be:

- Generation of analogous case.
- Analogous case triggers/activates schema A (for thinking about pulleys).
- Schema A is "run" to produce prediction for base and confidence in this prediction is sufficient to continue.
- The validity of the analogy relation is somehow confirmed.
- Schema A assimilates the target case and is run adaptively to provide an imagistic simulation and prediction for target.

In this view the analogy is helping the subject "project" a new image onto the target image to "see it in a new way" in terms of a new schema. That is, in projective analogies there is a "transfer of imagery" that goes beyond inferring a discrete symbolic relation.

In both types of analogy considered here the subject can have a schema as the source of knowledge driving an imagistic simulation of the analogous case. Thus the new knowledge in many analogies (and some extreme cases) is the connection to an intuition schema whose relevance to such a situation had not been seen before. For example, the above analogy above allows knowledge in the form of a runnable intuition about pulleys to be applied to a problem about a wheel.

However, just because the subject has a confident intuition about the outcome for the base case does not guarantee that this confidence will transfer to the target case. That takes confidence in the validity of the analogy relation as well as in the outcome of the base case. This makes reasoning by analogy more complex than using an intuition in an isolated thought experiment.

(Part 3) *Running an explanatory model.* The process (2b) of projective analogy above and in Fig. 15.6 foreshadows the more complex situation in (3) where a runnable schema is incorporated into an invented explanatory model, which can then be projected and run as an imagistic simulation of the mechanism operating in the target. Differences between explanatory models and "mere expedient analogies" were discussed at the end of Chapter 14. An example is: in Fig. 14.7a, a twisting

15.4 Imagistic Simulation in More Complex Reasoning Modes

schema is incorporated into the "Twist in an element of the spring" model. This model is projected onto a quarter turn of the spring and run. This establishes for the first time a model incorporating twisting that can actually be run to envision stretching in a circular coil. Running the twisting model helps evaluate the model by showing that it can explain the behavior of the target. A second historical example is a schema for elastic collisions being incorporated into a model of bouncing molecules being projected into a gas. In these examples I hypothesize that each model is run to yield predictions about the target case, allowing the model to be evaluated. In this view an explanatory model combines the characteristics of being invented, plausible, and projectable.

How do explanatory models utilize thought experiments and imagistic simulation? Another example that, although it turns out to be incorrect, also illustrates admirably these characteristics, is the bending model applied to the spring in episode 9 discussed previously. Hand motions in that episode gave evidence for simulation occurring, and the simulation led to a novel prediction of increasing slope. One can imagine ways in which running the bending model can trigger processes in Fig. 15.3 that answer the following questions about the model:

- Does it "fit" and is it spatially coherent when projected into the target and run? (Yes, small bending elements could fit into the spring.)
- Does it explain the targeted features of the phenomenon at hand? (Yes, bending might produce overall stretching.)
- Is it coherent with other features of the target? (No, running it leads to recognizing asymmetric stretching with unequal distance between coils.)
- Does it form other coherence relations with other theories? (No).

Thus Fig. 15.3 begins to provide a mechanism for theory evaluation via running a new model.

One can consider two ways in which imagistic simulation can be involved in explanatory models. During model construction the isolated simulation of an analogue source schema such as the twisting rod in episode 12b, may serve to confirm and establish a prediction for that simpler case. In the second way, once the source schema has been "built into" the model, the model can be run to yield predictions about the target. In another example, in episode 11A the simulation of twisting is projected into the square coil model as indicated by hand motions in an appropriate orientation and position. That is a more unfamiliar and difficult simulation than simply twisting a rod, and may be an extended or compound simulation, or may involve tapping implicit knowledge as shown in Table 15.4. It may be well outside the domain of normal application of the twisting schema. That is, running a new model can be an untested thought experiment in it own right. In fact, assuming we are talking about the subject developing a new model of a hidden mechanism that they have not observed, running it will always be a kind of untested thought experiment. Since a new explanatory model is the most important "prize" to be sought in scientific investigation, this view raises the incentive considerably for maintaining a research interest in untested thought experiments.

This illustrates how the sources of conviction in elemental thought experiments can also be sources of conviction for an explanatory model. As a historical example, if I have a model of an electromagnetic field that involves gears and the gears are in a configuration that jams and will not turn (as Maxwell did), then the model lacks internal coherence (Nersessian, 2002). In repairing this by adding more gears, I use my convictions about gears to build confidence in the internal coherence of the new model. Referring back to Fig. 15.3, a competent spatial reasoning system that can coordinate actions (of more than one gear or schema) in space in compound simulations would seem to also play an important role as source of conviction in running the model, and thereby helping to evaluate it.

Although Part 3 in Fig. 15.6 shows both model construction and running the model, I have focused in this section on running the model, since the construction process is potentially much more complicated, is not a simple reasoning process, and will be the topic of several later chapters.

(Part 4) *Evaluative Gedanken experiments*. Whereas I have defined an untested Thought Experiment (in the broad sense) as the act of making a prediction for an untested, concrete situation (exemplar, case, or "experiment") I defined the term "evaluative Gedanken experiment" in a narrower way as the act of making a prediction for an untested, concrete situation designed or selected by the subject to help evaluate a scientific theory (i.e. evaluates an explanatory model and/or its mathematical elaborations.)

Earlier I introduced the case of the "band" spring in Fig. 14.3 involving the vertically oriented strip instead of a wire wound into a spring. This imagined device resists vertical bending but allows twisting. This appears to be a strategy for evaluating the plausibility of the bending mechanism. By using a vertical sheet or strip of material rather than wire, he removes the mechanism of bending in order to see whether it changes the spring's behavior. But this doesn't appear to make the spring stop stretching – the spring still stretches nicely. So bend is not the only source of stretch. Thus, the subject appears to be removing one possible mechanism for producing stretch (bending) to see if it produces much less stretch. Since it does not in his estimation, it reduces the credibility of bending as a necessary mechanism. The hand motions and imagery reports in this segment provide evidence that imagistic simulation was involved here. This is also a "controlled" experiment in the sense that the subject has selectively removed bending without removing torsion as a possible source of stretching.

15.4.2 *Evaluative Gedanken Experiments as the Most Impressive Kind of Thought Experiment*

Such Gedanken experiments appear to be one of the most dramatic and sophisticated uses of simulations, because they have the power to shed serious doubt on a valued theoretical model and, among thought experiments, they are potentially the closest to a critical experiment in science. One designs a special case (the "band" spring) where the hypothetical model (most stretch comes from bending, shown as M1 in

15.4 Imagistic Simulation in More Complex Reasoning Modes

Fig. 15.6) yields a prediction (eliminating bending should remove most of the stretch). But one also has some other independent source of information that can confirm or deny that prediction. As shown in the diagram, the experiment works via a sort of "pincer movement," where the subject ingeniously invents a case that can be predicted via both the model and an established independent schema. The most interesting case occurs when both sources generate imagistic simulations (as with the subject's strong intuition about the band still being "stretchy," shown as S6 in Fig. 15.6). Each of the schemas can then be run on the Gedanken case independently and compared, allowing the subject to discount or support the model.

In another Gedanken experiment in episode 10a, the subject has already generated a microscopic explanatory model of shear forces and displacements in the wire that shows one element being displaced downward relative to the previous element. He draws wire element displacements as a kind of slowly descending, spiral staircase. However, by running a Gedanken experiment he discovers a contradiction: this model predicts that stretch in a loosely coiled up wire and a tightly coiled wire of the same length should be the same. But this is counter to his intuition since in running this comparison macroscopically without analyzing it, he imagines that the tightly coiled wire would stretch less. This case is not the same as the original problem because the wire length in this case is the same for both springs. It is a Gedanken experiment that pits the shear model against strong intuitions about the behavior of loosely and tightly coiled material, and it yields a striking and decisive contradiction to his model. Multiple instances of depictive hand motions in this episode provide evidence for imagistic thought.

A third beautiful example of a Gedanken experiment from episode 13a is the torsionless coil made of small cylinders with bearings that can turn "with perfect ease" (Fig. 14.2). It appears to have been designed to evaluate the torsion model by testing the necessity of torsion as a causal factor in springs. The subject's torsion model predicts that there will be little or no restoring force in a torsionless coil. The protocol here is sparse, and there are no quotable imagery indicators for this Gedanken experiment (by the same subject that generated the first "Band spring" experiment described above), but if its prediction was generated in the same way, we can explain it as a second example of an imagistic simulation generated by applying a familiar physical intuition scheme. One can run a simulation, starting at the top end of the torsionless coil, repeatedly using a physical intuition schema (independent from the model) to imagine that large segments of the coil will rotate downward until the spring collapses into a (somewhat crooked) vertical chain of elements (this is easiest to imagine in my own case with a torsionless bearing placed every 180°, or 90°, along the spring coils). Here the intuition schema being used is the prior knowledge that an object that is free to turn on an axle will naturally rotate so that its mass hangs down below the axle at the lowest possible point. In contrast to the bending mechanism in the first Gedanken experiment above, this suggests that torsion plays a very important role, by showing that when it is eliminated, it destroys the power of the spring to resist stretching.

Presumably the predictive power of this prior knowledge schema comes in turn from the factors listed in Table 15.4. If the prediction from running this schema conflicts with the prediction from the model being tested, it will count as evidence against the model. If the predictions coincide, then it can count as supporting evidence for the model. This is why I call such Gedanken experiments "evaluatory." Another important plausible source of knowledge, shown in Table 15.4 and Fig. 15.3 but not shown explicitly in Fig. 15.6 is the subject's basic spatial reasoning system and naive physical knowledge of how objects can be placed and oriented in space. I assume this knowledge can contribute to each simulation shown in Fig. 15.6 and is an important resource present in any thought process involving imagery. Here it contributes to imagining the spatial configuration of the coil elements after they have collapsed. So during the "running" of the experiment, both schema Sc6 and spatial reasoning are sources of confident knowledge, independent of the model, that allow the subject to check on the plausibility of the model when applied to a specific case. The torsionless coil is a classic example of a Gedanken experiment, novel and untested, designed to evaluate the torsion model by pitting it against a more primitive intuition schema for how the torsionless coil will behave.

Imagery enhancement to make the mechanism directly "perceivable." Another type of Gedanken experiment attempts to make the hidden mechanism of the model directly "perceivable" (amenable to detection in imagery). An experiment used for evaluating the presence of twisting is the stretching of a flat circular ribbon out into a straight vertical ribbon in episode 14 so that a full twist can be seen in the ribbon in Fig. 14.4c. This thought experiment provides further evidence at a macroscopic level for the twisting effect, by using a special case. This thought experiment appears to use the imagery enhancement techniques of reshaping the wire into a ribbon and stretching it to an extreme. In this case the experiment happens to support the model.

A final evaluative Gedanken experiment in Fig. 14.7 and episode 17 that appears to work via simulation and imagery enhancement occurs when a small segment of the spring is "softened" in order to localize and examine whether there are twisting effects. This serves to make the twisting effect on the wire imageable and to confirm its presence. Similar experiments in Figs. 14.9a, b use the "softened segment" technique again to show how torque can operate on all parts of the spring uniformly. These invented situations appear carefully designed to confirm these features.

The first three of the five Gedanken experiments discussed in this section, the vertical band spring, the torsionless coil, and the recoiled wire appear to constitute a somewhat different species of Gedanken experiment from the softened segment and twisting ribbon experiments. The first three share the characteristic of comparing the predictions of a model to intuitions about the behavior of a particular case. The last two appear to work by imagery enhancement that makes the mechanism more "perceivable." Both types, however, appear to be designed to evaluate a proposed model.

Gedanken summary. In summary, intuition driven imagistic simulations can play a role in evaluative Gedanken experiments as well. Evaluative Gedanken experiments can be impressive because their purpose is to evaluate exploratory or mathematical models, one of the most important tasks in science. They are an advanced form of reasoning because they can involve multiple simulations, because they come late in the model development process to evaluate a model after it has

15.4 Imagistic Simulation in More Complex Reasoning Modes

been proposed, and because some of them have a special "logic" or form as shown in Fig. 15.6,[2] part 4. Because they may provide the most sophisticated examples of thought experiments with high conviction I will give a further analysis of several of these Gedanken experiments later in the section on thought experiments and reasoning.

15.4.3 Multiple Types of Reasoning Processes that can Utilize Thought Experiments Run Via Imagistic Simulations

Figure 15.6 helps in highlighting differences between analogies, model construction, and evaluative Gedanken experiments as reasoning modes. For example, a thought experiment like comparing an extremely narrow spring to a wide one is a direct simulation of the target problem (Type Ia) whereas imagining the new case of a bending rod (Type IIa) is an analogy. Both may involve imagistic simulation but the analogy is a less direct use because a relation must be inferred or transferred from one case to another. Secondly, Fig. 15.6 helps one distinguish between a familiar analogous case and an evaluative Gedanken experiment case. Whereas an analogous case tends to be similar to the target case in an important way, this is not necessarily true for a Gedanken case, depending on the type of experiment. For example, a Gedanken designed to test the situation where the mechanism in the explanatory model has been destroyed (such as the torsionless coil) should be dis-analogous to the target. Thirdly, explanatory model construction is different than Gedanken construction. The anchoring case introduces a schema that was adapted and incorporated into the explanatory model during its construction, whereas the evaluative Gedanken case is designed after model construction in order to test the model.

New knowledge from old knowledge. We have been asking about the roles thought experiments using imagistic simulation might play in scientific theory construction. Four different answers have been given in Figs. 15.5 and 15.6 concerning the form of more complex plausible reasoning processes that can involve untested thought experiments. Each of the numbered processes is more complex than an isolated thought experiment using an elemental imagistic simulation. But all of them can involve an imagistic simulation driven by a runnable prior knowledge schema(s) as a source of new knowledge, as shown in Fig. 15.6. That is, each type of reasoning generates new knowledge from old knowledge. Thus the set of processes illustrates both the diversity and unity of creative expert reasoning.

---

[2] Fig. 15.6 does not purport to be an exhaustive framework – i.e. to show *all* ways that analogies or evaluative Gedanken experiments can take place. For example, certain analogies can involve inference from factual information in a familiar source case rather than inferring new knowledge from simulating a newly designed case. Rather, the figure focuses on a framework explaining how these forms of reasoning can take place for new untested cases via imagistic simulation.

15.5 Are Imagistic Simulations Operating in the Mathematical Part of the Solution?

Earlier I concluded that a significant part of the analysis of a new system can actually be qualitative in nature and that this analysis can involve imagistic simulations. Did the processes generating the more abstract mathematical relations use imagistic simulations as well? I will cite two reasons for an affirmative answer.

1. It is possible to view the causal diagram and equations in Fig. 14.12 as the ultimate outcome of the long process of reasoning in this protocol. However, what is not captured in that diagram is the spatial model that the subject developed so painstakingly, a model that embodies a dynamically runnable mechanism, and that is the foundation or source for developing the representations in Fig. 14.12. Figure 15.7 is then a "hybrid" diagram showing links between spatial and descriptive causal structures. It shows an explicit network of named, isolated variables being constructed in its upper portion as one moves from the starting point of an anchoring conception on the left, to an explanatory model that incorporates this conception on the right. However, the findings from a large variety of episodes described in this chapter argue that the crucial underpinning for such networks of relationships between variables (and also for more precise mathematical relationships that are in turn built on top of these networks) are the

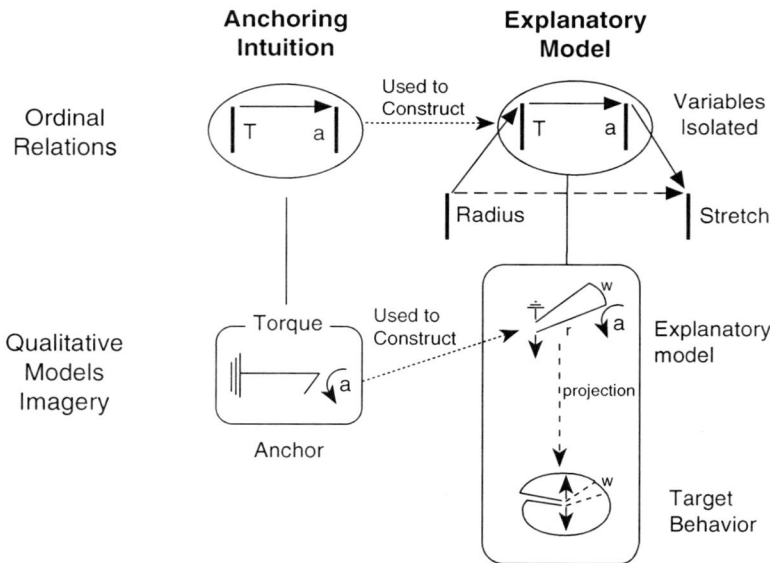

Fig. 15.7 Source analogue is used to construct model at both imagistic and ordinal levels. T = torque (roughly, "twisting force"); A = resulting twist (angular displacement)

images of actions and movements of objects shown in the lower half of the figure. In the view taken here, these are the origins of such variables and relationships and the source of their meaning. As argued throughout this chapter, these origins are difficult to explain via an appeal to processes involving static, discrete symbol structures alone.
2. In a number of the examples in Chapter 14, processes involving the use of imagistic simulations were identified that were at a higher level of precision than the qualitative level in analogies, model construction, and Gedanken experiments. For example, these occurred in episodes 20, 22, 25, 26, 28, and 29. Although these were not from external subjects, they do provide initial evidence for the plausibility of the role of imagery in quantitative arguments. This argues that imagistic simulations were operating in the mathematical part of the solution as well as the qualitative part.

15.6 How Thought Experiments Contribute to Model Evaluation

This question was raised at the beginning of this subchapter on thought experiments in the following form: the paradigmatic tool for evaluating theories in science has always been the laboratory experiment. Can a scientist also detect problems in a model such as conflicts, inconsistencies, or anomalies via an untested thought experiment? If so, how?

15.6.1 Evaluation Strategies

Three mechanisms were identified:

Running an explanatory model: The earlier example of bending causing an increasing slope in the spring illustrated the idea that a thought experiment could lead to the recognition of a hidden conflict implied by an incorrect model. This is an example of evaluating a model via a thought experiment by simply running the model mentally on the target. This is a simple technique, but it would seem to be very important for evaluating the initial plausibility of the model when it can be used.

Gedanken experiments that expose the mechanism: The twisting ribbon thought experiment in Fig. 14.4 appeared to use the *imagery enhancement technique* of reshaping the wire into a ribbon. Like the experimental technique of a "stain" for exposing hidden structure in tissue, subsequent enhancement techniques of making a ribbon of the right size and proportions may improve the inspectability of the imagery even further. Devising an apparatus to study Brownian motion provides another possible parallel with an example from experimental science. This and the stain technique are ways of making an invisible mechanism visible.

The experiments involving a softened segment of the spring to exaggerate twisting effects in Fig. 14.7 provided another example of using techniques to make the mechanism visible. Since they are presumed to access no additional schemas in memory, how can they add conviction? This is a difficult phenomenon to explain using discrete symbolic representations of knowledge and it is explained more plausibly by enhanced imagistic simulation for these cases of "envisioned mechanism."

Gedanken experiments that examine the effects of changes in key variables: Examples like the vertical band spring illustrate that convictions developed during imagistic simulations can be powerful enough to cast real doubt on an explanatory model. In that case the stretchiness of the band spring where bending was impossible argued strongly against a mechanism for stretching that depends solely on bending. In such an example a second simulation of the case can be used as an independent source of conflict, separate from the running of the model itself. Similarly, the visualized collapse of the spring in the Gedanken experiment of the torsionless coil argues strongly against a model of the source of a spring's elastic force that does not include torsion forces.

A scientist using such a Gedanken experiment makes model evaluation into an art form by deftly selecting, or more likely generating, a case that can force new conflict or coherence comparisons to a head – comparisons that speak to a tightly focused key issue. Cases like the "vertical band spring" and the "torsionless coil" also exemplify how some of the same principles that apply to laboratory experiments, such as control of variables, can be applied to Gedanken experiments. As illustrated by the experiments above on the bending and twisting hypotheses, it is also possible for *comparative evaluation of two competing mechanisms* to take place via thought experiments.

In cases like the vertical band spring where the Gedanken experiment leads to conflict rather than coherence, the power of a confident simulation in a Gedanken experiment that conflicts with a candidate model is larger, because of the power of a single counterexample to weigh heavily against a theory claiming to be general (stretching cannot *all* be due to bending). This kind of Gedanken experiment therefore may lead to the most dramatic or powerful cases of the application of convictions from a thought experiment. Also, because two simulations can be involved by being run and compared, it can be one of the most impressive and creative uses of imagery in advanced cognition.

It is interesting that each of the experiments above could in principle have actually been performed given the right idealized materials and resources. Gooding (1992) has specified this as an important characteristic of thought experiments. With these examples there is a strong parallel between thought experiment techniques and laboratory techniques. Of course, thought experiments become even more useful when the laboratory experiment is unfeasible, but still imaginable in principle.

15.6.2 Summary

The question asked at the beginning of this section was whether, in addition to real experiments, thought experiments that utilize imagistic simulation can help one evaluate a scientific theory. In summary I concluded on the basis of examples from transcripts that they could do so in at least the following ways:

- Running an explanatory model on a new case type was deemed to be an untested thought experiment.[3]
- This can produce new coherence or dissonance relations with prior knowledge about target behavior. I cited an example where a new dissonance relation was strong enough to dethrone a favored model.
- An evaluative Gedanken experiment where imagery enhancement is used to expose the mechanism operating in the target.
- A second type of evaluative Gedanken experiment where a special experimental case is designed that can pit an independent source of prediction against the model. In two of the three examples presented here, there was evidence that the independent source was an imagistic simulation.

15.6.3 Combining Reasoning Processes into a Model Construction Process

Figure 15.5 helps to clarify the evaluatory role of running an explanatory model described above. It also foreshadows how the reasoning processes in Fig. 15.6 will be described in the next chapter as serving a larger investigation process. The model construction process at the top of the figure contains a basic GEM cycle that alternates between model generation or revision and model evaluation. The primary role of running the model (or running a Gedanken experiment) is shown to contribute to model evaluation, whereas the primary role of analogy is for model generation or revision. The diagram provides a picture of how imagistic simulation can be foundational for various reasoning processes and thereby for the entire model construction process.

---

[3] In the case of inventing a new mechanism to explain an observed experimental result, there are some fine points to consider in identifying thought experiments. The experimental result itself cannot be a TE because it has been observed. However, the model used to explain it (e.g. an image of gas molecule behavior) could be a TE because one has not observed molecular behavior. Rather one is predicting how the molecules are behaving at the microscopic level in the experimental condition to produce the observed result at the macroscopic level. In this view the prediction about the microscopic activity of the hidden mechanism is part of the explanation of the macroscopic behavior. This stretches the use of the term "thought experiment" in the broad sense to the nonexperimental area of explanatory models. I think this is proper because the thought experiment paradox applies there perfectly well. However, the narrower use of the term "Gedanken experiment" is not intended to include such examples.

Figure 15.8 illustrates how these processes can work together over time by showing in column 1, a single simplest path through most of the processes in Fig. 15.5 along with the structures they produce (column 2). Figure 15.8 shows how imagistically expressed knowledge in a schema at the bottom might be transferred "up the ladder" in that figure over time to develop an analogy, an explanatory model, and a Gedanken experiment to test the model, in that order. Figure 15.8 also separates design processes in column 1 from application or running processes in column 3. One can orient to the figure by considering the explanatory model in row 3 to most often be the scientist's ultimate goal. Other processes are engaged in to serve this goal. Row 2 can be thought of as a subprocess that is "called" by part of the cycle at the left of row 3.

Most of the design processes in column 1 will be described as cycles of generation, evaluation and revision. Thus in a more complete figure there would be return arrows going from the running process in the right column back to the design process in the left column. The design processes can utilize imagistic, constructive transformations to modify a design (discussed in Chapter 16). The application or running

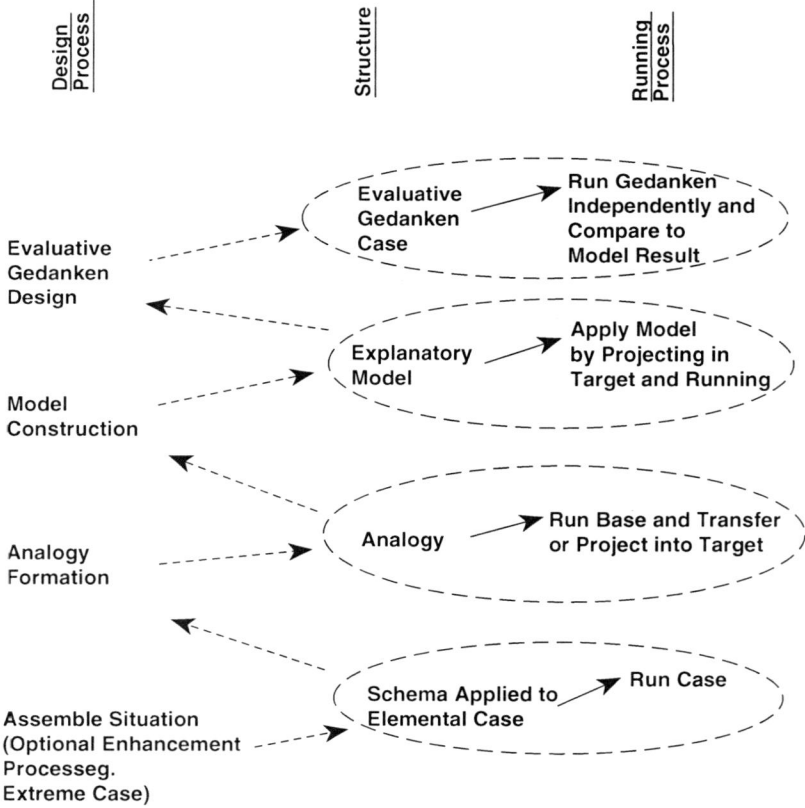

Fig. 15.8 Simplest path through all four major model construction processes

processes in column 3 have been described as (usually compound) imagistic simulations that are driven by perceptual/motor schemas.

Relations between the major processes in Fig. 15.8 are shown by the dotted arrows which show flow of information, not flow of control. These each produce a corresponding structure in column 2. Most of the diagrams in this book have focused only on processes but this diagram also shows structures in column 2 as outcomes of those processes. In the right-hand column, running each structure, once it is constructed, involves activating the structure and allowing it to generate dynamic imagery for the case. By showing how structures feed processes that form further structures, Fig. 15.8 illustrates how new knowledge structures can be constructed from prior knowledge schemas.

15.7 Chapter Summary

We saw earlier that some thought experiments can produce predictions with a high level of conviction. I have provided examples indicating that their value is not limited to providing a simple prediction. The right-hand side of Fig. 15.3 shows four different possible benefits of an isolated untested thought experiment. In addition evidence was provided that all three higher-level complex reasoning patterns in Fig. 15.6 can utilize thought experiments via imagistic simulations as a subprocess. The three figures just mentioned represent three levels of analysis of thought experiments, corresponding to the mechanisms by which they work (Fig. 15.3), the reasoning operations they can be involved in (Fig. 15.6), and the way those operations are combined to serve model construction (Fig. 15.5). Combined, these patterns of use constitute an initial model of how thought experiments work and can lead to convictions, how they participate in reasoning, and how they in turn can contribute to the creative construction and evaluation of explanatory models in scientific discovery.

On the basis of multiple examples from case studies, this chapter also documented the following findings concerning the role of untested thought experiments:

Functions for thought experiments were wide-ranging. Thought experiments can be used to help generate or refine as well as to evaluate models. That is, they can contribute to all three elements of the model Generation, Evaluation and Modification cycle. The examples discussed here indicate that thought experiments can be a potentially powerful and complementary alternative to real experiments at both qualitative and quantitative levels.

Thought experiments can be a source of divergence. Thought experiments are a potential source of divergence because several of the processes in Fig. 15.3 are to a large degree spontaneous and uncontrolled, such as the activation of other schemas, and the emergence of new image features.

Gedanken experiments. I also introduced the concept of Evaluatory Gedanken Experiments: their argument structures can be more complex than that of a simple untested thought experiment, but like them they can also be accompanied by evidence for imagistic simulation (Clement 2006).

Diverse and unified modes of reasoning. The thought experiment mechanisms in Fig. 15.3 can work within the larger reasoning modes in Fig. 15.6. The figures show how these modes are tied together by an impressive set of common similarities with respect to mechanism. Figure 15.6 on the other hand, also highlights the differences between these higher-order plausible reasoning modes and the *way* that they can use imagistic simulation as a subprocess. The finding that imagistic simulation is used as a subprocess for this wide range of reasoning operations adds to the evidence for its central importance in scientific investigation.

Widespread applicability of the fundamental paradox. The finding that thought experiments occur within such a wide range of reasoning operations means that the Fundamental Paradox applies to a broader set of processes than is commonly realized. Being able to make this point justifies the utility of the broad definition of an untested thought experiment used here, which is designed to encompass those situations that raise the paradox.

15.7.1 Addressing the Fundamental Paradox of Thought Experiments: Sources of Conviction

A final set of findings speaks to the question of the fundamental paradox of thought experiments: how can they lead to convictions? The findings on the central role of both prior knowledge schemas based on experience and rationalistic processes for extending that knowledge via imagistic simulation elaborate on and provide evidence for Gooding's (1994) and Nersessian's (1991) view that thought experiments derive their power from a mixture of both an empirical and a rational source. In this chapter an attempt was made to describe a number of such sources explicitly based on case study examples for each source. Convictions from thought experiments appeared to have their origins in imagistic simulations with sources that included the extended application of prior knowledge, implicit prior knowledge, compound effects from more than one schema, and spatial reasoning or symmetry operations, as shown in Fig. 15.3 (Clement, 1994b, 2002, 2003). This theory of the origins of conviction in thought experiments was supported by observations from transcripts. It was also supported by its ability to explain the phenomenon of the *imagery enhancement* strategies attempted by subjects. These were each identified as enhancing one or more of the mechanisms shown on the left and center of Fig. 15.3. These sources and roles for thought experiments have not been sorted out clearly in the past and it is hoped that this analysis will make it easier to study them in the future. Thus an attempt has been made in this chapter to develop a more detailed model of what it means for a concept to be "intuitively grounded" by being "embodied". In this model perceptual motor schemas not only provide a locus for meaning but also have inherent properties of adaptability and modest generality that are of significant potential value in science.

15.7 Chapter Summary

Limitations and next steps. Simulation may not be the only means for operating with thought experiments; some may also use qualitative or mathematical deduction from principles. However, simulations are an extremely interesting means and appeared to be ubiquitous in the present examples. Although schema-driven imagistic simulation can explain a major source of new knowledge in an untested thought experiment, it does not explain how the experiment was generated or *designed*. This comment applies to analogies, explanatory models, Gedanken experiments, and even extreme cases. In the above sections, I have focused instead on the process subsequent to design of *running* of a given experiment so that I could propose mechanisms and outcomes for that process. But in most of the above uses of imagistic simulation, the choice or design of the case to be simulated is extremely important. I cited examples where certain cases appear to enhance the use of schema-based simulation, and presumably part of the art of this kind of thinking is to design a case that maximizes the mind's potential for this process. And the most important task facing the subjects in this chapter was to design an explanatory model that can explain target behavior. Larger design processes for explanatory models will be discussed in the next chapter that utilize many of the smaller reasoning processes described in the present chapter.

Chapter 16
A Punctuated Evolution Model of Investigation and Model Construction Processes

In this chapter, I attempt to develop a higher level model of investigation processes which can account for the different stages of investigation shown in Table 14.1 and other major features observed in the solution in Chapter 14. It will describe three major subprocesses: one for generating initial descriptions at an observational level, one for constructing qualitative explanatory models at a more theoretical level, and one for mathematical modeling. As a first approximation, these are roughly matched to the early, middle, and late stages in the protocol. Direct simulations of the target and expedient analogies will be used primarily within the first description subprocess. In the second subprocess, explanatory model construction, mechanistic plausibility will become an important criterion for evaluating models. In the third, geometric and quantitative schemas will be applied as a way to add quantitative precision onto the qualitative model. However, I first need to discuss a core concept to be used in describing the middle process of explanatory model construction: *abduction*. This concept is used so broadly in our field that it needs to be split into at least two concepts with different labels to be useful.

16.1 Abductive Processes for Generating and Modifying Models

16.1.1 Defining Abduction

16.1.1.1 Evolutionary View

I have argued that a central feature of the learning process engaged in by these subjects was a series of explanation cycles of generation, evaluation, and modification. The cycle is "evolutionary" in the sense that it responds to a criticism from a discrepant event or internal criticism by modifying the model ("species") and then re-evaluating it. Different attempts at modification can lead to the "survival of the fittest model" in the face of critical evaluations. In this section my first purpose is to examine how the "generative" processes of model generation and modification might be better understood theoretically. Model evaluation will be discussed later.

Peirce (1958) and Hanson (1958) used the term abduction to describe *the process of formulating a hypothesis which, if it were true, would provide an explanation for the phenomenon in question.* This definition is a very open one. In their view the hypothesis could even be a guess about a hidden mechanism at work in the system as long as it explained the observations. In this section, I will argue that the core of the model generation process for the protocol in Chapter 14 is a conjectural abductive design process. Abduction is often considered to be a weaker inference process than deduction because it can involve "guessing." However, a second purpose is to show how this weakness of abduction can be compensated for by the presence of an intelligent evaluation and revision cycle. Magnani (1999) notes that different authors have used two epistemological meanings for the term "abduction":

- A narrower sense: the formation of explanatory hypotheses (explanatory models). (I will call this "generative abduction" since it refers only to the act of hypothesis generation or revision.)

and

- A broader sense: including generation (generative abduction), evaluation, and revision cycles for developing a single explanatory model, and later, evaluative comparisons between rival models. (When needed I will use the terms "model evolution" and "model competition" for these two aspects of this process. Like many others I will use the unmodified term "abduction" in the broad sense to include all of these processes – others have used the term "inference to the best explanation" in this way.)

I will begin by focusing in the rest of this section on the narrower process of generative abduction within a single cycle of model generation (or revision), because this process is seen here as being at the core of model construction. As conceived by Peirce and Hanson, the possibilities are rather open for how generative abduction might occur, and the term has come to indicate the possibility of an explanation construction method that could be different from traditional logical inferences such as deduction or induction by enumeration (where by the latter I mean finding a pattern of a common feature(s) in a set of observable events). The main requirement, however, is that, if true, the abducted hypothesis would explain the phenomenon. Details about how abduction may actually occur in humans have been poorly understood.

16.1.1.2 Generative Abduction Refers to Both Generation and Revision

I will assume for purposes of this discussion that there are enough similarities between model generation and model revision that I can treat them as both possibly utilizing generative abduction; both are concerned with the production of a viable

16.1 Abductive Processes for Generating and Modifying Models 327

Table 16.1 Some candidates for instances of generative abduction in the present protocol

- Proposing bending as the mechanism underlying stretching (episode 9)
- Proposing shear as the mechanism underlying stretching (episode 10a)
- Proposing torsion (via the analogy of the hexagonal coil) as a mechanism underlying stretching in the helical spring (episode 11a)
- Proposing "unbending" as an additional mechanism contributing to stretching and restoring force in the spring (episode 15a)
- The addition of elongation of the wire as an additional possible mechanism contributing to stretching and restoring force in the spring (episode 15c)

model under certain constraints. Thus "generative abduction" simply refers here to a method of "constructing or revising an explanation," but with an emphasis on the idea that it can be an act of creative design and that it is not necessarily a more formal act. Instead, we are opening the door to its using a variety of possible sources, including rough analogies or even guessing.

Candidates for instances of generative abduction in the present protocol are shown in Table 16.1. There I have limited my attention to episodes where a new element of a mechanism was identified that was considered to be a candidate for a process actually operating within the spring wire. Patterns in the above episodes include:

- The subject considers a single specific case like the original spring, square coil, hexagonal coil, or drastically stretched spring.
- A new type of deformation that may be going on in the wire in that case occurs to the subject.
- For the related cases (the last three cases above), the new type of deformation is conjectured to also occur in the original spring (possibly along with other types).
- Subjects express some confidence, but not complete confidence, in these results.
- In the last two cases the subjects does not appear to be intentionally seeking a new mechanism; rather it emerges from the case being considered for another purpose.

Generative abduction is thought of as being a different and complementary process from hypothesis *evaluation*, to be discussed later.

16.1.2 Construction Occurred via Generative Abduction Rather than Induction or Deduction

An initial question raised by the abduction idea is whether it can be reduced to deduction, or induction by enumeration. It would be convenient if subjects had a reliable algorithm for inferring models automatically from the information in front of them. But in the present case I will argue that the process is considerably less automatic and more inventive and tentative in the form of a design or educated guess under constraints.

16.1.2.1 Is the Generation Process a Deductive One?

It is difficult to *prove* deduction is not taking place "behind the scenes." But one can support the hypothesis that generation is operating in another way with the following observations:

- The lack of certainty in the reasoning of the subjects
- The origin of model elements in schemas activated by analogy
- The lack of syllogistic reasoning forms in the statements of the subjects

A more traditional approach to the spring problem would be to find a set of general physical principles that could be used to analyze the system. Such principles, if available, could be used to deduce properties explaining why the wide spring stretches more. But because the subjects were not mechanical engineers this was not a possible option for them. Rather subjects often conjectured that the spring involved bending as a mechanism, often using an analogy to a bending rod. However, the subjects did not indicate complete confidence in this conclusion. Rather they spoke of it as a very plausible conclusion, usually based on the analogy. And several said they were sure a longer rod would bend more, so the uncertainty was not due to an uncertain premise.

It is worth considering whether some of the candidate cases of abduction cited in the previous section are instead instances of the simpler process of applying an established principle from prior knowledge. This is the simplest potential explanation of the process the subjects used to understand the behavior of a bending or twisting rod, for example. But to apply bending or twisting to the wire in a spring is a much more unfamiliar and uncertain affair, as one is outside of the normal domain of application of the bending and twisting schemas.

In a deduction, results are derived via logical rules that combine statements assumed to be true to produce a new statement that should be true. We did not observe subjects speaking about such formal inferences in the instances of generative abduction listed above. Their explanatory models appear to be constructions that can explain events, not formal deductions from prior principles. Furthermore, the hypotheses in question were all initially quite tentative when posed, and certainly did not carry the confident sense of validity one hopes for from a deduction. S2, on further analysis and criticism of his bending model, proposed that the spring involved torsion as well. This also appeared to be an educated guess, both because he had never observed this effect (and could not during the interview) and because it was originally seen to apply to the analogous case of a polygonal coil, rather than the spring itself. Although the subject appeared to understand the physical principle of torque causing torsion, and it was eventually seen to apply to the polygonal coils, thinking that it might apply to the circular coil was still a conjecture at this point. And he did not specify how to apply it at a very high level of precision, e.g. where to assign the center of leverage, lever arms, etc. Thus, it is an example of a tentative hypothesis or abduction

16.1 Abductive Processes for Generating and Modifying Models

in contrast to a logical deduction. The same can be said for "unbending" and "extension" (tension) as applied to the spring – these are also tentative attributions.

Evaluating and criticizing these hypotheses was a more convergent process, and there some might describe the subjects as exploring deductions from their current model, although I have usually described it as running a simulation using the model. "Running a simulation" in a thought experiment is seen here as a substitute for deduction that may lack formal validity but that can nevertheless carry conviction, as discussed in the previous chapter.

On the other hand, when a model has become a formal mathematical model in the form of an equation and predictions are made from it according to logical rules, one can refer to deduction. But the observations above argue that there was a lack of the use of formal principles or syllogistic forms in large parts of the protocol that do not deal with equations. I will continue to analyze most of the pre-quantitative, and even some of the mathematical reasoning without using the concept of deduction, and this may contribute to changing the view of the expert scientist as a "logic machine." The reasoning appears less formal, certain, controlled, and convergent than in deduction.

16.1.2.2 Inductive Generation?

Were the model generation processes observed primarily inductive? Simplified diagrams for contrasting deduction, induction, and abduction are shown in Fig. 16.1. By induction here (formally, induction by enumeration) I mean a process by which a

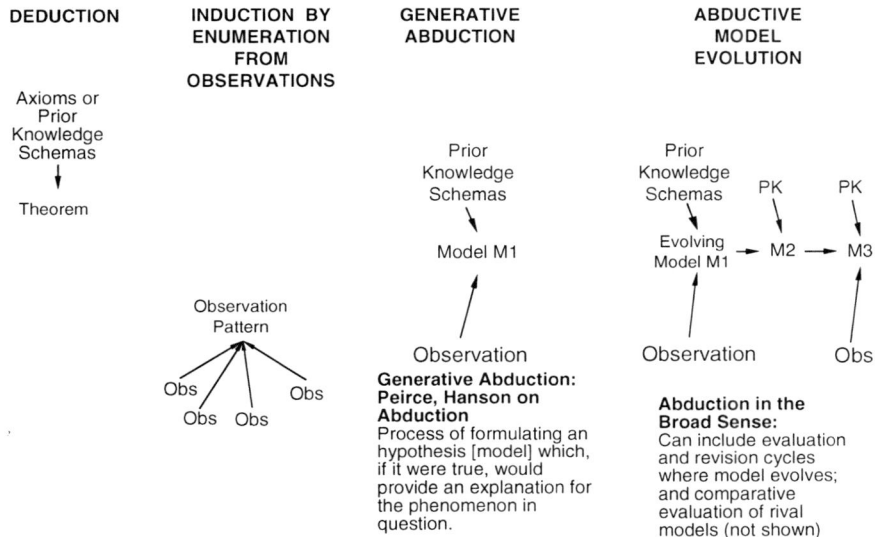

Fig. 16.1 Contrasting reasoning processes

more general principle is abstracted from a set of empirical observations. The principle then serves as a more abstract summary of a pattern in the observations. (The vertical dimension in Fig. 16.1 is roughly associated with that of Fig. 6.10.)

In the present cases the models developed were clearly not distilled via induction from a set of new observations of spring behavior since no such observations were made; rather, the subject appeared to construct a theoretical model at a different level by assembling nonobservable elements (generative abduction). Second, it is highly unlikely that most of these elements were observed in the past by the subjects; torsion bending, and tension are not observed in springs in daily life – these are inventions created at a theoretical level to explain observations at an empirical level. This fits the idea that an explanatory model is not just a pattern in or summary of observations – it is a hidden mechanism not initially subject to direct observation, – and that generative abduction can generate *unobservable* mechanisms, which induction by enumeration from observations cannot do. For example, it is very difficult to support induction as the mechanism for the introduction of the bending model. The inference is made in the context of a single example generated in episode 8. It was not based on observations of springs, and it describes a hidden mechanism, rather than a pattern of behavior. The above arguments appear to apply whether one is dealing with the initial generation of a model or a revision.

16.1.3 Generative Abduction: Basic Model

The simplest model of model construction presented in this book is the abductive process of generating an explanatory model in part 3 of Fig. 15.6. As depicted there, model generation is the construction of a somewhat general, runnable schema assembly within spatial and other constraints, including most basically the constraint of giving a plausible explanation for the target phenomenon. Prior knowledge elements in the form of schemas or concepts are used, but in new combinations to make a new model. In this view generation is more like designing an assembly of schemas and concepts to form an explanation, than it is like deduction or induction.

I have also argued that many of the models constructed during the protocol were capable of generating imagistic simulations. In the theory presented here generative abduction includes the process of forming an image of a mechanism that could be operating in the target, and this image may be new and novel. The new schema assembly then is able to "operate" on this image to generate imagistic simulations of the dynamic mechanism – to animate the mental image and then observe its operation through time. A somewhat more detailed model of generative abduction is shown in Table 16.2. There the subject designs a mechanism within known constraints that could produce the behavior observed, and does so by piecing together elements from available schemas and transforming them as needed. This process includes subprocesses for partitioning the observed system into causally connected parts. The small "increments" in the spring wire are one (rather standardized) example of this.

16.1 Abductive Processes for Generating and Modifying Models

Table 16.2 Basic elements of generative abduction process

Generate explanatory model of hidden mechanism M

1– If possible, partition system into quasi-independent or repeated elements

2– With compatibility with at least some of the constraints in mind, find by association viable candidates for model components: find imageable elements as well as applicable existing science schemas or practical schemas for modeling and explaining aspects of the target case

3– If necessary use an extreme case or analogy to suggest other explanatory model elements

4– Incorporate selected imagery from above elements into model and/or transform existing elements of model to form runnable, behavior-causing mechanism within known constraints

16.1.3.1 More Detailed Models: Design Under Constraints

As used here, a defining feature of generative abduction is that it provides an explanation of a phenomenon. However, a scientific abduction would aim to satisfy various additional constraints and desiderata for scientific theories; the most basic of these are that the model be plausible as a mechanism that could actually be operating in the system, that it not use occult powers, that it be coherent with previously developed elements of one's overall theoretical framework, that it not conflict with trusted explanations of previous phenomena that are in the "hard core" of a science, and that there be some attempt at precision in describing it (as opposed to being satisfied with a loose literary metaphor). This leads us to hypothesize along with Thagard and Shelley (1997), Nersessian (1992), Clement (1989b), and Darden (1991) that the process is one of design under constraints. That is, it is a "create the most plausible explanatory model that occurs to you" strategy – essentially a guess – but a very educated guess when informed by multiple constraints. These multiple constraints, as well as sources of ideas, are shown in Fig. 16.2.

16.1.3.2 Analogies and Extreme Cases can Contribute to Generative Abduction

Analogies or extreme cases seem to be involved to some extent in all of the primary examples listed in Table 16.1 above. This supports the idea that analogies can play an important role in abductive model construction. The authors above have also proposed that there may be a close relationship between analogy and abduction, but what exactly is the nature of the relationship? Is forming an analogy equivalent to constructing a model or completing a generative abduction?

A hypothesis consistent with the examples in Table 16.1 is that a central way that analogy can make a contribution to model construction is by suggesting a prior knowledge schema or the form of a model element to use as a building block. The analogy provides material to use in an "educated guess" about a component of the mechanism. Thus as shown in Fig. 16.2, analogies may not be the only process used within generative abductions, but they can act as a subprocess for abduction by suggesting relevant prior knowledge elements to use in constructing a model. Not all analogies are

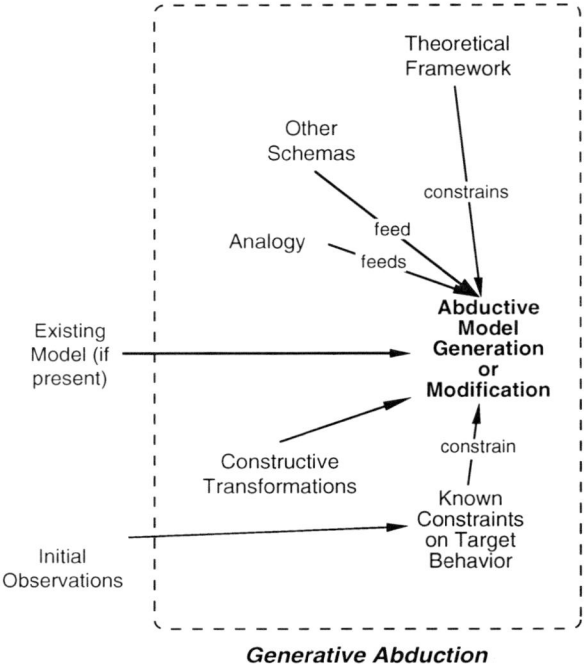

Fig. 16.2 Multiple inputs to generative abduction process

used in this way; the ones that are I will term "explanatory source analogues." This provides an initial model of generative abduction as a creative design under constraints process, as illustrated in Fig. 16.2, and this will become the core of the model of investigation processes being developed. To make further progress, the process must be expanded to include a way to evaluate the models it develops. And two other major processes, an initial description process and a mathematical modeling process, are needed to account for all the modes of investigation in the protocol. This is the topic of the next section.

16.2 Qualitative Investigation Processes

16.2.1 Introduction to Three-part Model of Investigation Processes

In this section, I attempt to specify a more detailed set of investigation processes which can generate many of the strategy types observed in the transcript as well as the five modes of investigation observed in the solution and listed in Table 14.1.

16.2 Qualitative Investigation Processes

I will propose a three-part process that alternates between emphasizing *description* and *explanation, ultimately progressing to mathematization*. A consolidation of the processes identified in the protocol in Chapter 14 is shown in Fig. 16.3. This figure should be viewed as a simplified outline of a set of cognitive processes that could produce the variety of types of conclusions and sequence of solution stages in Table 14.1. I have intentionally cast it in the form of a procedurally organized program for reasons of making it transparent for a large number of readers. Later I will discuss whether it could be made to be more realistic by using some other formalism, such as production systems. A more compact representation is given in Fig. 16.4. Note that inductive pattern recognition and empirical methods in general are not the area of focus in this book and are simply treated here as a black box.

I first give a brief overview of processes in Fig. 16.4. The first two cycles in the figure are the: (1) descriptive prediction process: methods for describing (predicting or accounting for) a relationship R; and (2) explanatory modeling process: constructing an explanatory model for explaining relation R. The distinction between them is that an accurate description, even a predictive one, need not include an explanation and does not imply understanding. (See for example, episode 12e.) The descriptive prediction process can be used to develop confidence and detail in a relationship, including the surface level relationship between independent and dependent observation variables. By use of observations, applications of descriptive schemas, expedient analogies and other techniques, this process can sometimes increase the subject's confidence in a predictive relation to a very high level, as was the case in Part I of the spring problem solution. However, no new relation need be identified or studied other than the initially specified one, and no explanation for it is given.

The Explanation Process II in Fig. 16.4 on the other hand generates a new intermediate causal relation at a deeper hidden level, that provides an explanation for the original causal relation or outcome, and this should increase understanding in the sense that the subject in episode 12c referred to. This cycle therefore begins to develop an explanatory model, not just a description of a phenomenon. Figure 15.5 gives a simplified overview of the major subprocess relationships that occur within the explanation process. There the cycle of model generation or revision on the one hand and evaluation on the other is shown in abbreviated fashion at the top, and other processes can be seen to feed these two main activities. Analogies can aid in generating a model, which can be evaluated by either running it in a thought experiment, or designing a real experiment.

The third major cycle generates a mathematical model for either a descriptive or explanatory relation coming out of one of the first two cycles. Here this cycle is engaged in the later stages of the solution, first to add geometric, then quantitative detail. One motivation for the distinction between this mathematical process and the explanatory modeling cycle is that the addition of geometrical or quantitative descriptive accuracy in the relations does not introduce new causal explanatory structure (unless it is part of applying a larger science principle). That is, mathematical modeling can take the precision of an existing causal relation to a mathematical level without necessarily modifying the explanatory structure. The level of mathematical detail sought in the model will depend on the goals of the subject.

Fig. 16.3:
MODEL OF SCIENTIFIC INVESTIGATION PROCESSES USED IN THE SOLUTION: SOME METHODS FOR INVESTIGATING A RELATIONSHIP R
(Focusing on Non-Empirical Processes)

INVESTIGATE RELATIONSHIP R

Given a current knowledge structure for a target case consisting of at least an initial description of the case and a targeted relationship R_1 between variables X_1 and Y_1; Investigate R_1 via the I. Description Process, and/or II. Explanation Process below.

CYCLE I. DESCRIPTION PROCESS: METHODS FOR DESCRIBING (PREDICTING OR ACCOUNTING FOR) A RELATION R_1 (X_i, Y_j)

A-Attempt to Generate a Predictive Description for Relation R.

   1-Direct methods (a, b, and c below may interact strongly).

      a-Apply any appropriate, descriptive prior knowledge schemas from (1) science principles or (2) practical knowledge.

         1.Try various alignments of the schemas to parts of the target system.

         2.Label newly observed or aligned features.

      OR

      b-Make additional "observations" (of real phenomena; or of images in a running simulation of R) (These may be naturalistic field observations or simply trying to run a direct imagistic simulation of the behavior of the whole target. If many observations are made, look for and summarize patterns in the observations.*)

      OR

      c-Partition the system under inspection into quasi-independent or repeated pieces and repeat Process I again for each subsystem.

   OR

   2-Indirect Methods- Use an Analogy or Extreme Case -this may provoke the activation of other new ideas from prior knowledge.

      a-Work from the Target: Transform Target containing R, via a transformation thought to be conserving, into an analogous system or extreme case that is more compatible with existing schemas. (Often produces a simplifying analogy.)

      OR

      b-Work from the Source: Find a possibly analogous system that is understood or feasible to analyze (This more remote type of analogy can be generated by association or via a transformation that is not known to be conserving.) Then try to confirm the validity of the analogy. (Often produces an expedient analogy.)

B- Evaluate Accuracy and Confidence Level for Method(s).

   1-Estimate confidence level for method. Compare different description methods; if agree then raise confidence in relation, if not, lower it.

   OR

   2-Design experiment to test description of behavior and gather support.

      a- Improve experiment if necessary.

C- If Confidence Level or Support is Inadequate, Improve Method in A and Retry. For example this could include using a transformation to modify the imageability of a direct simulation of the target (see table 10B.8), or the aptness of an analogous case.

D- If This Does not Yield Adequate Confidence, Try Another Method in A.

E- If Methods Above Fail, Try to Explain R in Process II Below. This may lead to a imagining a hidden mechanism that yields a prediction for the target.

F-If confidence level is adequate, one has the option to:

   1- develop more precise mathematical description of R using Process III below.

   OR

   2-Develop an explanation for R using Process II below.

(continued)

16.2 Qualitative Investigation Processes

Fig. 16.3 (continued)

CYCLE II. EXPLANATORY MODELING PROCESS: CONSTRUCTING AN EXPLANATORY MODEL FOR EXPLAINING RELATION $R_i (X_i, Y_i)$:
(Assuming there is no ready explanation for the target phenomenon accessed from memory.)

A- (Generative Abduction:) Generate or Modify Explanatory Model of Hidden Mechanism (by Generating One or More Mediating Causes M between X and Y in $R_i(X,Y)$) (See Figure 16.5 and accompanying text).

    1-Design mechanism within known constraints that could produce the behavior observed by piecing together elements from available schemas or analogies and transforming them as needed.

        a-Although partitioning of the target objects into subsystems may have taken place earlier, there can still be a need to attempt to partition further into quasi-independent or repeated elements to support modeling explanations.

        b-With compatibility with at least some of the constraints in mind (known constraints on the model and from the larger assumed theoretical framework, and constraints on the target) find by association viable candidates for model components: Find imageable elements as well as applicable existing science schemas or practical schemas for explaining relation R. Some of these will be Source Analogues since R will be well outside their normal domain of application.

        c-If necessary use an extreme case or a simplifying analogy to the target to suggest other elements of explanatory model.

        d-Incorporate selected imagery from above elements into model and/or transform existing elements of model to form runnable, behavior-causing mechanism M within constraints accounting for R.

    2-If possible, progress to fully imageable model. For schemas used: Align schemas to model and target images while attempting to integrate the explanation by forming complete spatio-temporal connections between elements in a runnable model.

        a-If necessary use an alignment facilitation analogy via conserving transformations of target to aid in alignment.

    3-Label and define image elements carefully.

    4-Attempt to generalize explanation internally for a part of the system to all equivalent parts.

B) Evaluate Explanatory Model.

    1-Evaluate plausibility of explanatory model.

        a-Run projected simulation of model on target case. Inspect for expected, contradictory, and new effects by comparing to behavior of target according to basic criteria developed in Ch. 14:

            1-Initial plausibility: when the model is projected into the target and run do the effects "fit" and are they spatially coherent?

            2-Do they explain the targeted features of the phenomenon at hand?

            3-Are they coherent with other features of the target?

            4-Do they form other coherence relations with other theories, including general principles such as symmetry?

            5-As the model is run, note any new spontaneous schema activations as possible inputs to the next round of model modification. Also accumulate newly emergent or highlighted image features of target as constraints on target behavior.

(continued)

Fig. 16.3 (continued)

  6-(For later stage of model development) Do the new relations provide a completely connected causal net? (Still later) A completely spatio-temporally connected mechanism?
 OR
 b-May design evaluative experiment to test model (as a real experiment or evaluative Gedanken experiment).
  1-Transform target case to eliminate (or reduce) specific mechanism; run model and inspect for decreased effects.
  or
  2-Transform target to exaggerate (or increase) mechanism; run model and inspect for increased effects.
  or
  3-Find a case that makes the mechanism directly "perceivable" in observations or in imagistic simulation.
 OR
 c-Other methods.
 2-If model is plausible evaluate model with respect to other desirable features for scientific models (see Chapter 18).

C) If Difficulties Identified do not appear to be Fatal, Modify Model in Light of Difficulties and New Effects Noted Above by Returning to (A) with these added as New Properties of Target or New Constraints on the Model. In addition try to pursue alternative models; If Difficulties appear to be Fatal, Reject Model and Generate New Model in A Above. If that fails, restart the investigation by generating more ideas in Process I.

 1-If evaluation positive for a piece of model, add weight to its priority in being retained as a constraint on further model modification and add support to overall model.
 2-As accumulate support for an overall model, decrease divergence of processes used.
 3-As accumulate "frustration" (time spent) with an unsolved anomaly, increase divergence of processes used.

D) Options: For Each New Relation R_{i+1}, R_{i+2}, etc. Added by the Model to the Causal Net.

 1-Investigate new R_{i+1} recursively via the investigation procedure in this figure.
 OR
 2-If mathematical level of precision is desired, develop a mathematical representation/model for R_i and R_{i+1} via Process III below.

E) Once Model Development has Stabilized, Compare Rival Models.

 1-On scientific criteria for the best explanation/description, including weightings in (IIC) above.
 2-On relative magnitude of effects to determine whether some have negligible effects.

CYCLE III. MATHEMATICAL MODELING PROCESS: CONSTRUCTING A MATHEMATICAL MODEL FOR RELATION R_i

A-Starting from the Existing Qualitative Description or Model (and any accompanying science schemas), Generate Mathematical Description of Relations at Mathematical Precision Level L (different geometric, or quantitative levels of precision: Levels 4 and higher in Table 14.2). Describe geometric relations, and/or quantitative functional relations between independent, intermediate, and dependent variables:

(continued)

Fig. 16.3 (continued)

1- Find relevant mathematical schemas (a, b, and c below may occur in any order and may interact strongly).
 a- Although partitioning of the target objects and model objects into subsystems may have taken place earlier, there can still be a need to attempt to partition the detailed behavior of the model elements (e.g. trajectories) into quasi-independent or repeated elements.
 OR
 b- Use symmetry arguments to find equivalencies and cancellations in system.
 OR
 c- Apply standard mathematical schemas to describe elements involved in R at level L.
  1- Look first for standard matches to standard objects, e.g. for the circle of wire, all radii are equal. Make simplifying transformations to facilitate matches for less familiar objects or relations, e.g. a spring coil is a circle with break.
  2- Run the imageable qualitative model underlying R, noting the direction and shape of movements and their connections to other movements.
  3- To attempt to apply geometric or quantitative schemas at level L make visual matches to fundamental "standard" visual geometric or mathematical function models (e.g. right triangles, or multiplication of x line segments of length 1 strung together).
   a- For example, to determine proportionality, one can try to run a simulation of variables X causing Y, then imagine doubling X and see if it is "obvious from the picture (or simulation)" or from known geometric relations, that Y should double.
   b- If necessary, develop alternative representations of the system using diagrams, vectors, graphs, etc.
2- Try to find the best alignment of the schemas to parts of the target system.
 a- Attempt to make visual and functional matches to parts of the target.
 b- Imagistic Alignment Analogy: Transform system containing R via a transformation thought to be conserving (or reversible) into an analogous one that is simpler or more compatible with existing mathematical schemas, so that the imagery and features of the target and the mathematical schema can be aligned correctly.
  1- As precision levels become higher, use less drastic transformations so that they conserve relations at the higher level of precision. Make minor transformations of system to exaggerate independence of or increase imageability of parts or changes in order to envision alignment of mathematical schemas to each part.
  2- For quantitative relationships, this means transforming the system containing R via a conserving transformation that (a) isolates a subsystem so that the relation between its inputs and outputs and its approximately independent contribution to the overall phenomenon can be calculated; or (b) makes it easier to apply standard quantitative mathematical schemas.
  3- If necessary use other heuristics described in Chapter 11 to simplify quantification such as:
   Partitioning and reassembly
   Embedding
3- Label newly observed or aligned image features carefully as mathematical entities.
4- For quantitative models, use algorithms to make further inferences by performing calculations or combining and simplifying formulas.
 a- After such manipulations, interpret and reconnect results to the visual model and target case.
5- Attempt to generalize a description internally for a part of the system to all equivalent parts.

B- Evaluate Accuracy of Mathematical Model.

1- Is the Mathematical Model consistent with existing constraints? (e.g. Are geometric predictions and symmetries consistent with known properties of the target system?)
2- Is the model predictive at the level of precision attempted? Design experiment to test description of behavior:

(continued)

Fig. 16.3 (continued)

> Empirical testing (if possible)
> OR
> Mathematical evaluative Gedanken experiment, e.g. use extreme case.

C-Modify Model in Light of Difficulties and New Effects Noted Above by Returning to (III A) with these Added as New Properties of Target or New Constraints on the Model, or Reject Model and Regenerate in A Above.

D-If no Difficulties are Detected at Precision Level L and Greater Precision is Desired, Return to IIIA Above and Develop Mathematical Model of R at Precision Level L + 1.

Fig. 16.3 Model of scientific investigation processes used in the solution
(Note: The processes of analogy, simulation, and Gedanken experiment that appear in the above investigation process have their own subprocesses outlined in Table 17.1)

16.2.2 GEM Cycles

Two loops are shown in each process in Fig. 16.4 for a total of six. In each process one loop is a GEM cycle of generation, evaluation, and modification. Many aspects of the protocol in Chapter 14 can be accounted for by assuming that the solver made several passes through each of the cycles in Figs. 16.3 and 16.4.[1]

16.2.3 The Explanatory Depth and Precision of Description Dimensions

The product of these investigation processes will be a knowledge structure containing different elements at both observable and theoretical levels. The dimensions of these differences are most easily described by treating them initially as dichotomies, then expanding them later to larger spectra. The simplified dichotomies for explanatory depth (explanatory model vs. observables levels) and precision of description (qualitative vs. quantitative) are shown as the two dimensions of Table 16.3. Here the dichotomy is illustrated by examples from the elastic particle theory of gases. The table can be thought of as depicting four parts of a final knowledge

---

[1] Readers may recognize section 1 of Fig. 16.3 as related to Table 5.2: Some methods for prediction and/or understanding of the analogous case. In this chapter that set of methods is raised to a significantly higher level of generality. Here, instead of understanding being a subprocess for analogical reasoning, the situation is reversed, and analogies are thought of as a subprocess of the more general process of investigation for understanding. I will retain the idea that these two processes can call each other heterarchically, but from now on will represent understanding via investigation processes as the higher order goal in science.

16.2 Qualitative Investigation Processes

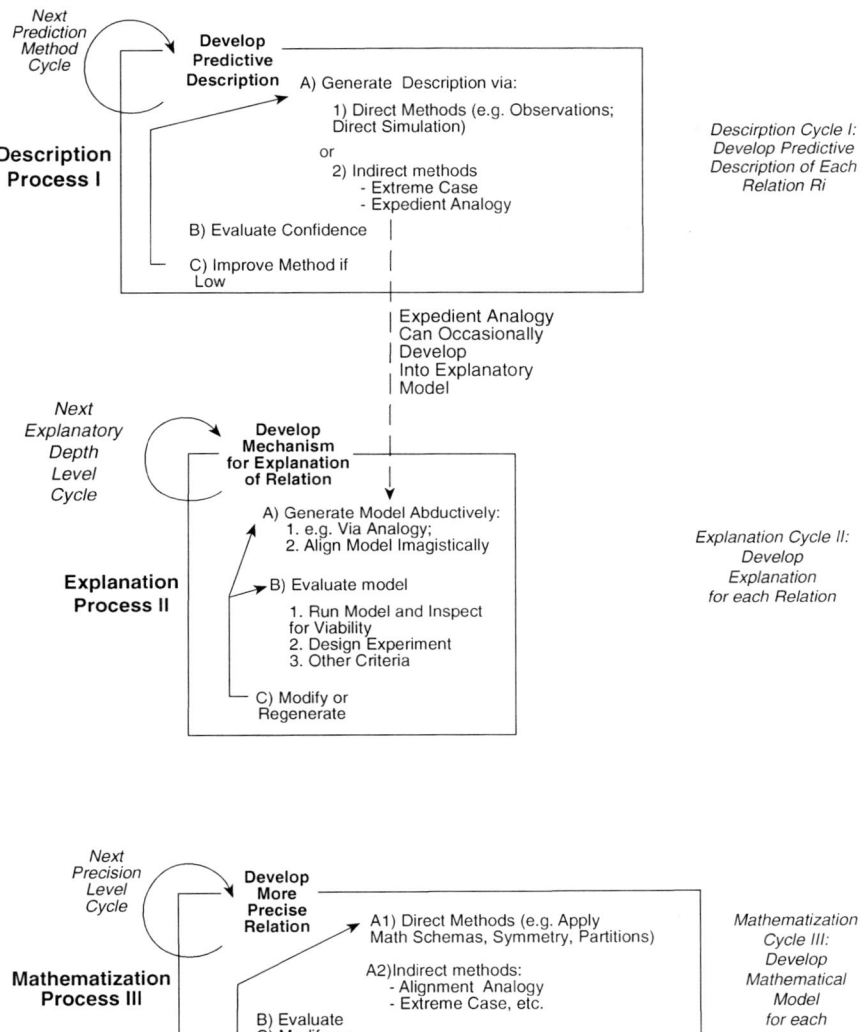

Solution Algorithm

Fig. 16.4 Cycles I–III in scientific investigation process

structure in memory resulting from an investigation. The traditional distinction between observational and theoretical knowledge is shown, with each having qualitative and mathematical levels of description. A parallel table for the spring problem is shown in Table 16.4. Note that with these distinctions, it is possible to have theories without mathematics (in the upper left corner) as well as mathematical

Table 16.3 Depth level vs. precision level in types of knowledge produced by scientific investigation: gas behavior example

Precision level	Qualitative	Mathematical
Explanatory model	Elastic particles model	$P \propto nmu^2/V$
Observable pattern	Higher temperature increases pressure	$PV = nRT$

Table 16.4 Depth level vs. precision level in types of knowledge produced by scientific investigation: spring problem example

Precision level	Qualitative	Mathematical
Explanatory model	Torsion model	$S = 2\pi F r^3/q$
Observable pattern	Greater coil width increases stretch	$S = cd^3$

descriptions without theory (lower right corner). The separation of explanatory model and mathematical model development processes in the present theory helps to account for this.

A major theme in the theory presented in Fig. 16.4 is that the explanation cycle II builds in an upward direction from the lower left cell to the upper left in Tables 16.3 and 16.4, while the mathematization process III attempts to move to the right from either left-hand cell. This can occur once the description process establishes an initial qualitative relationship in the lower left cell.

16.2.3.1 Mediating Causal Mechanisms vs. Levels of Precision

The distinction between explanation and description is also depicted diagrammatically in Fig. 16.5. Figure 16.5a shows an initial hypothesis that X causes Y. The causal relation R_1 might start as a purely qualitative, potentially observable relation that is predicted (e.g. higher temperature *causes an increase in* pressure, or greater spring width *causes an increase in* stretch). Methods in the description process (I) in Fig. 16.3 can be used to attempt to increase the confidence with which the relation $R_1(X,Y)$ is predicted (methods such as the use of multiple expedient analogies and extreme cases that appeared in stage 1 of the composite spring protocol.) Alternatively, the explanation process II in Fig. 16.3 can be used to deepen understanding by providing an explanation for the relation. This is done by hypothesizing a hidden mechanism or mediating cause M (e.g. twisting spring elements or increasing gas molecule velocities) between X and Y as in Fig. 16.5b; this produces two new relations $R_2(X, M)$ and $R_3(M, Y)$ as shown. Now instead of greater spring width *causes an increase in* stretch, one has, greater spring width *causes an increase in* torsion and greater torsion *causes an increase in* stretch. The hidden mediating variable M of torsion in this case has been inserted into the causal chain

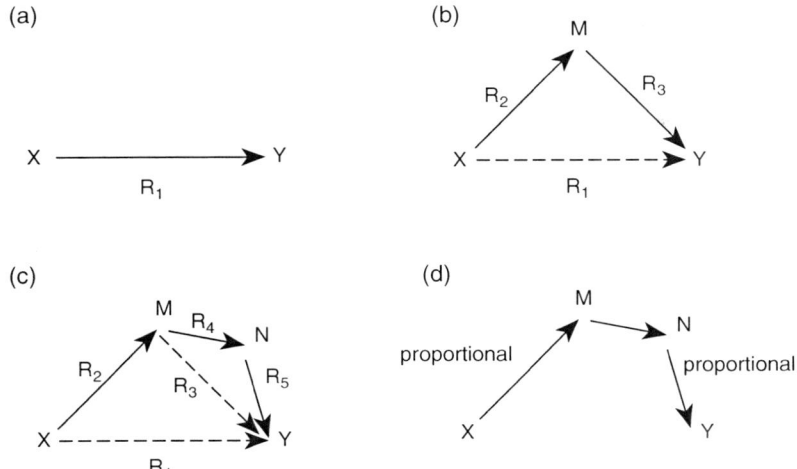

Fig. 16.5 Types of explanation for a causal relation. (a) qualitative observable; (b) qualitative explanatory; (c) deeper qualitative explanatory; (d) mathematical explanatory

as a mechanism that explains the original observable relation R_1. The explanation cycle II in Fig. 16.3 can be recursively applied to new relationships as they are added (e.g. where R_1 is originally explained by R_2 and R_3, R_3 can in turn be explained more deeply by R_4 and R_5, as in Fig. 16.5c). This shift is represented by the loop in the upper left corner of Process II in Fig. 16.4. (Note: This is the simplest possible model of causality, used here as the simplest way to illustrate how model construction might proceed to form deeper and deeper explanations recursively. Other more elaborate models of causality or interaction might be substituted here, but the same basic idea of recursive depth should apply.)

The mathematization process can be used to increase the precision of any of these relations from qualitative to geometric, or to quantitative function levels, as shown by adding in more precise relations of proportionality in Fig. 16.5d. The process illustrated in this chapter culminates in a structure that provides an explanation for the behavior of the target system in the form of a hidden, mechanistic model of torque and twisting (developed in the explanation process), in this case its precision enhanced by an accompanying mathematical model of those mediating variables at both geometric and quantitative-algebraic levels of precision.

Although it did not occur in the protocol in this chapter, the mathematization process could have been used alternatively to merely provide a mathematical model of the description of observable behavior provided by the description process. Such a model might simply reflect the cubic relation between diameter and stretch as determined by experiment, without modeling the underlying physics explaining the system. Or the following hypothetical episode could have occurred after episode 8c:

Now is there any way for me to predict some features of the quantitative function between width and stretch? By the extreme case of no weight, I know that this function has to go through the origin. A default hypothesis is that the relationship is linear, with stretch being proportional to coil width. To test this, I am going to imagine experimenting with hanging a 1 lb weight on a single-coiled spring made of heavy coat hanger wire. A pound is the weight of one pint of water. I imagine 1 lb on a 2" diameter coat hanger loop stretching less than 1/2"; then on 6" loop stretching more than 6". This tells me that the relationship between width and stretch is not linear and may go with the square or the cube of the diameter.

The above episode utilizes thought experiments to begin describing the target relation at a quantitative level directly without pursuing an explanatory model first, indicating that in theory, the mathematization process can operate on the results of the description process as well as the results of the explanation process.

The changes shown in Fig. 16.5 are only one of several types of conceptual change or expansion that can take place during the growth of an explanatory network (Thagard, 1992; Clement, 2008c). And complex real-life cases of moving to a deeper level of explanation may involve far more than an unpacking of a single relationship. However, the figure does serve to introduce the idea of depth of explanation as distinguished from precision of description, and serves here as a simplified model.

16.2.4 The Three Cycles in the Outlined Investigation Process Can Generate the Five Major Observed Modes of Investigation in the Protocol

Figures 16.3 and 16.4 then, describe a first order model to explain major aspects of the protocol in Chapter 14. The three types of cycles in Fig. 16.4 are designed to account for the five stages or different modes of investigation in Table 14.1. (As mentioned in Chapter 14, the empirical content of this table is reflected more in the existence of five categories of episode types or modes of investigation, rather than in their ordering, since the protocol is a composite one. The stages organize the protocol episodes in a plausible but idealized order. However, the present ordering does not seem far out of line with the protocols of most individual subjects. And the most central problem is to account for the processes being used in each stage as an observed set of similar episodes, rather than their order.) Stage 1 is accounted for by the description process I in Figs. 16.3 and 16.4. Much of stage 2 and all of 3 can be accounted for as a product of the explanation cycle (Process II), whereas stages 4 and 5 are accounted for by the mathematization cycle (Process III). I will review processes involved in each stage below. By reading the descriptions below along with Figs. 16.3 or 16.4 and referring to the transcript in Chapter 14 where necessary, the reader can make an initial evaluation of the viability of the outlines in those figures. I will then support or critique various specific features of the model in later sections.

16.2 Qualitative Investigation Processes

Stage 1. Efforts to Develop an Initial Qualitative Description or Prediction for the Targeted Relationship (Episodes 1–8b) (Primarily Modeled by the Description Process I)

The protocol begins with the subject passing through several processes in the descriptive cycle shown in Figs. 16.3 and 16.4, Process I. According to Figs. 16.3 and 16.4, it is hypothesized that a subject would first simply try to observe the targeted phenomenon; or attempt to apply standard scientific schemas or principles, similar to how a competent student might solve a physics homework problem. Lacking an opportunity to make observations of real springs, an initial attempt is made early in the protocol to find a scientific principle as a prior knowledge structure in memory that will make the prediction process easy. Failing this, the subject attempts a direct simulation of the target problem to develop a (weak) prediction that larger diameter causes greater stretching. Performing such an imagistic simulation is not always possible, but when it is, it can provide a starting point. Extreme cases of very narrow and wide coils, as well as expedient analogies to cases involving a double length spring, rubber bands, foam rubber, and a bending rod were then used to enhance and increase confidence in this prediction. This stage illustrates the possibility of extended periods of work in the descriptive cycle I of Fig. 16.3 with significant increases in confidence, but without engaging in real explanations or understanding of why the wide spring stretches more.

Although they will not be discussed at this point, it should be noted that the subprocesses of analogy, simulation, and Gedanken experiments that appear in Figs. 16.3 and 16.4 each have their own subprocesses outlined in Table 19.1 Also note that imagistic thought experiment methods may be used in the description process in addition to empirical methods. Although the topic of this cycle is always focused on predicting potentially observable behaviors of the target, the means used to investigate these are not limited to "observation in the narrow sense." Something like "observations" can also take place internally in imagistic thought experiments.

Stage 2. Searching for and Evaluating Initial, Qualitative, Explanatory Model Elements (Episodes 9–16) (Primarily Modeled by the Explanation Process II)

The solution then moves to a deeper question, namely, constructing an *explanation* for the predicted relation between coil width and stretch (Stage 2 in Table 14.1). The hypothesized process shifts to the explanation cycle II in Figs. 16.3 and 16.4. A unique characteristic of the explanation cycle is that it generates unobservable elements, i.e. qualitative theory. Here it is assumed that there is no familiar, ready explanation via established theoretical principles for the target phenomenon and that the initial search for such an explanation in memory has failed. In the spring problem, the bending idea began as a simple expedient analogy to a bending rod in Process 1. But this then suggests the theoretical conjecture that somehow bending is actually involved in the spring wire, an attempt at explanatory modeling that brings the subject into Explanation Process II. However, the bending idea leads to the anomaly of increasing slope and distance between coils – when the model is evaluated by actually being run in the spring as a simulation – stymieing the subject. (This ability to produce a negative result is why *running the model on the target in an imagistic simulation* is featured in Fig. 15.5 as the most basic method for model evaluation.) This negative result causes the subject to give up on explanatory modeling for the moment, and the subject returns to the description process to reevaluate the rod analogy. This leads him to generate the hexagonal coil analogy as a bridge between the bending rod analogy and the spring in order to help him evaluate the original analogy.[2]

But simulating running the hexagon then leads unexpectedly to the AHA phenomenon and torsion insight in episode 11, interpreted as the spontaneous activation of a schema during the simulation of the hexagonal coil stretching. This is an example of the "volatility" of simulations. It is an apparently unplanned process that takes place in spite of and in addition to, rather than because of, the overall investigation process in Fig. 16.3. At this point the hexagonal coil has become a "provocative analogy." The torsion possibility as a possible new explanatory mechanism then leads the subject to Process IIA again, where a simplifying transformation is used to generate a square coil. An attempt is made to analyze, explain, and understand the square coil by utilizing both a bending schema and a torsion schema. As described in Chapter 15, these are concrete, perceptual/motor schemas which enables running simulations of twisting and bending of the sides of square coil in thought experiments during episode 12, leading to the causal model in Fig. 14.1. Still within the explanation process but returning to the helical spring, at this point the subject has only the vague beginning elements of an explanatory model, but it is an extremely important start. It consists only of having identified certain *model elements that are undeveloped* – that bending and twisting could be involved and that stretching could somehow cause bending and twisting – but no detailed picture of *how* they might be caused in the helical coil.

In episodes 15a–c the inquiry continues at this level of identifying and evaluating other possible model elements (unbending, tension) in the explanation process. This is followed by model comparison work (part IIE) aimed at attempting to find out whether twisting plays a more important role compared to these other possible mechanisms. Unbending and tension are dismissed via imagistic simulation estimates as making a negligible contribution. Then three different Gedanken experiments are designed as a way of evaluating the bending and twisting models and are able to favor twisting over bending. The mechanisms of unbending and tension are considered but are deemed less important via imagistic simulation estimates as making a

---

[2] An alternative view here that would stay within the explanation process is that the hexagonal coil is an attempt to create an analogy facilitating imagistic alignment, in which he attempts to "see" bending in the spring. The latter interpretation is supported by the subject asking just before considering the square and hexagonal coils: "What could the circularity do? Why should it matter? How would it change the way the force is transmitted from increment to increment of the spring?" The difference is subtle and may not be entirely decidable from the protocol. Perhaps the hexagonal coil serves both functions.

negligible contribution. The explanation cycle in Fig. 16.3 is capable of accounting for most of these events.

In episode 16 a foray is taken into the mathematical modeling cycle, as the subject attempts to calculate the total quantitative twist in the polygonal spring as a model of the helical spring. Episode 16 does not fit into one stage because it is a condensed description of a modeling foray that would involve several stages and levels of investigation.

So far the torque schema is aligned to adjacent segments of the polygon. But since this alignment places the lever (torque) arm incorrectly, the leverage goes to zero in the limit as the sides become infinitesimal in the calculation of the twisting effect, and the answer given of zero torque and twisting is incorrect. This conflicts with several of the subject's previous arguments that there is something right about the torsion idea. So the subject proceeds to deepen the torsion model along another route.

Stage 3. Seeking a More Fully Imageable and Connected (Integrated) Causal Model Mechanism (Episodes 17–19c) (Also Modeled by Explanation Process II)

In episode 17 there is a shift to a new level of explanation within Process II concerning whether one can learn to actually "see" how twisting is caused in the helical spring (cf. "seeing as"). I interpret what is occurring in this episode as learning how to project an image of torque and twisting into the circular spring coil so that it is aligned with the somewhat complex shape of the spring, as described in process IIA2 in Fig. 16.3. An initial level of this kind of understanding is achieved when the subject answers the questions: How can one imagine how force, torque, twisting in an element, and stretch could be connected in the spring? Where could a force act to cause a twist? How would that translate into stretching? At this stage the model approaches a fully mechanical explanation – one where causal schemas are aligned to specific points in the model and target images and where there is a complete spatiotemporal contiguity between elements in the runnable model from original causes to final effect.

The analogy of a coil that has only one small twistable segment with the rest of the coil acting as "rigid handles" in episode 19 is then an example of an *imagistic alignment analogy*.[3] In this case it aids in determining how to align the imagery and features of the torsion schema (e.g. fulcrum, lever arm, twist deformation) with the spring in a more precise way. In this view, earlier in stage 2 the subject possessed isolated images in a causal sequence: (1) the weight is applied and the spring stretches, (2) the wire elements begin to twist and the torsion resists this. But there was no spatiotemporal continuity in – no imagery for – the connection between (1) and (2). After stage 3 the subject can carry out a plausible series of integrated imagistic simulations all the way down the causal chain from force to torque to twisting to stretch

---

[3] Defined as an analogy intended to aid in imagistically aligning features of an explanatory model, source analogue, target, principle, or mathematical schema, in a more precise way.

and "see or feel" how and where the stretch causes twisting. I call this a "fully mechanical" explanation.[4]

A series of problem transformation analogies and symmetry considerations then convinces the subject that this model developed for one segment applies everywhere in the spring. This corresponds to process IIA4 as an effort to generalize the explanation internally for a part of the system to all equivalent parts. The final result of this qualitative modeling process is shown in Fig. 14.15, where the anchoring conception of twisting and torsion is used to build up an explanatory model of twisting in small elements of the circular coil.

16.2.4.1 Explanatory Model Evolution via GEM Cycles

A basic feature of protocol Stages 2–3 then are that they illustrate *explanatory model evolution* via a GEM cycle as shown in Figure 6.8. This process generated a sequence of progressively more accurate models from shear, to bending, to bending and twisting, to twisting alone, and finally to refined twisting with correct alignment of torques. Added to this were lesser effects due to unbending and tension mechanisms.

16.2.4.2 Importance of Alignment and Use of Imagistic Alignment Analogies

Stage 4. Increasing the Level of Precision of the Model to Include Geometric Relationships (Episodes 20–22) (Primarily Modeled by the Mathematization Process III)

The subject spends this stage in the mathematization cycle III as the twisting model is reexamined at a geometric level of precision, with the following questions:

> Exactly where does the twisting occur in the geometry of the spring? Exactly where do the lever arms, through which forces or torques applied and transmitted, begin and end?

In episode 20 various analogous cases are used to argue that the torque is uniform everywhere. Symmetry arguments are used to find equivalencies within the system (process IIIA1B), but are still short of describing a quantitative function between width and stretch. Another question is: What are more exact geometric constraints on the spring as it stretches? (Episodes 21–22).

Here, no new causal factors are being identified. Rather, these processes help to increase the precision with which the twisting in various locations can be described.

---

[4] While one cannot always achieve fully mechanical explanations in science, e.g. "a magnet acting at a distance," one can still strive to make explanations as mechanical as possible, e.g. by introducing intermediate elements such as the magnetic field which connect the magnet spatially and causally to the attracted object in a somewhat more integrated way. And simply because mechanical explanations have not yet been achieved in a few areas of science does not mean that they are not preferred in most others. It may be that mechanical explanations are preferred because they make it easier to reason with the hand-eye manipulation systems that are so uniquely well developed in humans.

Thus, this is the stage where uniformities, geometric details concerning shapes, symmetries, and points of intersection, as well as pre-numerical details concerning equivalencies and relative sizes of lengths, forces and other measurable entities are determined. So this stage is concerned with increasing precision of mathematical description at a geometric level.

Stage 5. Developing a Quantitative Model Built on the Foundation of the New Qualitative and Geometric Models (Episodes 23–34) (Also Modeled by Mathematization Process III)

All of the preceding work makes it possible to construct detailed quantitative functions in episodes 23–34, and this is accounted for by the mathematical modeling process. This includes the application of established physical and mathematical principles at a quantitative level to the geometric model that has been built so far.

An initial conjecture in episode 23 involving reassembling the twisting segments of the spring into one long twisting wire in episode 23 appears to use the same *partitioning and reassembly* strategy from Chapter 11, except there it was static volumes that were being rearranged but conserved, and here it is dynamic twisting actions that are being rearranged and perhaps conserved. After this, by continuing to generate special cases and partitions of the problem, standard principles of torque and geometry are activated, adapted, and "fit onto" pieces of the spring. This appears to be successful even though most of the principles are for straight line geometries, and the spring is actually curved. These processes appear to rely on flexible mathematical models, analogies, and multiple compound simulations. For example, the "Clamshell" analogy in Fig. 14.13 was used to help the subject determine the correct alignment of the physical apparatus to the mathematical schema relating angle to arc length. It is therefore another example of an imagistic alignment analogy.

The above processes allow the development of algebraic expressions for details in the model. But these arguments all use the imageable models that were developed at a qualitative level as a foundation. In this case the move from stage 4 to 5 can be accounted for by a move within the mathematization cycle from geometric to a quantitative level of precision, both in the schemas being applied and the evaluation criteria being used.

16.2.5 Separate Explanation and Description Processes

Thus the more elaborate three part model in Figs. 16.3 and 16.4 improves on the earlier and simpler GEM cycle model from Chapter 6 in being able to generate the set of investigation modes depicted in Table 14.1. Its most prominent features is that it shows separate description (both qualitative and mathematical) and explanation processes. Stages 1, 4, and 5 (accounted for by Processes I and III) are dominated by description and precision increasing activities without adding basic new causal explanations (they do not add new qualitative relations to Fig. 16.5 or establish connections between the image of the explanatory model and the target), whereas stage 3 and most of 2 (accounted for by Process II) are dominated by such causal explanation work. The description and explanation processes are both at a higher level (of control) than the analogy processes described in Section I of the book, in fact, analogy processes can be called as a subprocess by each of the three major processes in Fig. 16.3. Yet all of these three processes and the analogy subprocesses are similar in that they all have generation, evaluation and modification processes at their top level. (GEM cycles may be nested, as when the entire cycle of an analogy improvement process occurs within the Generative portion of the larger, explanatory process that called it.) Thus the GEM cycle strategy is ubiquitous in this theory.

16.2.5.1 Partitioning

Partitioning appears in all three major process cycles in Fig. 16.3 and therefore one should ask whether this repetition is redundant. However, its repeated inclusion represents the different levels at which it was used. For example it was used at the descriptive level I to focus on a single coil of the spring, at level II to break the coil further into twisting segments, and at level III to break movement trajectories into quantifiable parts (episode 28). Thus it is used at finer and finer levels of detail as the solution progresses.

However, partitioning as a process can also be seen as much more ubiquitous – it can be thought of as involved in object recognition or formation at a perceptual level and at a higher level in the application of a schema that is assimilating some part of a perception or image to some structure. Thus partitioning may be involved in many places in Fig. 16.3. In fact the partitioning of the coil into segments might be seen as the result of the application of a schema. Thus the process is poorly understood and it may not be sufficient to think of it simply as a separate subprocedure in Fig. 16.3. It is an important topic for further research.

16.2.6 Computational Model of Todd Griffith

Todd Griffith (1999) conducted an extended study of the spring problem titled "A Computational Theory of Generative Modeling in Scientific Reasoning" in the College of Computing at the Georgia Institute of Technology. He and his advisors developed a computational view of model evolution processes, most of which correspond to those I have described above in Process I: descriptive prediction cycle (Griffith et al., 1997, 2000). Griffith and Nersessian asked permission to use our transcript data on the spring protocols as a database for their study. This led to Griffith writing a set of detailed procedures for accessing, evaluating, and transforming/revising analogies to the spring problem. They also give a preliminary outline for how many of the principles used in the expert protocol study might also be used to explain the model construction process used by Maxwell in developing theories of the EM field.

16.2 Qualitative Investigation Processes

In Chapter 3 it was found that a majority of analogies in the spring problem protocols were generated via transformations rather than by associations to a prior schema in memory Clement (1988). The Griffith studies focus on the role of heuristic, generalized transformations such as "approximate a circle by a polygon" in the spring problem. These are used to generate "improved" analogies once an original analogue is accessed from memory. They proposed that many aspects of this process are simulatable in principle by a computer program and that such a program can account for a considerable number of the transformational moves made across the ten subjects in my data. Some of the features that make this an impressive study are:

- The work makes a large number of connections to and builds on previous AI approaches to creativity, analogy, design, and simulation. It extends this work in a coherent way. This difference in the starting theoretical framework means that Giffith's interpretation and my own have evolved largely independently and partially in very different directions even though a major part of our data base has been the same. However, I believe the set of phenomena in this data base is so rich and so poorly understood that using several different theoretical frameworks to attack it makes excellent sense.
- Griffith's system is one of the most detailed descriptions of aspects of progressive qualitative model construction yet developed. (The term "model" in their study is used generically to cover both the system's knowledge about the target and about analogies to the target.) Like the present study, it develops an explanation of how scientists generate new models via a process of model evolution. It emphasizes the role of progressive transformations of analogous cases in this process.
- One reason Griffith's work is pioneering is because so little work has been done on geometric transformations as tools for analogy generation and evolution.
- Ablation experiments were performed to show how various pieces of the model are necessary to modeling subject behavior. Program performance is described as deteriorating when the system is run on paper without a modular component.
- In this sense the system has been developed to a high-enough degree of precision to be runnable in principle and to provide an initial sufficiency check for targeted aspects of the subjects' behaviors.
- That a cognitive model that was developed for five subjects (including S2) was supported by being able to account for most of the major moves observed in protocols from an approximately equal number of other subjects is rarely attempted in AI research and is impressive.

Griffith also admirably describes some limitations of his model:

- Visual reasoning is not as sophisticated as in humans; certain processes "are treated as atomic operations rather than as progressive simulations that are carried out in 'the minds eye.'" (Griffith, 1999, p. 268)
- The system "is primarily concerned with solving the problem, and will continue exploring until this goal is achieved. S2, on the other hand, is more concerned

with understanding. Even after he claims that he is 99% confident in his answer he continues to explore in order to increase his understanding. In this way, as well as in other ways, the...system falls short of being a general scientific reasoning agent, and can be best characterized as a scientific problem solver." (Griffith, 1999, pp. 74)

One can then describe the Griffith thesis as having achieved a detailed computational model primarily for aspects of Process I: the descriptive prediction cycle, shown in Fig. 16.4. This includes initially attempting to apply principles directly, analogical retrieval, evaluation by mapping differences, and modification of analogies by transformation to improve them. Such transformations can generate novel analogue cases. This means that this system can address issues of finding an answer via analogies and perhaps assessing confidence in it. Other interesting features treated as known facts, such as "constant slope" in the spring, can also be highlighted by the system by way of contrast with certain analogies.

Several of the above processes might be adapted in an expanded simulation system to provide some of the elements in Processes II and III in Fig. 16.4. The system also includes the accessing of physical principles that may be activated by a novel analogy, and this helps to explain the torsion insight in S2. Only in this last way does the Griffith program begin to move outside of Process I and begin to address an issue in Process II: the explanatory modeling cycle. The most important limitation, as Griffith points out, is the present difficulty in the computational approach with modeling the construction of visualizable models which utilize spatial reasoning and mental simulation. Thus in his system, the final output is the identification of a "best" analogous case with the answer to the problem mapped as an inference from discrete symbolic relations in that case.

In the present study of imageable understanding and its construction – the qualitative product is an elaborated, visual, schematic, explanatory model of the spring with new forces, torques, and microscopic deformations projected into an image of the spring system. The construction of such an output uses a variety of processes that are not contained in the Griffith system. This is to be expected given the difficulty of handling these representations within the current state of AI techniques. For example, some of the imagery-generating schemas used in the construction of the present model have been transferred from analogies, in a way that preserves the runnability of dynamic elements in imagistic simulations, including kinesthetic/visual representations of force and deformation. The present study also describes some of the processes needed to develop an output at the geometric or quantitative level, once a qualitative model is in place.

16.2.7 Evaluation Functions can Guide Control

Thus one of the challenges in the present study is to go beyond issues of evaluating predictive confidence in the description process to issues of evaluating qualitative explanatory understanding and mathematical modeling in the other two cycles in

Fig. 16.4. One's understanding of an explanatory model is evaluated initially on the criteria of model plausibility (including runnability), clarity (including imagistic alignment) and depth, in process II. Mathematical accuracy is evaluated and improved in Process III. This section describes how evaluation criteria might be used to control movements between the three major processes.[5]

It is hypothesized that the investigator has a goal to reach an acceptable level of "satisfaction," defined as follows. The goal of satisfaction has several subgoals: predictive confidence in an answer, explanatory understanding, precision or accuracy, and generality. Different individuals in different situations will weigh the importance of these four subgoals differently. What the problem solver will do at any point depends on whether their current solution has a high-enough rating on one or more of these subgoals. As work continues, new arguments or models are brought in, until the subject "is satisfied" with the estimated confidence, precision, explanatory understanding, and generality of their knowledge state. The ordering of these goals is not fully determined so they can operate to some extent in parallel. These subgoals can be described as follows:

(1) Predictive confidence can be generated simply by having rote knowledge from an authority or by having more than one nonformal method (such as expedient analogies in Process I) point to the same conclusion. (2) Explanatory understanding has two components: model plausibility and explanatory depth. (a) A model has plausibility when it is based on combining plausible prior knowledge structures (including intuitions with primitive plausibility) to explain the target phenomenon in a way that does not conflict with other intuitions and spatial reasoning. This means the phenomenon is explainable in terms of an imageable chain of events understood via a subset of prior knowledge schemas (perhaps in new combinations). Vague, disconnected imagery will count less than more precise causal imagery with complete instantiation and imagistic alignment of schemas and a complete spatiotemporal connection between cause and final effect. (b) Explanatory depth is the ability to say "why" something happens at a deeper causal level as defined earlier and illustrated in Fig. 16.7. (3) Precision is related to a qualitative/quantitative spectrum. Although in science we separate the concepts of precision and accuracy, here we will conflate them to say that the subject has a goal to be able to make predictions or "run" the model to some (personally determined) degree of precision and/or accuracy which may vary from qualitative to geometric to quantitative. (4) Generality is related to the number and diversity of situations that the model can account for. (Note that these are basic criteria: other goals may be possible, such as "coherence with other models." Multiple goals for evaluating models will be discussed further in Chapter 18.) Table 16.4 can be seen as the output of a relatively complete analysis of the spring problem and could guide a more complete specification of the output variables of the program in Fig. 16.3. This output has four possible parts as shown in the four cells. One

---

[5] I am indebted to Thomas Murray for helping to articulate the relationship between evaluation criteria and control here.

can also imagine an initial state of the solution with all but the lower left cell blank and a conjectured prediction in that cell. As an investigation proceeds successfully, various criteria for evaluation will show progress. Confidence as the characteristic associated with the descriptions in the lower cells will rise as more methods or data lead to the same prediction. On the other hand, a measure of "understanding" with the explanatory model in the upper left cells will increase with imageable plausibility, alignment/connectivity and depth of explanation. Mathematical precision will increase below and above as mathematical representations are introduced toward the right.

The following control structure is proposed. An investigator starts at Process I and tries to develop confidence. At any time, if an investigator places priority on having a higher level of understanding, then they can move to Process 2; if they place priority on having a more mathematically precise solution/model, then they can move to cycle III. Results in these areas can also add to confidence by contributing another method that confirms earlier results.

16.2.7.1 Production Rules

In pursuit of transparency at this level of development, the activity within each cycle has been modeled by a procedural net (with procedural steps and Boolean conditionals). On the other hand, the higher-level control structure, with competing goals determining transitions between major cycles is merely described above. But this highest level could also be modeled as production rules which "fire" opportunistically when their conditions are met, with a goal being more likely to dominate when the satisfaction level for its evaluation criteria is low. This would allow one to model the interruptions that allow another process to take over when a relevant schema is activated. In addition, processes that require more effort than other methods, could be weighted accordingly to reflect this cost.

Solutions that produce conflicts are tabled permanently unless further investigation does not yield any alternatives, in which case Tabled ideas can be returned ("back tracked") to. It may also be possible that multiple mechanisms will have to be constructed by the investigator to explain a complex phenomenon. (In the future it may be fruitful to attempt to model the activity within each cycle as a production system, as we have outlined here for the highest-level control structure, to allow for further flexibility and interruptability.)

16.2.8 Comparison to Griffith Study

I am now in a position to summarize some of the differences between the computational model of Todd Griffith and the model developed here, as shown in Table 16.5

A major achievement of Griffith, Nersessian, and Goel is that they show that the modeling of a number of these processes can be done at a very high level of detail.

16.2 Qualitative Investigation Processes

Table 16.5 Comparison between the computational model of Todd Griffith and the model developed here

Griffith's computational view	View developed in this book
Discusses analogies but does not discuss explanatory models as a special type of representation	Central distinction between expedient analogy and explanatory model projected into target
Transformations used to modify and improve analogous cases	Transformations used to modify and improve analogous cases, but transformations are also used to generate initial analogies (Clement, 1988) and to modify explanatory models; and sometimes to evaluate analogies
Initial analogies always retrieved from memory	
Source of transformations is a library of "generic" transformations	Source of transformations is a spectrum running from some transformations that are extremely generic (I call these spatial reasoning operations) to ones that are much more specific (I call these constructive transformations)
Transformations can generate novel analogue cases	Transformations can generate novel analogue cases or novel explanatory models
Novel cases can activate as yet untapped prior knowledge schemas as resources	Novel cases can activate as yet untapped prior knowledge schemas as resources
Explains episodes of "analogy improvement" as driven by finding and reducing differences between current analogue and the target	Identification of problems during analogy evaluation or explanatory model evaluation can inform modification process leading to "improved" analogy or model
Analogies evaluated by comparing (mapping) discrete features to target and examining differences	Several imagery-based mechanisms for evaluating analogies, some of which may not depend on mapping discrete features (Chapter 17)
	Several imagery-based methods for evaluating explanatory models
All analogue cases are considered as target cases (that can have analogies of their own) as well.	An analogue case can itself be analyzed recursively as a target, if the subject is unsatisfied with their understanding of the case
Access to generic scientific principles in LTM	Access to generic scientific and mathematical schemas in LTM, as well as physical intuitions of at least modest generality
Final output is the identification of a "best" analogous case with the answer to the problem mapped as an inference from discrete symbolic relations in that case	Final output at the qualitative level is an elaborated visual schematic explanatory model. (In the present case of the spring, this means new forces, torques, and microscopic deformations projected into an image of the spring system.)
	Also produces geometric and quantitative models
Solutions are grounded on discrete factual relations as "axioms"	Solutions are grounded on perceptual/motor schemas that embody physical intuitions as "axioms"
Procedural control structure	Production-like control at top level; procedural net at mid level, but no commitment to latter as best model for control
Investigator's model in detailed pseudocode	Model identifies major subprocesses

The opportunity in the present study is to develop models of other more elusive processes that must be given an initial description before refined models can be formulated.

16.2.9 Explaining Insight: Unpredictable Spontaneous Accessing of Subprocesses

16.2.9.1 "Cross talk" Between Cycles Via an Analogy

The three cycles in Fig. 16.4 are seen as generally operating independently and in the order shown, but there will be exceptions to this. I note first the possibility of a special kind of "cross talk" between the description and explanation cycles: sometimes an expedient analogy created within a description cycle can be a starting point for constructing an explanatory model within the explanation cycle. This seems to have been the fate of the bending rod analogy in Chapter 14. For many subjects this was merely an expedient analogy. But in Chapter 14, it was eventually taken as a model of what was happening to elements of the spring wire and at that point became explanatory. (This model was then criticized and refined further so that "unbending" was the surviving model element in the final model.) Thus, an analogy that starts as an expedient analogy can sometimes grow in status to play the role of an explanatory analogy that provides ideas for constructing elements of an explanatory model. In this case then, the descriptive cycle I "fed an idea" to the explanatory modeling cycle II, as shown by the dotted line in Fig. 16.4.

16.2.9.2 Spontaneous and Opportunistic Insights can Arise from New Features Recognized in Thought Experiments

An attempt to evaluate the rod analogy via the bridging analogy of the hexagonal coil led to the sudden recognition of torsion in the coil in episode 11a in Chapter 14. Thus a thought experiment intended for one purpose can lead unexpectedly to the generation of a different model for a different purpose. As another example, extreme stretching of a coil made of a flat horizontal ribbon (episode 14) was done to enhance the imagability of a twisting effect in the wire, and yet this led to the unexpected recognition of a component of stretching coming from unbending of the coils (episode 15a). The more general pattern is that a thought experiment designed for one purpose can sometimes end up serving a new unexpected purpose. This suggests that subjects can have multiple goals close at hand "in the wings" in a semiactive state, and that insights can occur opportunistically when an image created for one purpose brings to full activation one or more schemas that serve another purpose, interrupting the original activity.

This is one of the implications of Fig. 15.3, which depicts the hypothesis that any untested thought experiment operating via imagistic simulation is "volatile" in the sense that it can at times generate new image features and/or boost the activity of other schemas spontaneously and divergently (shown as a dotted line in Fig. 15.3). This effect will be revisited later in this chapter as an added way that evaluation processes can affect generation processes. Although Fig. 16.3 is expressed as a procedural net for simplicity and transparency, such an effect could be modeled by using a production system architecture that handles multiple parallel goals competing for dominance and the possibility of interruptions. In that view one would think of the model generation step in Fig. 16.3 part IIA as "watching for" such new effects that might aid in model revision, even while the subject pursues other major processes.

In summary, this constitutes the explanation provided for the insight behaviors mentioned above, and the resulting overall pattern of punctuated evolution (evolution with an occasional mini-revolution) in the protocol. Some aspects of the ordering of subprocesses in Fig. 16.3 appear to be predictable, but other effects, such as the volatile secondary effects of imagistic simulations, are unpredictable and add divergence and goal shifts to the system.

16.2.10 Generality

A dimension not shown in Tables 16.3 and 16.4 is that of generality – whether the knowledge refers to a specific case or apparatus in the laboratory, or whether it has been generalized to a wider domain of cases. Cycles II and III in the theory of investigation processes in Fig. 16.3 incorporate a separate subprocess for moving toward a more general observation pattern or model, and these reflect, for example, places in the protocol where the subject makes generalizations from a specific part of the image of the system he is considering to other parts. Further increases in generality would occur if the subject were able to generalize to different types of springs (or say, types of gasses, for the elastic particle theory) by analyzing new data or making the model more schematic, but I have not collected data on this issue.

16.2.11 Levels of Explanation and Precision

The four types of resulting knowledge from the investigation shown in the cells in Tables 16.3 and 16.4 are abbreviated in Fig. 16.6. Downward arrows there mean "explains," while rightward arrows mean "provides a foundation for." In this section, I make a number of conjectures that go beyond the data in the present work but that may be applicable to future research:

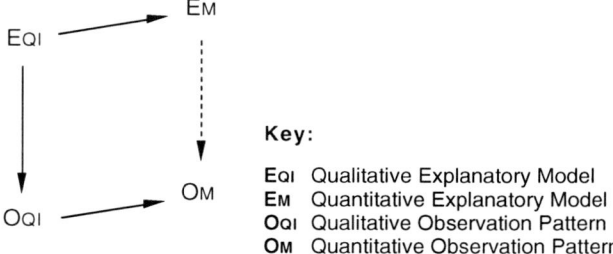

Fig. 16.6 Basic dimensions of explanatory depth and mathematical precision in modeling

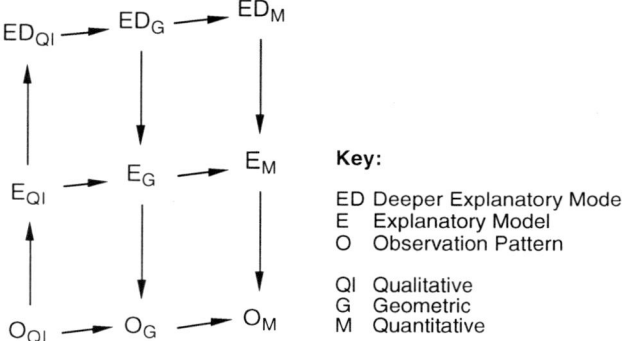

Fig. 16.7 Expanded dimensions of explanatory depth and mathematical precision in modeling

- Inferences can take place in a downward direction from a prediction for any item in the upper row of Fig. 16.6 to a prediction for one below it. However, if we do this in a meaningful way for, say, mathematical model E_M, it may be more appropriate to think of the prediction "originating in E_M and flowing through E_{QL} and O_{QL} to O_M," as the basis for the inference of translating E_M at the models level into a prediction for O_M at the observational level.
- This same C-shaped path in reverse would appear to be very useful for the process of explaining a quantitative data pattern O_M, and in many cases may be the only route. For example, explaining measurements taken on gas compression experiments would comprise developing a qualitative model of the gas first before developing an explanation at the mathematical level.

In the present model, explanations generated at level I explain observations at level 0. A more complex version of the kinetic theory that adds another level of explanation depth and an intermediate level of precision (geometric precision) can be envisioned as shown in Fig. 16.7. level I of explanation for gasses simply states that gas molecules each carry a certain amount of energy. Level II is also explanation, labeled E_D for "deeper explanation"; in this case, it would be a model of *how* a gas molecule can carry energy: translation, rotation, and "vibrations" between bonded atoms. The intermediate level of descriptive precision shown with G subscripts in the figure is the geometric level, including considerations of how molecules bounce at the wall, molecular geometry, etc. This leads to several more conjectures:

- Formation of a second level explanation is seen as essentially a Cycle II process; the deeper level II explanation should explain the level I explanation.
- Mathematization can again take place at this level (Cycle III process).
- An overall theme of Chapter 16 with regard to Fig. 16.7 is that the upward development of the left-hand column provided an essential foundation for quantitative work. Only by doing a great deal of qualitative work on E_{QL} was the process able to progress to a deep mathematical model via E_G and E_M.
- Empirical testing (or testing against an accepted model at a lower level of explanation) can be called for in either Process II or III in Fig. 16.3. Whenever a new explanation is formed, if possible it should be tested for plausibility against knowledge from the item below it.

16.2.12 Limitations of the Model Presented

The basic pattern of generation, criticism, and revision as a cycle that can operate without new empirical input is portrayed by Lakatos (1976) in the context of the history of mathematics, and by Clement (1989b), Nersessian (1991), Holland et al. (1986), Falkenhainer (1990), and Darden (1991) among others in the context of science. The more detailed description of investigation processes in this chapter concentrates on methods for which there is evidence in the present protocols. Some researchers have proposed additional investigation processes; see for example the lists of methods for theory development and modification compiled by Darden (1991) in the context of the history of genetics.

16.2.12.1 Positive and Negative Elements of the Model

The individual processes or "moves" in each small segment of the protocol described in Part I of this chapter, such as analogy evaluation or the running of a thought experiment, are "descriptive" of the thinking of these expert subjects rather than "prescriptive," to the extent allowed by the data. The same cannot be said for the overall structure of the investigation process on a larger time scale shown in Fig. 16.3 however. That structure is a possible way to explain the rationally reconstructed protocol in this chapter, and as such it is prescriptive, idealistic, and probably overly procedural in character. As with any analogue model, this model of cognitive processes has both positive and negative elements (Hesse, 1966). That is, there will be both similarities and differences between such a set of procedures and the cognition of scientists. Certainly the explicitness with which these procedures are described as named algorithms with relatively well-defined control structures is not a property being ascribed to all of the subjects' processes – i.e. we would not expect the subjects to be able to articulate all of these procedures as a plan. Thus, I do not claim that any subject necessarily had all of these top level procedures in mind as conscious and examinable entities. However, the explicitness of description and transparency of this theory makes it possible for us as investigators to make progress in modeling and communicating about the processes.

16.2.12.2 The Need for a Composite Model

A second problem is that no one subject gave evidence for all of these processes. Therefore an amalgamation of data from different subjects into a single protocol was used to promote the development of a fuller theory. This means that the overall process represented by the theory may be more extensive than that used by any one subject. However, it is possible that this gives it some value as a step toward a prescriptive theory for use by scientists.

16.2.12.3 Hierarchical vs. Distributed Control

As in models in other sciences, there are also features for which we are currently uncertain as to whether they are representative of the thing being modeled – in this case human theory construction (Hesse, 1966). This last "neutral" category includes items such as whether the loops in the procedure are governed by an established (although possibly unarticulated) control structure or whether they are an emergent behavioral feature that is a spontaneous collective result of the activity of more autonomous cognitive structures. As it stands, the procedures are more loosely defined than most computer programs, with a multitude of procedural (as opposed to logical) "or's" indicating several alternative methods for accomplishing various goals. But scientists' processes are probably even more loosely structured with respect to flow of control than is shown here, as is allowed in production systems. Further work attempting to recast the model as a production system would allow for sudden shifts in goals in response to new developments, such as an opportunistic shift from description (or problem solving) to explanation goals (see Karmiloff Smith and Inhelder (1975) for a discussion of this phenomenon in 5-year-olds!) In the concluding chapter of this book I will argue that an intermediate level of partly centralized and partly decentralized control allows for interruptability and multiple methods of model generation and evaluation. This intermediate level of control will be seen as a key feature for fostering effective creativity.

However, actually building a running computer simulation of the Investigation Process specified in the theory would be quite difficult if not impossible at this time, given its fundamental reliance on presymbolic imagistic simulation processes. Rather than speculating on the exact nature of the control system and detailing every process to a runnable level, I have concentrated here on identifying specific presymbolic heuristics, while giving an initial description of a prescriptive goal structure that is compatible with the protocol data. Therefore it is important not to take the order of the steps in Fig. 16.3 too literally; subjects can be at different stages on different sub-problems at a given time, e.g. they can alternate between developing the explanatory mechanism and increasing precision or pursuing other goals.

However, taken as a simplified model, the hope is that this description as a procedural net provides a more transparent outline of a way to account for a number of features of scientific inquiry, including the stages of model growth shown in Table 14.1. It also gives an explanation for how a wide variety of nonformal reasoning methods are used, and describes some of the generative and evaluative power of processes which use imagistic simulation. It represents an attempt to combine selective aspects of the computational and embodied paradigms in a composite model.

16.3 Mathematical Modeling Processes

I laid some groundwork for the analysis of mathematical modeling by analyzing nonformal mathematical reasoning processes in external subjects in Chapter 11. The data on mathematical modeling in this chapter is less diverse and independent because it is based only on the author's self-protocols on the spring problem. Nevertheless the patterns of thinking show many correspondences to the qualitative work from other subjects on the spring problem. A mathematical model is shown as a separate entity on upper right in Fig. 16.6. However, it is thought of as being built on top of the qualitative explanatory model. It starts from a schematic qualitative simplified picture or diagram of an explanatory model that can be drawn as an external diagram but that can also be represented internally as an image (or at least parts of it can at any one time). As part of the diagram is assimilated to a more precise geometric schema, it becomes a geometric model. Then as quantitative relationships are connected to these from one's storehouse of standard quantitative and algebraic schemas, it can become a quantitative model. (For some situations, such as word problems, this can happen without a geometric step in the middle.)

16.3.1 Cycle III: Mathematical Modeling

The process of mathematical modeling as depicted in Figs. 16.3 and 16.4 begins by looking for simplifying partitions and symmetries in the system, using analogies and other heuristics as needed, e.g. to pin down variables that remain the same throughout the spring. The Mathematization Process then attempts to apply standard mathematical schemas to describe static and dynamic relationships in the causal explanatory model, first at a geometric level, and then at a quantitative level. This is achieved by looking first for standard matches to standard objects, e.g. $C = 2\pi r$ for the circle of wire, and by making simplifying assumptions if helpful to facilitate matches for less familiar objects or relations, for example, a slightly helical spring coil can be treated as a circle with break. These assimilative matches are assumed to be primarily visual ones. Correcting adjustments can be added to these approximations later if necessary.

A false solution was generated early on in episode 16 by applying the torque schema lever arm to the wrong part of the coil. This episode exposed the importance of imagistic alignment of a schema like torque with the situation to which it is being applied. Imagistic alignment appears to be as important for quantitative modeling as for qualitative modeling. Imagistic alignment analogies can be used as an important tool for this alignment task.

16.3.1.1 Evaluating a Mathematical Model by Running It

Here one might ask what it means to run a mathematical model, and whether one can evaluate a mathematical model by running it. In the geometric case an example is the "opening the clam" diagram in Fig. 14.13a with a softened segment at w. The investigator envisions that as the vertical distance between a and b is increased, the angle theta will increase proportionately (for small angles). This relation can be expressed mathematically, leading to a mathematical description for the relation between stretching, s and twist in a segment, theta. It therefore qualifies as running mathematical model in the sense of coordinating two motions known to be related mathematically.

However, by running this model the subject also evaluates it, because he or she sees a discrepancy with an earlier assumption, namely that the orientation of elements of the wire is going to remain horizontal. At point < a > the orientation is no longer horizontal, as shown by the "handle" there, so something is wrong with the model and it must be amended in Fig. 14.13b. Thus, by paying attention to features at a geometric level of detail (angles, horizontality), the subject is able to evaluate the plausibility of a mathematical model by running it.

Other ways that a mathematical model can be evaluated are via mathematical Gedanken experiments (described below) or empirical testing if that had been available. There is less independent evidence here than in the qualitative episodes with external subjects, but most of the mathematical processes above are conceived of as involving imagery, because for one they appear to involve projecting movement into diagrams. This extends the imagery-based theme to the mathematical domain. In the next section, I elaborate on the plausibility of thought experiments being involved in many of the processes of mathematical modeling.

16.3.2 Untested Thought Experiments at Higher Levels of Precision than Qualitative Modeling

In addition to their import in developing qualitative models, untested thought experiments can contribute to the refinement of a model to reach higher levels of precision than the qualitative level. Consider the following examples:

16.3.2.1 Determining Significant and Insignificant Effects

Thought experiments were used to determine the relative sizes and insignificance of the unbending and tension effects kinesthetically (in episodes 15b, 15c). I interpret this as running different explanatory models in thought experiments to compare the relative sizes of their effects. However it is somewhat controversial whether to count this kind of comparison as a "higher level of precision than the qualitative level."

16.3.2.2 Determining the Form of a Mathematical Model

1. Determining the linearity of a mathematical relationship between twisting and length by imagining adding the twists of two identical coupled twisting rods in series (episode 25). I interpret this as a prediction from a compound imagistic simulation for twisting and symmetry to predict equal contributions in the same direction, assimilated by a mathematical schema for adding angles.
2. Developing a mathematical model by calculating the amount of twisting in a coil by using the "frozen handles" techniques in Fig. 14.13 (episode 28). There the softening technique allows the generation of the quantity of twisting introduced by a given stretch within a familiar geometry. I classify this as a simpler equivalent imagery alignment analogy which uses imagery enhancement techniques to partition the stretching action into two pieces runnable as a thought experiment and "measurable" via the application of metric geometry principles.

16.3.2.3 Mathematical Gedanken Experiments

Finally, the use of a mathematical evaluative Gedanken experiment was described in episode 32 which tests the final equation in the solution against an extreme case like the spring with a coil width of zero. This is a special kind of thought experiment and has the same logic as the qualitative Gedanken experiment shown in Fig. 15.6 part 4. A prediction of the model is compared to a prediction from an independent source for a specially designed case, here an extreme case where the prediction could be made from an imagistic simulation.

All of the cases discussed in this short section on higher levels of precision still appear to rely on untested thought experiments. In examining the cases, they appear to exemplify most of the elements in Fig. 15.6, only now they are at a higher level of precision than the qualitative level; they are moving toward a mathematical level of precision. This suggests that each mode of reasoning in that figure can take place not only at the qualitative level but at higher levels of precision as well. I conclude that untested thought experiments can be involved in a variety of ways in mathematical as well as qualitative modeling.

16.3.3 *Mathematics and Explanation*

The relationship between explanation and mathematics is vexed, and I have only scratched the surface of it in this document. However, I will provide some conjectures about the connection. The scientific principle of torque schema used here in the solution is both scientific and mathematical – it is both a qualitative/causal model and the geometric and mathematical details connected to that qualitative model. However, such schema assemblies do not in general contain sufficient knowledge about how to apply them to a novel problem, as was illustrated by the misapplication of this schema in episode 16. The hope of applying a mathematized science schema like torque may be an incentive for the subject to add realigned, connective precision to the model during the course of the protocol, in order to see how to apply the schema. As a mathematical schema is activated as possibly applicable, it can motivate simplifying modifications in the qualitative model, such as being sure to apply force at the center axis of the spring so that torque is the same everywhere. Thus the goal of applying an existing physico-mathematical principle can push qualitative modeling in fruitful directions. In this sense, the mathematical schema and qualitative model can "move toward each other". This is a different view than saying that a given qualitative model is simply "re-represented in" mathematical terms. A mathematical model results as mathematical schemas are accessed, selected, and aligned but also as the qualitative model is realigned, simplified, or sharpened. This may be a more accurate picture than simply saying that the qualitative model is "translated" into a mathematical one.

This allows one to revisit the theme developed in Chapter 14 concerning the existence of a gradual transition rather than a sharp line between qualitative and quantitative modeling. It was *not the case that mathematical modeling simply consisted of applying equations*. The initial stages of building toward a quantitative work model involved an intense focus on expanding the imagery of the qualitative model. That is, additional qualitative, geometric, and eventually quantitative precision was added to existing qualitative model simulation elements. For example, features of lever arms that twist a segment of the coil were described in increasing precision: exact location of the lever arm, presence and direction of movement, geometric trajectory of movement, and finally measurable displacement and angular displacements generated by the movement. In this enterprise, *expanded precision in imagery* to a geometric/mathematical level appeared to be at least as important as increased precision in linguistic or mathematical-symbolic description.

The present theory generates questions for future research. In the present theory, scientific schemas are applied or invented qualitatively in the explanation cycle, whereas quantitative, geometric, and descriptive/observational schemas are applied in the mathematization cycle. But if qualitative explanations and mathematical descriptions are developed in separate processes, how then do *mathematically based explanations* occur, such as explaining the difference between a nuclear reactor with an unstable vs stable design? I would hypothesize that neither mathematics nor science can do this alone, and that it is the qualitative model, enriched with mathematical schemas to give descriptions of the relations at a mathematical level of precision, that is able to provide a meaningful and satisfying explanation of say, why a reactor can be stable. In other words it is a combination of explanatory and descriptive processes, using both scientific and mathematical schemas, that produces the explanation at this level of detail. Indeed it is a strong test for a qualitative *model*, if increasing its precision via the mathematization cycle to a mathematical level, while independently increasing the precision at which *observations* are gathered to a mathematical level, (both horizontal motions to the right in Fig. 16.7) leads to two sets of mathematical conclusions that match within an expected range of error.

I have only begun to analyze the nature and construction of mathematical models in this book. Their full analysis must await further research, but the present analysis illustrates their connection to and grounding in qualitative models and imagistic, nonformal reasoning methods. Overall, I am surprised at the voluminous amount of qualitative preparation that it took to understand the spring at a level that could be finally quantified in an equation. This reinforces the importance of qualitative analysis, and illustrates how it can support, and be tightly coupled to, quantitative analysis.

16.4 Abduction II: How Evaluation Processes Complement Generative Abduction

Although each of the three major subprocess in the investigation process in Fig. 16.4 are important in their own right, in the protocols analyzed here the process of explanatory model construction in cycle II is the one most responsible for Aha phenomena and scientific insights or breakthroughs. I will spend the rest of this chapter attempting to further unpack its subprocesses, starting by returning to the process of generative abduction. In this section, I reexamine the concept of generative abduction and discuss how even though it is a relatively weak strategy, it is complemented by evaluation processes that can make it powerful. I also discuss the question of the degree of divergence in abductive processes that might be most effective, and how that might be achieved.

16.4.1 Multiple Sources of Ideas and Constraints for the Generative Abduction Process

As discussed earlier, Fig. 16.1 contrasts generative abduction in the third column schematically to deduction and induction by enumeration in the first two columns. In the fourth column it depicts the process of Abduction in the Broad Sense which includes evaluation and revision cycles and a consequent evolution of the model. The term "generative abduction" as used here is still rather broad even though it is narrower than "abduction"; as it stands the concept does not specify a precise mechanism for generating models. I think it is still useful, however, to keep the term until we have a consensus on such a mechanism. It is associated here with a probable set of functions and features that such a mechanism will have to satisfy.

In Fig. 16.8, I take the next step in making these functions more explicit. It is based on the simpler but very important pattern of model evolution shown on the right-hand side of Fig. 16.1 with inputs to the evolving model from both observations "from below" and prior knowledge schemas "from above." Figure 16.8 shows the flow of information over time for a pass through the explanation cycle II (in Fig. 16.4), with the basic alternation between model generation (or revision) and evaluation processes during the construction of an explanatory model. Generative abduction is modeled in Fig. 16.8 by the processes within the dotted box. Multiple sources of constraints on the model revision process are shown, including known structural and behavioral characteristics of the target system, one's general theoretical framework, and input from the most recent evaluation of the last iteration of the model. Which of these multiple constraints is actually kept in mind during a generative abduction as opposed to being examined during a later evaluation phase is a question for further research and may in fact vary from case to case.

The figure helps clarify the two concepts of abduction proposed; the evaluation process is not considered a part of the generative abduction process, but it is considered a part of abductive model evolution in the large sense. Generative abduction in the narrow sense then refers here to the process that drives a single act of model generation or revision. Hypothesized subprocesses involved in generative abduction are shown in procedure IIA in Fig. 16.3. An abduction may be "fed" directly by a prior knowledge schema. Analogies can play different roles within the generative process: providing concrete elements, activating schemas, aiding in the application of schemas.

16.4.2 Model Evaluation can Provide Inputs to the Next Abduction Cycle

A more traditional view of model evaluation focuses on the empirical testing of the model. But here, running the model on the target in an imagistic simulation is thought of as the first means of evaluating a new or revised model for adequacy in

16.4 Abduction II: How Evaluation Processes Complement Generative Abduction

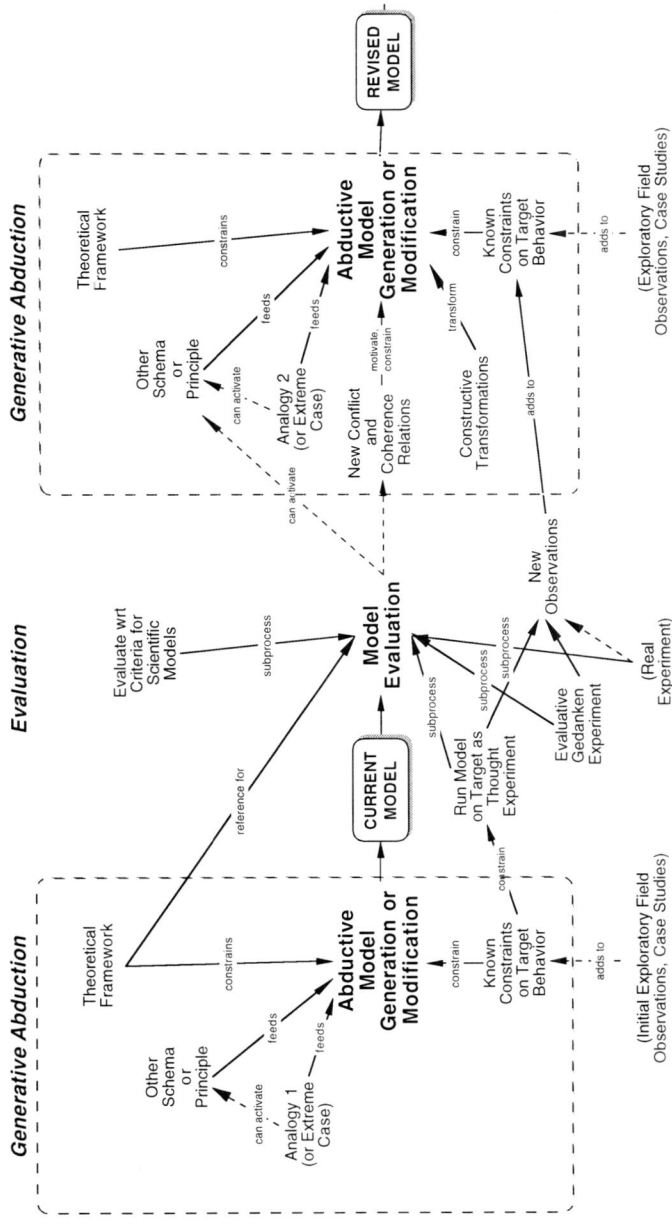

Fig. 16.8 Flow of information over time for explanatory modeling cycle II

providing a viable mechanism that can produce the behavior of the target. The subject can also proceed to use other more detailed criteria for evaluation, as indicated in procedure IIB in Fig. 16.3. Empirical contributions are shown in the bottom row of Fig. 16.8 as an indication of their potential contribution, but since there were virtually no empirical inputs in the protocols they are shown in parentheses. Although exploratory field observations and case studies are shown here as primarily contributing constraints and material to be explained by model generation and revision, they can also provide data for model evaluation.

Figure 16.8 also captures the idea of informed revisions: first, there is an arrow from evaluation to model modification indicating that criticisms of the old model are to be addressed as new constraints in producing a modified model. This makes them "intelligent" revisions that contrast with a pure "series of random guesses" process such as that used in a simple model of biological evolution. For example, when the initial bending model was applied to the spring, I hypothesized that the increasing slope idea emerged from running the bending idea while applying it to the spring in an imagistic simulation. This generated a new conflict relation and constraint on the *model* – that the model could not include bending – which needed to be resolved via model modification. Secondly, it brought to the foreground the issue that the *target* (spring) obeyed the constraint of having coil separations equally spaced apart. The latter restriction is shown as a second arrow from new observations to known constraints on system behavior in Fig. 16.8. Tracking of constraints on target behavior and observations of it takes place at the bottom of the figure, while tracking of constraints on the model and simulations of its behavior take place across the middle of the figure. Thus running the model on the target can be a powerful initial evaluation strategy for generating constraints on both the model and the target. In a study of student responses to anomalous data, Chinn and Brewer (1998) identified certain students who did not treat anomalies as something needing to be explained. Thus one cannot take this "passing of criticisms back to the abduction process" for granted and this serves to highlight it as a skill.

This figure also indicates that model evaluation can provide inputs to the next abduction cycle, not just in the form of constraints, but as a way of activating new ideas. That is, evaluation can play a role in generation or revision! When a model is run on the target, this simulation can trigger a variety of other processes, as indicated in Fig. 15.3, some of which can suggest new ideas. An example of this was the evaluation of the twisting model via stretching a circular ribbon in episode 14. Imagining the ribbon fully stretched in the vertical direction was originally intended only as an evaluation that would allow one to "see" twisting in the spring. But it triggered two ideas for model revision in episode 15, "unbending," and extension of the wire in the spring, as new ideas for sources of stretch.

As another example, if one considers the initial generation of the hexagonal coil as part of an attempt to evaluate the bending model of the spring, then its role in triggering the recognition of twisting is also an example of how an evaluation process can trigger a new idea for the next round of model modification. Thus it appears that because the imagistic simulations that are used in model evaluation are volatile processes, model evaluation can trigger new ideas for model revision.

16.4 Abduction II: How Evaluation Processes Complement Generative Abduction   367

Thus we have the possibility of a doubly effective intelligent modification process, not just because a criticism of the old model is taken into account, but because evaluation processes can trigger new ideas for modification. Although the latter are heuristic strategies that are not guaranteed to work, this is still a sense in which the GEM cycle is "doubly intelligent" in contrast to a blind evolution strategy. Overall, this dialectic alternation between generation and evaluation is still at bottom a "guess and test" process, but what makes it a very sophisticated one is: the multiple sources of suggestion for generation; the tracking of multiple constraints on generation; plus the multiple criteria and means for evaluation as in Fig. 16.8.

Some aspects of this analysis are rooted in those of Clement (1989b) and inspired by Nersessian's (1991) insightful work on model development in the case of James Maxwell and his process of developing electromagnetic field theory. These aspects include model revision, multiple contributing analogies, and novel thought experiments. Nersessian also hypothesized developing multiple constraints on the model and target. Here, in addition to these features, an attempt is made with the help of Fig. 16.8 to present as transparent a picture as possible of the model construction process – one that separates processes, structures, and labeled information flows between them. Other additions on the generative side are to: separate out two uses of the term abduction, at least three roles of analogy, and the important role of constructive transformations in model modification. On the evaluative side the contribution attempted is to delineate: the central evaluative process of running the model on the target, a separate role for evaluative Gedanken experiments, and, in addition to these basic strategies for attaining explanatory plausibility, the possible activation effects from running simulations, and a separate process of evaluation with respect to more esoteric criteria for scientific models.

16.4.3   Role of Transformations in Model Modification

A number of examples of constructive spatial transformations being used to generate analogies and extreme cases were provided in the protocols in Chapters 6 and 14 (e. g. transforms rod into a polygonal spring in episode 10b in Chapter 14; transforms the circular coil by "softening" part of it and "hardening" other parts to form a close analogous case in episode 17; transforms spring to a case with extremely narrow coils in episode 6). These examples prefigure the idea that constructive spatial transformations are also an obvious candidate for a model modification process during abduction in the form of changes in an element of a model or the shape of an element. We saw the model of twisting in the spring change subtly from twisting in an element due to torsion forces applied by adjacent segments as lever arms, then whole quarter or half sides of the coil acting as lever arms, and finally "handles" from the middle of the spring coil acting as lever arms. We can think of these as small constructive transformations in the model that produce model modification. They appear in step IIA1d of the model construction process in Fig. 16.3, part II.

In summary, any particular instance of abductive model modification might then occur by any of several mechanisms: accessing and adding an element from a prior knowledge schema directly by association and adding it to the model, doing this via a provocative analogous or extreme case, or changing some aspect of the model via a constructive spatial transformation.

16.4.4 Distinctions Between Constructive Transformations, Running a Schema in an Imagistic Simulation, and Basic Spatial Reasoning Operators

I will maintain a general distinction between running a schema that is able to simulate the behavior of an object or system (such as the twisting schema), and constructive transformations that are actions which form, or modify, a new object or system (such as forming/adding the "handle" gc in Fig. 14.9c or "softening" the segment w to make the twisting effect macroscopic and imaginable). The distinction between constructive transformations and the process of running a schema in an imagistic simulation can also be seen by means of Fig. 15.3. Constructive transformations are not shown in that figure because they are used to construct models or cases, not to run them. It makes sense that the activity of domain specific schemas (such as torque or acceleration) would be more involved in running a model, whereas more general constructive transformations for fabricating, assembling, and modifying objects would be more involved in generating or revising a model. This corresponds to the form of Fig. 16.8, where revising the model ordinarily involves a constructive transformation, whereas running the current model (an assembly of schemas) to evaluate it does not.

More formally, I refer to an imagistic simulation when, given a fixed system with certain variable features, the subject applies one or more schemas to the system to simulate autonomous operation of the system, leading to a prediction about the system's behavior. The latter operation may be represented by the subject analogically by thinking of physically acting on the system. This means that constructive transformations and imagistic simulations can be partially similar in that they can both utilize action schemas. However, the purposes of these two processes are different. In a constructive transformation, the goal is to produce a new (improved) case or model. In an imagistic simulation the goal is to produce a prediction from an existing case or model.

16.4.4.1 Spatial Reasoning vs. Constructive Transformations

I should also clarify another potentially confusing issue around various uses of the term "transformation." Minor perceptual operators such as zooming, and rotating are also sometimes referred to as transformations. In models of thinking

16.4 Abduction II: How Evaluation Processes Complement Generative Abduction

Table 16.6 Some processes involved in thinking via dynamic imagery

Running a model or schema in an imagistic simulation
Schema generates a predictive simulation, such as: applying torque, acceleration of a car, rolling a ball

Constructive spatial transformations
Such as cutting, deforming, or reshaping an object, joining multiple objects – involved in constructing models, analogous cases, Gedanken experiments, or extreme cases

Basic spatial reasoning
General visual operations, e.g. zooming, scanning for features
Comparing size, shape, color, distance
Imagining basic manipulations for arranging object positions, e.g. rotating, moving, rearranging, stacking, nesting, abutting, aligning, etc. and inferring the resulting new spatial relationships

that involve imagery, such spatial reasoning operations are used so ubiquitously that to analyze them here would take one to a much finer level of detail than that of the main focus of this book. The transformations under analysis here are usually more specialized constructive ones involved in generating models (or analogies, or extreme cases), e.g. joining, cutting, deforming objects. Therefore when I use the unmodified term "transformation" or "spatial transformation" in this book I am in general referring to such constructive transformations, as shown in Table 16.6. When I use the term "spatial reasoning" this refers to the more ubiquitous basic perceptual operations such as rotating and zooming. Thus it is useful to distinguish basic spatial reasoning operators from both constructive transformations and the process of running a schema in an imagistic simulation.

16.4.5 Coherence and Competition Between Models

A feature not shown in Fig. 16.8 is that the subject developed several competing models of the spring and weighed them against each other, such as the simple bending vs. twisting models of the spring. There were many possible coherence or dissonance relations between each model and various constraints: the subjects prior experience, intuitions, and other physics knowledge, in addition to the other criteria for evaluating scientific models listed above. This means that deciding between two or more competing models is a multi faceted task involving many weighted factors. This can be considered a third-, higher-order subprocess of abduction. Therefore, in addition to the process for developing generating and improving a single model shown in Fig. 16.8, one needs a larger "meta-evaluation" process that compares competing models (abduction to the best explanation), and this was modeled in Fig. 16.3 as the last process in the explanation cycle (Section II-E). Chapter 18 will discuss multiple criteria for evaluating and comparing scientific models based on the work of Darden (1991), Kuhn (1977b) and others. Thagard (1989) has examined one aspect of this problem by exploring various parallel algorithms for computing complex networks of coherence relations that "settle" on a dominant model.

16.5 Seeking an Optimal Level of Divergence

16.5.1 The Problem of Accessing Relevant Prior Knowledge: An Ill-structured Problem

The most "standard" method for solving a problem or giving an explanation in science is to activate relevant schemas or principles by learned rules of relevance. This is not possible however for unfamiliar cases for which there are no familiar rules. In the spring problem, subjects quickly ran out of direct associations to traditionally relevant principles, and that method failed, leaving the subject with no "standard" moves to make.

Thus the model construction problem as characterized here is inherently ill-formed in that one is not just finding a strategy for ordering a given set of moves, rather one has to first find or invent the relevant states and operators or "moves of the game" themselves. In this situation it is not clear what prior knowledge may be relevant – any prior knowledge schema that is used is likely to be outside of its normal domain of application so that normal routes to activating such a schema will not be present. This means that the process of abductive model generation needs to be a divergent and creative one. On the other hand, it should also be one that does not violate too many of the constraints on the problem so as to retain relevance. This is the problem of accessing relevant prior knowledge creatively in order to construct explanations and predictions for a new situation.

Classical problem solving theory deals with how to order moves within a fixed problem representation or problem space – a fixed set of goals, operators and possible states. Here, however, each time a new analogy or schema is activated, one has changed the problem space. Thus we are dealing with cognition that is not working within a fixed problem space; in fact, the task itself – explanation – would seem to be to invent a new problem space in the form of a new representation (a model) for solving prediction problems.

There are additional sources of difficulty in creative model construction:

- The sheer difficulty of homing in on a key variable among innumerable possibilities.
- As discussed earlier, an Einstellung effect can heighten the access problem by keeping the subject locked into a limiting view of the problem. This can be caused by the coherence of a previously developed view.

16.5 Seeking an Optimal Level of Divergence

– The problem can be heightened further if the subject is using a persistent misconception he or she feels confident in. The extreme version of this problem is the problem of breaking out of a scientific paradigm.

Thus the problem of creative model construction is inherently ill-structured and requires the use of divergent methods.

16.5.2 Need for an Effective Middle Road with Respect to Creative Divergence

Now I want to summarize the advantages and disadvantages of the generative abduction process and argue that it needs to be one of moderate rather than extreme divergence. One advantage of having an abductive generation process is that it is more divergent than induction by enumeration of a pattern in the phenomena – one can create models of invisible mechanisms that are not visible in any pattern of phenomena. To do this however means that some of the ideas for an explanatory mechanism must come from a prior knowledge source other than the target phenomena (although any such mechanism must fit the constraint of explaining the phenomena).

The factors above provide important motives for model generation processes to be abductive at least part of the time and allow for divergent and creative generation of model elements. I will discuss a number of possible subprocesses that could have served this purpose in the present protocols.

16.5.3 Analogies and Extreme Cases Appear to be a Fruitful Source of Divergence

One of the positive roles analogies and extreme cases can play is to help a subject break out of an Einstellung pattern. In Chapters 6–7, I mentioned the paradox that an analogy appears to take the subject away from the problem at hand – how can one move closer to a solution by moving away from the problem? However, when one is stuck in an Einstellung pattern, moving away from the current context may be just what is needed. Extreme cases and analogies appear to be useful in this way. This was seen in the discovery of "unbending" in the extreme case of the drastically stretched spring in episode 15a. It was also seen in the case of the hexagon analogy that triggered the torsion idea. These examples suggest that analogies and extreme cases may be effective methods for stimulating the discovery of new variables. However, their successful use may depend on generating analogies that are not too different from the target. The question remains as to how an analogy might be any better for this purpose than free association.

16.5.4 Dangers of Divergence: The Need for Optimal Divergence

In fact, some analogies are too divergent. Several different analogies can be generated for the same target. The volatility of imagistic simulations adds to this complexity. The generation of a new analogous case may activate one or more new schemas; when the schema is run, it may in turn suggest other schemas, analogies, or dissonances. A combinatorial explosion of ideas to pursue is possible. There is a premium on useful ideas, not just quantity of ideas. There is a question of just how far away from the original target one wants to wander, in order to stimulate ideas that are actually useful. So the idea of "optimal," or "contained" divergence may be a useful concept for thinking about scientific discovery, and possibly for use by problem solvers as a heuristic.

16.5.5 Some Methods for Reducing Einstellung Effects Via "Contained" Divergence

I will discuss several methods for reducing Einstellung effects while maintaining one's focus on ideas that are relevant to the problem. (Starred items below indicate those items for which there is evidence in the present data from one or more interviewed external subjects.)

16.5.5.1 Social Processes for Reducing Einstellung Effects

The social processes described by Dunbar (1997) and others may serve to enhance model construction by reducing Einstellung effects through individual differences in approach. They may also amplify the evaluation and modification processes that have been described here. Social processes go beyond the scope of this book, but are of obvious potential importance. Here I will concentrate on aspects of individual thinking that may help lead to optimal forms of divergence.

16.5.5.2 *Model Evolution Via Evaluation and Modification is a Convergent Influence

As discussed earlier, one answer to the problem of excessive divergence is the "intelligent evolution" pattern of a cycle of evaluation and modification where faults are targeted for revision and new constraints are passed forward to the next cycle of revision. This serves as one convergent influence to compensate for creative processes that are overly divergent.

16.5.5.3 *Imagery May Support Modification Under Multiple Constraints

One of the advantages of imagery is as a way of representing multiple constraints simultaneously as a context for the above modification processes. If abductive model modification is achieved via a creative transformation under multiple constraints, then this advantage would be a very important kind of *constrained divergence*.

16.5.5.4 *Analogies May Foster Contained Divergence: Analogies are a Conservative Strategy in Two Ways

First, because an analogy by definition seeks a case that is structurally similar to as well as different from the target problem, this gives it the potential to be optimally divergent in the above way. However, far analogies may often be too divergent to be relevant. Since the torsion breakthrough occurred in an analogous case generated by a "close" transformation of the original target problem, this raises the question as to whether that kind of close analogy might be optimally divergent in the sense that it is particularly effective for generating insights. A close analogy might be more likely to be effective than a far analogy to a remote context because with the far analogy the chances are good that there might not be enough overlap between the systems to allow for transfer. That is one reason why I have shied away from the classical definition of analogy (cases that share higher order relations but few surface features) which seems to exclude near analogies where only one feature may be changed. I will examine this issue further in the section on divergence modulation below.

16.5.5.5 *Extreme Cases

These are often generated by changing only the value of a variable in the target problem, and are therefore arguably even closer to the original problem than an analogy. This does not guarantee relevance to the target problem but it should make it likely. Extreme cases may therefore have a fairly good chance of being *optimally divergent*, and since they are easy to generate, I give them a high recommendation as a general heuristic, along with close analogies. The power of extreme cases to encourage students to see new variables (not just to resolve the influence of a known variable) has been documented by Zietsman and Clement (1997).

16.5.5.6 *Perceptual Activation as a Useful Alternative to Verbal Association

It is also possible that schema activation mechanisms that depend on perceptual mechanisms are an important alternative to verbal associations, and that these can be triggered by concrete analogous or extreme cases. Such perceptual activations might be especially important when there are spatial constraints such as shape or size to be conserved. This was most likely the mechanism leading to the critical activation of the torsion schema in episode 11a.

In summary analogies and extreme cases may have the potential to inject a "contained" or focused level of divergence into the construction process. More detailed strategies for fostering this are discussed below.

16.5.6 *Mechanisms for Modulating Divergence*

16.5.6.1 *Temperature Control

An interesting question arises as to whether subjects can control or modulate the level of divergence in their thinking. Mitchell and Hofstadter (1989) proposed a mechanism of "temperature control" in their simulations of analogy use whereby the degree of similarity required to activate an analogous case could be changed for different stages in problem solving. It makes sense that strong divergence is more important at the beginning of a problem. This suggests an investigator who is very divergent at the beginning of an investigation when he has little to go on, then less divergent later once a viable direction for modeling has been found. The protocol in Chapter 14 appears to mimic this pattern to some extent, and so specific elements were placed in the process model to model this strategy. It also makes sense to increase divergence when one feels "stuck" in a fruitless approach; e.g. some evidence can be found for this in Chapter 6, transcript line 57.

16.5.6.2 *Conserving Transformations of the Target Situation, were Effective Triggers for New Ideas in the Protocol

Mitchell and Hofstadter's method above appears to rely on analogies generated in a traditional manner via an association. Another possible strategy for modulating divergence is to use other methods for analogy generation. The strategy of creating an analogy via a transformation of the target case, if limited to transforming a small number of features, would appear to be a natural way to generate "close" analogies, with a small amount of divergence. For example, one could pick one or more factors deemed central as ones that should be conserved, then try transformations on other factors in order to retain essential problem constraints. Many of the analogies in Chapter 3 that were generated via a transformation are possible examples of this strategy.

16.5.6.3 *Constrained Analogy Generation

On occasion, there was some evidence in the transcripts that a subject was attempting to generate analogies via association within particular constraints. For example,

16.5 Seeking an Optimal Level of Divergence

just prior to generating the hexagonal coil, the subject says: "What else could I use that stretches, instead of a spring?" indicating that he is constraining his search to include only cases with stretching as a property.

16.5.6.4 *Bridging Analogies May Further Optimize Divergence

The hexagonal coil seems to have been intended as a bridging analogy for evaluating the earlier analogy to bending. The bridging nature of this analogy may be relevant also within the context of optimizing divergence. A bridging analogy is an attempt to find a case that splits the distance between an analogous case and the target with respect to similarity. The bending rod analogy had proven to be too "far" (as indicated by the anomalies the bending schema was generating when used within the spring model), and the hexagon as a bridge is somewhat "closer" to the target of the circular coil. Therefore it may have reduced the divergence of the case enough to be more relevant, but still capable of activating a different schema than had been used up to that point. The hexagon case is just different enough to trigger a new schema; but not so different as to have a mechanism fundamentally different from that of the spring, as did the bending rod. The bridging case thus may have acted to "fine tune" the degree of divergence or "distance" of the analogy, to optimize its chances for triggering an important idea. Bridging analogies were discussed in Chapters 4 and 6 as a strategy for evaluating the validity of a *previous* analogy (that function fits other protocol evidence, as was seen in the case of the "wheel problem" in Chapter 4). By contrast this newly identified role involves the generation of a new "improved" and more provocative or fruitful analogy and appears to be a second possible purpose for bridging analogies. I call this type of bridging analogy a "provocative bridging analogy." By having the effect of modulating divergence strategically, a provocative bridging case may increase the chances that a relevant schema for model construction will be activated. This constitutes the most fully developed model that will be discussed in this book of a mechanism by which the torsion insight occurred.

16.5.6.5 *Analogical Transformations can be Specialized Further to Modulate Divergence

In the model in Fig. 16.3, analogies can be generated with different levels of divergence. For example, in episode 17a the subject uses the small transformation of a "softened segment" to align the twisting idea with the coil. And in the later quantitative parts or stages of the solution, the subject uses specialized analogies in the form of target transformations that are very close to the target (such as the "frozen handles" transformation in episode 28), presumably so as not to destroy its fine structure. I hypothesized details in the mathematization cycle in Fig. 16.3 step IIIA2b as follows:

b-Imagistic alignment analogy: Transform system containing R via a transformation thought to be conserving (or reversible) into an analogous system that is simpler or more compatible with existing mathematical schemas, so that the imagery and features of the target and the mathematical schema can be aligned correctly.

1. As precision levels become higher, use less drastic transformations so that they conserve relations at the higher level of precision. Make minor transformations of system to exaggerate independence of or increase imagibility of parts or changes in order to envision alignment of mathematical schemas to each part.
2. For quantitative relationships, this means transforming the system containing R via a conserving transformation that (1) isolates a subsystem so that the relation between its inputs and outputs and its approximately independent contribution to the overall phenomenon can be calculated; or (2) makes it easier to apply standard quantitative mathematical schemas.

Thus an imagery alignment analogy is one that aids in determining how to imagistically align features of an explanatory model, source analogue, scientific principle, mathematical schema, or target in a more precise way. The latter heuristic is a specialized analogy that uses a very small transformation assumed to be conserving in order to establish equivalencies and make it more possible for schemas – in this case a mathematical schema – to assimilate and apply to the situation. The transformations used are thus modulated so as not to distort the system at the particular level of precision being described. For example, only affine transformations would be used if it is important at one stage to preserve proportional relationships between parts of the model. The above references to modulating the degree or "drasticness" of divergent processes raise the interesting question of whether the idea of temperature control could be applied to perceptual association mechanisms as well, depending on the level of development of and confidence in the current model.

16.5.7 *Summary for Section on Divergence*

I concluded that divergence could be produced intentionally, as in "brainstorming" by free association, but that it is also produced unintentionally by the naturalness of association processes and in particular by the volatility of imagistic simulation processes that activate other schemas. However, I have also hypothesized several methods for containing or even modulating the divergence of methods to a level appropriate to each stage in the solution. The following methods appear to be designed to maximize the chances for finding useful model elements and even scientific insights by producing *contained divergence*:

- Intelligent model modification in which faults are targeted for revision and new constraints are passed forward.
- Using imagery to represent models, which may support the latter process of modification under multiple constraints.

- Analogies by nature are cases that are partly different but partly the same as the target.
- Extreme cases are very similar to the target.

In addition several processes were seen as mechanisms for *modulating or fine tuning divergence* to an optimal level at different stages of the solution:

- Associative "temperature" control.
- Analogies that are generated with constraints in mind.
- Bridging analogies that move back toward the target from an initially too distant analogy.
- Analogies generated via conserving transformations.
- Within analogies there is a spectrum from very divergent associative analogies to very constrained minor transformations of the target situation, and these can have different purposes (early and late in the solution, respectively).
- In particular, a conserving transformation can be matched to the highest mathematical level of precision that must be conserved in order to retain important features of the target.

Thus it is possible to envision a system that is guided enough to converge on solutions, but is unconstrained enough to overcome Einstellung effects. Analogies and extreme cases seem especially well adapted to provide appropriate levels of divergence. The methods above are ways that I see creative theory generation walking a fine line between methods that are too convergent or too divergent, with the contingency that the line may be adjustable as well. Although they are not guaranteed, these appear to be effective ways of "maximizing the chances for insight."

16.6 Chapter Summary

16.6.1 Diagrammatic Summary

The theory of model construction developed in this chapter is represented in several of the later figures and tables, starting with Fig. 16.8. It shows the basic alternation between generative and evaluative modes as a model is evaluated and revised over time. It also shows how the analogy process can feed the abductive model generation or modification process, and how the evaluation process is served by the subprocesses of: (1) running the current model on the target and (2) running an evaluative Gedanken experiment. In addition, it shows how evaluation findings are used to constrain the next modification, and possibly to trigger new schemas for use in the modification.

The larger view of scientific investigation processes developed here was summarized in Fig. 16.4. Figure 16.8 provides a "blow up" picture of the explanatory modeling cycle II in Fig. 16.4 and shows how it impacts on the structure of the model over time.

16.6.2 Multiple Cycles and Goals in the Overall Investigation Process

Figure 16.3 unpacks Fig. 16.4 further by specifying a quasi-algorithmic procedure for each cycle. This model is capable of accounting for the different modes of investigation in the solution listed in Table 14.1. At the top level, the description process was used in stage 1, the explanation process was used in stages 2 and 3, and the mathematization process was used in stages 4 and 5.

Further hypotheses were advanced in this chapter concerning: the separation of explanation from description cycles in Fig. 16.3; the separation of these cycles from the analogy processes described in Section I (those analogy processes are called as a subprocess by each of the cycles in Fig. 16.3); and the similarity, at the top level, of generation, criticism, and revision processes in these three cycles. These cycles are capable of generating a rich model of a phenomenon at several levels of depth and levels of precision, as depicted in Fig. 16.7. These results indicate the power of nonformal reasoning and imagery-rich techniques, even when description at a mathematical level of precision is the goal.

16.6.2.1 Cycle I: The Descriptive Prediction Cycle

This cycle attempts to generate an initial prediction and description of the behavior of the target phenomenon without attempting to explain it. It first uses direct methods by looking for schemas the subject already has that might describe behavior, and attempting to make new "observations" via real observations or an imagistic simulation of the phenomenon. Failing these, it tries indirect methods such as expedient analogies and extreme cases. The object here is a correct prediction for *what* the behavior of the system is, not a deep explanation of *why* it behaves that way.

16.6.2.2 Cycle II: The Explanatory Modeling Cycle

This cycle attempts to generate an explanatory model that can explain why the phenomenon occurs. It is responsible for repeated cycling through the processes of model generation or modification alternating with model evaluation (GEM cycles) shown in Fig. 16.8. I have characterized the generative part of the model construction process as generative abduction within GEM cycles – a process that is less structured than deduction, but more structured than random guessing. This process was depicted in the dotted box in Fig. 16.8. Features of the present protocols were explained by hypothesizing the following features of the generative abduction process:

- Analogies can act as a subprocess for generative abduction by suggesting relevant prior knowledge elements to use in constructing a model.
- I speculated that in addition to using analogy, the remaining core of the generative abduction process is an act of design under constraints leading to an "educated guess."

16.6 Chapter Summary

- As such, it is often considered more conjectural than deduction, or induction over a sample; but I have portrayed it as being complemented by a cycle of evaluation and revision that makes up for this weakness to form a powerful learning process. I view this as a fundamentally different process from deductive proof, and it is at the core of the most productive episodes in the protocols.
- It makes sense that ideas for abduction should be creative and divergent in order to break out of Einstellung patterns, but they will not be applicable if they are too divergent. So there is a need for "optimally divergent" strategies.
- Generative abduction can be used to modify a model while incorporating new constraints generated by a previous model evaluation process, making the entire process more intelligent than a random evolution process.
- A third intelligent mechanism comes from various methods for modulating divergence during analogy generation.
- Imagistic simulation may have certain advantages as the internal representation for the design process inherent in model evolution by abduction, for several reasons, but in particular because of its ability to represent many interacting constraints simultaneously.

A final source of divergence documented in Chapter 15 is the volatility of thought experiments, which can lead to spontaneous associations (schema activations), or image feature recognitions. In particular, when a developing model or analogy is run in order to evaluate its viability, the simulation produced may trigger new ideas directly that can be considered during the next model modification. This adds a less predictable source of creativity to the mix.

I conclude that it is possible to envision a system that is guided enough to converge on solutions, but is unconstrained enough to overcome Einstellung effects.

16.6.2.3 Evaluative Modeling Process Compensates for Generative Abduction as a Weak Process

I take the quotations from expert interviews in Chapters 6, 7, and 10 as providing initial case material for grounding the following hypothesis:

> Given a trusted set of axioms or principles, generative abduction is a weaker form of inference than deduction from those principles for domains in which the principles are known to apply. But it can be quite powerful when used in conjunction with an evaluation and revision (evolution) process.

Although abduction led S2 and others into some "blind alleys," it also led to powerful results in the end. In this view evolution via abduction is a mode where the individual inferences are weak, but the overall strategy is strong because repeated evaluations and revisions can "home in" on a good model despite elements of trial and error and the possibility for missteps. When used successfully, the cycle can make up for the possible missteps in any particular generative abduction. For example, the model of bending for the spring could be considered a misstep, but then its evaluation did lead to the discovery of torsion and it is not clear whether torsion would have been discovered without this "misstep." Thus compared to deduction, abduction is a "weak" method for analyzing a system. It may ordinarily be used only when the subject is exploring new territory where established principles cannot be found that apply. In this view a strong evaluation process is an essential complement to generative abduction. I conclude that generative abduction, although weaker than deduction, can be a powerful process when combined with other processes. It can generate unobservable mechanisms, in contrast to induction by enumeration. Somehow the combination of processes shown in Fig. 15.5 is powerful. I will attempt to say why this combination is powerful in Chapter 21.

16.6.2.4 Cycle III: Mathematical Modeling

The process of developing a mathematical model was viewed as increasing the detail of description of variables in the existing qualitative model until relations between the variables could be described as mathematical functions. The process utilized subprocesses such as finding partitions and symmetries to simplify the problem, followed by applying standard mathematical schemas, first at a geometric level, and then at a quantitative level. Since the matching process appears to be primarily visual, this reinforces the idea of geometry as a giant connecting link between qualitative and quantitative models.

The importance of schema alignment became apparent, as a false solution was generated early on in episode 16 by applying the torque schema lever arm to the wrong part of the coil. Imagistic alignment analogies were identified as a tool for this alignment task. Because these analogies tend to be carefully designed bridges generated by a small conserving transformation, they are an important example of using controlled or modulated divergence in analogy generation. Accuracy of the model was assessed using an evaluation process that variously included processes such as running the model and using mathematical Gedanken experiments.

The fact that both qualitative and quantitative modeling processes use similar subprocesses, such as GEM cycles, partitioning, analogies, schema alignment, imagistic simulation, thought experiments, Gedanken experiments, and generalization, and the fact that there are several intermediate modes in the solution between qualitative and quantitative models, argue that the developmental distinction between these two types of modeling is not so sharp as has been traditionally described. Yet the two ends of the spectrum provide a strong contrast, with the initial qualitative models being extremely rough, when compared to the elaborate and carefully connected mechanism envisioned in a model with quantitative precision.

16.6.3 *Four Subprocesses at the Core of Complex Model Construction: Generative Abduction, Model Evaluation, Schema Alignment, and Mathematization*

I have attempted to construct a framework that takes visualizable mental models seriously as the central kind of knowledge being developed by an expert in a new

16.6 Chapter Summary

area and as the organizing goal for theory construction. The framework outlines how abductive model generation and revision, model evaluation via imagistic simulations, and repeated cycles of these last two processes in manageable step sizes can produce scientific explanations. Each of these utilizes the reasoning processes described in Fig. 15.6 and each of those reasoning processes in turn were analyzed as using imagistic simulations in Chapter 15. The latter three levels of analysis were anticipated in Fig. 15.5. Thus the goal of this chapter was to show how it is plausible that reasoning processes that can operate via imagistic simulations can be combined together to generate scientific theories.

Generative abduction is viewed here as a design process operating under multiple constraints for producing runnable models, and this is what makes it difficult and error prone. But repeated cycles of revision can make up for the weaknesses in any one abductive revision, as shown in Fig. 16.9. And small step sizes during revision can make possible the difficult task of making productive abductive revisions under multiple constraints. Intelligent revisions are possible when the revision process responds to prior evaluations, and this differentiates the process from a random trial and error process. Figure 16.9 is unusual, in that it shows the relation between a larger process on the left, and a subprocess of it on the right. Thus it is not a flow chart, but a description of relations that make the combination of processes at two levels successful.

As precision increases further, accurate schema alignment becomes important, and the subject may move toward the possession of a fully integrated, spatiotemporally connected model. At this point a fourth major process, the mathematical modeling cycle, can serve to add new levels of geometric and then quantitative precision to each causal relation in the model. Some of these processes can use analogy and extreme case analysis as a subprocess, and all of them can use imagistic simulation as a subprocess. Each of these is a source of creative divergence for generating new ideas. Part of the "art" of creative model construction lies in coordinating the many strategies documented for controlling or containing divergence so as to home in on productive ideas.

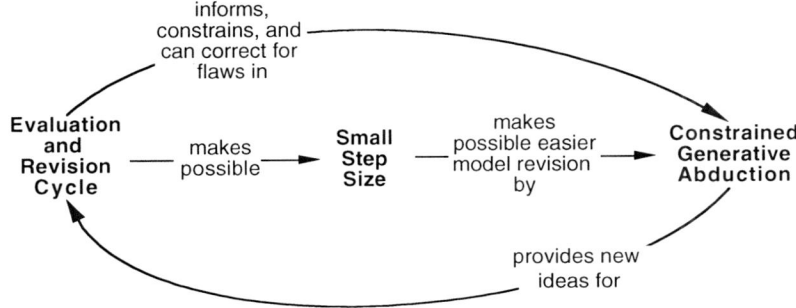

Fig. 16.9 Abduction plus GEM cycle strategy for complex model construction

Since there were virtually no observations to provide the subjects with an empirical input to the generation or evaluation processes described in this chapter, one can view the last three chapters as an exploration of the use of nonempirical processes to generate and evaluate models. Although I certainly do not wish to deny the power of empirical processes, this study advances new reasons for recognizing the impressive power of nonempirical processes. The concluding chapters will expand on this theme, as well as examining the extent to which the investigation model is "uncontrolled" or "controlled" (algorithmic). This will impact on the question of how to achieve a delicate balance between divergent and convergent processes for productive creativity.

Chapter 17
Imagistic Processes in Analogical Reasoning: Transformations and Dual Simulations

In Chapter 4, it was argued that criticizing and evaluating an analogy relation is just as important a process as generating the analogous case in the first place. In other words, even if the analogous case is well understood and has yielded a prediction for the target problem, one must establish confidence in the validity of the analogy in order to have confidence in the prediction. I discussed a well-known strategy for analogy evaluation, that of mapping discrete features, and presented some evidence that experts exhibit it. I also introduced an additional evaluation strategy detected in the protocols called generating a bridging analogy. Now that there has been a discussion of imagery-based processes in Chapters 12 and 13, four more new analogy evaluation strategies discovered in the protocols can be introduced: *conserving transformations*, *imagery alignment analogies* (a special type of bridging analogy introduced in chapter 16), *dual simulation comparisons* used to detect perceptual/motor similarity, and *overlay simulations* (a special type of dual simulation). Accompanying the appearance of these strategies in the protocols are observations that strongly suggest the use of dynamic imagery. These suggest the methods involve imagery and may be imagery based. These findings add to previous evidence (Casakin and Goldschmidt, 1999; Clement, 1994, 2003; Craig, Nersessian and Catrambone, 2002; Croft and Thagard, 2002; Trickett and Trafton, 2002) for formulating the general hypothesis that many analogical reasoning processes can be imagery based. I will also discuss evidence for imagery being involved when subjects use *transformations* as a method for *generating* analogies, partitions, extreme cases, and explanatory models. Throughout this chapter, I use evidence from external subjects only.

17.1 Two Precedents from the Literature

17.1.1 Structural Mapping and Evaluation

In a pioneering article on the mechanism of analogy evaluation, Forbus et al. (1997) discussed mapping-based methods for evaluating analogies. In that model, each correspondence in a mapping has a score, and an evaluation metric for the mapping is the sum of those scores. They also proposed a method for evaluating strength of an inference made about the target as result of the analogy. In their method, the strength of the inference is based on the number of connections to supporting elements in the original mapping. Thus they described some plausible methods of analogy evaluation based on an extension of structural mapping algorithms for determining a best mapping.

17.1.2 Wertheimer's Parallelogram

One of the strategies for analogy evaluation I will consider as an alternative to mapping is what I term a conserving transformation. Although the inference resulting from mapping and a conserving transformation can be identical, I will argue that the processes themselves are not identical. An example that can be considered a paradigmatic case of a conserving transformation is Wertheimer's (1959) method for determining the area of a parallelogram shown in Fig. 17.1. Is the rectangle shown there an analogous case in that it will always have the same area? Developing a theory for how subjects understand this in a way that is different from mapping discrete features introduces the main theme of this chapter. For most literate adults, cutting and rearranging nonoverlapping pieces of area in this case are transformations known to conserve total area. Thus these conserving transformations, combined with the spatial/geometricknowledge to recognize the formation of a rectangle after the transformation, serve to confirm a conjectured analogy relation between the rectangle and the parallelogram. Transformations were identified in Chapter 3 as the most common analogy generation method observed in this book, but those were not always found to be conserving transformations. In a conserving transformation, the subject has reason to believe that the transformation conserves one or more critical sought relationships when the target is transformed into an analogous case and vice versa.

The parallelogram example *starts* as a conjectured analogy. Alternatively one might give a formal proof, because with the use of enough geometric knowledge, a deductive inference can be made showing an equivalence. However, many of the subjects cited by Wertheimer who understood the equivalence were young students who did not have formal geometrical knowledge; therefore he appeared to be dealing with nonformal

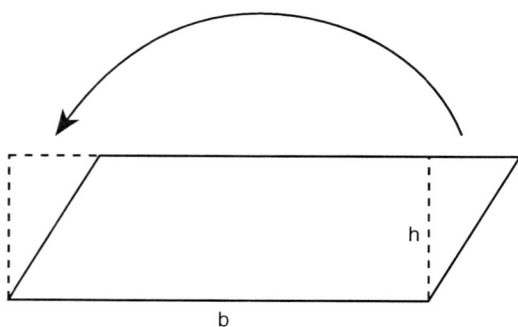

Fig. 17.1 Wertheimer's parallelogram problem

reasoning in his subjects. Their recognition that a rectangle of equivalent area was formed was nonformal and arrived at without use of formal geometric rules.

17.2 Conserving Transformations

The transformation can be broken down into three conserving transformations: cutting the parallelogram, moving the triangle to the other side, and joining it with the larger piece. To illustrate the process I will focus on the cutting transformation – does it conserve area? As the scissors completes the cut, most of us have very strong feeling that it is a conserving transformation. And if one is confident of that, there is no need for additional evaluation. One does not need to do a careful mapping of multiple features to know that the area of the two pieces is going to be the same after the cut as the area of the original. Thus this conserving transformation process appears different from the process of mapping discrete features as a method of analogy evaluation. The cut rectangle is considered analogous to the uncut rectangle here, not because there are multiple high-level features that are identical in target and base, but because a single action that leaves the area untouched has been found. A similar conserving transformation process can be described for the second transformation of moving the cut piece to the other side of the remaining trapezoid, and the third one of joining it.

17.2.1 *Transformations are not Equivalent to Mapping Symbolic Relations*

From this point of view the Wertheimer cut parallelogram case is not an example of mapping discrete symbolic relations. The *result* may be a mapping of lengths and areas and the relations between them; but the *means* to determining those equivalencies is not feature mapping, but rather a strong spatial intuition that the transformation is a conserving one that does not change those elements. In this regard, one can define the "targeted relationship" as the one for which an explanation or prediction is sought within the target situation (e.g. the relation between the dimensions and area in the parallelogram). Here, the crucial steps are: (1) find a transform that conserves area and (2) determine the relation between b and h and A in the target, respectively b and h and A in the base – in this case they are all conserved by the transformation). This guarantees that the targeted relationship is conserved. More specifically, as the scissors completes the initial cut, the fact that it does not change the area is known from the nature of the transformation. Similarly when the triangle is moved to the other end of the figure, one knows from the nature of that action that it will not change the area or the total length of the base when the pieces are rejoined. In this view these conclusions are derived from spatially embedded conceptions of length and area rather than a weighted sum of mapped elements.

17.2.2 Are Conserving Transformations Just Memorized Rules?

It may be tempting to propose that a conserving transformation is simply a memorized rule which gives a known result. And for transformations associated with an advanced principle like conservation of energy, the conservation knowledge may be mostly a memorized constraint in many subjects. For a conserving transformation like cutting an area, however, it is likely to be more than that, since conservation experiments done by Piaget and others have shown that many adults have a strong sense that cutting will not change area (or volume), even if they have not been taught that explicitly, although young children may not yet have this sense. Cutting as an area-conserving transformation carried out in adult thinking may be supported by strong, self-constructed, spatial reasoning operations and conceptions that involve basic notions of surfaces and space, and that carry a high degree of necessity for the subject. In this sense a transformation that is known or intuited to be conserving can be used to confirm a tentative analogy relation in mathematics. The exact character of these intuitions about particular transformations is an important area for future research, and goes beyond the scope of this study.

17.3 Conserving Transformations in Science

Subjects' think-aloud reports of modifying and manipulating shapes of objects are the main source of evidence for spatial transformations discussed so far. In these cases the idea that they are based on imagistic representations has the status of a plausible assumption. In the following cases from science problems other indicators from Table 12.3 will be used as evidence that spatial transformations can involve imagery.

17.3.1 Wheel Problem

Do conserving transformation processes occur in scientific thinking as well as in mathematical problem-solving? One possible example of a conserving transformation used in scientific thinking comes from the Wheel on the Ramp (Sisyphus) problem introduced in Fig. 4.1. Recall that Subject S7 first changed the problem to an analogous one involving an almost-vertical cliff with gear teeth:

> 101 S: Suppose it were tilted steeply and you did that; so steep as to be almost vertical (see Fig. 17.2a).

> 103 S: It seems like it would skid out from under you the other way [down along the cliff]. This – (moves hands as if turning an object clockwise) would get away from you here [at point of contact with cliff]. Let's assume it's gear toothed [gear teeth on the wheel and the cliff] and that it won't slip or that the friction is strong enough here that it's not going to slip under you.

17.3 Conserving Transformations in Science

This is actually a double transformation consisting of two parts: the change of slope, and the addition of gear teeth. The change in slope transforms the problem to one in which forces act mostly along one dimension: upward and downward, and it therefore simplifies the problem. S7 also uses a "traditional transformation" commonly used in physics: adding gear teeth to the wheel and the cliff to eliminate slipping. In his further work on the problem S7 never questions the validity of these combined transformations, and assumes that the targeted relationship in the problem situation is not affected by them. In fact he does find the extremely steep case easier to solve and these transformations are part of his path to a correct solution to the problem, to be discussed later in this chapter. The hand motions over the drawing here provide one source of evidence on the use of dynamic imagery. Although the drawing can be an external support for a static visual representation, it does not depict movements, so it is reasonable to hypothesize that the subject is performing a mental imagistic simulation of the wheel's motion. In summary, here S7 uses conserving transformations to generate an equivalent analogous problem and his positive evaluation of the validity of this analogy is immediate.

One can surmise that this is because one of the transformations is a standardized one in physics and both together are intuited to be irrelevant to the outcome of the problem. The origins of this kind of intuition have been studied since Piaget's early conservation experiments but are still poorly understood.

(Note: It may be objected here that all of the preceding examples of transformations in this chapter exemplify an analogy generation strategy, not an evaluation strategy. My response is that here it plays both roles. It is true that the preceding cases are also interpreted as being generated via a transformation; but if it is also immediately apparent to the subject that those transformation(s) are conserving, then analogy generation and evaluation take place very close together. In the present section, I will concentrate on the evaluation role. Sometimes these two processes take place separately, however, as illustrated in the next example.)

17.3.2 Spring Problem

The cliff and the gear teeth are examples of "minor" transformations; a more substantial transformation is illustrated by the following passage from S2's spring problem solution. Earlier he has considered the bending rod seriously as an analogy for the spring wire and in this later section he evaluates it by speaking of rolling up the bending rod:

> 102 S: You can imagine a spring...and you know...there's no difference between the top and the bottom. It's a symmetric situation.
>
> 105 S: You take your [straight, horizontal] wire, you say 'OK, you think it's the bending that does it. Well, then let's bend it. And then let's roll it up [around a vertical axis] to make the spring. And

106 S: You get a spring which stretches more and more at the bottom. The loops are wider apart!

107 S: Stretch it [the spring itself]: you don't get this increase of the distance between the loops toward the bottom. You just get a uniform stretching. And therefore the stretched spring cannot be understood as a rolled up bent spring.

Note the imagery report in line 102. He concludes that the spring wire cannot be bending. Here, the subject describes a very explicit transformation between the spring and the rod. The sequence is: he generates the rod analogy; he simulates bending in the rod; he evaluates the analogy by transforming it back into a spring; there is a conflict with a known property of the spring, and he discounts the bending rod analogy. This evaluation is extremely valuable in that it gives him information arguing that the conjectured mechanism of bending is invalid. Later, he makes this more explicit by examining what amounts to the inverse transformation in a cylindrical coordinate system:

113b S: Suppose I describe a helix in a ... cylindrical coordinate system so you have the helix spiraling up the cylinder...

113c S: then unwinding the helix is simply the same as unwinding the cylindrical coordinate system...slitting it down the side and laying it flat....And that certainly yields a straight line...

These passages suggest the use of spatial, visual transformations to evaluate the validity of a tentative analogy. The evaluation is very effective in that it leads to discounting the validity of the analogy. In this case the bending rod analogy was generated earlier, so the evaluation process is separate from the generation process, in contrast to the previous example of the vertically pulled wheel with gear teeth.

Prior to these sequences the subject had generated not only the analogous rod case, but what appeared to be a fully adequate mapping of symbolic features between the rod case and the spring case. Bending, length, and slope, in the rod were mapped onto stretching, width, and slope in the spring. The relation of <greater length causes greater bending in the rod> had been mapped to the sought-after relation of <greater width causes greater stretch in the spring>. And the mismatch of increasing slope in the rod but not in the spring is identified early on in the protocol before the passages reproduced above. Therefore the transformations reported in the above excerpts do not appear to be adding any new elements to the earlier mappings. Rather, these transformations seem to be increasing the subject's confidence that he has found an important mismatch in the slope feature. They are new ways to arrive at the same result, thereby supporting them. That is, the transformations are a *means* to determining a match or a mismatch as the outcome; they are not just the notation for a mismatch as read off from two different lists. The notion that the transformation should be conserving is quite plausible. If the main mechanism is bending, this "winding up" transformation is locally perpendicular to the bending; therefore it could very well be a conserving transformation.

The basic behavior pattern identified for a conserving transformation is: the subject describes a transformation that is a single change in a system (or a small number

17.4 Dual Simulation

of changes), followed by an immediate increase or decrease in confidence in its validity, with little mention of mapped features. This pattern suggests that the conserving transformation strategy is a process that can work independently from a strategy of mapping discrete features.

17.3.3 Newton's Canon

Evidence for the usefulness of conserving transformations can also be found in the history of science. Recall that Newton's famous thought experiment, involving a cannon on a mountain shot horizontally with more and more powder in it, was described as a series of bridging analogies in Chapter 4 (Fig. 4.10). The analogy relations between each of these bridging cases can be evaluated at a finer level of detail via conserving transformations. By reducing the difference between the falling apple and the moon to a series of conserving transformations, Newton provides a convincing argument that the principle cause affecting the motion in both cases is gravity. The purposes of this example are to: explain orbiting by connecting it to more familiar motions like dropping a ball, throwing a ball, and firing a gun; and to show that gravity is the primary cause in all these examples. The transformation between cases in Fig. 4.10 consists of gradually increasing the initial horizontal velocity until the ball goes into a circular orbit. Since this transformation does not change the gravitational force acting on the projectile, that force is conserved as the qualitative cause of the subsequent acceleration. The final case shows how a central gravitational force can be the continuing cause of a circular motion. This is the essential qualitative underpinning of the explanation of the motion of the moon (and the planets) for which Newton and Robert Hooke are credited. The series is convincing because the only element that is transformed is the initial horizontal velocity, and that is orthogonal to the force of gravity.

A traditional mapping approach to analogy evaluation focuses on determining that multiple similarities between the base and target are sufficiently *important*. In contrast, a conserving transformation strategy need only focus on determining that a single transformation from base to target is sufficiently *unimportant* (irrelevant to the targeted relationship). This may mean that confirmation of an analogy via a conserving transformation can require considerably less work than confirmation via mapping. This suggests that conserving transformations can be powerful in science as well as in mathematics.

17.4 Dual Simulation

There is evidence in the protocols for a sometimes imprecise but very direct strategy for analogy relation evaluation termed "dual simulation comparison," or "dual simulation" for short. This strategy uses the process of imagistic simulation

390   17 Imagistic Processes in Analogical Reasoning

discussed in Chapter 13, as follows. Imagistic simulations of the target and the analogous case are run in as much detail as possible. The dynamic images of the behavior of each system are then compared; and they may be inspected for certain aspects. If their behavior "appears" to be the same, the analogy relation receives some support, depending on the level of certainty in the comparison.

Example 1. A brief example that hints at this possibility follows. S2 says:

(Line 23) "surely you could coil a spring in squares, let's say, and it would behave more or less the same."

There is not very much data in this statement, but it is plausible that the subject created an image of a square spring and simulated the effect of hanging a weight on it. This could have been compared to the earlier image of stretching a normal spring, indicating that they were roughly similar in behavior. However, the resolution of the perceived similarity appears to be at a low level of detail. I call the comparison of overall behavior of the two systems in the protocol a "global comparison statement."

Here, one can build on the earlier analysis of extreme case reasoning in Chapter 15. Extreme cases cited with imagery reports there appeared to allow the subject to compare two imagistic simulations with greater contrast, thereby increasing confidence in a prediction markedly. If subjects can compare simulations in this manner in order to *contrast* them with respect to a particular relation between variables, it is reasonable that they may be able to compare simulations in order to see whether they are *similar* with respect to such a relation.

Example 2. Later this subject makes a statement similar to the one above:

S2: 119 Clearly there can't be a hell of a lot of difference between the circle, and, say, a hexagon...(draws hexagon)

S2: 121 ....I mean, surely springwise that [hexagon] would behave pretty much like a circle does.

Here, the subject somehow senses that the differences between the hexagonal and circular coils will not effect the qualitative behavior of the spring in an important way. It is clear that his conclusion here is not gained from "looking up a fact in memory," because of the novelty of the hexagonal case. Dual simulation of gross behavior at a low level of detail is a plausible way that this evaluation could have been made and the most probable explanation for the global comparison statement in line 121.

Example 3. More data is present in the following example of a zigzag spring discussed in Chapter 6. After seeing a problem with his first zigzag spring in Fig. 6.4a, the subject modifies it as follows:

(S2: 23) The problem with this idea is that...the degree to which the stretch...has to do with...the springiness of the joint. But the springiness of the spring – the real spring – is a

17.4 Dual Simulation

distributed springiness;....I wonder if I can make the spring....which is a 2 dimensional spring...but where the action...isn't at the angles...it's distributed along the length. And I'm going to do that; I have a visualization.... Here's a stretchable bar (draws modified zigzag spring in Fig. 6.4b) a bendable bar, and then we have a rigid connector.... And when we do this what bends...is the bendable bars...and that would behave (moves hand over weight in drawing and then back toward himself repeatedly) like a spring. I can imagine that it would... it would stretch, and you let it go and it bounces up and down (waves l hand up and down). It does all the things.

Here, the conjunction of the dynamic imagery reports, hand motions, and the global comparison of the two systems gives more support to the hypothesis that a dual simulation and comparison is occurring. The dual simulation appears to establish that a newly constructed analogous case is relevant and plausibly analogous in that its behavior is similar, at least at a gross level of qualitative behavior, to the target. But this does not tell the subject whether the two systems exhibit the same relationship between width and stretch. Thus in the above cases dual simulation appeared to serve only as a check on the initial plausibility of the analogy. One needs to be clear that dual simulation, as an analogy evaluation strategy, does not necessarily mean confidently simulating the targeted relationship independently in both base and target. In that case there would be no need for an analogy because the target could have been directly simulated on its own. However, when the targeted relationship cannot be simulated directly for the target case, as the examples indicate, dual simulation can still help one determine whether the target and base are similar with respect to other important behaviors, thereby increasing one's confidence that the analogy is sound (or eliminating the analogy from consideration).

The simpler static version of a dual simulation comparison would be a static image comparison, such as would be used to answer the question, "Is a man of average height taller than the back of a pony?" Kosslyn (1980) has documented that people report using dual image comparisons in such problems.

17.4.1 *Do Dual Simulations Differ from Transformations?*

The zigzag spring appears to originate in the earlier case of the square-coiled spring, and from transforming that idea into a two-dimensional zigzag spring. If we assume that the zigzag spring arose from a transformation, why place it in a new category called dual simulation? Such confusions come from not being careful to distinguish between analogy generation and evaluation processes, and between analogous cases and analogy relations. The question here is not how the case was generated, but how it is used. Thus with the zigzag case one can propose that:

1. The zigzag case was generated via a transformation process.
2. The zigzag case was generated for the purpose of forming an intermediate bridging analogy between the spring and the bending rod (larger purpose for generating it).
3. The analogy relation between the zigzag case and the spring initially received some support from a dual simulation (an initial evaluation strategy).

4. This analogy relation between the two cases may finally be disconfirmed however, by simulating the zigzag in more detail, realizing that there will still be bending in the elements, and not being able to reconcile this with the lack of bending imaged in the original spring (further evaluation strategy).
5. This may be described by the subject verbally as a mismatch.

Steps 1–3 are the main focus here, and they have some support from the transcript passage quoted above indicating that the zigzag case was generated via a transformation and then evaluated via a dual simulation. This example illustrates how dual simulations can be treated as a separate process from transformations.

17.4.2 Dual Simulation for the Square and Circular Coils

I hypothesize that a very important example of a dual simulation occurred in episode 11b in Chapter 14. There S2 constructed a spring with square coils as a thought experiment and determined that twisting and torque in the wire would be critical variables in determining the stretch. Major support for the validity of this analogy was found when his thinking about the square spring indicated that this spring would *not* show the property of increasing slope. His earlier simulations of the original normal helical spring had predicted that it too would not show this property. This was an important likeness in the predictions from two simulations for S2, and it apparently helped him put faith in the applicability of predictions from the square coil analogy. One could argue that this correspondence is just one piece of a mapping of discrete symbolic features. However, even if that is the outcome, one can hypothesize that this single correspondence in an eventual mapping is the *result* of a dual simulation, not the reverse.

17.5 Overlay Simulation

There is a possible exception to the proposal that dual simulations allow an indication of similarity only at a fairly gross level of precision. There is some evidence for the existence of a more precise type of dual simulation that I term "overlay simulation" where the image of one simulation takes place "on top of" a second image.

17.5.1 Examples of Overlay Simulation

17.5.1.1 Lever Overlay

As a first example recall S2's solution to the wheel problem in Chapter 4. S2 drew his lever analogy (Fig. 4.2b) directly on the wheel (Fig. 4.2a) and compared the movement of the wheel and the lever. This meant that the arrow symbolizing the application of a force by pointing to the top of the wheel was also pointing to the top of the lever. When two separate systems are represented as overlapping in the same external diagram with salient features aligned I call this an *overlay diagram*. Since the diagram does not move, it suggests the hypothesis that his internal dynamic images of the wheel and lever were intentionally of the same size and overlapping as well. One can hypothesize this made it easier for him to compare the expected movements and resistances of the wheel and the lever in a dual simulation. I call the internal process here an *overlay simulation* as a special type of dual simulation.

17.5.1.2 Spokes

Another overlay simulation may be responsible for the power of S2's "spoked wheel without a rim" analogy discussed in Chapter 4 and shown in Fig. 4.3C. This subject spoke of a tireless, rimless wheel. This is shown separately in Fig. 4.3C for clarity, but in fact the spokes were drawn within the subject's original wheel drawing. This may make it easy to sense that they behave in approximately the same way as the wheel when a force is applied. In particular, the way the rimless wheel moves on each spoke over a short distance can be seen as similar to the way the original wheel rolls. That is, it may appear to have the same kind of motion and therefore be amenable to the same type of analysis with respect to the causes of motion. Although such arguments must be bolstered mathematically to make them rigorous, as a form of heuristic reasoning, this type of qualitative argument can be quite compelling.

17.5.1.3 Pulley Overlaid on the Wheel

A third example of overlay simulation appears after the initial steps in S7's solution shown in Fig. 17.2. Working from the extreme case of rolling the wheel up an extremely steep hill in Fig. 17.2a, he goes on to propose an analogy to a pulley, as shown in Fig. 17.2b. This allows him to predict that it would be easier to push the wheel on the outside, since he knows that the pulley would cut the required force in half. S7 imagined the case of a pulley on a vertical ramp, as if the image of the pulley were "overlaid" on the image of the wheel. In this case he used the same drawing (Fig. 17.2b) to represent the wheel and the pulley, referring to it differently as one or the other in alternate fashion. (I have shown the pulley case separately in the figure only for clarity of exposition; the subject drew only Fig. 17.2b.) Presumably it is easier in an overlay simulation to switch rapidly back and forth between simulations of the two cases (or perhaps even to simulate them simultaneously).

> 105 S7: What it feels like is the weight of it [the wheel] – is pretty close to parallel with what you've got if you go roll it with a complete vertical. It now begins to feel like a pulley…(see Fig. 17.2b). What the vertical is over here no longer matters perhaps but we'll say it's er, gear-toothed again.

> 106 S:….And you're over here pulling like this [at x in 17.2b]. That feels like you're on the outside of a pulley pulling up.
> (Stares at Fig. 17.2b).

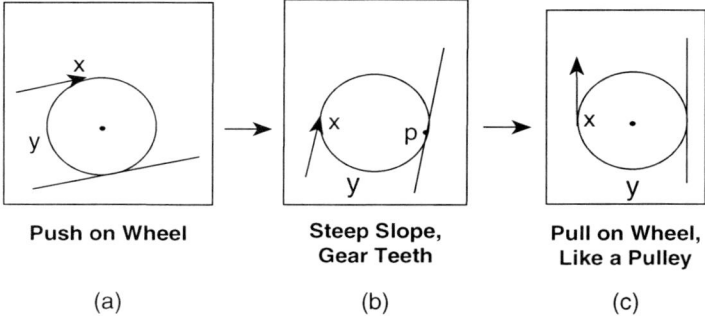

Fig. 17.2 Analogies for wheel problem (**a**) wheel on steep slope, with gear teeth; (**b**) pull on wheel, like a pulley; (**c**) push on wheel

> 107 S: And since you say it doesn't slip, then this thing over here (points to line in upper right of Fig. 17.2b and adds upward pointing arrowhead to it) must be providing the other half of it, something – it feels, in which case it's a classic pulley; no, it can't be classic pulley. But it's, like a classic pulley in which now you only need half of the force. If the weight of the thing is 10 lb. here, it feels like 5 would work here (writes 5 on upper left of 17.2b) and 5 over here (writes 5 on upper right) as though it were a pulley....So let's imagine it is a pulley.

> 108 S: [In] this new point of view, it feels like working at X [on the edge of the wheel] is better [than at the center].

Alternating references (seven alternations) to both the wheel and the pulley systems while staring at the same overlay diagram in Fig. 17.2b provide initial evidence for an overlay simulation here that compares the system of rolling the wheel straight up a vertical cliff to the pulley system. Again, although the imagery is probably assisted in this case by the drawings, the drawing cannot be providing perceptions of forces or motions involved, and so I hypothesize that these are imagined via imagistic simulations. Some evidence for kinesthetic imagery is indicated by phrases like *It now begins to feel like a pulley* and "feels like you're on the outside of a pulley pulling up" in the transcript, and such imagery clearly cannot already be depicted in the drawings.

Later he seeks further confirmation of the pulley analogy by adapting it to the original gently slanted ramp. Figure 4.5B shows the pulley laid on the ramp and pulled by a rope running up to the right from the top of the pulley. He then (1) generates another bridging analogy in Fig. 4.5C; and (2) uses an overlay simulation to confirm the analogy between the two cases on the right-hand side of the bridge in Fig. 4.5. Therefore this final example is more complicated because it combines the two strategies of overlay simulation and bridging.

> 162 S7: [Looking at Fig. 4.5A of the wheel on the ramp] Seems clear that – (silently holds both hands out in front of him as if pulling a rope for 4 s)....So we attach a rope to one of the teeth [as in 4.5C on the top of the wheel]. Now it becomes more like the pulley problem (holds r. hand out as if pulling a rope for 3 s) which I was thinking before. Anyway, the teeth at the bottom are playing the role of – the pulley doesn't look so bad after all. And

17.5 Overlay Simulation

you hang on for all you're worth up there, to keep it from rolling; to keep it balanced (see Fig. 4.5C).

The case of the rope attached to the edge of the wheel can be seen as an intermediate bridging case between the original wheel pushing problem and the pulley lying on the slope case.

> 163 S7: Seems a lot easier than getting down here behind it [at "Y"] and pushing. Why? Because of that coupling pulley effect. It seems like it would be a lot easier to hold it here [at "X"] for a few minutes (Holds hands in "pulling" position) than it would be to get behind it or even to attach a rope here [at "Y"] and – yeah, my confidence here is much higher now, that it's right. [easier to push at X].... And so the pull – it just felt right with the pulley feeling. Now pushing (lays extended finger on paper pointing up slope at X in Fig. 4.5A and moves it toward X) uh,... it's got to be the same problem...
>
> 178 I: OK. And do you have a sense of where your increased confidence is coming from? Is it this example?
>
> 179 S: It's the pulley analogy starting to feel right.

In this case the subject has been led to a second prediction that it would be easier to push the wheel on the outside. The depictive hand motions and personal action projections in references to feeling and pulling on the rope in the above passage can be seen as evidence for the involvement of kinesthetic imagery and imagistic simulation in this episode. Since the same drawing (Fig. 4.5A) is used to represent three systems: the wheel, the pulley, and the rope attached to one of the teeth on the wheel, I hypothesize that the drawing is supporting an overlay version of the dual simulation process, whereby the alignment between trajectories and forces in imagistic simulations of different cases can be more easily made. Focusing on the two cases on the right-hand side of the bridge in Fig. 4.5, evidence for dual simulation is then triangulated from several sources:

– Global comparison statements about the pulley and attached rope cases
– Evidence for kinesthetic imagery and imagistic simulation from depictive hand motions in both cases C & B in lines 162 and 163
– Attention alternates rapidly between cases using the single overlay diagram in Fig. 4.5A

Thus this last example illustrates the combined use of two analogy evaluation strategies: overlay simulation (as a special kind of dual simulation) and bridging.

17.5.2 Connection to Model Construction: Overlay Simulations and Model Projections May Involve Similar Processes

Overlay simulations may allow a more precise evaluation of the similarity between not only static figural shapes, but also motions, trajectories, and dynamic influences such as the size and direction of forces, within constraints imposed by both the target and source problems respectively. Thus this imagistic method may be especially well suited for the task of comparing complex, dynamic, spatially embedded relationships in the base and target.

I have argued that model evaluation techniques described in Chapter 16 can be different from the analogy evaluation techniques described in this chapter and Chapter 4. But it is well to ask whether there are also any similarities between these techniques. Overlay simulations provide the most graphic examples of *projective analogical transfer*, as described in Fig. 15.6 in Chapter 15. I hypothesize that *analogical projection* is exemplified by the last case discussed: toward the end of the protocol the pushed wheel can be *seen as* a pulley (although it is not a pulley – this is not an explanatory model for the system), and when the image of the pulley is projected or overlaid onto the image of the wheel, the intuitions about the pulley can be transferred by projecting them into the wheel.

17.5.3 Model Projection

Model projection, on the other hand, was described in Chapter 15 as the process of applying a model to a target case by projecting an image of the non-observable explanatory mechanism into or onto the target in order to imagine it running in parallel with observable actions in the target. For example, one might project an image of moving molecules onto or into the image of a moving piston in a cylinder. That process shares some characteristics with overlay simulation in that (1) the image of the explanatory mechanism and the actions of the target are run along with another one (2) the images may be "overlaid" in the same space, and (3) part of testing the model is a comparison of the behavior of the model projected into the target – yielding a prediction by thought experiment – with the previously known behavior of the target to see if they are compatible. Thus, something like overlay simulation may be used in model projection and testing as well. (This also suggests that, pedagogically, having students make overlay diagrams may be a good training exercise for developing model projection skills.) The identification of overlay simulations as a special case of dual simulation adds to the number of observations of dual simulation episodes, adding to the evidence for dual simulation as a significant form of analogy evaluation.

17.5.4 Imagistic Alignment Analogies

In Chapter 16, I defined an imagistic alignment analogy as an analogous case that aids in imagistically aligning features of an explanatory model, source analogue, target, principle, or mathematical schema, in a more precise way.

Spokes. The wheel made of spokes without a rim discussed earlier can also be interpreted as an imagistic alignment analogy. In working on the wheel problem, once the principle of the lever is activated, there is a difficult alignment problem for some subjects in determining where the fulcrum of the lever should go – at the center or at the bottom of the wheel? The bridging case of the wheel made of spokes without a rim serves to evaluate and affirm the relevance of the lever principle and to argue that the proper location of the fulcrum is at the point where the wheel touches the ground. It can help one perform a dual simulation to see that "Running" each segment of the wheel rolling appears to be extremely similar imagistically to running the movement of the single lever. Thus in addition to being a bridging analogy, the spokes without a rim appears to be an imagistic alignment analogy that aids the subject in determining how to imagistically align features of the lever schema, such as the fulcrum, with the wheel in a more precise way. In these cases the role of the analogy is not so much the standard one of a new source of main ideas or relationships; because that has already been provided by the activated source schema (e.g. the lever concept for the wheel) Rather, it is the details of how and where the imagery from the lever aligns – how to "see the wheel as" a lever – that become clarified by the new spokes analogy. Thus we can interpret the spokes analogy as designed to facilitate imagistic alignment and dual simulation for evaluating the lever analogy. In doing this it also plays a "domain expanding" role in that it stretches the domain of the lever schema to apply to wheels.

Square Coil. Late in the game, S2 may have reused the square coil case as an imagistic alignment analogy for seeing torsion in the spring coil. His first success in actually drawing a source of twisting in a circular coil comes immediately after revisiting the square coil, as seen the transcript in Episode 17 in Chapter 14. He first speaks of how one side of the square operates at 90 degrees to the next side in order to twist it. He then speaks of how pulling down at a point on the circular spring coil can twist a segment 90° away from the point on the coil. Depictive hand motions suggest the use of imagery in both cases. Here, the square coil analogy appears to be facilitating the difficult process of aligning the imagery of torque and twisting actions with the imagery of a circular spring. (Recall that torque was easily misaligned with the spring coil in Episode 16 in Chapter 14.) It makes sense that imagistic alignment analogies will often be bridging analogies, as is the case here. In this case the analogy helps to align a standard torque and torsion model, ordinarily applied to straight objects, and a target (the circular coil).

Since both of the above examples were initiated as bridging analogies, this suggests another common use of bridging analogies is to help provide an imagistic alignment.

17.5.5 Dual Simulation vs. Compound Simulation in Modeling

How does dual simulation differ from compound simulations? Compound simulation was described earlier as the coordinated action of two perceptual/motor schemas operating on the image a single case jointly to produce a single simulation. Dual simulation on the other hand is a process of comparing two separate simulations to evaluate their similarity. Although I do not have enough fine grained protocol data on this issue to support it fully, I can conjecture at this point the possibility of a progression from dual to compound simulation during some cases of model construction. This process could occur as a subject begins to "see the wheel as" a lever; or "see the gas as" a swarm of moving particles. She or he no longer questions the model, but runs it together with the target in a smoothly aligned way. Take say, the historical case of inventing an elastic particle model for gases, where one has a target situation of heating the gas in a cylinder to move a piston. One starts by asking: when one runs the imagined system of a swarm of tiny moving ball bearings in the cylinder does it appear to act the same way as a second system – compared to running the experiment of heating of the gas and piston? This question could be answered by a dual simulation that compares the simulations. But once the model is accepted one learns to run the particle model within the cylinder system. At this point it becomes a single (coordinated) compound simulation instead of a (comparative) dual simulation.

In the wheel problem, one may start by asking: when one runs the lever does it appear to act the same way as the spoked wheel without a rim and the original wheel? This question could be addressed by dual simulations that compare the simulations, serving to evaluate the lever as a model. Once the lever model is accepted one can run multiple levers within the wheel, to see the wheel as a set of levers in a coordinated way. Again at this point it becomes a single compound simulation instead of a dual simulation. I hope to evaluate this conjecture in future research.

This interpretation would fit the final state of modeling in the spring problem, which involves projecting a simulation of torque causing twisting causing stretching into elements in a coil of the spring wire (the behavior of the microscopic model), while simultaneously imagining the spring stretching (the behavior of the macroscopic target). As described, this is an advanced form of compound simulation.

17.6 Summary and Discussion of Types of Evaluation Processes: Contrasting Mechanisms for Determining Similarity

In summary, rather than a single process for mapping discrete elements in a symbolic representation, a number of additional processes for evaluating the analogy relation have been identified, namely:

Conserving transformations
Dual simulations to detect dynamic similarity
Overlay simulation
Bridging analogies

There was some evidence that these can involve imagistic representations. The hypothesized way in which these processes can be coordinated is summarized in the idealized algorithm shown in Table 17.1 for evaluating an analogy. The algorithm defines three procedures which can call each other. "Fail" here means "failure to confirm the analogy relation." The algorithm shows bridging as a higher level strategy,

17.6 Summary and Discussion of Types of Evaluation Processes

Table 17.1 Strategy using multiple methods for evaluating an analogy relation between a target case (T) and anchoring case (A)

- Evaluate analogy relation (T, A)
  - Direct evaluation (T, A)
  - If fail, Bridge (T, A)
  - If fail, quit
  - Report (T, A) confirmed
- Direct evaluation (T, A)
  - Dual simulation comparison (T, A) (or Static image comparison) to evaluate perceptual/motor similarity (may only provide initial plausibility) and/or
    - Overlay simulation
  - and/or
  - Find Conserving Transform from A to T
  - and/or
  - Map discrete symbolic relations in A and T
- Bridge (T, A)
  - Generate bridging case B
  - Evaluate (T, B); if fail, Bridge (T, B) recursively, or try a different bridging case B, or quit
  - Evaluate (B, A); if fail, Bridge (B, A) recursively, or try a different bridging case B, or quit

T = Target case; A = Anchor case; B = Bridging case

whereas the other evaluation methods are more direct ways of determining similarity and analogical validity. In this section, I contrast the above methods to each other, and speculate on some possible imagery based mechanisms for them.

17.6.1 Mechanisms for Dual Simulation (Including Overlay Simulation)

17.6.1.1 Dual Simulation Mechanisms

Dual simulation appears to be a sometimes rough but very direct method allowing the subject to evaluate whether the base and target have some similar behaviors. I have assumed that the subject was not able to do this initially for the targeted relationship itself, but that dual simulation can play a role in determining whether or not there are other informative similarities between base and target behavior in general.

17.6.1.2 Precedents in the Literature

Dual simulations can be contrasted most sharply with mappings of discrete symbolic elements if they involve analog processes of a perceptual/motor nature. Other processes that appear to have this analog property are: direct imitation of bodily movements; and face, object, and shape recognition. These processes can occur before the development of language in very young children. A valuable precedent here is the process of inspection and high-level perceptual similarity detection for familiar static images identified by Kosslyn (1980). A similar process has also been studied for novel static images by Finke (1990). One can speculate that a comparison made in a dual simulation might use a similar type of similarity detection process. except that it operates on images of dynamic actions and events. This can be referred to as "dynamic perceptual/motor similarity." Dynamic comparisons would involve each simulation generating a certain event, and the simulations being compared. The comparison could involve an interrogation for aspects that might affect the targeted relationship the same way in both cases as well as an openness to "observing" unexpected dissimilarities.

17.6.1.3 Observations Relevant to the Nature of Dual Simulations

In addition to imagery indicators, the transcripts quoted for dual and overlay simulations are characterized by their references to an undifferentiated "sense of being similar." Examples appeared in lines 23 and 121 of S2's zigzag spring protocol above and in lines 105–106 of S7's protocol. In these, rather than breaking the actions down into features such as force and torque components or angular displacements, subjects appeared to be simply comparing some skeletal version of the overall visual and kinesthetic "perception" of the two events. This may be a distributed and parallel type of perceptual comparison, rather than a mapping of discrete symbolic features; it could lead to estimates of the degree of similarity rather than a dichotomous decision between "identical" and "different" on a particular feature. Aspects of the two images that were aligned and identified as similar may then be given the same verbal label in a second stage of the process, e.g. a "force" at point A on the wheel and on the pulley. In that case the process of dual simulation would precede and contribute to the process of mapping of discrete symbolic elements.

17.6.2 *Mechanisms for Conserving Transformations*

One can also propose a possible imagery based mechanism for conserving transformations. The idea of using mental transformations builds on the findings of Shepard and Metzler (1971), who found that adults could compare an image of an object to say that it was identical with another that was transformed from it (e.g. rotated).

Whereas most previous descriptions of analogical reasoning in science have focused on strenuous efforts to find relevant relationships that are central to the causal operation of both systems, the most unusual feature of a conserving transformation strategy is that it focuses on aspects that are *not* central to the operation of the system by establishing with some level of certainty that the transformation does not affect the central causal mechanisms in the target. In some cases this may be known as a fact, but in others it may have to be inferred.

17.6 Summary and Discussion of Types of Evaluation Processes

17.6.2.1 Transformations vs. Mapping

Feature mapping can be contrasted with conserving transformations for which it is sufficient to know the behavior in the base corresponding to targeted relation plus knowing: (1) that the transformation applied to the anchor case yields the target case or vice versa (which can often be determined by a perceptual comparison); (2) that the transformation does not affect (conserves) the targeted relationship. Beyond these, *there is no need to map other entities explicitly between the target and the base* in order to confirm the analogy relation. It suffices to know that the only difference between cases A and B is that formed by the transformation and that that difference is irrelevant to the behavior of interest. This contrasts to the situation in Fig. 3.4c, where the differences between A and B may be unknown and may need careful mapping.

In conclusion, findings from the protocols analyzed in this chapter suggest that there are alternative methods for evaluating analogies in addition to mapping of discrete symbolic elements.

17.6.3 Bridging is a Higher-order Strategy Compared to Others

The separation of bridging from the other strategies in Table 17.1 is appropriate for two reasons. First, bridging in itself is an incomplete strategy for analogy evaluation, since *each half of the bridge must itself be evaluated.* The bridging method is a higher order strategy which breaks the problem of confirming a "larger" analogy into the problem of confirming two "smaller" analogies. Therefore bridging is most useful in conjunction with other evaluation methods and for facilitating their use. The other three evaluation processes are more fundamental than bridging in this sense, as shown in Table 17.1. Thus, whenever bridging is used, one should expect other methods of analogy evaluation to play a role as well.

Secondly, bridging adds to, rather than reduces, the number of tasks to be performed by creating two new analogy relations to evaluate, and so should ideally be used only when other less-time-consuming methods have failed. However, this does re-raise the problem of why experts bother to consider bridging cases, since they seem to create more work.

17.6.4 Combinations of Evaluation Methods

17.6.4.1 Square Torus

The bridging case of the square torus discussed earlier (Fig. 4.8) can now be understood more fully as involving both bridging and a conserving transformation in its development. The initial analogy proposed was between the torus and a cylinder of length equal to the torus' intermediate circumference. The subject then generated the bridging case of a "square torus." After some additional work, the subject was able to determine that cutting and reassembling the cylinder of length L into the square torus of the same volume would in fact produce an intermediate mid-perimeter of length L. At that point he had found a volume-conserving transformation (or sequence of transformations) between the cylinder and the square torus, as diagrammed in Fig. 4.8. This served to confirm the analogy relation for one half of the bridging relationship. He then appeared to realize that the same arguments could be made for a hexagon, or any polygonal torus. In this example the apparent role of the bridging analogy is to create a situation where the two cases are "close" enough to each other so that a confirmed conserving transformation between them can be found. Thus bridging is a supplementary method that "sets up" situations where more basic processes like conserving spatial transformations can occur, and the methods can be used together.

17.6.4.2 The Concept of Limit

These methods may provide us with a new perspective on the nature of mathematical arguments which use the concept of a limit. The interesting question in, say, the "stacks of wedges" transformation strategy for the torus problem shown in Fig. 11.4 is whether the "lumpiness" of the sides of the stack will throw off the volume calculation or whether the wedges will approach the volume of the smooth cylinder in a well-behaved manner as n, the number of wedges used, goes to infinity. Lakoff and Nunez (2000) theorize that mathematical thinking in such a case depends on a leap via a conceptual metaphor, where an infinite sequence of states is treated as though it had a final state – the state where n reached infinity. One can provide a more detailed hypothesis about how people think about the basic concept of a limit, by using the vocabulary just developed. This states that the lumpy cylinder with n pieces has the status of a promising bridging analogy between the cylinder and the torus. A converging sequence is then a sequence of bridging analogies, since the progression from one to the next produces another bridging case that moves closer and closer to the target. This exemplifies the potentially recursive nature of the bridging process as shown under "Bridge" in Table 17.1. The relationship between the torus and the lumpy cylinder on one side of the bridge is that of a confirmed analogy via reliable conserving transformations for volume (cutting and rearranging pieces), and this can be repeated as each new bridging case is constructed with more and more wedges. But the relationship on the other side of the bridge between the lumpy cylinder and the normal cylinder is that of a tentative analogy relation. It can be evaluated roughly as plausible via the simpler static version of a dual simulation comparison – a static image comparison. The question is whether the final links to the target (cylinder) are valid, in the case where there is an infinite sequence of bridges. Other mathematical arguments for the target side in the limit are needed to confirm that tentative analogy relation with mathematical precision. (This can be important, as in the case of attempting to calculate the area of a hemisphere by cutting it into quasi triangles meeting at the pole and then rearranging them into a rectangle, which does not work!) Prior to such a demonstration however, the plausibility

17.6 Summary and Discussion of Types of Evaluation Processes   403

that the sequence will converge, as supported by the increasing perceptual similarity detected in sequential static image comparisons, can become rather high. Again, this last evaluation mechanism is a rough one, but as a first heuristic it may lead the subject in a fruitful direction that can later be examined more precisely. Thus reasoning about limits can be seen as involving recursive bridging analogies.

17.6.4.3 The Wheel as a Pulley

The intermediate bridging case of the rope attached to the wheel from S7's wheel problem solution was part of a "triple" overlay drawing. The use of an overlay diagram and references that the wheel problem solution "felt right with the pulley feeling" supports the hypothesis that dual simulations were being used to evaluate these analogies. Thus there is evidence that the generation of the bridging case was an attempt to create a pair of more tractable dual simulations, between it and the original problem and between it and the pulley analogy. This suggests that another role of intermediate bridging analogies is to generate chains of cases that are "close" enough perceptually to compare and evaluate with confidence in a more perceptual way, via dual simulations.

17.6.4.4 Imagery Based Evaluation Methods can Explain the Effectiveness of Bridging Cases

Thus subjects are sometimes observed to go through the effort to construct a bridging analogy in conjunction with other strategies. Evidence was presented indicating that in these cases, the purpose of the bridging case was to facilitate an imagistic evaluation strategy such as a conserving transformation or dual simulation. Subjects are apparently willing to go through an effort to support such imagistic processes, and this argues that they are valued by these expert subjects. The explanations above for the effectiveness of the higher-order bridging strategy in terms of the theory of imagistically based analogy evaluation methods are an example of a successful application of the theory and therefore add some support to it.

17.6.4.5 The Relationship Between Imagistic Methods and Discrete Symbolic Methods is Unclear

I have hypothesized here that the analogy confirmation methods of conserving transformations and dual simulations (including overlay simulations) operate on depictive, imagistic representations rather than on discrete symbolic representations. Unfortunately, these methods will sometimes be hard to detect, since spontaneous imagery reports are not common in our mode of speaking. Also, imagistic methods may result in or be used in conjunction with the mapping of discrete symbolic relations. For example, global recognition of perceptual/motor similarity in dual simulations could lead to mapping discrete symbolic relations more explicitly in a second stage. The latter is easier to report during think-aloud problem-solving because of its verbal nature. Nevertheless, cases such as the ones discussed in this chapter can be found where there is some evidence for imagery-based methods. In some cases there is more evidence for these methods than there is for mapping. On the other hand, it is clear that mapping discrete symbolic elements is a valuable strategy for some purposes. The relationship of the imagery-based strategies to mapping is still unclear, but when subjects can articulate such mappings, that may add another important kind of precision to the process of analogy evaluation. Subjects who can use such a combination of methods, involving both perceptual/motor and symbolic representations, would presumably have the greatest advantage.

17.6.5 Comparison to Structural Mapping of Images

Previous studies describing evaluation processes from a structural mapping perspective have virtually all been based on the use of discrete symbols as an underlying theoretical framework. I have drawn the contrasts with imagery-based methods above based on this assumption. However, a "liberal" structural mapping researcher might claim that mapping processes are still valid for all analogies even if the elements being mapped are imagistic representations or elements thereof. Does the present data then indicate that there are evaluation processes that are different from the mapping process no matter whether the elements being mapped are symbolic or imagistic? This is a harder claim to support than the previous ones because it concerns a fine-grained process and because some minimal mapping (e.g. of the targeted relation to the target problem) will probably be involved in any evaluation. However, the following considerations provide some initial motivation for proposing the theoretical possibility that evaluation processes can involve more than mapping correspondences between finite sets of discrete elements, even if the elements in the mapping can be imagistic:

- A theoretical argument can be made that it would be inefficient to bother with a complete mapping when a subject thinks they have found a conserving transformation, as in the parallelogram problem and in the rotation of the wheel problem to a vertical cliff with gear teeth. This can receive some weak support from transcripts where there is immediate acceptance of such analogies with no references to mapping discrete imagistic features.
- Global similarity comparisons such as the wheel "beginning to feel like a pulley" (lines 10–106) convey the idea that the subject has not broken down or analyzed the cases into elements to be mapped, rather that they are making a global comparison of two molar simulations.
- Such a perspective would take seriously the idea that some imagistic simulations for a system's behavior can be molar events that are not broken down even imagistically into consciously heeded "pieces" so that the pieces can be compared one by one to yield a similarity "score". Rather, as the target is run as a simulation, it is "seen as" the base holistically–as when the pushed wheel is 'seen as' a pulley or the rimless spoked wheel is "operated as" a lever.

17.6 Summary and Discussion of Types of Evaluation Processes

- One purpose of 'overlay simulations' may be to facillitate the running of a dual simulation via a kind of holistic imagistic alignment.

It is by no means possible to settle such a fine-grained issue by using a limited number of protocols from case studies, but they do serve to raise the pointed question, at least for me, of whether methods other than mapping of discrete elements can be used for analogical inference and analogy evaluation.

Contrary in spirit to the latter perspective are certain types of 'imagistic alignment analogy'. It was thought that the rimless spoked wheel, for example, could help one fix the proper place to think of the lever's fulcrum in the wheel, whether at the center or at the point of contact. This sounds like imagistic mapping, and seems a potentially important process. To rectify these two perspectives, it is possible to envision an initial, more primitive process of imagistic global comparison in a dual simulation that yields a feeling that the cases are analogous, followed by a more careful imagistic mapping process which aligns aspects known to be important in each system.

17.6.6 *Conclusion on Evaluation: Four Main Analogy Evaluation Methods, Not One*

This chapter concludes with the proposed possibility that, in contrast to most of the previous literature in this area, we have four analogy evaluation methods rather than one: bridging (including imagistic alignment analogies), mapping of discrete features, conserving transformations, and dual simulations (including overlay simulations) (Clement, 2004). Evidence was presented indicating that conserving transformations and dual simulations can be imagery based. I have hypothesized that bridging is a higher-order strategy whose purpose is to make it easier to conduct one of the other evaluation processes. In cases where a bridging analogy serves to make the use of one of the above imagery-based methods possible, or plays the role of an imagistic alignment analogy, it too can be termed an imagery-based evaluation mechanism. Questions have also been raised as to whether imagistic processes can act prior to and feed symbol mapping processes or whether they can act in conjunction with them in a complementary way, and these are very interesting topics for future research.

17.7 Use of Imagistic Transformations During the Generation of Partitions, Analogies, Extreme Cases, and Explanatory Models

Up to now this chapter has been concerned primarily with analogy *evaluation*, but at this point to expand the theme of imagery-based analogical reasoning I will shift to consider processes of analogy *generation* or *modification*. In Chapter 3, I cited transforations as the most commonly observed analogy generation method for the spring problem, such as modifying the springs to have equal widths and shortening one of the springs relative to the other. Chapter 11 documented many spatial transformations in the mathematical Torus Problem. If such generative transformations are associated with changing a feature of an object or a system, there is a question as to whether the change is symbolic or imagistic. A large proportion of cases of transformations in the present protocols were accompanied by a new or modified drawing generated by the subject. This suggested that an imagistic transformation was involved. These processes occur very quickly, and observations of accompanying imagery indicators from Table 12.3 were not as common as were the drawings, but they did occur, as discussed below (using cases from external subjects only).

Partitioning: Transformations can generate partitions by cutting a system into parts. In episode 4 in Chapter 14 a subject partitions the spring into repeating coils and then focuses on one coil, thereby transforming the problem into a simpler one. Evidence for imagery during this episode was provided by hand motions.

Analogy Generation: Torus Problem: There is evidence that the cylinder analogy was generated by S6 using a transformation:

> 001 S: It's probably pretty close to a worm, er
>
> 002 S: I mean a cylinder. Where you know, if you laid out the doughnut on the ground, uh, if you cut it open and laid it out, it would basically be the area of the base times the length around the middle.... I'll just turn it into a cylinder....
>
> 005 I: ....When you thought about the cylinder, do you know how that arose?...
>
> 006 S: Well I mean I in fact...just imagine the knife cutting it open and you know, laying it out...
>
> 008 S: ....I mean to say I thought of cutting it at one edge and it sort of flopping down and then the uh, the doughnut becomes a cylinder.

Here, the imagery report in line 6 suggests that the transformation being used here is generated via an imagistic representation.

Spring Problem: (1) A second example of imagistic transformation occurs in the generation of the second "zing zag" spring discussed in the section above on Dual Simulations. There the subject appears to modify or transform his first zigzag spring into an "improved" one, and says that he has "a visualization" as he does this.

Spring Problem: (2) The analogy in Episode 7a in Chapter 14: "what if I recoiled the spring and made the spring twice as long – instead of twice as wide?" appears to be generated by a transformation. This is immediately followed by a simulation of the new case with multiple imagery indicators, giving indirect but still significant support to the idea of the transformation being imagistic.

Wheel Problem: Similarly in the wheel problem, S7's generation of the case of rolling the wheel up a vertical wall in line 5 (in this chapter), as well as the generation of the case of attaching a rope to the top of the wheel in line 162, are immediately followed by depictive hand motions.

Extreme Case Generation: The Spring (2) episode above is followed by a transformation to an extreme case with accompanying imagery indicators: "I'm imagining (moves hands together slowly in 5–6 small movements) that one applies a force closer and closer to the origin of the spring." This quotation contains depictive hand motions as well as a dynamic imagery report as imagery indicators. Extreme cases are usually accompanied by language that suggests generation by a transformation, and this makes intuitive sense theoretically.

Another extreme case in episode 5 is accompanied by an imagery report: "if you...imagine shrinking the coils to a very small diameter."

Explanatory Model Modification: The attempt to imagine the twisting mechanism operating in an actual spring coil in episodes 17–18 is more subtle example of a transformation. Here, the subject transforms his model of the spring by softening an increment of the spring and carefully choosing the place from which a torque is applied to that segment. The change in the position of the force applying torque here is an example of transforming an explanatory model that involves a drawing and has accompanying nearby imagery indicators.

The examples above appear to be most parsimoniously explained by assuming that the subjects were transforming the images of concrete objects in order to generate an analogous case, partition, or extreme case. In the case of analogies, the process appears to be a very different one from retrieving an analogous case from memory via a linguistic association, and the model presented here is therefore different from most previous models of analogical access.

17.8 Conclusion

On the basis of think aloud protocols, and associated imagery indicators such as depictive gestures, this chapter has proposed three new types of imagery-based analogy evaluation mechanisms: transformations, dual simulations (including overlay simulations), and bridging analogies (Clement, 2004). Evidence for transformations, as an imagery-based mechanism for *generating* analogies, partitions, and extreme cases was seen in other cases. These findings provide additional evidence for formulating the general hypothesis that many analogical reasoning processes can be imagery based. In the next chapter, I will extend this theme further by examining evidence for the transfer of imagery and runnability from source analogues to models and argue that this is an important source of model flexibility in science.

Chapter 18
How Grounding in Runnable Schemas Contributes to Producing Flexible Scientific Models in Experts and Students

18.1 Introduction: Does Intuitive Anchoring Lead to Any Real Advantages?

This chapter takes a more global point of view and attempts to explain why grounding models in runnable schemas appeared to be so desirable for both experts and students in the previous chapters. Sections I and II of this book show that for experts, analogical reasoning can be difficult, risky, and time consuming. This raises the question of why the subjects who used analogies went to the trouble to do so. A simple answer might be that it was their strategy of last resort – that they could not think of any other way to make a prediction. However, in the case of S2's solution to the spring problem in Chapter 6, the subject already has a prediction with a high confidence level at the point where he complains that his "understanding is still quite low" and continues to pursue analogies. He goes on to eventually ground his solution in runnable schemas of twisting and bending. But he did not do this primarily to increase confidence in his solution, which was already quite high. The hypothesis to be developed here is that such subjects were attempting to construct new understandings that were as deep as possible by grounding these attempts in prior understandings. In particular, the subjects generated analogies in order to find relevant anchoring schemas constituting their prior understandings. This paid off when a schema was found that was runnable with conviction and that could be used as a grounding element in the construction of a model for the target situation.

In the case of S2, this type of anchoring appeared to eventually increase his understanding of the system considerably. Finding the analogy of a twisting rod allowed him to apply his intuitive knowledge of twisting to develop a model for the source of stretch in a spring, as shown in Fig. 14.15. It is evident that he went to a great deal of trouble to attain this increased sense of understanding since he had already reached a high degree of confidence in his prediction and since it took him an additional period of real conceptual struggle to attain that understanding. The expert pattern of developing an intuitively grounded model by anchoring it in runnable schemas (and here in particular, in physical intuition schemas), is one of the major findings of this study. But why did these experts go to this extra trouble? Why is this kind of grounding or anchoring so desirable?

18.1.1 Review of Findings on Imagistic Simulation and Runnable Schemas

In this chapter I discuss some potential benefits of anchoring a model in a prior knowledge schema, the most important benefit of which is labeled "transfer of dynamic imagery" or "transfer of runnability" from anchor to model during model construction. Throughout this chapter I use evidence from external subjects only. In Chapters 12–15, transcripts were examined which seemed to show subjects "simulating systems or events in their heads." At a minimum, this means that they were able to produce predictions of future states of an unfamiliar system. In addition, those examples provided other kinds of evidence supporting the hypotheses that the subjects were:

(a) Using dynamic imagery
(b) Using concrete knowledge structures with modest generality that could generate simulations (runnable schemas)
(c) Using these structures in a way that was important to their solution process

For instance, some of the observables supporting hypothesis (a) above were dynamic imagery reports and corresponding depictive hand motions. Most of the examples in Chapters 12 and 13 that supported hypothesis (b) were "elemental" cases where the subject focused on a single causal relation in the simulation. Thus the question in those chapters was not how to combine such elements into a complex mechanism, but how to account for the ability of the subjects to simulate a single causal relation that they may never have attended to before.

To account for these data, a schema-driven *imagistic simulation* process was hypothesized wherein a schema assimilates a particular image of a system and produces expectations about its subsequent behavior in the form of a dynamic image that can predict or account for an event. Figure 13.3 gives an overview of this process. It was assumed that an expectation about how an object behaves is embedded in a permanent and somewhat general schema. A central assumption of the imagistic simulation hypothesis is that the schema produces a prediction by in some sense regenerating or generating and running through an action or event rather than simply recalling a list of static facts about it.

A *"runnable" schema* was defined in Chapter 15 as being a schema which has this capability of generating imagistic simulations. A runnable schema can assimilate a static image of an object or system and perform mental actions on that image to generate a dynamic image anticipating possible future states of the system.

18.1.2 Transfer of Runnability Hypothesis

I began by asking the question: "Why did experts go to so much trouble to find anchoring examples? What are the potential benefits of intuitive anchoring?" This chapter proposes the following hypothesis:

18.1 Introduction: Does Intuitive Anchoring Lead to Any Real Advantages?

Transfer of Runnability Hypothesis

Part 1: Transfer from Source to Model

> Using one or more runnable schemas to provide contributing elements to the construction of a complex explanatory model can result in the model itself having the ability to generate and run simulations. This is referred to as "transfer of runnability." More specifically, parts of a runnable source schema can be transferred from source to model so that there are similarities in the dynamics of the simulations generated by them.

Part 2: Model Runnability Facilitates the Flexible Application of the Model to New Transfer Problems

> Runnability in turn helps provide the model itself with the crucial property of flexibility in access and application to unfamiliar problems, as well as other valuable properties.

These hypotheses provide an explanation for the motivation experts exhibit toward finding anchoring analogies and examples, although they may express this only vaguely as a search for a "way to understand it at a gut level." To make the above into a clearer hypothesis one must first consider the nature of the process of model projection. This question was dealt with in Chapter 14 by using the concept of image projection illustrated by the following example. Twisting became part of an explanatory model where the image of small twisting elements was taken seriously as a mechanism which could actually contribute to stretching in the spring wire. This culminating image was projected into the spring itself, as shown in Fig. 18.1.

This example illustrates the way in which an *explanatory model works by generating an imagistic simulation and projecting it into the target case.* This suggests that it is extremely valuable for an explanatory model to support imagistic simulations. In what follows I consider the question of how such a runnable explanatory model might be constructed from runnable intuitions.

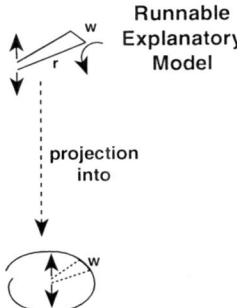

Fig. 18.1 Twisting model projected into spring target

18.1.3 Models Can Inherit the Capacity for Simulation from Anchors

The rest of this chapter follows the themes in Fig. 18.2, working from left to right. In this Section I propose that a runnable anchoring intuition schema can be used in the construction of an explanatory model; and that this can contribute to the simulation capabilities of the new model. This hypothesis has been implicit in much of the discussion in Chapters 14, 15, and 16, but it is time to make it explicit and to examine the possible benefits for model use.

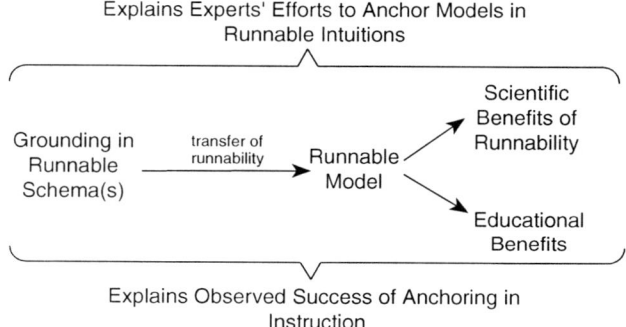

Fig. 18.2 Form of global arguments in Chapter 16; Key: → = "makes possible."

18.1.3.1 Runnable Schemas are Incorporated and Adapted into a Model as It is Constructed

The diagonal arrow in Fig. 14.15 foreshadowed the idea that a schema activated in thinking about anchoring examples could be incorporated into an explanatory model during its construction. The schema may be modified or simply incorporated to become part of the model, as shown in Fig. 18.3.

18.1.3.2 Models Can Inherit Runnability

The runnable schema or source analogue transfers its runnability to the more complex model. For example, as the model of the square spring coil constructed by S2 is developed, he appears to tap a runnable schema for twisting as a source analogue and use it to think about the square spring. The square coil serves as a simplifying model of the spring. It appears that the twisting action and its consequences are projected into the square coil by the subject. The square coil appears to inherit the runnability of this schema, evidenced not only by the way he talks

18.1 Introduction: Does Intuitive Anchoring Lead to Any Real Advantages?

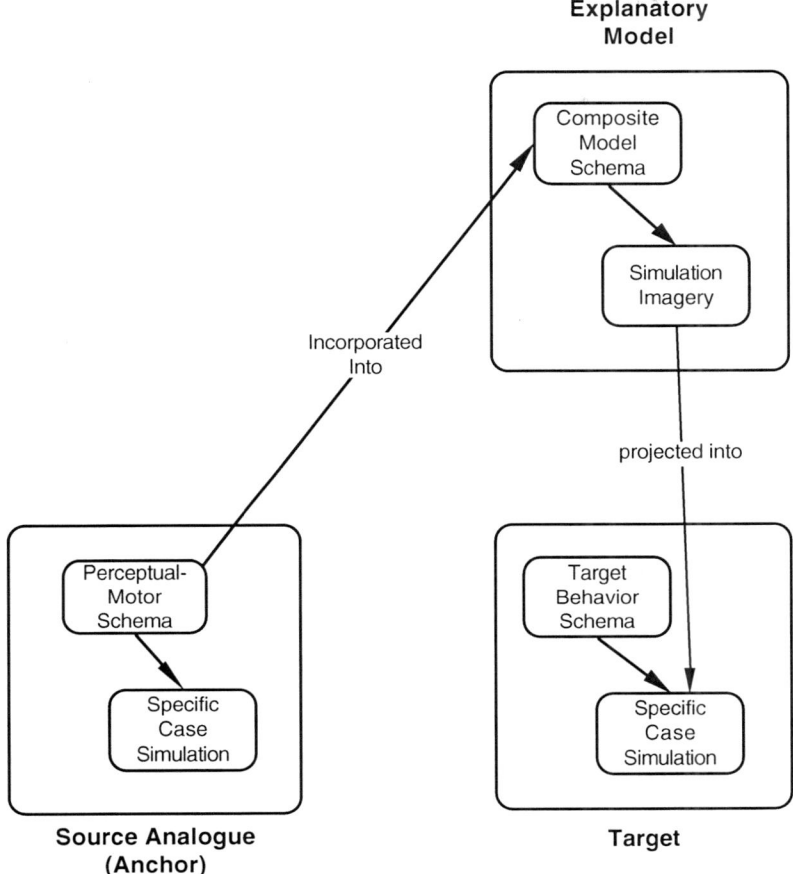

Fig. 18.3 Transfer of runnability to model and application to target

about the square twisting, but also by the way he produces similar twisting motions with his hands for the rod analogy and the square coil, as he makes predictions from these. These appear in sections 12a and 12b in Chapter 14. The similar hand motions suggest that the form of the imagery is similar in both cases. I have hypothesized that the runnability of these elements in the square spring model, along with the imagistic "summing" or canceling of the effects of these applications via spatial reasoning, is what allowed him to "interrogate" the model to generate the prediction that a wider coil will stretch more. In this view he is generating information by running the model within its spatial and geometric constraints, rather than making a set of deductions from previous facts. Thus there is evidence for transfer of runnability from the twisting schema to the square coil model. Presently I will argue that this kind of runnability is one of the most valuable properties a model can have.

18.1.3.3 Transfer of Runnability to Circular Coils

A further culminating transfer of runnability for S2 was described in Chapter 14, section 17a. In this episode he has just reviewed the square coil model and asks whether one can actually see stretching forces causing twisting in a helical spring. By drawing a helix with two points 90° apart in Fig. 14.7a, he is able to imagine a downward force at point b exerting a twisting force on the segment at point a, so that the twisting deformation in point a allows point b to drop as a contribution to stretch. I interpret this as the first occasion where S2 has actually been able to envision a mechanism for twisting occurring in the spring coil itself that is aligned so that it captures how forces are causing the twisting. Here, he again makes similar twisting motions over the drawing, indicating that some similar imagery has been transferred from the earlier case of the twisting rod.

The entire process is referred to as perceptual-motor grounding or anchoring of the model. The vocabulary I will use to describe this process is shown in Fig. 18.3. There an anchoring, source schema on the left (e.g. twisting) is incorporated (or adapted) into the explanatory model on the upper right. Recall that both a schema of modest generality and the image of a particular example are involved when an imagistic simulation of the anchor is "run." When aspects of the source schema are incorporated into the explanatory model, there is a transfer (or inheritance) of the capacity to generate simulations (runnability) for this part of the model. (This does not guarantee a correct model or even the ability to run the entire model if the schema provides only a piece of the model, but it does provide an important part of what is needed to run the model.) Once a runnable model has been assembled, episodes of running the model can be projected into an image of the target. This transfer of runnability view is supported by the parallel hand motions for twisting of the (a) rod and (b) $^1/_4$ coil of the spring shown in Fig. 18.4.

The subject may also have an older knowledge schema in the form of a target behavior schema which incorporates his or her prior general knowledge about the target and its behavior as in Fig. 18.3. If the projection from running the new explanatory model fits the separate simulation generated by the target behavior schema, then the model can receive initial support as a plausible model.

The overall argument structure supporting the interpretation given here is shown in Fig. 18.5. The presence of similar hand motions during the subject's investigation of the source case of twisting the straight rod and during his investigation of

Fig. 18.4 Transfer of runnability from twisting schema to model of spring: parallel hand motions for twisting a rod and twisting one quarter of a spring coil

18.1 Introduction: Does Intuitive Anchoring Lead to Any Real Advantages?

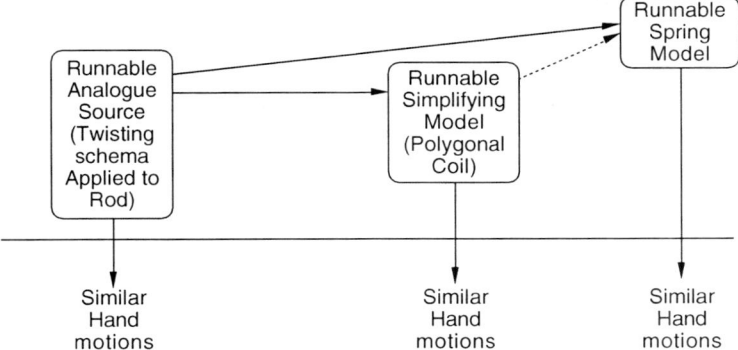

Fig. 18.5 Transfer of runnability during model development

the two models provides evidence for a transfer of runnability from the source schema to S2's model of the spring.

18.1.3.4 What is the Mechanism of Transfer?

The simplest model for how runnability is transferred is to assume that the anchoring schema is simply "applied" directly to or incorporated directly into the model in progress, then modified as necessary. This will ordinarily be an act of applying a schema outside its normal domain of application, so there will be issues and questions about where to apply it and how to align the action of schema imagistically with the target. The resolution of these questions will be part of the abduction-under-constraints process that is involved in any change in the model and it will need to be evaluated and possibly modified further by the subject.

18.1.4 What, Exactly, is Transferred?

There are several possible aspects that could be transferred to a model including:
- Static imagery forms
- Runnability: schema elements that generate dynamic imagery
- Confident runnability, local conviction about how aspects of the model will behave

The transfer of runnability hypothesis stated earlier in this section includes the first two ideas above. In the case of the twisting wire this included a static image form (an elongated wire-like object) and a schema (twisting) that can generate dynamic imagery. Is one justified in posing a stronger hypothesis here that includes the transfer of some *confident* runnability (conviction concerning the behavior of the model)? Because S2's sense of understanding rises sharply after he applies his torsion insight to the spring, there is some justification for proposing the second, stronger hypothesis, as follows:

> Using one or more runnable, *intuitively grounded* schemas to provide contributing elements for constructing an explanatory model can contribute to giving the model itself an ability to generate and run confident simulations. This is referred to as "transfer of runnability with conviction."

By assuming that this kind of runnability comes ultimately from an intuition, I mean that it comes from a schema that has some self-evaluated conviction. (Or by saying that it comes from an intuitively grounded schema we allow for the possibility of recursive grounding back through a chain of conceptions to an underlying intuition.)

I call a model that is derived from an intuition in this way an *intuitively grounded model*. Theoretically one would expect intuitive grounding to be a desirable property for a model because it means one has been able to make a model coherent with basic self-evaluated convictions developed from practical experience and other firmly established sources. This explanation serves here as a hypothesis for why S2's rating of his own understanding (as distinct from his confidence in the answer) rose sharply after he applied the twisting idea to the square coil. While they may not provide perfect and certain access to the structure of the world, since such intuitions are based in perceptual motor components that share such a close interface with the world, they are inherently "in touch" with the world. Their perceptual motor nature also means that they are compatible and closely linked to other commonly used physical intuitions and spatial reasoning operators that aid in making inferences from them.

Thus, I have generated a hypothesis at two levels:

> Transfer of runnability or imageability from a perceptual motor schema can occur during the formation of a runnable model.
>
> Transfer of runnability with conviction from a special kind of schema – runnable intuitions – occurs during the formation of a runnable, intuitively grounded model.

The second hypothesis implies the transfer of runnability contained in the first hypothesis and is stronger and therefore more difficult to support. Colloquially, the second hypothesis provides an initial theory for how an explanation can "make sense" to a subject. Instances of the second are also instances of the first and the first therefore has a broader scope. As will be seen there are differing transcript episodes in the data base that support each of these variations of the hypothesis to differing degrees.

18.1.5 *Example of Transfer of Imagery and Runnability in Instruction*

Clement and Steinberg (2002) analyzed the protocol of a student being tutored in an electricity curriculum which used analogies to attempt to construct models of electric potential (voltage) and charge flow (current) that could be anchored in the

18.1 Introduction: Does Intuitive Anchoring Lead to Any Real Advantages?

student's intuitions about (perceptual motor schemas for) pressure and air flow. The teacher provided 8 hours of interactive tutoring over 5 days. Care was taken to develop an imageable model by working from concrete analogue examples (e.g. a leak in a tire) and using student generated drawings that had incorporated a color coding scheme suggested by the curriculum for different levels of "electric pressure." This subject was able to map and apply an air pressure and flow analogy to electric potential and current as her tutor helped her build a model for electric circuits. The success of this process led the authors to hypothesize a transfer of runnability from the analogue air pressure conceptions to the electric potential model, as shown in Fig. 18.6 (Clement, 2003). The hypothesis is supported by evidence from the subject's spontaneous use of similar depictive hand motions over drawings during the original air analogy, and again during the instructional circuit examples, indicating that she was using a similar type of imagistic simulation in both cases. Gestures for both pressure and flow appear during the tire analogy and then during her work on instructional problems on circuits. Furthermore, the subject's spontaneous use of similar depictive hand motions during a later posttest provided evidence that the instruction fostered development of a dynamic mental model of fluid-like flows of current caused by differences in "electric pressure," that could generate new imagistic simulations for understanding relatively difficult transfer problems (Clement and Steinberg, 2002). This led the authors to describe the core of her new knowledge as runnable explanatory models at an intermediate level of generality. This study suggests that the model runnability achieved by grounding a new model in runnable prior knowledge schemas (here physical intuitions) during instruction may foster a type of model flexibility that aids in its use in transfer problems.

Transfer of runnability may be related to what Hesse (1966) called "material similarity" (as opposed to simply formal or abstract similarity) and what Brown (1993) called "reattribution of agency." In this view, the target model is enhanced

(a) (b)

Fig. 18.6 Similar depictive hand motions as evidence for transfer of runnability from analogue case to model; (a) air pressure in tire: presses hands together around imaginary entity; (b) charge "under pressure" in upper capacitor plate: presses hands together around imaginary entity, then pulls them apart

with new concrete material features (Brown and Clement, 1989) that come from the base and enable the construction of a new explanatory model. This new explanatory model then enables a different attribution of agency than the earlier view of the target situation.

18.2 Cognitive Benefits of Anchoring and Runnability for Models

In this section, I attempt to develop a framework to explain why the runnability derived from grounding a model in one or more runnable schemas might be a very useful cognitive property for an explanatory model to have, shown in Fig. 18.7. The figure moves well outside the data base in this book in attempting to connect the idea of grounding on the left with desiderata for theories from history of science studies on the right, although I will make some connections to expert case study data in this section for the sequence of items from Runnable Schemas on the left to Predictive Testability and Explanatory Adequacy on the right.

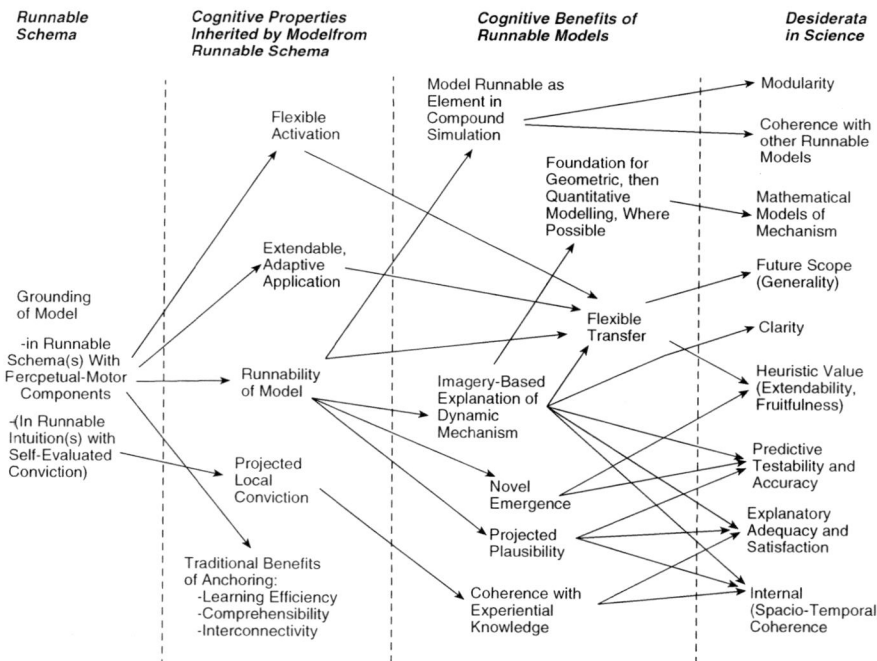

Fig. 18.7 Special benefits of inherited runnability for scientific models

18.2.1 Traditional Benefits of Building on Prior Knowledge

There are traditional reasons for "starting from what the student knows – from confident prior knowledge" (referred to here as "anchoring") that may not depend on runnability, such as the following ones.

1. *Efficiency*. An anchoring idea may serve as a starting template for constructing a more complex idea. This should save the time involved in reconstructing that piece of the idea. Thus anchoring should contribute to the efficiency of learning.
2. *Comprehensibility*. The familiarity of an anchoring idea may help increase the comprehensibility and level of understanding with which the more complex idea is held.
3. *Interconnectivity*. Using anchoring conceptions as building blocks may increase the interconnectivity of one's knowledge network since it creates links between old and new conceptions. This in turn should increase external coherence and retention.

The three benefits above should apply to any type of confident prior knowledge used in instruction, including factual knowledge, as shown at the bottom of column two in Fig. 18.7. But as shown in columns 2 and 3 there are additional benefits that may accrue from runnability, as described below, that go beyond there.

18.2.2 Benefits of Transferring Runnability from a Schema to an Explanatory Model

When an explanatory model is constructed from one or more runnable schemas, I hypothesize that the runnability inherited by the model should contribute to the usefulness of the model in the following ways.

18.2.2.1 Runnability Enables Flexible Application to Unfamiliar Problems

One of the most important characteristics a model can have is the capability for flexible transfer. Runnable schemas are assumed to have perceptual activation mechanisms and adaptive tuning capabilities (as described in Chapter 13), and if inherited by the model, these properties should support the flexible application of the model to unfamiliar transfer problems as described below:

(a) *Flexible activation*. Knowledge is useless if one cannot access it at appropriate times. In the present view, runnable schemas should be accessible by association via imagery or perceptions currently in the attention of the subject via a perceptual recognition process. If this property were inherited by a model grounded in an runnable schema, it would mean that the model's involvement in a problem may be triggered by parallel and distributed perception-like recognition processes separately from explicit feature analysis via linguistic symbols.
(b) *Adaptive tuning and application*. Once incorporated within the construction of a model, a runnable schema's perceptual motor components should also contribute to

flexible application of the model in a new situation. This derives first from a runnable schema's ability to "stretch" the domain of its application. Piaget and others have documented the process of early perceptual motor schemas assimilating and adapting to more and more new situations from their naturalistic observations of infants, so in this view this capability is built into our earliest mental structures.

The capability to tune themselves and adapt for new conditions has also been attributed to generalized motor schemas by Arbib (1981), Schmidt (1982), Rosenbaum (1991), and others, as discussed in Chapter 13. This raises the possibility that a schema could adapt its performance to new conditions as the schema performs, or is made ready to perform either a real action or an internal imagistic simulation. One of the original intentions in putting forward the construct of a runnable schema was to distinguish the schema-driven simulation process from "reading from" a list of memorized facts or a single figural episodic memory of an event. The schema somehow *generates* imagery that is "customized" to the specifics of the immediate context to some extent. That is achieved by a schema that is schematic (it does not carry specific details), and that can adapt to generate events for a range of situations. This is what gives runnable schemas in general (and, it is hypothesized, physical intuitions in particular) their general applicability. However, such a schema is still concrete in the sense that it operates on images of concrete objects. Thus runnable schemas have the somewhat paradoxical property of being both general and concrete. Perhaps this is less surprising though when one considers that our concrete action schemas used in everyday actions also have such properties of generality – they are flexible in their application to new situations. If this property were transferred to a model constructed from the runnable schema, it would also contribute to the ability of the subject to apply the model flexibly in a new situation.

18.2.2.2 Adaptive Assimilation and Application in a Student Transcript

The transcript below from a tutoring session on the direction of surface friction forces illustrates this type of adaptive assimilation. About 1 week earlier, the subjects had been taught in class to apply their runnable intuitions about the "springiness" of springs to solid objects like tables, in order to explain how solid objects can exert normal static forces (via the same instructional techniques as those used in Chapter 10). This includes the idea that as a force is exerted on the object, it deforms the object and causes it to push back with an equal and opposite force. Thus the intent of the instruction was that their runnable intuition about "springiness" would be incorporated into their model of elastic normal forces between solids. This new elastic force model is apparently applied spontaneously by subject B in line 7 to the friction case of the "bumps (on the table) pushing back" depicted in Fig. 18.8.[1]

---

[1] This is an oversimplified model of surface friction since a large part of the force can also be caused by chemical bonding or plowing mechanisms. Nevertheless it is a useful initial model for making the horizontal direction compelling (this is quite counterintuitive for many students who say that friction is a directionless effect or a downward force).

18.2 Cognitive Benefits of Anchoring and Runnability for Models

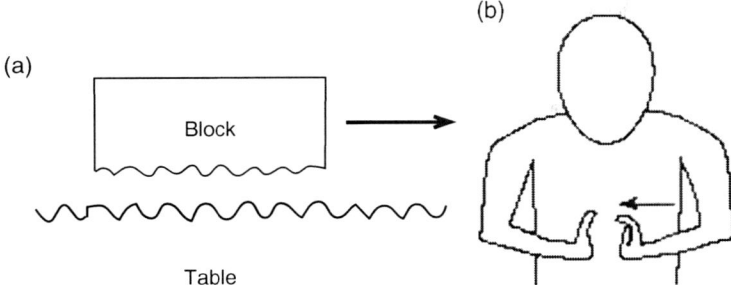

Fig. 18.8a Block sliding on table, **b** "You hit the bump and the bump pushes back"

Question: Given a cart or block sliding to a stop on the surface of a table, in what direction does the frictional force act?

1 SA: The block's going this way [to her right], and the friction from the table's trying to stop it from going in that direction. So in order to stop something that's going this way, [left] you're gonna – pull that way,
[This is correct, but does not explain how the table is able to push.]

2 I: OK. The, the question arose this morning, how can the table push –

3 SA: Up?

4 I: – push that way [left] on the cart?

5 SA: That's a good question! Is the table really pushing that way? [Puzzled tone]

6 I: Well, the model we use is that, if you take a microscope, and blow it up, so that the block is very big (draws Fig. 18.8a), the block has a rough edge, and the table's kind of rough, too. And, some of these bumps on the table will hit bumps on the block. So that's how the table can push that way. So as those bumps hit each other as the block slides over the table, the table pushes back.

7 SB: Oh, I see. It's just, um, is it kinda like the same way you said about pushing down (pushes down on table with right hand) and it, (moves same hand up 5 inches above table) it comes up again? Do you know what I mean? The spring?

8 SA: Yeah. (Presses down on table with both hands repeatedly) Like the springy movement of the table. (Referring to a discussion the week before about the source of a table's elastic vertical force)

9 SB: Now (places hands as shown in drawing) now, like, now you've got a thing coming up like this and you're pushing against it (left-hand fingers push on right hand) like, these are the bumps, really big (right hand), and these are the bumps (right hand) and it, it, um – (pushes fingers on left hand against fingers on right hand)

10 SA: You (her left fingers push right fingers back as in Fig. 18.8b) hit the bump and the (right fingers push) bump pushes back.

11 SB: Yeah.

Here, the subject takes a macroscopic elastic force model and applies it to a very different-looking microscopic situation. This is a case of transfer to a new and different topic that is impressive for a beginning student. I hypothesize that their elastic normal force model in this case involves the perceptual motor activity of imagining a force – deformation – reaction force sequence and that an extended application of this perceptual motor idea allows them to interpret the new friction situation. They appear to be able to transfer the runnability of their schema for static normal elastic forces in a vertical direction to the context of dynamic frictional elastic forces in a horizontal direction, even though the geometry and other aspects of the situation are very different. The evidence for this view is more complex here because the subjects are triggering ideas for each other and reasoning collectively in a very effective way. Subject A's "pushing" hand motions in her statement of the connection in line 10 are the culminating expression. She simulates the frictional forces by pushing her right fingers against her left fingers, saying "you hit the bump and the bump pushes back," and then pushing her fingers against each other in the opposite direction, representing the elastic reaction force of the bumps on the table. These indicate a kind of imagery where a force causes a deformation that produces a reaction force in the opposite direction. They are similar in exactly this way to Subject B's earlier hand motions representing an object pushing down on a table and the table pushing back up. This similarity in hand motions provides some evidence for transfer of runnability. Apparently their schema for the normal reaction force in the table developed a week earlier was still runnable at this later date, and it was able to become active and apply itself to the transfer situation of a frictional force in a flexible manner. Thus, flexibility of activation and adaptive application, as shown in column two in Fig. 18.7, for novel situations appears to have been a benefit of having a runnable model in this case.

The intent of starting from a hand pushing on a spring in the instruction a week earlier was to ground their model on a familiar perceptual motor intuition and build on the resulting imagery. It is plausible that the grounding of this model in a perceptual motor intuition of springiness is what allowed the model to assimilate and apply to the new situation in such a flexible manner. This leads us to a view of runnable source schemas being used to construct runnable models, and the models, themselves recursively becoming source schemas for building more advanced models.

18.2.2.3 Projected Plausibility and Spatiotemporal Coherence

Runnability should also play an important role in helping a scientist to evaluate the plausibility of a model. By this I mean evaluating whether the model is (in the words of Harre, 1961) a "candidate for reality." This should occur when the subject projects the model into an observable target situation. Such factors as whether the working mechanism portrayed by the model could actually *fit spatially and temporally* within the observable target situation, and whether it could provide a plausible chain of hidden causes to connect the observed conditions and behavior, would enter into determining the plausibility of the model. In unsuccessful cases, running the projected model could lead to emergent incompatibilities. Such a case has been discussed in Chapter 6 where S2 appeared to project the mechanism of bending into the spring and judged it to be highly suspicious. This eventually inspired further work and significant breakthroughs. In addition, if more than one

18.2 Cognitive Benefits of Anchoring and Runnability for Models

runnable schema were involved in constructing a model, those schemas would have to be plausibly compatible with each other in spatial and physical ways during the running of mental simulations – another requirement for internal coherence of the model.

In this view, when a model inherits the ability to run imagistic simulations as a major mode of representation, it *inherits a set of very important constraints* which depend on the spatial and temporal character of that representation. The subject may be able to use these constraints as the model is "run" to criticize and eliminate the model as a realistic view of the world. On the other hand, in cases that pass such tests it can be hypothesized that this kind of evaluation builds into the model an important form of self-evaluated plausibility. Another way to describe this outcome is to say that the model is spatially, temporally, and causally coherent with important aspects of prior knowledge. Thus inherited runnability should give one a powerful foundation from which to evaluate the spatiotemporal plausibility or coherence of a model.

18.2.2.4 Novel Emergence

From the imagistic runnability of a perceptual motor schema, the model may also acquire some of its potential for generative processes that can produce new information when the model is run. On occasion this may lead to future improvements in the subject's theories. Generative processes could happen in several ways:

(a) Since simulations of physical phenomena are spatiotemporal, when a model is grounded in this way it makes it compatible with spatial reasoning operations, and these may be used to extend inferences from running the model.
(b) When the model is run in a new context, it may lead to new or novel predictions that are based on *extended application* of the assembly of schemas comprising the model (Clement, 2003).
(c) The hidden sources of *implicit knowledge* that reside in perceptual motor schemas (as discussed in Chapter 13) could provide emergent knowledge in a simulation produced by a model.
(d) Many models will be assembled from more than one source schema. Running two or more runnable schemas in parallel in a compound mental simulation in such a model can lead to the recognition of *unanticipated emergent interactions*. For example, S2's act of running multiple instances of the bending schema in the orthogonal sides of the square coil led to the emergent prediction of no cumulative bending in the spring.

This kind of feature emergence from a runnable model is seen as a contribution to the testability of the model, as shown in Fig. 18.7. In other cases it might contribute to the heuristic value of the model (the potential for stimulating further scientific discoveries) by enhancing its potential for generating new and unexpected findings.

18.2.3 Recursive Runnability of Models As Thought Experiments Explain Many of These Benefits

The reader may note that the properties I have been discussing for runnable models are very similar to properties of runnable schemas that were used to explain sources of new knowledge and conviction in untested thought experiments in Fig. 15.3, and this is no accident. What is being proposed is that models can inherit these properties by being constructed out of runnable schemas. (Moving one more step backward in the development of this argument, several of these properties were described as "elemental" sources of knowledge from imagistic simulations in Chapter 13.)

To take this argument one step further, some unity is achieved by assuming that once a model inherits runnability from a schema, the model itself is eligible to participate recursively in Fig. 15.3 as a runnable schema. The act of a model being applied for the first time, or to a new and unfamiliar situation, qualifies as an untested thought experiment as defined in section 15.1.1. That is, the model should be able to generate such experiments by running imagistic simulations on cases outside its normal domain of application, and utilize spatial reasoning in these simulations. It should be able to run in coordination with other schemas or models in a compound simulation which gives it *modularity*. And these may generate new unanticipated effects which give it heuristic value for generating new theory. In fact, it should be able to participate in any of the plausible reasoning operations in Fig. 15.6, increasing its generativity and heuristic value, as it may ultimately serve as a runnable component in a larger mode of use.

18.2.4 Transfer of Conviction

Can conviction be transferred along with runnability from a source schema to a model? Having some self-evaluated conviction was used as an important defining feature of the construct of "intuition" defined in Chapter 13. I have suggested that runnable intuitions such as the twisting schema for confidently simulating cases like the twisting rod can be transferred to a model of the spring. Here, I examine how tapping an intuitive schema with self-evaluated conviction can be one source of confident knowledge for an explanatory model. It most likely would not be a complete source, since the conviction that is transferred would be a conviction about how that element of the model will operate, *if* it is part of the mechanism operating in the target. That in itself does not constitute support for whether it is actually part of Nature's mechanism. In this book, that last kind of support or confidence in validity is thought of as coming from a different source – from the whole model surviving repeated evaluation and criticism cycles.

But the hypothesis stands that valuable convictions can be transferred from a prior knowledge schema to a model – convictions that inform how that part of the model should behave. When S2 envisions twisting occurring in the square coil, he is quite confident about how twisting can occur in each side, even though the context is quite different from the isolated case of a single twisting rod. Apparently the twisting schema can be used adaptively here, during a simulation of the whole square coil. Chapter 15 proposed the theory that conviction could in some cases be maintained at a reasonable level, even during the extended application of a schema outside of its normal domain of application. This then outlines a way in which an anchoring intuition held with some conviction might contribute to confidence in an explanatory model. Conviction can be transferred from a runnable intuition schema to an element of the explanatory model as elements of the intuition schema are applied to the model. This does not guarantee the correctness of the model, but it allows confident inferences to be made from the model that can be tested.

I have added this idea as a separate "track" at the bottom of Fig. 18.7 to indicate how adding conviction to the properties of a runnable source schema may in turn add to the conviction with which the resulting model can be run. This in turn adds to coherence between the model and the prior experiential knowledge of the subject, assuming that the intuition is derived from prior experiences. This constitutes an increase in the internal coherence of the model, and that should increase the explanatory satisfaction of the model. However, the precise relationship between runnable schemas, those which are also intuitions with experientially based conviction, and the explanatory models that utilize them is a complex one, and one that requires much more research to sort out.

Earlier, I cited evidence from the square coil transcript for a case of increased understanding that fits the idea of a transfer of conviction. In addition there is some evidence for transfer of conviction in the case discussed in the previous section where two students are being tutored on the friction example. The two students give some evidence for increased understanding when student B says "Oh, I see" and the tone of the students' statements changes from puzzlement to confident affirmation. These observations are sparse, but alert teachers who can pick up many clues from facial expressions and tone of voice as well as verbalizations use them effectively to check the level of understanding during instruction.

18.3 How Runnable Models Contribute Desirable Properties to Scientific Theories

In this section, I illustrate the possible extendibility of a theory of imagistic simulations and runnability by discussing how the benefits of runnability listed above may add power to the role that a model plays in a scientific theory.

18.3.1 Scientific Theories and the Role of Runnability

Components of scientific Knowledge. As shown in Fig. 6.10, advanced areas of science use several types of knowledge, including:

(a) A body of shared observations
(b) Summaries of patterns in these observations
(c) A theoretical core consisting of an explanatory model or hidden mechanism (and sometimes several layers of models) which can explain the observation patterns
(d) Formal and preferably mathematical descriptions of these models which can add detail and precision to the models and provide deductive tools for reasoning and prediction

This is a synthesis of the views advocated by Harre (1972), Hesse (1966), Nagel (1961), and Giere (1988). What I am attempting to add to this view is the idea that imageability, and in particular, runnability (or confident dynamic imageability) is an extremely important feature of explanatory models. I will make this argument by providing a description of the desiderata of scientific theories that should plausibly be supported by runnability. In this view runnable explanatory models are only one component of scientific knowledge, but they are a very important and central component.

Nagle's functions of visualizable models. Nagel (1961) provided an initial outline related to some of the points that I will make here. Among the possible functions that Nagel identified for visualizable analogue models are that they can:

1. Help to articulate newly constructed theories
2. Suggest key questions for the refinement and extension of theories
3. Allow the application of theories to concrete physical problems by suggesting points of correspondence between theoretical elements and observable variables
4. Contribute to the achievement of inclusive systems of explanation by providing links between theories

I will build on this early philosophical assessment in the present discussion and combine it with the present protocol-inspired constructs to argue that runnable models can be even more central than Nagle proposed in several ways, e.g. they can: provide added flexibility and generality; yield unexpected insights or conflicts in new situations; provide a foundation for building mathematical models; and yield additional explanatory adequacy by providing clarity, grounding, and spatiotemporal coherence.

Runnability contributes to desiderata for scientific theories. Table 18.1 shows a list of desiderata for scientific theories discussed by Kuhn (1977b) and a more detailed list of strategies for theory assessment discussed by Darden (1991). Figure 18.7 shows a condensed set of these desiderata for scientific theories in the right-hand column. Lines connecting these desiderata to column 3 indicate cognitive benefits of runnability that could support the desiderata, as follows.

18.3 How Runnable Models Contribute Desirable Properties to Scientific Theories

Table 18.1 Criteria for evaluating theories. Phrases (except those I have added in brackets) are quoted from: Darden (1991), p. 257, and Kuhn (1977b), pp. 321–322.

Kuhn's characteristics of a good scientific theory
    Accurate – within its domain… in demonstrated agreement with the results of existing experiments and observations
    Consistent – not only with itself, but also with other currently accepted theories (I refer to internal and external coherence for these two senses)
    Broad scope – beyond the particular observations, laws or subtheories it was initially designed to explain
    Simple
    Fruitful of new research findings

Darden's strategies (criteria) for theory assessment
    Internally consistent [internal coherence] and nontautologous
    Systematicity [and] modularity
    Clarity
    Explanatory adequacy
    Predictive adequacy
    Scope and generality
    Lack of ad hocness
    Extendibility and fruitfulness
    Relations with other accepted theories [external coherence]
    Metaphysical and methodological constraints testable
    Simple
    Quantitative when possible

- Taken as a whole, the case study of S2 documents how the *projected plausibility* of a model being run on a target can lead to increased *explanatory adequacy* (as indicated by the subject's perceptions of increased "understanding" of the target situation). Such explanations via imagistic simulation in the spring problem led to descriptions of a whole new level of causal mechanisms underlying observable spring behavior. These mechanisms also had the potential to produce *predictions via imagistic simulation* for cases not yet observed (e.g. lack of increasing slope in the square coil case, presence of stretching in the vertical band spring). Thus, a runnable model can provide a central source for generating both testable predictions and satisfying explanations of physical phenomena. Furthermore, in Chapter 16, progressively more detailed imagistic simulations of deformations in the spring were used to develop a runnable mechanism with internal coherence in being fully connected spatiotemporally – where one could "see" (and "feel") precisely how force could cause torque which causes twisting which causes stretching. This was seen as the culminating form of *explanatory adequacy* and *clarity* at the qualitative level.
- Flexible transfer is a natural mechanism for contributing to the *future scope* of a theory, because the scope of a theory is the breadth of new phenomena that it can explain upon examination.

- The *heuristic value* of a theory is its ability (identified by Nagle) to raise questions and trigger new lines of inquiry. Flexible transfer contributes to heuristic value by raising questions about the boundaries of application of the theory. In addition, Hesse (1966) pointed to the way that *neutral elements* in models – elements which were neither clearly applicable nor inapplicable to a target – can be effective for raising questions. Exploration of neutral spatial aspects of models that were inherited from runnable schemas could generate new avenues for investigation.
- Novel emergence (e.g. emergence of increasing slope in the case of envisioning the bending model in the spring; and emergence of "unbending" as a causal factor in the spring in the case of imagining extreme stretching) contributes to potential *heuristic value* (extendibility) of the model. It also contributes to potential *testability*, by generating new predictions for testing. Viewing a formed explanatory model recursively as another runnable schema eligible to participate in the process of a thought experiment, as depicted in Fig. 15.3, allows one to envision many possible sources of divergent ideas that might emerge from running the model by projecting it into a new case. From this point of view, the idea that a model can generate new imagistic simulations for new cases gives it the potential for generating new ideas, coherence relations, conflict relations, and questions – models too can be "volatile" in this sense. In addition, the idea that a runnable model can be run with others as an element in a compound simulation gives it a desirable kind of *modularity* and the potential for *coherence with other runnable models* (see Fig. 18.7).
- Projected plausibility, as described in a previous section, would also contribute to *testability*, as well as to *explanatory adequacy* and the spatiotemporal and thematic aspects of *internal coherence*.
- Finally, a runnable model that produces imagery-based explanations can provide a core *foundation for geometric and then quantitative modeling*, as illustrated in Chapters 14 and 16, contributing to the attainment of mathematical models.

18.4 Conclusion

A central property of any explanatory model, as defined here, is to provide a hidden causal mechanism. A runnable model that can generate imagistic simulations offers a natural means of formulating explanations and predictions for events in dynamic, causal terms. An explanation is produced by imagining a set of hidden events (the mechanism) which cause the phenomenon. The same mechanism should also produce predictions for cases not yet observed. This chapter argues that runnable schemas can provide the imageable pieces or building blocks for constructing a model that can be run to provide such an explanation. That is, the ability to run the parts of the model is inherited from the runnable schemas. The recursive runnability of models as macro schemas that can participate in the thought experiments shown in Figs. 15.3 and 15.6 explains many of the properties in Fig. 18.7, columns 2, 3, and 4.

18.4 Conclusion

18.4.1 *Initial Support for the Runnability Hypothesis*

The main argument structure of this chapter was illustrated in Fig. 18.2. The theme of the chapter is that models can inherit runnability when they are constructed from runnable schemas, and from physical intuition schemas in particular, and that this runnability can provide important scientific and educational benefits. This theme links many of the findings in this book together. It might be objected that "transfer of runnability during model construction" is a vacuous concept which merely makes a tautological inference of the form "models can be run because they are runnable." I believe this objection is false because the above theme builds on the findings and hypotheses first developed from case studies of elemental runnable schemas and imagistic simulations in Chapters 12 and 13; and the "transfer during model construction" concept is based on studies of how such elemental schemas are assembled together during model construction as developed in Chapters 6 and 10. This kind of origin in case studies gives the hypothesis its initial empirical grounding and plausibility. The theory receives its initial support from its ability to explain several findings:

- It helps explain how S2 was able to obtain new emergent predictions from his novel square coil model and how bending and twisting schemas contributed to components of this model.
- It explains the presence of similar hand motions in the source case of twisting the straight rod and in the two models of the spring: the square coil model and the quarter turn of the coil twisting an adjacent element of the coil.
- It helps explain how students in the transcript of the "Direction of the Friction Force" problem were able to apply an earlier elastic force model, and the form of hand motions in that episode.
- It explains parallel hand motions in analogical and target model domains in other settings – such as learning of electric circuit theory described at the beginning of this chapter (Clement and Steinberg, 2002).
- It helps explain why the anchoring strategy was successful in the tutoring studies described in Chapter 10.
- It explains why the effortful tasks of finding analogies to runnable anchoring examples was important to expert subjects discussed in this book.
- It helps explain the fact that many scientists prefer visualizable mechanisms as a major mode of explanation in science, a preference cited by historians of science such as Hadamard (1945), Harre (1972), Hesse (1966), Nagel (1961), Miller (1984), and Giere (1988), as well as by Einstein (in Hadamard, 1945). I have hypothesized more specifically that simulation capability enables one to have a dynamically imageable causal mechanism.

The above findings on transfer of runnability and the arguments for the value of runnability in scientific models add support to the more general hypothesis developed in this book that imagery based methods are important in scientific thinking. They also support the general theme that significant elements in science are embodied in the sense of being rooted in perceptual – motor schemas and can be seen as an extension of intuition – of practical prior knowledge structures. The importance of prior knowledge to learning in students is a topic of continuing central interest in education, and this application of the above findings will be discussed in Chapter 21.

Section VI
Conclusions

Science is not just a collection of laws, a catalogue of unrelated facts. It is a creation of the human mind, with its freely invented ideas and concepts. Physical theories try to form a picture of reality and to establish its connection with the wide world of sense impressions. Thus the only justification for our mental structures is whether and in what way our theories form such a link.

Albert Einstein and Leopold Infeld (1967, p. 294)

Chapter 19
Summary of Findings on Plausible Reasoning and Learning in Experts I: Basic Findings

In this chapter, I begin a summary of basic findings on analogy, model construction, and imagistic simulation primarily from Sections I, II, and IV of the book, respectively. Chapter 20 will focus on more advanced findings on these same topics from Sections V and VI, and Chapter 21 will focus on philosophical, psychological, and educational implications, including Section III on student thinking. Those who desire an advanced summary of the theory developed in the book at four levels of diagrammatic models will find it at the beginning of Chapter 21. Here, I will begin with a simplified and somewhat polemical overview of theoretical findings as an advanced organizer for the conclusions to be presented. Connections to data sources will be made in later sections.

19.1 Brief Overview of Theoretical Findings

19.1.1 Model Construction in Experts

Think-aloud protocols from the experts studied in this book revealed a hidden world of plausible reasoning and learning processes based on imageable models – a world whose arguments are quite different from the formal arguments we often associate with experts. A theoretical framework was developed depicting scientists constructing deeper understandings of the world not just in terms of equations or formal principles, but also in terms of dynamically imageable models. They appear to be creative inventors of these imageable models rather than simply being logical manipulators of linguistic symbols.

The largest questions addressed by the book are: How do scientists develop new theories? How do creative insights occur? In simplest terms, the book puts forward the following framework in response to these questions, outlined in Fig. 15.5. Theory construction has at its core a process for constructing qualitative explanatory models. Explanatory models were identified in Table 6.10 in contrast to three other types of knowledge used in science: data, observation patterns, and theoretical principles. (The last category can include the development of quantitative explanatory models as well, but only after the full development of a qualitative explanatory model.)

Explanatory models are generated by a cycle of generation, evaluation, and modification (GEM cycle) that is usually evolutionary rather than revolutionary. However, the construction process was found to have divergent as well as convergent subprocesses, and it could lead to rapid scientific insights occasionally, as well as long periods of frustration or foundation building, yielding a pattern of "punctuated evolution."

A striking feature of much of the expert thinking examined here is its generative character. Analogies and models that are new and novel can be *generated* rather than recalled or derived. This fits the idea that the reasoning at the core of the model generation processes appears to be primarily abductive rather than derived from induction or deduction. Model construction utilizes a number of subprocesses, and three major ones are shown in bold in Fig. 15.5. These three important reasoning and learning processes were the major topics of three early sections of the book:

- Section I. The use of *spontaneously generated analogies*
- Section II. Explanatory *model construction cycles* of generation, criticism, and revision
- Section IV. The use of *schema-driven imagistic simulation* where the subject imagines a device and attempts to "run" its behavior

The model construction cycle at a top level can use analogies as a subprocess for model generation. Both analogy and model construction can in turn use schema-driven imagistic simulation as a subprocess.

The analysis of imagistic simulation began with the initial case studies of imagery use in Section IV, which identified a large variety of observation types that could be considered imagery indicators. Evidence was presented that physical intuitions can be thought of as perceptual/motor schemas that can generate imagistic simulations, and that these concrete processes are not just peripheral ones, but play an important role in the thinking of expert subjects. These processes can also be used in unfamiliar thought experiments. What is remarkable is that thought experiments can be a source of new convictions even though they involve novel systems that have not been observed before. I characterized this as the fundamental paradox of thought experiments. An attempt was made to understand the sources of conviction in thought experiments in order to address the paradox; the identified sources included compound simulations, implicit knowledge, spatial reasoning, and extended imagistic simulations where an intuition schema is "run" on a situation outside of its normal domain of application.

In Chapter 20, I will summarize findings on the role of imagistic simulations in higher-order reasoning processes. For example, analogies and models can be run using extended imagistic simulations. These two processes in turn play a central role in the generation and evaluation, respectively, of explanatory models, providing two major links for understanding how scientific theory construction can depend on imagistic simulation. The most advanced use of analogies documented here was to make possible the transfer of runnable simulations from elemental schemas such as physical intuitions to a newly constructed explanatory model that then becomes runnable (capable of generating imagistic simulations). This leads to the view that models can inherit runnability from the source analogues used to construct them, giving the models the valuable characteristic of flexibility in application. Thus the overall view developed was that discovery and learning in science can involve the punctuated evolution via GEM cycles of new models expressed as imagistic simulations. One major source of model generation is the use of analogy. Outside the realm of empirical testing, two major sources of model evaluation are running the model on a target situation, and designing evaluative Gedanken experiments.

19.1.2 Model Construction in Students

The first set of findings of this book, then, is the documentation in think aloud protocols of the existence of powerful uses of nonformal reasoning by experts, including analogy, model construction, and imagistic simulation. The second set of findings concerns the documentation of the successful use of these three processes by students as well. Some student uses occur spontaneously, and others can occur if assisted by tutoring. Whereas most prior research has concentrated on expert–novice differences, this finding supports the idea that there are important expert-novice similarities in the area of plausible nonformal reasoning and learning. This finding – that students are capable of learning science in many of the same ways experts do – was discussed in Chapters 8, 9, 10, and 18. It has important implications for the design of science instruction, including the need to foster subprocesses for imagistic simulation, analogical reasoning, and explanatory model construction. It was proposed that these are three major sources of conceptual understanding in science for students as well as experts.

19.1.3 Summary Table of Expert Subprocesses

Table 19.1 shows a condensed and simplified summary of the construction cycles involved in generating viable models, analogies, Gedanken experiments, and simulations. This table serves as a review of many of the major subprocesses described in the book. In the model construction section, for summarization purposes, I show only the simpler view of model construction developed in Chapters 5 and 6. A more complex view was developed in Chapter 16, as depicted in Fig. 16.3. Findings on Gedanken experiments will be summarized in Chapter 20.

19.2 Analogy Findings, Part One

19.2.1 The Presence and Importance of Analogy in Expert Thinking: Significant Analogies

In this section, I will review the initial findings on expert use of analogy discussed in Part One of the book. Are analogies actually used on the way to an attempted solution or explanation, and not just afterwards as a pedagogical

Table 19.1 Summary of expert subprocesses[a]

1. Model construction process (see more-detailed algorithms, including mathematization of the model, in Figs. 16.3 and 16.4)
   (a) Generate initial model
      1. Gathering elements
         (a) Analogy(s)
         (b) Other elements
      2. Abductive design of a model that can explain behavior of target
   (b) Evaluation
      (1) Rationalistic evaluation, e.g.
         (a) Running the model
         (b) Generating an evaluative Gedanken experiment and/or
      2. Empirical evaluation
   (c) Revision or rejection
      1. Transform model to fix criticisms
         a. Repeat evaluation and revision
      2. Reject and repeat generation
2. Analogy process
   (a) Analogy generation: Find or generate analogous case that has tentative analogy relation to target
      1. Via association or
      2. Via transformation or
      3. Via principle
   (b) Alignment and evaluation of analogy relation. (See detailed evaluation algorithm in Table 17.1.)
      1. Via mapping or
      2. Via conserving transformation or
      3. Via bridging analogy or
      4. Via dual simulations
   (c) Evaluation of understanding of base
      1. Try direct methods
      2. Break into parts or
      3. Use extreme case or
      4. Use extension analogy or
      5. Construct explanatory model
   (d) If case is inadequate under (b) or (c) above, revise and reevaluate it
   (e) Application as
      1. An expedient analogy with transfer or inference directly from source analogue
         – Infer properties or relationships in target, or
         – Transfer problem action or procedure or
      2. Provocative analogy activating a latent schema, or
      3. Source analogue (proto-model) as a building block for constructing an explanatory model or
      4. Bridging analogy (for Domain expansion or Imagistic alignment) or
      5. Extension analogy or
      6. One arm of an induction
3. Elemental simulation (see Fig. 13.3)
   (a) Activate dynamic schema(s) and assimilate or generate image of particular situation
   (b) Run simulation by applying schema(s) to image
   (c) While running, "interrogate" simulation by examining the experience for specific properties, or:
   (d) While running, allow the experience to activate new schema(s)
   (e) If fail, enhance simulation case and run again

Table 19.1 (continued)

4. Evaluative Gedanken process
   Given a hypothesized model (mechanism) M and an independent source of knowledge K:
   TYPE A
   1. Find or generate simple case C that can be both
      – analyzed for a prediction via M
      – predicted via another source of knowledge K such as a direct imagistic simulation
   2a. If prediction or comparison is not clear, improve C
   2b. If multiple mechanisms MA, MB are possible, constrain case to isolate particular MA as cause, eliminating MB; and vice versa
   3. If M and K conflict, detract support from M and consider modifying or rejecting M
   4. If M and K agree, add support to M
   OR
   TYPE B
   1. Modify target to make hidden mechanism M detectable
   2. Run target and inspect for evidence of mechanism
   3. Add or detract support for M accordingly

[a] I do not claim that the processes in Table 19.1 must be conscious, explicit plans in the subject, necessarily. But their rational reconstruction as procedures that call each other here is the simplest and most transparent way to outline a plausible model. Nor do I claim that this description is complete; it is intended to reflect some of the most important subprocesses. Furthermore, whether these patterns are explicit, consciously applied program-like sequences of heuristics or whether they are thought patterns that emerge from the interaction of less conscious, relatively autonomous cognitive entities operating in parallel remains to be determined. They are first-order models of the cognitive processes at work

communication tool? One of the first findings reported in this study was the observation that: experts solving problems and giving explanations can generate spontaneous analogies where the subject initiates and forms the entire analogy. This contrasts with most psychological studies, which have focused on presented analogies where elements of both sides of the analogy are provided by the investigator. In addition, almost all previous studies of analogy have tended to focus on novices rather than experts. (An exception is Dunbar, 1997.) Thirty-eight spontaneous analogies were observed in the study of 10 experts described in Chapter 3. Overall, these were impressive in their number and variety. Thirty-one of the observed analogies were "significant" – playing a substantive role on the way to a problem solution or explanation attempt. Thus, spontaneous, significant analogies produced by experts were perhaps the most prominent feature initially identified in the protocols for study and they constituted a starting point for the investigation. The variety of analogies presented suggested a more creative face of expert problem solving than is normally described. In the remainder of this section, I summarize four major subprocesses identified in spontaneous analogical reasoning, identify some new features of these processes, and highlight some of differences between these initial findings on analogies and those in most previous studies.

19.2.2 Literal Similarity and the Problem of What Counts as an Analogy

A "spontaneous analogy" was defined as occurring when the subject is solving a problem and: (1) refers to another situation B where one or more features ordinarily assumed fixed in the original problem situation A are different; (2) indicates that certain structural or functional relationships (as opposed to surface features alone) may be equivalent in A and B; and (3) describes B at approximately the same level of abstraction as A. This definition is somewhat different from the classical concept of analogy. The pioneering structural mapping theories of Gentner (1989) and Holyoak and Thagard (1989) were described in Chapter 2. For Gentner, analogies can be contrasted to what she terms a literal similarity between two cases: in a literal similarity, the two cases share many concrete surface features as well as perhaps some limited abstract relations, whereas in an analogy, the cases predominantly share abstract relations and very few surface features. I have argued for a somewhat broader definition of analogy. Building on Clement (1988), in Chapter 3 I argued that some "close" analogies such as the hexagonal coil analogy, that some would call "literal similarities," are indeed difficult, creative achievements that deserve to be called analogies. The "closeness" of the hexagonal coil did not imply "weakness" since it was a powerful idea that led to a genuine scientific insight. Thus "considering a situation B which violates one or more fixed features of A" was taken as central to the definition of a spontaneous analogy. I consider this a more important criterion than requiring case B to have many surface features that are different from A's features, since such "close" analogies appear to be one of the most fruitful and powerful types of analogies observed.

19.2.3 Analogy Subprocesses

Major subprocesses identified in case study examples in Chapter 2 for using analogies include those outlined in Table 19.1, part 2, and I will discuss the first three of those here.

19.2.3.1 Analogy Generation

Chapter 3 analysed generation methods used in the 31 analogies generated by 7 of 10 subjects. Analogy generation sometimes occurred via the classic mechanism of an association to a somewhat similar and familiar case in memory. On the other hand, many were novel cases generated via a transformation of the target problem. The phenomenon of generating novel analogies via transformations of the target casts analogy in an even more creative and generative light than in the previous literature on analogy, because in previous views the base of the analogy was familiar, not novel. Since theory construction in science can involve the creation of novel models these more generative methods of novel analogy formation laid the groundwork for and foreshadowed methods of scientific model construction, to be reviewed later in this chapter.

19.2.3.2 Analogy Evaluation

Analogy evaluation was seen to be just as important, however, since analogies can be misleading. Previous work has emphasized the role of feature mapping in evaluating analogy relations. The discovery of invented, intermediate bridging analogies as a new type of evaluation strategy was documented in case study examples Chapter 4, not only in expert protocols, but also in historical writings of Galileo's predecessors and Newton. The evaluation side of analogy use might be expected to be more convergent than creative, but the phenomenon of invented bridging analogies demonstrates that analogy evaluation can be a creative process as well. The suspicion that a bridging analogy may somehow help one to "see the target as" the analogous case prefigures the later analysis of imagistically based analogy evaluation methods to be discussed in Chapter 20.

19.2.3.3 Developing the Analogous Case

The case studies also indicated that an analogy is only as good as its source case (base); and especially when the source case is created anew in analogy generation, it may need to be analyzed until it is firmly understood. But firm understanding is what we had subjects set out to find in the first place, so we have come full circle. Only if the base is easier to understand or make predictions about than the target, does the analogy succeed. Thus development of one's understanding of the base may be necessary as a subprocess in analogical reasoning. Case study evidence was presented in Chapter 5 that understanding the base can at times involve the recursive use of analogy by generating a second analogous case (termed an "extension analogy") used to analyze the first analogous case. Another form of multiple analogy use is the "improved analogy" which occurs when a second analogy is generated to improve on an initial but faulty analogy. Thus, analogies can go through cycles of evaluation and revision or replacement and this makes their use more dynamic or evolutionary than is often portrayed.

19.2.4 *Initial New Distinctions and Findings on Analogy*

Table 19.2 outlines some differences between analogical processes identified in previous studies and new additional processes identified in Chapters 1–7. By looking for patterns in actual expert protocols that contain spontaneous analogies, one finds a much wider variety of analogical reasoning processes, as summarized in Tables 19.1 and 19.2. With respect to higher-level questions, the prominence of analogical reasoning in the present data supports the idea that expert thinking is less algorithmic and more divergent than is commonly thought. Analogies appear to be a nonempirical source of hypotheses, and this contributes to arguments against a purely inductive view of hypothesis formation. Further analysis of other analogy generation, evaluation, and

Table 19.2 Comparisons between initial findings on analogical subprocesses in the present study and those in most previous studies

Most previous studies: analogical processes identified	Present study: additional processes identified in Chapters 1–7
Primary context: Presented analogies.	Primary context: Spontaneous analogies
Emphasis on accessing analogous case in permanent memory via an association	Emphasis on generating analogous case via a transformation of the target case
Analogous cases are familiar	Generation of novel invented analogous cases
Retrieved analogous case itself does not require development	Efforts to improve or develop greater understanding of the analogous case; sometimes this occurs via a second extension analogy
Evaluation of an analogy relation via mapping	Use of bridging analogies for evaluation of an analogy relation

application methods that depend on imagistic simulation was presented in Chapters 17 and 18. I will summarize those findings in Chapter 20 and compare them to the classical view of analogy, and these will add to the contrasts in Table 19.2.

19.3 Model Construction Findings Part One and Initial Connections to General Issues in History/Philosophy of Science

19.3.1 Extraordinary vs. Natural Reasoning

How are hypotheses and theories formed in science? The history of ideas concerning scientific hypothesis/theory formation is rich and varied. Approaches to this question have ranged from: (1) proposing that hypothesis formation is largely the slow inductive formation of generalizations from multiple empirical observations (from Carnap, as described in Suppe, 1974) to (2) proposing that hypotheses are generated nonempirically through a super-normal form of reasoning or "Eureka" mechanism (Polyani, 1966). The data in the present study supported an intermediate view that is more complex than either of these more extreme points of view. This issue is part of a larger set of interrelated long-range questions that continue to torment and motivate researchers interested in scientific thinking. One way of parsing these is via the hierarchy in Table 19.3, where questions lower in the hierarchy are used to speak to higher-level questions. A number of the questions are stated intentionally as strong dichotomous choices simply to heighten the issues. Findings on analogies in the previous section contributed by speaking primarily to questions IB1 and IB3. In this section, I will primarily focus on question IA concerning insight, based on the case study of insight and basic methods of model construction in the S2 protocol in Chapters 6–7. I will wait until Chapter 21 to revisit the other questions after reviewing imagery-based findings and a more detailed theory of model construction in Chapter 20.

19.3 Model Construction Findings Part One   441

Table 19.3 Hierarchy of questions about scientific thinking

I – Do hypothesis formation processes involve natural reasoning processes or supernormal, extraordinary ones?
   A – What is the pace of change: incremental, revolutionary, or somewhere in between? Can scientific reasoning contain Eureka events that involve sudden reorganizations, or does it progress smoothly in an incremental manner?
   B – What are the primary mechanisms for theory and model construction? If they do not use formal logical methods of reasoning, is there an informal but rational method of reasoning and discovery?
      1 – How are theories generated – using the unconstrained, divergent, distributed imagination, as in an artist's thinking, or are they the product of tightly constrained, convergent, algorithmic operations? Do scientists use nonformal reasoning processes like analogy, or formal ones like deduction?
      2 – How are theories evaluated – How are problems detected in a theory?
      3 – Are some of the above mechanisms nonempirical as well as empirical?
II – Is scientific knowledge a set of formal propositions or an extension of nonformal intuitions?
Is it symbolic or imagistic?
Is it abstract or concrete?
How can it be rigid enough to be stable and yet flexible enough to apply to new situations?

19.3.2 Extraordinary Thinking?

The Gestalt psychologists used concepts like "Einstellung effect," "(perceptual) transformations," "good forms," and "incubation period" to give general descriptions of mental correlates to the phenomenon of insight. Although this work pointed to some general problems and features of insight, it did not propose detailed mechanisms of insight or discovery. Eventually work died down on this problem, partly because of the difficulty of the problem and the shift of interest to more tractable problems of memory and cognition that were not at such a high level of processing.

In his classic work in philosophy earlier in this century, Popper may have sensed these same imposing difficulties as he avoided the nature-of-hypothesis-formation question by focusing on the evaluation of scientific hypotheses rather than their generation. However Kuhn (1970) reopened this issue when he showed how difficult some types of creative hypothesis generation can be by documenting the resistance of the scientific community, and the individual scientist's own conceptual systems, to certain changes in theory, dubbing them "revolutions" that can involve a "gestalt switch" in conceptions. The term "revolution" and stories about insights such as Einstein's thought experiments and Darwin's Malthusian economics analogy revived the insight issue by raising question I above: Are the hypothesis formation processes used by scientists super-normal in the sense of being different in kind from normal processes of everyday thinking?

19.3.3 Eureka vs. Accretion Question

A significant subquestion labeled IA in Table 19.3 concerns the "revolution" vs. steady accumulation (accretion) issue concerning the pace of change during

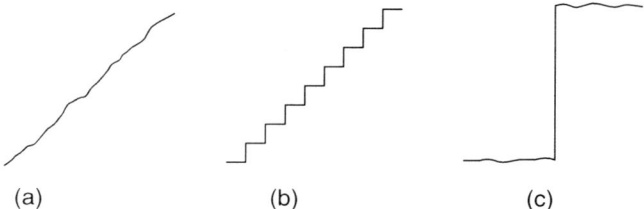

Fig. 19.1 Some possible types of change in perceived levels of "understanding" over time. (a) Smooth accretion; (b) evolution; (c) pure Eureka event

hypothesis formation: Were these inventions "sudden bolts from the blue" or did they arise gradually? Presented in its historical form, one can ask, are such creative new theories products of a sudden "revolutionary" change or of slow accretion of small elements? This turned out to be a complex issue.

The graphs in Fig. 19.1 offer one way to envision the contrast between a instantaneous Eureka event (graph c) and smooth accretion (graph a), as well as an intermediate "evolutionary" pace (graph b). I will use the term "evolutionary pace" to mean a process with a number of medium-sized intermediate steps; and the term "accretionary" to mean a smooth trajectory of very small steps. One possible response is that the answer depends on the timescale: over the scale period of an hour, the process may look like graph a, whereas the same events viewed over a scale of years will look like c. This idea is likely to contain some truth in many cases, but using protocol data one could still reask the question with regard to what happens within a critical hour. Does the hour-long process really look like a, or can it look more like c, with critical events taking place over minutes?

The above questions are interrelated as expressed in Table 19.3, because a very rapid pace might be one sign of an extraordinary process.

19.3.4 A Case Study of Scientific Insight

Chapters 6 and 7 examined a study of insight behavior where a sudden breakthrough in a solution appears to occur, allowing me to begin to address the pace of change question. S2 used an analogy to create an initial bending model for the spring but characteristics of this model were dissonant with other facts he knew about springs. After a long period of work on the spring problem, in which he becomes increasingly discouraged, he suddenly sees a new variable in the system, which turns out to be a key insight. This appears to be a strong "Aha" episode of the kind that Wertheimer and the Gestalt psychologists were trying to understand, where his pace of speaking suddenly increases, and his tone of voice and body language change from a discouraged tone to one of positive surprise and excitement. The insight, in which he recognized the new causal variable of torsion at work in each element of the spring, allowed him to view the system in a new way. It gave him a mechanism which explained *why* the wider spring would stretch more than the narrow one, involving the recognition of a new causal variable in the system. I termed this an insight, defined as a relatively sudden, impressive breakthrough leading to a significant structural improvement.

19.3.4.1 A Pure Eureka Event?

Surface level support for the instant Eurekaist view was provided by the strong Aha reaction in S2's protocol indicating that there was a relatively fast insight that opened up important directions for new thinking. But I also argued that it was not an example of a "pure Eureka event" in the sense that there was not evidence of it involving extraordinary reasoning of a completely different genre than that used in everyday thought. Instead it was explained by the processes of dissonance, analogy, and schema activation.

19.3.4.2 Punctuated Evolution

My response to the pace of change question was more complex, since the pace is relative to the timescale one is viewing from, and also because the pace seems to be uneven. When one looks at this thinking-aloud case study microscopically over minutes on a small timescale, one sees an arduous dialectic process of conjecture, evaluation, and modification of hypotheses that precedes the breakthrough, as opposed to a pure "Eureka" event that takes place quickly and effortlessly. On the other hand, progress does not take place as a completely smooth, incremental accretion of new knowledge either. Progress appears to be blocked when the subject is "locked into" his current conceptualization of the problem for long and sometimes frustrating periods. It is argued that an analogy generated by the subject then leads to a "reorganizing insight." Once this occurs it seems to trigger a rapid chain of further ideas. Thus, insight processes are found which support the view of science as involving significant rationalistic reorganizations in a relatively short period of time. When we add in the data from the composite protocol in Chapter 14, more of the protocol looks evolutionary, as it passes through the five stages of investigaton identified there. But the pace of progress is still uneven, with one or two "revolutionary" periods of work, one of which is illustrated qualitatively and hypothetically as segment 4 in Fig. 19.2. (I would count the recognition of "unbending" in Episode 15a as a second insight in the Chapter 14 protocol, involving something approaching a rapid "Gestalt switch" from bending to the opposite deformation of unbending.) Thus we have a process that is "evolutionary" (as opposed to either revolutionary or accretionary) for much of the time (segments 1, 5, 7), but that has some periods of insight, and some periods of stagnation where little progress is made, or where regression occurs. This means that the conceptual changes along the way varied in size, from a very large change in adding torsion to the model to

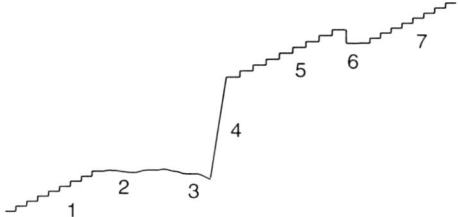

Fig. 19.2 Punctuated evolution with periods of gradual evolution (Sections 1, 4, and 6), stagnation (Section 2), and insight (Section 4)

much smaller changes as well as unproductive periods and setbacks. I call this a *punctuated evolution* view of theory change, by analogy to Gould and Eldredge's (1977) concept of intermittent as well as smooth progress in biological evolution. This is an intermediate position between Eurekaism and accretionism.

19.3.5 Initial Exploration of Mechanisms of Hypothesis Generation

19.3.5.1 Sources of Insight

As initial contributions for question IB1 in Table 19.3, in Chapters 6 and 7, I concluded that some of the sources of S2's insight were:

- Creative problem transformations which generated many analogous cases and the exploration of these analogies
- Attempts to apply an invalid analogy (bending) which led to dissonance.
- Further analogizing which triggered the new activation of an established but inactive schema
- The latter, as S2's Aha! insight, appeared to be an instance of suddenly applying the new schema to the problem

These sources indicate that a considerable amount of creativity was involved in this solution. There was a rapid addition of new structure to the subject's conception of the problem, so I did consider it to be significant scientific insight. However, the subject had to try many analogies and false models first, and go through frustratingly unproductive periods of hard work in his solution, so the insight did not come easily, and the subject had done a considerable amount to prepare the ground for it in this case. The processes used during the insight appear to be describable as natural extensions of everyday reasoning, but unusually creative and effective ones.

S2's case study suggests several general hypotheses:

19.3 Model Construction Finding Part One

- Some subjects can generate many analogies as part of a model generation process, although many can be discarded as not useful.
- Dissonance can play an important role in model criticism.
- The more evolutionary GEM cycles can be interrupted by a sudden insight caused by an unplanned schema activation.

Dissonance and GEM cycles also provide part of the answer to question IB2 in Table 19.3: how are theories criticized or evaluated? It was hypothesized that the ability to generate theories and the ability to criticize one's own theories stringently may interact: knowing that one can criticize later allows one to generate more freely; and knowing that one can always generate more ideas allows one to criticize more aggressively.

19.3.5.2 Analogies vs. Persistent Misconceptions

The persistence of the subject's initial model and the tension between it and the perceived anomaly may be partially analogous to the persistence of a paradigm in the face of anomalies in science, since paradigms create cognitive as well as social barriers to change. An important function of the strategy of searching for analogous cases is that such divergent thinking may help the subject to break away from such a stable persistent model. This helps to explain the observed presence of intermittent periods of negligible progress and rapid insight in such protocols.

19.3.6   Section Summary

19.3.6.1   Insight and the Pace of Change

Thus a type of "guided trial and error" process was documented that can converge on a new model, leading to an intermediate position between Eurekaism and accretionism: "punctuated evolution" with periods of slow and rapid progress. I gave a preliminary answer of no to the question of whether S2 exhibited extraordinary reasoning, by describing a series of plausible reasoning moves that could allow S2 to make his discovery without a pure Eureka event. They appear to be describable as natural extensions of everyday reasoning, but unusually creative, and effective ones. I will revisit this question in Chapter 21.

19.3.6.2   Preview of Educational Implications

Section III also documented the use of analogies and extreme cases by students, and showed that they could construct explanatory models with these tools, especially when the process is scaffolded by an instructor. This indicates that many of the learning processes used by experts for model construction are ones that nonscientists can use. This fits an overall view of science as involving reasoning that is an extension of everyday reasoning and intuition. The use of creative reasoning and the complimentary relation between generative processes and convergent evaluatory ones are issues that are too seldom recognized in science education, and I will discuss this in the conclusion section on education in Chapter 21. To make further progress on analyzing model construction methods requires the analysis of imagistic thought experiments and more detailed views of analogical reasoning, taken up in the next sections.

19.4 Imagistic Simulation Findings, Part One

The main goal of Section IV was to construct a plausible initial framework proposing basic relationships between physical intuitions, perceptual/motor schemas, visual imagery, kinesthetic imagery, and simulations. Introducing and expanding the concept of imagistic simulation in Section IV was a foundational move in the book because it went on in later sections to propose that imagistic simulation can play a role as a subprocess in each of the other major reasoning processes shown in Fig.15.5. The following paragraph gives an overview of the general perspective developed in Section IV.

A considerable number of imagery indicators, such as spontaneous imagery reports and depictive hand motions, were observed in the protocols. Subjects also gave evidence for the use of intuitions–knowledge structures that were self-evaluated and concrete but somewhat general. These observations were explained by hypothesizing that rather than using only equations and words to think with, scientists can use dynamic imagery as a representation with which to "run simulations" of new models and events, and perform spatial reasoning operations. Hand motions and other indicators have the potential to provide evidence for such imagistic thinking. The studies also highlighted the importance of selected *actions* on images that embody causal relationships. For example, many subjects appeared to have the physical intuition that the action of applying more weight to the end of a rod may make it bend further. We can think of such intuitions as permanent action schemas that generate temporary dynamic images to think with in imagistic simulations. Such a concept of schema-driven imagistic simulation helps explain (a) how expert subjects were able to make predictions for novel situations as well as (b) the widespread use of "anchoring intuitions" as sources for instructional analogies. Attributing action schemas to the subjects also allows one to use motor theory as an initial foothold to begin to formulate a theory of how subjects can generate imagistic, mental simulations. This contrasts strongly with a view of experts as only manipulating abstract symbols and allows one to describe the sense in which expert knowledge can be embodied. I will now summarize several subtopics from the above in the form of hypotheses proposed on the basis of the case studies in Chapters 12 and 13.

19.4.1 Imagery Indicators as Observational Concepts

Starting from the work of Pavio, Shepard, Kosslyn, and Finke, there has now been a long history of findings on imagery, and I made the assumption that imagery is a viable mode of representation in some circumstances in most normal adults. A remaining question is whether it is used in advanced cognition and whether it can play a non-superficial role there. Assuming people are capable of using it, are there any indicators for when it is being used? Is there a way we can use protocol data to speak to this issue?

The following are some of the new types of *observational* concepts and vocabulary developed here as potential indicators for the use of imagery: dynamic imagery reports, personal action projections, imagery enhancement reports, and corresponding depictive hand motions. In Chapter 12 case study evidence was presented suggesting that depictive hand motions observed during expert reasoning could be taken as an indicator for and partial reflection of thought processes involving dynamic imagery. These findings were coherent with previous literature on gestures reviewed there that has also suggested that they are tied directly to thought processes rather than being simply a means for communication. Chapter 13 presented additional cases where the subject exhibits depictive hand motions along with other co-occurring imagery indicators. An expanded list of imagery indicators was given in Table 12.3. To my knowledge these have not been used together before, and many of them are new indicators. They made it possible to propose models of the role of intuition and dynamic simulations that were grounded in case study observations.

A finding that surprised me that supports the *importance* of imagery-based processes was evidence for imagery (and simulation) enhancement, wherein the subject transforms a situation so as to make it easier to imagine (e.g. S10 adding "paint dots" to his image of the spring wire in order to help him imagine whether it was twisting). This example from Chapter 13 is difficult to explain via an appeal to discrete symbol structures alone because it is hard to explain how there is anything to be gained from adding the dots except the enhancement of imagery. A number of other indicators of the importance of imagery based processes to the subjects were listed separately in Table 12.3. These allow one to provide evidence that imagery is not an epiphenomenal "side effect" in these scientific contexts.

19.4.2 Mechanisms for Imagistic Simulation

In Chapter 13, I tried to speak to the question: where could these images be coming from? How are they connected to the actions depicted in the hand motions? A major problem was how to represent permanent knowledge of spatial and physical information, given that experienced images are clearly temporary. This was done by hypothesizing the involvement of perceptual/motor intuition schemas.

In common usage the term "intuition" can be used very loosely, and pains were taken here to give it a much more precise meaning. I chose to use it for a type of knowledge rather than for a type of reasoning. Most of the examples studied were "physical" intuitions. The nature of a physical intuition is difficult to analyze because it is often treated by both subjects and researchers as a primitive whose internal structure cannot be examined. Nevertheless, by doing microanalyses of a number of case studies, it was possible to make progress on outlining characteristics of these processes. On the basis of clusters of observations, elemental intuitions were hypothesized first, to be perceptual/motor schemas as knowledge structures that are concrete and have modest generality. Second, other hypothesized characteristics of intuitions are that they are self-evaluated, and have some level of conviction or plausibility without the need for further explanation or justification. The last five characteristics are in some cases each inferable from transcript observations that can be considered intuition indicators. The cognitive hypothesis proposed to explain instances with combined intuition and imagery indicators was to think of an intuition as a schema that produces expectations through an imagistic simulation. For elemental physical intuitions, these would often represent a single causal relationship, in which case I called the process an "elemental simulation."

19.4.3 *Terminology for Imagistic Simulations*

At this juncture I need to review the way in which I have used terminology associated with "simulations" in the following hierarchy from Table 15.5 (each category includes those below it).
[Mental] Simulation (prediction of future states of or changes in a system)
    Imagistic simulation (simulation with use of imagery)
        Schema-driven imagistic simulation (SDIS) (imagistic simulation generated bya runnable schema with perceptual and/or motor components and modest generality)
            Intuitively grounded or intuition-based simulation (self-evaluated SDIS, with at least some conviction present)
Collins and Gentner (1987) used the term "mental simulation" at the first level above to refer to cognitive entities, which allow one to make new inferences by anticipating a sequence of multiple future states of a target situation. There they do not mention imagery or specify a detailed mental mechanism underlying this activity and I have tried to retain the noncommittal use of the phrases "mental simulation," or simply "simulation" (that is agnostic in particular as to whether imagery is involved) in this book. At the point in Chapter 12 where some evidence had been assembled for imagery use, I began to switch to the term "imagistic simulation" at the second level of specificity above in order to label the subset of cases of mental simulation where the involvement of imagery was hypothesized. At the third level, in Chapter 13, I hypothesized that permanent schemas with perceptual and/or motor components were often involved in generating these temporary imagistic simulations, calling them "schema-driven imagistic simulations." An alternative kind of imagistic simulation would be an episodic, figurative image memory or "replay" of an individual event that is more specific than the range of events covered by a general schema. But since the focus of this book is on new and often novel imagistic simulations that cannot be merely episodic memories, I continued to use "imagistic simulation" for short to mean "schema-driven imagistic simulations" generated by

schemas that have at least modest generality. At the fourth level, most but not all individual schemas discussed in this book appear to be primitive intuitions, defined as a schema having the additional feature of being self-evaluated with some conviction, and standing without need of justification. (Examples of schemas that are not intuitions for most people include facts such as the shape of the earth as a sphere, and the formula for the volume of a cylinder or the derivative of the sine of an angle.) More specifically most examples here are *physical* intuitions, defined as an intuition containing knowledge about a concrete physical phenomenon or system.

19.4.4 Imagery During Simulation Behavior

I will now flesh out the above overview by stating the findings of Section III as hypotheses. I refer to simulation behavior below when a subject makes predictions for a change in a system or a series of states of a system operating over time. Evidence from the above observations and others shown in Figs. 13.3 and 13.5 and Table 12.3 supported the view that:

(H1) Simulations can involve imagery, not just linguistic or mathematical representations.
(H2) Simulations can involve dynamic, not just static imagery.
(H3) Simulations can involve the motor system in many cases, not just the perceptual system.

Taken together, the observations in Chapters 12 and 13 seemed to be most parsimoniously explained by assuming that the subjects were operating on dynamic images of concrete objects (in some cases imagining them being manipulated by their hands) and then "inspecting" the resulting objects and relations between them for new properties.

19.4.5 Image-Generating Perceptual Motor Schemas as Embodied Knowledge

Do these simulations appear out of nowhere? Assuming that something like "predictive imagery" exists, what is the source of knowledge or inference that generates it? The proposed answer to this central question was the following:

(H4) Schema-based simulation hypothesis: an elemental imagistic simulation involves a relatively stable schema with perceptual and/or motor components that is activated from long-term memory, not just a temporary image. This schema can generate a temporary simulation involving dynamic imagery in working memory.

The initial example given for this process was subject S2 imagining whether it was easier to twist and deform a long rod vs. a short rod shown in Fig. 13.2. Although there have been a number of theoretical studies of how networks of causal relationships might yield inferences, there have been almost no analyses of the nature of individual causal relations. This separation of a permanent schema for generating the action of twisting, and an image in temporary memory (the particular twisting rod), means that the theory separates what is old, familiar, permanent, and more general (the schema) from what can be new, novel, temporary, and more specific (the image of particular objects and movements) in an elemental simulation.

Figure 13.4 shows a physical intuition schema generating an imagistic simulation. This is of interest partly because these intuitions appear to be rather concrete in nature, while the traditional characterization of expert knowledge is rather abstract. Evidence from experts thinking aloud about individual causal relations was cited suggesting that:

> (H5) Schema-driven imagistic simulations (and physical intuition schemas in particular) can play a central role in problem solving and explanation in experts that is more than simply a "start-up" role – the use of perceptual motor schemas and imagery and intuitions was not just a side effect, but was part of the central argument.

However, it is likely that many parts of the theory developed would apply to other kinds of schemas with perceptual/motor components, not just physical intuitions. Hypotheses 4–10 are framed in this section more generally at a higher level in the taxonomy shown in Table 15.5 (reproduced earlier in this section), where they are stated in terms of perceptual/motor schemas instead of in terms of physical intuitions. This is because I see no reason for why the hypotheses should depend on a schema being fully intuitive – being completely self-evaluated and unjustified. The hypotheses would appear to apply just as well to more elaborate schemas that are, say learned partly from authority, as are the schemas involved in running a ball and stick molecular reaction model. These schemas are based on images of concrete perceptual/motor elements: the breaking and forming of attachments between spherical objects of different sizes. Therefore the hypotheses in this section are framed at what appears to be the most reasonable and probable level for the theory. However, most of the evidence presented in this book comes from physical intuitions, and therefore a number of related hypotheses were initially framed more narrowly and conservatively in Chapters 12 and 13 in terms of intuitions.

Types of evidence from protocols that can be used to support hypotheses 4 and 5 are summarized in Fig. 13.5. The fact that such schemas can be applied to different situations, including some new situations not previously encountered, argues for what may appear to be a paradoxical combination:

> (H6) Concrete Generality Hypothesis: Subjects can use perceptual motor schemas of a schematic nature (in the sense of a schematic diagram that is stripped of specific detail features) that are both *concrete* (imaginable as perceivable events) and somewhat *general* (applicable to a range of situations).

This hypothesis conflicts with the widespread assumption that the major difference between expert and novice knowledge structures is the central role played in expert knowledge structures by abstract components not found in novice knowledge (Chi et al., 1981). Although the separation between experts and naive subjects is significant, the finding that concrete perceptual/motor schemas can play a central role in expert reasoning makes it less sharp.

19.4.6 Sources of New Knowledge in Imagistic Simulations

The process of generating new information from imagistic simulations poses an alternative form of reasoning and learning in contrast to more traditional forms such as induction and deduction but how this happens is not well understood. Three hypotheses were put forward as to how perceptual/motor schemas might be an emergent source of new knowledge when operating on an unfamiliar case in an imagistic simulation:

> (H7) An extended application of a schema can occur when it is applied to an unfamiliar case (e.g. in a transfer problem) not originally considered to be in its domain of application. The subject applies an existing perceptual motor schema to an image of an unfamiliar situation, and the schema can assimilate and adjust its expectations about the situation, e.g. via tuning mechanisms. This takes advantage of the natural flexibility of perceptual motor schemas. In this case new knowledge is generated as the image of a new event is generated and the schema is "stretched" both in criteria for activation and in adapting its actions to the details of the case. This may permanently change the future domain of application of the schema. (Piaget called this "accommodation".)
>
> (H8) Imagistic simulation can be used to derive new knowledge by converting implicit knowledge in a perceptual motor schema to explicit knowledge.
>
> There are implicit relationships and constraints that are built into action schemas, which can become explicit when the schema is applied to an image of a particular situation with a particular question in mind. The concept of tuning parameters in motor schema theory provided one precedent for this idea.
>
> (H9) Two or more schemas can assimilate the same situation and operate jointly in a compound simulation for more complex cases.
>
> An impressive example of such a complex simulation was S2's novel construction of a spring made of square coils, with both bending and twisting occurring in each side of each square. The subject was able to correctly predict that the spring would have no net increasing slope from top to bottom, and that a wider spring of this kind would stretch more.

Hypotheses 7, 8, and 9 describe some of the possible root sources of emergent knowledge in plausible reasoning via imagistic simulation. These are ways *new knowledge can be derived, and they do not depend on inferences from chains of word-like symbols.* These were viewed as "fundamental" sources of knowledge from simulations and can be used as foundations for more complex sources such as thought experiments. They provide an important part of the response to the paradox of how a scientist can "run" an experiment in the head that has not been done before, to be discussed in Chapter 20 in the section on models and thought experiments.

19.4.6.1 Pinker

The above processes provide an outline of specific mechanisms that can implement a basic idea proposed by Pinker (1984) when he made the speculations in the quotation at the beginning of Section IV of the book. In the present case I am generalizing from "objects" in Pinker's statements to "events," on the basis of think aloud case studies. Hypotheses 1 through 9 above were discussed in Clement (1994a), and most are summarized in Clement (1994b). Chapters 12–13 in this book expand on these themes.

19.4.6.2 Barsalou

Barsalou (1999) has described a theory of perceptual symbols which represent schematic elements of perceptual experience and that can be integrated to produce simulations and combined in productive clusters. Barsalou emphasizes the application of these ideas to categorization, concepts, and language, and he is therefore attempting to extend the central role of simulation to many parts of cognition, in effect extending it to most knowledge structures. His position, like mine, is controversial, but if he is right then many of the ideas about simulation in this book should apply more broadly to reasoning with many types of knowledge structures. It should be very interesting to follow his progress in this regard.

In summary, an explanation for protocol observations was given by proposing a description of knowledge schemas that are image-generators rather than simply discrete symbol generators, but which are permanent and can represent causal relations in a way that is more general than specific temporary images. One can view this kind of knowledge as a set of permanent schemas that generate temporary images to think with. Thus I speak of *schema-driven* imagery and simulations, and these can also be used in extended applications for novel cases or compound simulations for complex cases.

19.4.7 *How Perceptual Motor Schemas are Useful in Scientific Thinking*

19.4.7.1 Use of the Motor System in Thinking

At this point I need to reflect on the way simulation examples in this study go beyond the use of perceptual imagery by involving the motor system. A system that utilizes imagistic representations is only as powerful as the reasoning operators that utilize those representations. Rather than only postulating spatial reasoning operators for this purpose (even though they are also important), one possibility is that everyday practical actions for manipulating objects with the hands can provide a large number of transformational operators for reasoning via simulations. In Chapters 12 and 13, I introduced an approach to imagery that placed equal weight on perceptual and motor components of imagery. Basic action schemas like "bending" were thought of as coordinating perceptions and actions together to produce bending in objects via the hands. This perceptual/motor idea could then be extended metaphorically to other situations where external forces are doing the bending, as in a rod bent by a weight. I called the anthropomorphic version a "personal action projection" or "personal analogy." This view was supported by a number of transcript episodes where the subject begins to think about the problem as one where he or she is doing an action on an object in the problem, even though in the original problem another object is the origin of the action.

In these cases the subject's motor system appears to act as an analog device for modeling elements of the systems in question. It is plausible such analogue models allow access to visual and/or kinesthetic intuition schemas as a form of prior knowledge, or to the spatial reasoning supported by the systems for controlling the hand-arm manipulative system, and that this is one reason that people go the trouble of using such analogies. This hypothesis is lent support by the fact that these manipulative models were not mentioned in the problem statements but rather were introduced through the efforts of the subjects.

These episodes of self-projection appear to happen so effortlessly that they give the impression that this is a natural and perhaps preferred mode of thinking for the subject. It may be objected that since this data comes from solutions to a problem that involves forces, this pattern could be narrowly restricted to that domain. However, very similar patterns were also observed in Chapter 12 in the Rotation of the Earth problem, in purely geometric subproblems in the spring solution in Chapter 14, as well as in the "doughnut" problem of finding the volume of a torus. There (Chapter 11) S6 made statements like: "I just imagine the knife cutting it open and you know, laying it out." Apparently the use of everyday action schemas is not restricted to force problems. I do not wish to rule out other possible modes of thinking that depend more one-sidedly on perceptual patterns, but my sense is that the perceptual/motor mode has been given far too little attention. Treating a perceptual/motor action schema as a natural unit of cognition, rather than separating cognition into a perceptual stage followed by a cognitive symbolic stage followed by a motor output stage, provides an approach to cognition alternative to the traditional symbol processing approach, if one thinks of these processes as heavily imbued with perceptual and motor processing over time during the central part of the thinking. This provides an initial description of what it means for important aspects of expert thinking to be embodied.

19.4.8 Intuitive Anchors

The findings on perceptual/motor schemas also provided a foundation for the concept of "anchoring" in analogical reasoning – the use of a base whose behavior is familiar and predictable. As a self-evaluated source of knowledge and conviction, intuitions (in the specific sense used here) may be a natural source of anchors as a solid base for analogical reasoning. Building on the work of diSessa (1983), this allowed me to formulate hypotheses that:

> (H10) Intuition schemas can provide a source of conviction and grounding for coherence and a feeling of sense-making through schema-driven imagistic simulations.
> (H11) This can make them useful as the base of analogies used for scientific reasoning.

These themes were introduced in Chapters 6 and 13.

19.4.9 Role of Perceptual/Motor Schemas in the Construction of Model Assemblies

The concept of schema-driven imagistic simulations was used later throughout the book in a number of ways and I will review some of these here to foreshadow topics to be discussed in the rest of this book. In section 16.1.3 it was hypothesized that scientific explanatory models are based on assemblies of image producing schemas. Coactivated along with a well-developed, scientific schema are precise verbal labels for key components in the imagery. This view then includes imagistic and verbal components of knowledge. Thus it is not that scientists necessarily have only well-developed images or verbal principles, but that they have well-developed, general schemas (explanatory models) that can assimilate particular situations and generate general, schematic, labeled images that selectively represent crucial variables of those situations, along with actions that allow them to "run" the schema as a model of the situation and labels which allow them to describe it. This comprises a view of scientific knowledge as based on image generating, perceptual/motor schemas. When these schemas are also intuitions and include the property of self-evaluated conviction, they contribute to one's conviction in a model.

> (H12) Intuition schemas may provide basic building blocks for constructing imageable models in science. Certain self-evaluated intuitions that are selected and refined can become scientific conceptions and these can be assembled into scientific models (Clement, 1989b).

The theory was further expanded in Chapter 18 to include "transfer of runnability and conviction" from anchor to model as one explanation of the origin of generality and conviction in explanatory models (discussed further in Chapter 20).

19.4.10 Connection to Experiments and Situated Action

The important cognitive role played by actions involved in scientific experimental practice has been documented by Tweney (1986) and Gooding (1992). For example, in discussing the case of Faraday, Gooding uses a method of diagramming similar to one developed by Tweney and Hoffner (1987) to map out steps in the development of critical experiments. He shows for the case of Faraday how experimental work can begin as informal, unclear choices between actions, how the experimental actions alternate with conceptual operations, and how work proceeds with irregular progress and with procedures and observables only gradually made unambiguous and reproducible. His findings suggest that, experimental work can be situated in perceptual/motor actions in complex ways. This is also true of non-experimental work in the present study. Connecting this important work to the present theory of the role of actions in nonempirical thought is an important topic for future research.

19.4.11 Section Summary

Hypotheses 1 through 9 summarize the sense in which schema-driven imagistic simulations have many properties of interest to scientific thinking. The schema is termed an intuition when it has the further properties of self evidence and self-evaluation. Intuitions can provide grounding for convictions in an investigation. In Chapter 13, I chose to introduce all of the properties in hypotheses 1–10 using examples which involved physical intuitions. However, on reflection, and in formulating the general theory in the hypotheses in this chapter, I believe most of the key properties are properties more generally of embodied perceptual/motor schemas running imagistic simulations. So I have expressed those properties here first in hypotheses 1–9, and then added properties from physical intuition schemas in hypotheses 10–12. This chapter has reviewed basic findings from the first half of the book on imagery and intuition, analogy, and model construction. In short, experts can ground understandings on perceptual motor intuition schemas. These can be used as the base for an analogy and ultimately as a component of an explanatory model. Clement (1993a) provided evidence for students having useful intuitive preconceptions that can be used as anchoring analogies to ground qualitative scientific models. In Chapters 8–10, I argued that significant educational benefits can be derived from making full use of these intuitions in students, and this theme will be discussed in Chapter 21.

Chapter 20
Summary of Findings on Plausible Reasoning and Learning in Experts II: Advanced Topics

In this chapter, I provide a summary of findings on more advanced issues in the same three areas discussed in Chapter 19: analogy, imagistic simulations in thought experiments, and model construction. This is followed by discussions of evidence for the importance of imagery in the expert investigations, and the importance of transfer of imagery from source schemas to models, and lastly by a section on the methodology of this study.

20.1 Analogy Findings, Part Two

20.1.1 Comparison to Classical Views of Analogical Reasoning

20.1.1.1 Major Subprocesses in Generating and Evaluating an Analogy

Gestalt psychologists such as Wertheimer (1959) and Duncker (1945) studied the role of analogy in higher-level problem solving and were primarily interested in it as a process that might generate holistic reorganizing insights. They initiated work on the important role of analogies and transformations of representations in problem solving by doing interpretive analyses of interviews. However, in large part they did not break down analogical reasoning and examine its subprocesses. Interest in the issue of analogies was largely lost within psychology until the late 1970s and 1980s, when pioneering researchers such as Gentner (1983) and Holyoak and Thagard (1989) rekindled it by developing information processing theories of subprocesses used in analogical reasoning. I have termed the theories they developed in this vein the "classical view" of analogical reasoning. In this section, I will compare and contrast findings on analogy from the present study to this classical view.

Structural mapping theory is a highly developed, sophisticated theory that is predictive and that has explained a large body of experimental evidence. Much of it has been refined to a high level of precision by representing it in a computational form that can be run as a simulation and compared to data from human subjects. The classical view can be described as involving four major subprocesses. These processes are described as operating on sets of explicit, discrete propositional descriptions of the base and target as follows:

- The analogous case is *accessed* by being activated associatively and retrieved from permanent memory.
- An *alignment* or *mapping* is generated between corresponding elements in the base and the target, and the structural soundness of the analogy is *evaluated*.
- One or more inferences are projected from the base to the target.

20.1.1.2 Extending the Set of Subprocesses for Analogical Reasoning

In the studies described in this book, it was found that when experts use spontaneous analogies, they exhibit a wider variety of creative behaviors that adds to those discussed in the classical view (summarized in Table 19.1, part 2). I will try to paint these differences as sharply as I can in order to promote discussion and further work.

A. *Generation*. Whereas the primary mechanism of analogy formation is usually taken to be access via associations, the primary mechanism observed here was generation via transformations. For example, one subject in Chapter 3 transformed the spring problem by mentally uncoiling the spring into a straight horizontal rod with the same weight hanging on the end. Another took the wide and narrow springs mentioned in the problem and hooked them horizontally on opposite sides of the same weight, creating a "tug-of-war" situation. Such actions lead to the conclusions that:

1. Analogies are not just accessed in permanent memory by association from the target; they can be generated by transforming the target case.
2. Many of the cases generated in this way turn out to be newly invented, rather than familiar cases.

For this reason I have referred to this first stage as analogy "generation" in order to cover both accessed and invented cases. By this I mean generation of the new possible similarity relations between target and base (and in some cases generation of the base itself). This is shown as the second subprocess in Table 19.1 part 2a, and reflects an alternative to the classical associative view.

B. *Alignment and evaluation of the analogy relation*. In the classical view, alignment means sorting out similarities and alignable differences by mapping discrete symbols. A weighted mapping of identities, and correspondences between discrete symbols in the representations of the target and base is the basis for evaluation and determining the soundness and relevance of an analogy; In contrast, the case studies in this book indicate that other mechanisms exist for evaluating analogies:

20.1 Analogy Findings, Part Two

1. A prominent pattern observed here was the effort to find a *conserving transformation* indicating subjects gave evidence for using spatial and other transformations to produce analogous cases, and when these were simulated, the results could simply be transformed back to the original problem (e.g. Section 17.3.2, line 105). This allows one to hypothesize that in some cases alignment and inference may come almost "for free", without necessarily using a process of mapping discrete features.
2. In the present study, alignment means *imagistic alignment*, not symbolic mapping. Other alternative analogical methods of evaluation and alignment, called *dual simulations*, and *overlay simulations*, were observed, to be discussed at the end of the next section on imagery. There I will review the idea that the above methods may all rely heavily on imagery.
3. In some cases an *intermediate bridging analogy* was used to achieve the above goals – a "secondary" analogy created in order to evaluate an original analogy. By "bridging the gap" between an uncertain analogy and its target, such cases can provide a novel pathway for analogical inference. Some of these were termed "imagistic alignment analogies" because their purpose appeared to be to help in aligning the original base and target imagistically.

Evidence for individual elements of a feature mapping process was sometimes observed in the transcripts but not always. In constructing the analogy section of Table 19.1, I have placed the "mapping" and "evaluating soundness and relevance" processes under one concept: "Alignment and evaluation of the analogy relation," where aligning can mean imagistic alignment. This leaves open the possibility of having strategies for evaluating the analogy relation that do not involve mapping of abstract propositional features.

C. *Understanding the base*. A step not ordinarily mentioned in the classical view is evaluating (and possibly extending) one's understanding of the source or analogous case. Most previous studies assume that the subject has a firm understanding of the knowledge in a source conception that is pertinent to the target. However, some subjects in this study focused on an analogous case (base) that they realized was insufficiently understood to be used. In these instances they often used one of several strategies shown in part 2c of Table 19.1 to improve their understanding of the source (Burstein, 1988; Clement, 1983).

D. *Application*. In the classical view, the basic idea behind an analogy is that some candidate inference can be made about the target, based on the structure of the base. Transfer involves making inferences about the target that extend the initial correspondences generated in the mapping. I call this a method for direct inference from an expediant analogy.

In the present view, Table 19.1 shows a variety of other possibilities in the application stage of using an analogy. For example, they can play a more provocative role in activating an essential schema that has never been applied before to either the target or the base. They can also contribute to the larger purpose of the construction of an explanatory model that provides an explanation of the target at a deeper level. I will discuss these possibilities later in this section.

20.1.1.3 Example of Alternative Subprocesses for Analogical Reasoning

The most important example of a scientific insight in this book, the torsion insight of S2 in the spring problem, exemplifies several of these alternative subprocesses for analogy. First, the hexagonal coil analogy from which the insight arose was generated via a transformation rather than accessed in memory. Second, S2 attempted to deepen his understanding of this analogous case – it did not come "packed with information," ready to be used as a prior knowledge source. This makes sense, since it was just newly generated! This raises a paradoxical question: how can any information be gleaned from a new analogous case if the case has never been thought about before? Thirdly, in answer to this question, the transcript suggested "provocative" triggering of a prior knowledge schema (torsion).

The torsion idea was not a relationship transferred from within his "hexagonal coil conception," it was a schema used here to analyze not only the target (spring) for the first time but also the base (hexagonal coil) for the first time. Thus, S2 attempted to deepen his understanding of this analogous case. Fourth, the torsion schema appears to help generate a case in the form of a simple straight twisting rod that can be viewed as a source analogue. The function of the twisting rod analogy is to be a proto-model for the spring in that material and concrete aspects of it (e.g. twisting material and concrete forces and/or motions) are eventually incorporated into a separate explanatory model for the spring, as opposed to simply directly inferring a projected abstract relation from it to the spring. On the basis of parallel imagery indicators like hand motions, it was argued that some runnable schema elements capable of generating imagery, not just static propositional structures, were transferred to the subject's spring model.

It was argued that the torsion idea can be considered a scientific insight that identified a new causal variable in the phenomenon being explained and broke the subject out of a strongly held prior view. Thus, it is possible for these new subprocesses of analogy to play a powerful role in expert thinking. One is motivated to study analogy at a detailed process level by the finding that it appears to play a key role in creative scientific thinking by helping the subject break out of previous frameworks and assumptions. This more generative, transformative, and provocative view of the subprocesses perhaps explicates some of the enigmatic themes of transformational and focused creativity involved in using an analogy that were emphasized by Wertheimer and the Gestalt psychologists.

The hexagonal coil example also illustrates how these new subprocesses fit together – it makes sense that a newly invented case would be generated by a transformation rather than an association process (how does one associate to something that is not yet in memory?) and that it would more often contribute by provocative triggering of a previous idea during an attempt to better understand the new analogous case, rather than by inference from information stored with the analogous case (because the case itself was not stored). Thus this appears to be a different mechanism for analogical reasoning from the classical one, using a set of alternative subprocesses that can work together with each other.

20.1.1.4 Analogy Characteristics

Several of the above features as well as other comparisons to previous work can be seen in Table 20.1, which outlines several dimensions for categorizing analogies. Asterisks in the table indicate areas that are underrepresented or absent in previous studies. Whereas much of the previous research on analogy in problem solving has focused on analogies that access an existing source that is strong in prior knowledge with low or medium surface similarity to the target, this book has focused on invented analogies having a source that is either strong or weak in prior knowledge, with high or at least medium similarity to the target. In the case of S2, the hexagonal spring was an invented, close, simpler case which provoked the activation of the torsion schema and led to the development of an explanatory model. The extent to which the dimensions in Table 20.1 are truly independent, or have dependencies or correlations among them, has not been fully explored and is an important task for future research. These dimensions combined with those alternative processes listed under analogy in Table 19.1 form a complicated space of possibilities for characterizing different kinds of analogies.

20.1.1.5 Roles of Analogy in Scientific Thinking

Table 20.2 outlines the different roles for analogy that have been identified in this book, and is an expansion of Table 5.3. (There are at least three other roles of analogy that I did not observe or discuss in the book but that could be added to this table. The first is as a *descriptive metaphor* which is intended merely to illuminate rather than develop a scientific prediction or causal similarity. The second is as *one arm of an induction* leading to a new generalization; in that mode, a successfully

Table 20.1 Characteristics of different types of analogy

A. Knowledge of source
   1. Strong
      Utilize known knowledge stored with source case
   2. Weak*
      Source lacks development: source case may act as provocative trigger to access other knowledge not stored with source, e.g. for newly invented analogous cases, virtually no knowledge is stored.
B. "Distance"/Degree of source–target similarity
   1. "Far" or remote or low surface similarity
   2. Medium
   3. "Near*" or close or strong surface similarity
C. What is inferred or transferred:
   1. Abstract relation: nonconcrete factual or relational element or strategy
   2. Projective material: imagistic, concrete/schematic material (and imagability or runnability)*
D. Role or purpose served (see Table 20.2)

*Areas that are underrepresented or absent in previous studies

confirmed analogous case may focus the subject on a common relational abstract pattern that can be induced as a generalization from the two instances in the base and target. Central features for the new category will be among those shared by the source and target (Gick and Holyoak, 1983). A third is a *formal analogy* such as that between an LRC circuit and spring oscillations, where the similarity is captured in the form of an abstract equation or principle.)

Within Table 20.2, the first role, expedient analogy, is the predominant view in previous work on analogy. In that mode, analogies were useful only for making a prediction for the behavior of a target. This mode can make efficient use of relevant prior knowledge directly for target inferences and prediction. The other three basic uses suggest different mechanisms:

- An analogy can help through new indirect provocative triggering of a principle, schema or method, as the analogous case itself is analyzed. This type of analogy is often a simplifying transformation of the target problem.
- Analogies may serve as proto-models, that is, as starting points for the development of a scientific explanatory model. Examples cited here were the bending rod and twisting rod cases for the spring. Other examples have been proposed by Clement (1981, 1989), Holland et al. (1986), Falkenhainer (1989), Nersessian (1992), and Gentner et al. (1997) (in the latter chapter on Kepler's use of analogy, the authors go beyond the classical view of analogy). I theorized in Chapter 18 that models constructed via projective transfer of runnability from an established source schema inherit desirable properties such as flexibility, and analogous cases that facilitate access to such source schemas are therefore extremely valuable. I have therefore placed an emphasis in this book on the role of analogy as a *source analogue*, or *proto-model*, concrete aspects of which are incorporated into a larger and more complex model. This leaves open the possibility that more than one analogy can be involved in the development of a single model.

Table 20.2 Different roles for analogy in the case studies

1. Expedient analogy with inference or transfer directly from source analogue
   - Infer properties or relationships in target, or
   - Transfer problem action or procedure
2. Provocative analogy*
   Case activates a schema not stored with the case as it is analyzed or run and this is applied to the target or explanatory model for the target.
3. Proto-model (source analogue)*
   Prototype case and/or associated schema incorporated as element of explanatory model for target. The analogous case may be a starting point that is later revised and elaborated to become part of a model explaining the target.
4. Bridging analogy (Chapters 4, 17). Typically these stretch the domain of application of an existing intuition or model. Some bridging analogies can serve as an imagistic alignment analogy when the new case imagistically aligns features between an explanatory model, source analogue, scientific principle, mathematical schema, or target.
5. Extension analogy. Analogy that develops the understanding of a previous analogous case that is poorly understood (Chapter 5). (This last role can be subsumed if one views a poorly understood analogous case recursively as a target to be understood)

- Bridging analogies can serve to expand the domain of a relevant schema or principle; or they can aid in the imagistic alignment of two other representations.
- Extension analogies can develop the understanding of a previous analogy.

It should be noted that when an analogy like the bending rod analogy is generated, the subject may not always know ahead of time what its eventual purpose will be; it may simply appear to be "an interesting case to pursue".

One possible reason for why these three roles have been underemphasized in the classical work on analogies is that much of that work focused on tasks of finding a problem solving action or comprehending a story whereas the present work focuses on conceptual explanation tasks in science. In the former the process often involves comprehending a comparison or analogy between a presented pair of cases; whereas in the present research the process can start from only the target and can involve the subject both generating an analogous case and using it to stimulate new ideas for the larger process of building an explanatory model. In summary, the asterisked items in Tables 20.1 and 20.2[1] (to be extended in Table 20.3 later) signify differences from the classical view that add to those presented earlier in Table 19.2. Clement (2007a) has noted that such different types of analogy may be used to support different types of conceptual change. If so, an awareness of these different types may have important implications for teaching strategies.

20.1.2 Analogies and Imagery

The classical work on analogy does not examine the role of imagery. Chapters 15, 16, and 17 of this book began to examine the question of whether there was a link between analogy and imagery. In this section, I will examine that link further by reviewing findings on several analogy subprocesses that are explained parsimoniously by hypothesizing that they involve imagery and spatial reasoning. These include: spatial transformations used in analogy generation and evaluation, and dual simulations, also used in analogy evaluation.

20.1.2.1 Role of Spatial Transformations

A theme in this study that connects to the Gestalt psychologists is that of the importance of spatial transformations in advanced thinking. Chapter 11 explored their role in mathematical thinking. Section 17.7 documented the role of transformations

---

[1] In developing Tables 19.1 and 20.2, I considered including "simplifying cases" as one of the roles of analogy. However, this is often an effective substrategy for generating analogies, via a transformation, that can serve any of the above roles, especially by activating new schemas for analysis (cf. Figure 16.3). I also have not included 'projective analogies' as a type in Tables 19.1 and 20.2 because their ultimate purpose appears to devolve into either expedient analogy or an explanatory proto-model.

in analogy generation in science problems, and linked these processes to imagery indicators providing evidence that analogies could be generated via spatial transformations using an imagistic representation. Chapter 17 also discussed examples of evaluating an analogy via a transformation including: "rolling up" the bending rod to evaluate it as an analogy to the spring in Chapter 16; as well as cutting the torus in wedges and reassembling it in Chapter 11.

Since spatial transformations are recognized in the literature as a major process for perception and for reasoning via imagery, they signal an important way in which analogical reasoning can involve imagery. Cooper and Shepard (1973) found that subjects perform mental transformations such as spatial rotations over times that correlate in a linear way with the size of transformation required. Kosslyn (1980) has described a variety of transformations that play an essential role in perception. He reports on a large number of experiments that support the claim that these same transformations can be used with images as well. More constructive transformations where forms are modified and combined were performed by subjects in studies by Finke (1990). In the present study constructive transformations were observed as part of the process of using analogies. These were described by subjects as transformations of objects or actions in space, and almost all occurred as the subjects stared at a drawing of their own making. Several were accompanied by imagery indicators.

Wertheimer was prescient. These findings on transformations suggest that Wertheimer was on an important and precocious trail when he proposed that spatial transformations were important to cognition in mathematics and science. His solutions to the "area of the parallelogram" problem appear to me to provide paradigmatic cases of the conserving transformation strategy for analogy evaluation. In Chapter 17 evidence was presented for the use of conserving transformations to confirm analogies, not because there are high-level features that are identical in target and base, but because a conserving spatial transformation has been found that is recognized immediately as producing the base from the target, with the key targeted relationships conserved. This is hypothesized to be a different mechanism from the mapping of many discrete symbolic features. In summary, spatial transformations appear to be used in science as well as in mathematics and to be a first way in which analogies can utilize imagery.

20.1.2.2 Role of Dual Simulations and Imagistic Alignment in Imagery-based Analogies

The strength of the classical structural mapping approach is that it specifies procedures for elaborating and expanding a structural match between source and target by analytically breaking it down into individual relations and examining possible matches between those relations. For certain domains and situations this may be the best approach. The present study, on the other hand, examines the possibility of a more holistic approach to analogy evaluation based on imagistic matching and fit. This is more like the process of motor schema alignment, where

20.1 Analogy Findings, Part Two

for example, in order to grasp a large, very strangely shaped pitcher, one may have to explore different orientations and placements of the hand in order to grasp and lift the heavy pitcher successfully. This is a matter of "fitting" one's own schema for grasping to the new situation at hand. This is something like the process of fitting a schema for, say, the idea of torque and torsion in a straight rod of metal, to the deforming forces in a curved spring. Where and how to apply the forces and resulting deformations is not apparent at first, but by imagining different configurations for twisting, one eventually finds one that "fits." These examples carry a sense of imagistic alignment rather than 1-1 mappings of many discrete elements. I called this process the use of "dual simulations." This explains why the earlier list of major steps in analogical reasoning in Chapter 4 had "evaluating the analogy relation" as the second step, rather than "development of a mapping". Now I would label the step, "Aligning and evaluating the analogy relation" often imagistically.

Dual simulations were described in Chapter 17 as involving *perceptual/motor similarity comparisons* between two separate simulations, an anchor and a target. Dual simulations contrast most strongly with comparisons of discrete symbolic relations when they involve global, analog comparisons of a perceptual/motor nature. For example, S2 generates a "zigzag spring" as an analogy in the spring problem and makes an initial evaluation of it as follows:

> S2: (23) "And that would behave like a spring. I can imagine that it would... it would stretch, and you let it go and it bounces up and down. It does all the things."

Here, the conjunction of the dynamic imagery reports and the comparison of the two systems provided evidence for a dual imagistic simulation and comparison. The ability to compare two static images and identify similarities and differences is considered to be a fundamental process. A dual simulation is made up of two imagistic simulations, and perhaps utilizes a similar process.

Overlay simulations. Other data supporting the presence of dual simulations came from a phenomenon that puzzled this researcher for some time – the presence of "overlay diagrams" in which one system is drawn on top of another system. For example, S7 drew a pulley rope on top of a circle originally drawn to represent the problem about pushing a wheel up a hill. He proceeds to alternate between talking about the wheel and talking about the pulley while staring at the same "overlay" drawing. I accounted for this by assuming that the overlay drawing supports dual simulations; the drawings of the two cases overlap so that a perceptual comparison of their behaviors can be made easily. The subject imagines pulling on each one; and then imagining whether each will "work in the same way." Here, intuitions about the pulley appear to be projected and transferred onto the wheel.

In proposing a description of the process of dual simulation, I started from the process of inspection and perceptual similarity/matching for static images identified by Kosslyn (1980). A comparison made in dual simulation might use a similar type of matching operator, except that it operates on images of dynamic actions and events. This can be referred to as "dynamic perceptual/motor similarity." Like other methods it is not infallible, but in some cases it can support or reject hypotheses with confidence. Thus I hypothesized that this type of evaluation may be based on a distributed and parallel type of perceptual matching process, rather than a mapping of discrete symbolic features.

Imagistic alignment analogies. A final analogical reasoning process where imagery appeared to play a role was imagistic alignment analogies, often appearing as a type of bridging analogy. For example, the spoked wheel without a rim analogy in the wheel-push problem was interpreted as helping the subject align the placement of the fulcrum on the wheel in order to apply the original analogous case of a lever. In doing so this expands the domain of application of the lever schema to include wheels, so it also plays a "domain expander" role. In a similar way the bridges used between the book on the table and spring cases in Chapter 9 expanded the domain of the "elastic force (springiness)" schema to include rigid objects.

20.1.2.3 Coherence from Explaining Bridging and Overlay Drawings

In Chapter 17 the credence of the overall theory of imagistic analogy evaluation methods was supported by the realization that the somewhat enigmatic mechanism of the bridging strategy discussed in Chapter 4 could be explained by seeing it as an auxiliary strategy for enhancing the more basic process of imagistic alignment, or the imagistic evaluation processes of dual simulation and/or conserving transformations. That is, one of the functions of bridging analogies is to generate chains of pairs of cases that are "close" enough to compare in these more perceptual ways. Table 17.1 ties these methods together into a possible idealized process for evaluating analogies that coordinates the various strategies.

20.1.2.4 Imagistic Alignment, Evaluation, and Inference Mechanisms

In this perspective there are two fundamental mechanisms involved in analogical evaluation and inference: conserving transformations and dual simulation via imagistic alignment. It is possible that these could operate prior to mapping via discrete symbols, although the relationship between these processes and mapping is still unclear. Thus in the present view, imagistic alignment, conserving transformations, and dual simulations are not equivalent to mappings between discrete symbolic elements. Arguments were also presented in Chapter 17 that these evaluation processes can involve more than mappings of elements, even in an approach that assumes that some elements in a mapping can be considered to be imagistic. While at this stage there is certainly not a body of competing evidence that can compare in size to the extensive empirical work that has been done within the mapping symbolic features paradigm, the present studies highlight cases that raise the question of whether concepts different from mapping are needed to provide full and parsimonious explanations of the behavior.

20.1 Analogy Findings, Part Two

Table 20.3 How newly described analogical reasoning processes that can utilize imagery differ from the classical theory

Most previous studies and classical view – analogical processes identified: processes are explained via manipulations of a propositional representation	Present study – additional processes identified: processes are explained via imagistic simulations and transformations	Protocols provide initial evidence for imagery involvement?
Previously presented analogy is accessed by association as a familiar case	Subject spontaneously generates (sometimes novel invented) case	Yes
	Cases can be generated via transformation	Yes
Case needs no development	Sometimes case is developed further	Yes
Evaluation of an analogy relation via: mapping of discrete features	Evaluation of an analogy relation via:	
	Bridging analogies	Yes
	Conserving transformations	Yes
	Dual simulations (including those evidenced by "overlay representations")	Yes
Existence or purpose of bridging analogies not specified	Purpose of bridging analogies is to enhance conserving transformations, dual simulations or imagistic alignment	Yes
Transfer of or inference from abstract, symbolic relations	Transfer of concrete, schematic images or image generating schema elements for system behavior in projective analogy or for behavior of hidden mechanism in explanatory model construction	Yes
Computer simulation of theory possible	Current theory identifies major subprocesses	

20.1.2.5 Summary: Alternative Subprocesses of Analogical Reasoning

The main analogy subprocesses accounting for the case studies in this book are shown in Tables 19.1 and 17.1. Asterisked items in Table 20.1 and Tables 20.2 and all of 20.3 summarize dimensions along which the newly described analogical reasoning processes appear to differ from the "classical view." For each such dimension protocol data examples have been analyzed that contain initial evidence for imagery involvement. It was also hypothesized that several of these alternative analogical processes were also involved in the most productive insight documented in this book, S2's torsion insight in the spring problem. These findings suggest that further investigations of the above alternative mechanisms are an important problem for research.

20.1.3 *Analogies and Model Construction*

In this section, I examine linkages from findings on analogy to findings on the construction of explanatory models. Claims about the powerful role of analogies in scientific theory development have been made by notable figures such as Campbell (1920), Einstein and Infeld (1967), Polya (1954), Wertheimer (1959), Hesse (1966), Harre (1967), Gentner et al. (1997), and Nersessian (1984). The present study supports this view and proposes a variety of mechanisms by which such development can take place.

20.1.3.1 Expedient Analogies vs. Explanatory Models

Expedient analogies. Although simple expedient analogies provided a powerful tool for these subjects, it was apparent in the interviews that some of the subjects were seeking a stronger understanding by pursuing an explanatory model – a hidden process or mechanism that could be actually be operating in the target system. An explanatory model is more than an initial rough analogy. For example, in Chapter 14 we saw that the expedient analogy of the bending rod gave some predictive power, but limited explanatory power. It was not viable as a mechanism that could actually be operating in the target. Eventually the twisting idea was used to form the explanatory model of torques producing twisting in elements with precise orientations in the circular coil. Figure 14.15 illustrates this as a three-element view of explanatory model construction (as a particular example of the general relations shown in Fig. 14.16): (a) the anticipated behavior of a target case; (b) the anticipated behavior of an anchoring analogous case (also referred to as a source analogue or proto-model); and (c) the anticipated behavior of an explanatory model. In the present framework explanatory models are the ultimate goal, so expedient analogies have become lower in importance as an outcome than explanatory models, but the various types of analogies are still seen as enormously important tools for developing those models.

A simpler, two-element view is that the source of the analogy *becomes* the new model of the target and therefore the model need not be distinguished from the source analogue. The model *is* the source analogue, including its abstract relational structure. That symbolic structure is then shared by source and target. Prediction of the target phenomenon occurs via inference rules from higher-order predicates.

By contrast, in the present three-part view, aspects of the source analogue may be incorporated into part of the explanatory model being constructed, but the model is usually at least a modification of, and often a more complex structure than, any single anchoring source. The multiple anchors idea has support from the success of the "multiple analogies" strategy used in instructional contexts by Spiro et al. (1989), Steinberg and Clement (2002), and Glynn (1991). The imagery indicators cited in episodes of model construction in Chapters 13–16 support the sense in which the present view also contrasts with or complements language-based views of models as a semantic net or set of verbal descriptions. This is complemented by the present section on how analogies can utilize imagery and involve processes of simulation, projection, and imagistic alignment. In contrast to the symbolic structure approach described above, in imagistic modeling, explanation occurs via a compound simulation wherein an image of the mechanism operating is seen to cause or result in an image of the target operating in a way corresponding to the observed or given phenomenon being explained.

Returning to the legacy of Georg Polya (1954), analogy was one of the heuristic devices that he highlighted in writing about methods for solving hard problems in mathematics. We can use Table 20.2 to speak to the question of whether analogies should be thought of simply as one of many possible "heuristic devices" or tricks for jogging thinking into a new path in solving a specific insight problem – or whether they have more serious roles. The table implies that either can be true. A playful *expedient analogy* could remind one of some relevant schema or solution strategy for a puzzle, and this would be a one-time-only specialized use for that analogy. On the other hand, when analogy plays a role as a *proto-model* (source analogue) in the evolution of a scientific model (such as the twisting rod analogy for the spring), and aspects of the analogy become part of the model, this is a central and lasting role, even though the analogy does not constitute the entire model. This can enrich the model with new material components that form an imageable mechanism. In this case aspects of the analogy can become part of the core meaning of the explanatory model, and it becomes more than a heuristic used to suggest a new operator for the problem. At the end of this chapter, I will summarize the argument from Chapter 18 that analogies that act as proto-models can be a source of flexible runnable knowledge schemas that can make a scientific model powerful. In this way analogies can provide a pathway for grounding scientific knowledge in confident and runnable prior knowledge schemas. Thus Table 20.2 indicates that the relationship between analogies and models is more diverse and complex than many previous studies have indicated, since any of the roles of analogy listed there can play a role in developing an explanatory model, or laying the groundwork for one. Many of these roles were also represented in Figs. 16.4 and 16.8.

Going beyond the observational definition given in Chapter 3, one can then provide a condensed summary of the present view of analogy in cognitive terms as: Given a target problem that involves predicting and/or explaining a phenomenon, the subject creates or accesses a significantly different related case with the aspiration that it is still similar enough to the target problem to generate relevant ideas that can be applied to the target. Such a case can suggest prior knowledge or strategies or perceptual/motor alignments to apply to the target. When successful, this can yield a prediction for the target or it can feed a generative abduction process for constructing or revising an explanatory model of the target as a third entity. Subjects apply various evaluation processes to detect relational or imagistic similarity that is causally relevant in order to decide whether the analogy is valid or decide whether the resulting model is viable.

A major challenge in this study has been to improve the precision with which the term analogy is used. When one examines the spontaneous creative thinking of experts, rather than controlled or forced choice experiments in the setting of the laboratory, the variety of types of related cases produced is extremely rich. It is tempting to treat all of these as analogies but this would lead to an overblown and fuzzy concept of analogy. In retrospect, for me the concept of analogy has been narrowed in some directions but expanded in others while developing this book. The possible purposes of analogy have broadened greatly from my original conception of it as a heuristic device. Another major expansion of the concept of analogy has come from the inclusion of analogies resulting from target transformations that may change only a single feature of the target, but still produce a significantly different case. On the other hand, a major narrowing of the concept has come from explicitly differentiating analogies from partitionings, explanatory models, and Gedanken experiments.

Summary of analogy findings, Part Two. In summary a collection of new observed patterns of analogy use have been presented along with some new imagery-based processes designed to explain them. The new observations include novel analogous cases generated via spatial transformations, bridging analogies, overlay simulations, and others in Table 20.3. In some cases analogies can play a very central and substantive role in model construction as a source of essential content, meaning, and flexibility for a model by passing on the capacity to generate imagery. These findings provide an answer to the question of why expert subjects go to the trouble of using difficult-to-construct analogies when the effort is high and the payoff uncertain. However, less work has been done here on the role of language. Although an attempt has been made to identify important alternatives to mapping discrete features where they seem to be justified by protocol events, I do not wish to deny the additional power of naming imagined variables linguistically and mapping them to another linguistic description. This can help take a subject's analysis to a higher level of precision and explicitness, and can make it more amenable to further precise evaluation. Classical theories of analogy appear to be based exclusively on linguistic or propositional symbols. Eventually a synthesis will be needed to integrate the role of both imagery and language in analogy use and model construction. For example, an explanatory model might be fixed in memory by a labeled pictorial diagram, where the different labels help to differentiate closely similar imagistic features that might otherwise be confused. Conversely, carefully constructed schematic images may help disambiguate linguistic terms that might otherwise be confused. Thus my hope is that the constructs developed here will stimulate further research on how these types of representation may be complementary.

20.2 Imagistic Simulation Findings, Part Two: Thought Experiments and Their Uses in Plausible Reasoning

20.2.1 Overview

Part One of the book introduced the higher-level process of scientific model construction and the role of analogy therein. Part Two introduced the lower-level process of imagistic simulation and the role of perceptual/motor schemas and dynamic imagery therein. Beginning in Chapter 15, I attempted to merge these two themes to examine the role imagistic simulation might play within analogy and model construction. In these contexts, evidence for imagistic simulation often appeared within *untested thought experiments* (in the broad sense) in which subjects made predictions about the behavior of concrete systems they had not previously experienced. In this book most of the imagistic simulations I chose to focus on were thought experiments, because most occured in new and unfamiliar circumstances. This is a worthy test of the power of imagistic simulations since they must then produce real inferences in strange surroundings. A major conclusion of that section was that analogy and model construction often involve subprocesses that can be considered untested thought experiments (such as running and making predictions from a novel, invented analogous case, or from a novel, invented, explanatory model), and these in turn often depend on imagistic simulations.

Thought experiments were considered to be an interesting process in themselves because of a problem that continues to challenge cognitive science, which I call the fundamental paradox of thought experiments: "How can findings that carry conviction result from a new experiment conducted entirely within the head?" Where does this power come from? Some of these "experiments" can actually provide results powerful enough to strongly discredit an explanatory or mathematical model. Even thought experiments that do not generate this strong a result, are subject to the same paradoxical question.

20.2.2 Summary of Findings on Thought Experiments

Previous work on thought experiments has hypothesized that they may play an important role in scientific discovery and evaluation. Nersessian has theorized that they may involve the use of imagery and mental simulation. Although the historical role of thought experiments as a scientific tool have been discussed by Kuhn (1977d), Nersessian (1992), Gilbert and Reiner (2000), Brown (1991), Gooding (1994), Koyre (1968), Sorensen (1992), and others, what has not been previously examined systematically are the underlying cognitive processes nor the specific roles that thought experiments can play in analogy and model construction on the basis of analyses of think aloud protocols. In addition, the question of *how* thought experiments utilize imagery, mental simulations, or mental schemas has not been sorted out. Building on the model of imagistic simulation in Fig. 13.3, an attempt was made in Chapter 15 to provide initial empirical grounding for a set of distinct meanings for the concepts of "untested thought experiment," "evaluative Gedanken experiment," "analogy," and "explanatory model," and an analysis of how they each can depend in different ways on imagistic simulation. The theory of how untested thought experiments are run presented in Fig. 15.3 offers a description of the sources of conviction in such experiments that addresses the fundamental paradox of thought experiments. In particular, Chapter 15 attempted to:

- Base these findings on evidence from think-aloud case studies rather than an analysis of historical records, with the finer level of detail that makes possible.
- Connect the hypothesized presence of generalized perceptual/motor schemas to empirical evidence from hand motions and other indicators; and to previous work on generalized motor schema theory.
- Trace the origins of new knowledge and sources of conviction found within the individual simulations themselves. Imagistic simulations can generate new findings by way of mechanisms termed "extended schema application," conversion of implicit to explicit knowledge, spatial reasoning, symmetry, and (in conjunction with one or more additional schemas) compound simulations. Thus the origins of conviction in thought experiments can come from not one but several possible sources, as shown on the left-hand side of Fig. 15.3. This explanation provides an initial response to the thought experiment paradox. In retrospect, the sources of conviction represented on the left side in Fig. 15.3 are surprisingly diverse. They indicate that the answer to the thought experiment paradox may come not from a single primary source, but from many possible sources, including knowledge structures, reasoning processes, and prior learning processes.
- Provide additional supporting evidence for certain sources of conviction from the phenomena of imagery enhancement, where subjects actually make an effort to invent cases that increase the effectiveness of sources such as implicit knowledge and spatial reasoning. Instances of generating a simpler problem or extreme case with imagery indicators like hand motions were explained most plausibly and parsimoniously by the present theory of imagistic simulation.
- Distinguish between the sought prediction as the primary outcome of the simulation and important secondary outcomes such as new image features, coherence and dissonance relations, and schema activations, shown on the right-hand side of Fig. 15.3.

Individual occurrences of the latter additional benefits are somewhat unpredictable, and this means that TEs can be considered somewhat "volatile" in producing unanticipated effects. This was interpreted later as one source of creativity and divergence as well as criticism through dissonance, helping to explain how there can be sudden breakthroughs in an investigation.

Chapter 15 went on to place imagistic simulation in the context of a broader theory of a more complex reasoning by:

- Formulating distinctions between four different modes of plausible reasoning in which thought experiments can occur (shown in Fig. 15.6)
- Presenting case study evidence that each of these reasoning modes can derive part of its predictive power and conviction from one or more imagistic simulations
- Hypothesizing that these different modes have different mechanisms for utilizing the power of imagistic simulations to generate new knowledge

Thus I attempted to show how some major scientific reasoning processes for model building could be based on imagistic simulation as a subprocess. The finding that thought experiments occur within such a wide range of reasoning operations means that the fundamental paradox applies to a broader set of processes than is commonly realized.

20.2.3 Broader and Narrower Categories of Thought Experiments

I also identified a special kind of thought experiment in the narrow sense: evaluative Gedanken experiments. In all thought experiments the subject makes a prediction for an untested, fairly specific concrete situation, but an evaluative Gedanken experiment is also designed or selected by the subject to help *evaluate* a scientific concept, model, or theory (e.g. an explanatory model and/or its mathematical elaborations). I believe both the broad and narrow concepts of "thought experiment" as defined here are useful, and both can be analyzed in think-aloud protocols. The broad concept is appropriate for expressing the fundamental paradox. The narrower concept of an evaluatory Gedanken experiment encompasses some famous TEs in the history of science, impressive in that they can even contribute to eliminating an established theory. *Documenting such cases indicates that nonempirical creative methods can be used for theory evaluation, not just for theory generation* (Clement, 2006).

If we take evaluatory Gedanken experiments to be the kind of thought experiments that have been of most interest to philosophers of science, then there are interesting comparisons and contrasts here to the view of Kuhn (1977d). He theorized that thought experiments can work only in domains where one has relevant real-world knowledge and experience. This is compatible with the present view that emphasizes the role of prior knowledge schemas in such experiments, although I believe it is important for these schemas to extend beyond their normal domain of application, in order to count the case as an untested thought experiment. I have also pointed to how the coordination of multiple schemas in new combinations and with spatial reasoning can contribute to thought experiments going beyond the bounds of prior experience, a view compatible with that of Gooding (1993). In addition, it is unclear whether TEs that rely on symmetry arguments are empirically based. Thus in the present view, thought experiments can involve prior knowledge schemas that were built up historically from empirical experiences; but they may also involve reasoning that produces predictions that go well beyond empirical knowledge.

Kuhn also theorized that the role of thought experiments was to *disclose a conflict* between one's existing concepts and nature. This is compatible with some of the examples presented in Chapter 15 where evaluatory Gedanken experiments had a negative outcome, but others, such as the torsionless coil (episode 13a) and radically stretched ribbon coil (episode 14) experiments had *positive* outcomes – providing some support for the importance of torsion or existence of twisting respectively (see also Brown and Clement [1992] for historical cases of positive TEs). In sum, the documentation of thought experiments in expert thinking deepened the result from previous sections that important nonempirical processes occur in scientific thinking.

20.2.4 Can Thought Experiments Allow One to "Get the Physics for Free?"

Once a thought experiment is defined, in many cases the process of "running" the experiment to yield a result is observed to occur very quickly. This suggests that the process used takes surprisingly little effort and this may feel to the subject like "Getting the physics for free." In addition these experiments can yield results with high levels of conviction where the answer appears to be "obvious" with no evidence in the transcript for formal inferences via deduction or via induction by enumeration. In the literature on diagrammatic reasoning, Barwise and Shimojima (1995) describe the possibly related process of getting a "free ride" when drawing a new diagram for a problem makes subsequent inferences particularly easy (with the suggestion that the same thing can happen with internal imagery). What can the present theory say about this issue? The response is not so simple because the knowledge derived from an imagistic simulation in a thought experiment can have a variety of sources and outcomes according to Fig. 15.3.

It appears that the closest thing to "getting the physics for free" in that figure would occur in obtaining implicit knowledge from an elemental imagistic simulation driven by an intuition schema. For example, in the "twisting rod" experiment of Fig. 13.2, the subject seemed to be able to "consult his motor system" in a fairly direct manner, yielding an immediate answer of high certainty. Although he still needed to formulate the right question and design the right case to think about, during the actual running of the 'experiment' the result emerged with surprising speed and effortlessness. Perhaps running it took some minimal processing time, but for this part of the process the physics was impressively easy to get, if not totally "free."

"Virtually for free" or "very inexpensively." One should be able to generalize from this example involving implicit knowledge to other cases involving the other main sources of conviction in thought experiments identified in Chapter 15: the extended application of a schema to a case outside its normal domain of application, running compound simulations with two or more schemas, and spatial reasoning. These processes can also be fast and efficient once the subject has formed an image of a subproblem that is solvable via one of these processes. It is hypothesized that one has the sensation of getting new results for free when there is a good match between plentiful,

20.2 Imagistic Simulation Findings, Part Two

fast, nonformal mental resources available and the subtask at hand, e.g. when the problem can be solved quickly by spatial reasoning.

I conclude that one *can* "Get the physics virtually for free" during the running of a thought experiment in the sense that it can occur remarkably quickly and effortlessly by using plentiful, efficient, and therefore inexpensive resources.

There are a number of caveats to this "virtually for free" statement, however: In the case of the twisting rod, reducing the original problem to that question was an effortful and complex achievement – an aspect that was not at all for free. After all, that event occurred almost 40 min into S2's protocol. Thus preparing for the running of the experiment was expensive, even though running it was not. In general, the caveats are:

- Much preparation in formulating the right question and case can be involved, including designing the thought experiment itself.
- A good deal of prior work may have gone into building up the resource used to run the experiment, e.g. building up a motor skill or spatial reasoning skill during childhood.
- A good deal of effortful reasoning may have to occur *after* tapping into the "free" knowledge in order to apply or refine it, that is, the experiment may only address a problem analogous to the target.
- In some cases, the running of the experiment itself may require nontrivial effort, if this strains the capacity of the imagery system or involves the coordination of many schemas on the same image. Some subjects running imagistic simulations were seen to be working quite hard to apply spatial reasoning and new combinations of prior knowledge to make inferences about new situations – thereby obtaining the new knowledge for considerable effort rather than "for free," e.g. the "paint dots on the spring" case in line 22, Section 13.3.5. Thus although some thought experiments feel effortless to run, others feel quite effortful. (However, for effortful cases the subject may try to improve the design of the experiment so that this is not an issue and it becomes a "better" thought experiment.)

These caveats make examples of "getting the physics for free" more like a "surprisingly large dividend gained suddenly from an earlier investment" than something gained for free. And even in the cases involving the least work it would be more accurate to say "getting the physics very inexpensively" or getting "an inexpensive ride." Thus in some cases, during the moment one utilizes one of the sources of knowledge in imagistic simulations, such as converting the knowledge implicit in a perceptual/motor schema to explicit knowledge, it may be that subjects are getting the physics virtually for free. But preparing for, accessing, and then utilizing such knowledge may be quite effortful.

20.2.5 Section Conclusion

Chapter 15 presented evidence that thought experiments and imagistic simulations can be involved in expert scientific thinking in a large number of different ways, providing further support for the important role played by imagistic simulation in expert cognition. It also presented evidence for multiple sources of conviction within imagistic simulation that addresses the thought experiment paradox. Combined, these sources and patterns of use constitute the beginnings of a theory of how thought experiments and imagistic simulation work and how they can contribute to creative thinking/scientific discovery. Questions remain about details in this theory: exactly how spatial reasoning is used, what the limitations of simulation are, how new variables or image features emerge, etc. Thus the account given here does not fully resolve the paradox, but it does provide an initial explanation that can be evaluated further and revised. The origins of new knowledge in thought experiments is thereby seen to be a multifaceted, difficult, and fundamental topic, and further research in this area is very much needed.

20.3 Model Construction Findings, Part Two: An Evolutionary Model of Investigation Processes

20.3.1 *Top Level of Scientific Investigation Process*

In the previous major section of this chapter, I briefly described how Chapter 15 traced an upward path from basic to more complex processes, starting from the basic process of elemental imagistic simulation and summarizing how it was used within the four major nonformal reasoning operations in Fig. 15.6. In this section, I summarize how those reasoning operations are in turn combined to produce a larger scientific investigation. The earlier study of a single subject's solution in Chapter 6 had argued against an overly inductionist view of the source of new hypotheses in investigations. The analyses in Chapters 14, 15, and 16 focused on the same "Spring Problem" but used more protocol data from other subjects and included data on depictive hand motions and other imagery indicators. This enabled the development of a more comprehensive picture of noninductive investigation processes that could account for the following solution stages.

20.3.1.1 Stages

Five stages of analysis were described in Table 14.2 10a, showing a gradual transition from qualitative to quantitative modeling:

1. Efforts to develop an initial *qualitative description or prediction* for the targeted relationship
2. *Searching for and evaluating initial, qualitative explanatory model elements*
3. *Seeking an imageable model that was fully connected spatiotemporally*
4. *Seeking a geometric level of precision* in the spatial and physical relationships in the model
5. *Developing a quantitative model* on the foundation of the improved qualitative/ geometric model

The construction cycle shown in Fig. 6.3 was proposed earlier as a way to explain alternating periods of evaluation and construction in the transcripts. However, although it provides a core idea, a single process of this type cannot account for a pattern as complex as the one above involving five different stages.

20.3.1.2 Three-part Model of Scientific Investigation Processes: Explanation vs. Description

Therefore the following framework for a theory of explanatory model construction was developed to account for the composite protocol in Chapter 14. The three-part top level of organization for the framework is shown in Fig. 16.4, including description (accounting for stage 1 above), explanation (accounting for stages 2 and 3), and mathematization (accounting for stages 4 and 5) Processes.

This helps explain one of the tensions I experienced in writing Chapter 6 on model construction from analogies. There I lumped together all analogous cases and explanatory models under the single undifferentiated concept of a "scientific model," defined broadly as a mental representation of a system that can predict or account for its structure or behavior and that has a basic level of plausibility that rules out, for example, occult properties, and requires that the model be internally consistent (not self-contradictory). A single generation, evaluation, and modification (GEM) cycle was used to describe the evolution of such models. Starting at the very end of Chapter 6 however, I began to distinguish between analogies and explanatory models as two different types of scientific models. This motivated the distinction between the first two of the three processes shown in Fig. 16.4.

An investigation in that figure starts when the question is raised about a targeted relation R between X and Y in a target case: Does X influence Y and if so, how? In essence, whereas the Description Process I simply aims to predict the relationship between factors X and Y as confidently as possible, the Explanation Process II aims to develop an explanatory model – one or more new causal relationships as hidden intermediate causes between X and Y, as shown in Fig. 16.5b. The Mathematization Process III then builds on the output from either Process I or II to produce more precise representations that can approach or reach a quantitative level.

Thus there is what appears to be a familiar distinction between handling observational concerns at a surface level within Process I and handling theoretical concerns at a hidden mechanism level with Process II. However, in Process I something like "observations" can also take place internally in imagistic thought experiments, so there are some ways in which these processes are *not* cut along traditional lines. In fact, one of the surprising results of the study is how much of the investigation[2]

---

[2] After developing the model of investigation processes involving separate prediction and explanation processes, I noticed that there is some similarity to a finding in a rather distant domain: Karmiloff Smith and Inhelder (1975) found that 5-year-olds appeared to shift between problem solving and explanation goals in working with toy blocks that exhibited novel behaviors. This may be analogous to the shifts between the Description Process I and The Explanation Process II in Fig. 16.4 suggesting a possible connection to roots of scientific behavior.

can be handled using imagistic processes as opposed to empirical ones. I will comment on each of these three processes.

20.3.2 Process I: Description Cycle

Using Cycle I, subjects could find multiple methods to make predictions for observable behaviors in the target via techniques like direct imagistic simulations of the target, extreme cases, and expedient analogies. The analogies used were subject to evaluation and improvement. Agreement between such multiple methods was described as a major source of confidence by Polya (1954) for making predictions.

As discussed in Chapter 16, Griffith et al. (2000) have written a set of detailed procedures for accessing, evaluating, and revising analogies to the spring problem, based on transcript data that we shared with them. The program accounted for a considerable number of the transformational moves made in generating analogies. Almost all of these fall into the Description Process I in Fig. 16.4. The Griffith study succeeded in showing that certain spontaneous analogical processes in creative thinking can be modeled and evaluated at a high level of computational detail. Their largest self-described limitation was the inherent difficulty in formulating a computational approach to visualizable modeling that included the construction of new understandings (as opposed to predictions) through mental simulation. The present study attempts to model these aspects, as well as thought experiments, other analogical processes, contained divergence strategies, and to some extent, mathematical models, as follows.

20.3.3 Process II: Explanatory Model Construction

Evidence for subjects seeking understanding by going beyond the Description Cycle I appears in data from the composite protocol that indicates that subjects sought explanations of underlying mechanisms, not just predictive descriptions of target behavior. Most of Chapter 16 focused on Process II: Explanatory Modeling. As pictured in Fig. 16.4, its central feature was a cycle of model generation, evaluation and modification (GEM cycles). This means the generation of partial or incomplete models on the way to a new theory should be a normal occurrence. (Cycles of empirical testing and modification are acknowledged in every description of scientific method. However, the present view also includes nonempirical GEM cycles using rationalistic forms of model evaluation where no new data is collected.) This discussion grapples with the question of how extraordinary creative scientific reasoning is – must it be a sweeping, unnatural, Eureka-style act of synthesis? The argument from detecting GEM cycles is no – a new scientific model can be evolved more gradually in many steps. Although this answer to the Eureka question was first given in Chapter 7, it now has a deeper meaning because protocol explained by Figs. 16.3 and 16.4 shows GEM cycles occurring in each of the three

20.3 Model Construction Findings, Part Two

major processes. Thus the idea of using an *evolutionary* GEM cycle as the foundational organizing principle for the investigation process has been generalized to explanatory and mathematical model development, in addition to analogy development, extending the argument against the pure Eureka view.

Two additional diagrams, Figs. 20.1 and 16.8, describe Process II in Fig. 16.4 in more detail, but they do so differently. Four levels of processes are shown in Fig. 20.1; each level draws on processes in the level below it:

Level 4: Runnable model application and comparison
Level 3: Explanatory model construction
Level 2: Advanced nonformal reasoning processes
Level 1: Schema-driven imagistic simulation, intuitions, imagistic transformations

These correspond to topics in major sections of this book. As a first-order model, Fig. 20.1 highlights distinctions in the theory between a number of concepts that

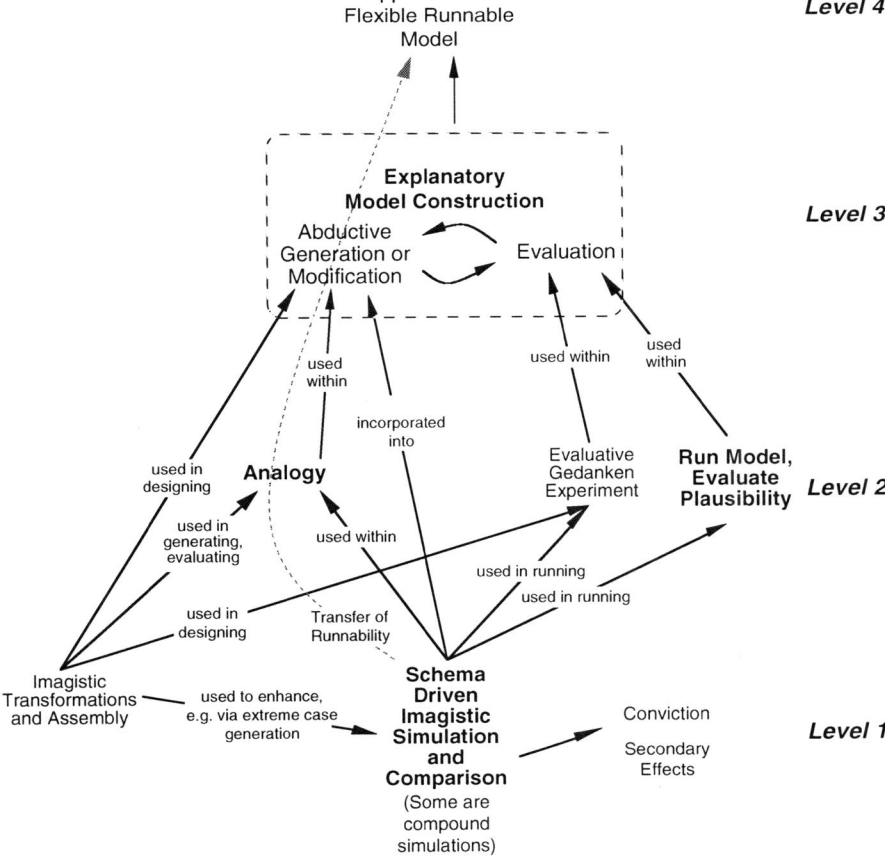

Fig. 20.1 Some major nonempirical subprocesses in the construction of flexible models

have been confounded in the history of the field: using imagistic transformations in design vs. using simulations, forming explanatory models vs. running the model, and using analogies vs. using evaluative Gedanken experiments. As shown there, each of the latter three processes can utilize schema driven imagistic simulation as a subprocess, as evidenced in the previous major section of this chapter and Chapter 15, particularly in Fig. 15.6. In this view, far from being an auxiliary technique or "frill," imagistic simulation plays a foundational role in the investigation process. Also, any simulation can also lead to unexpected secondary effects, such as activation of new schemas, or the formation of new conflicts or coherence relations with prior knowledge schemas, and some of these are shown in Fig. 16.8. General, constructive, imagistic transformations are shown as contributing to the generation or modification of models and analogies, and even to the enhancement of elemental simulations. They are involved in imagined actions like cutting, deforming, or reshaping an object, or joining multiple objects in order to form new elements. Relationships between these diagrams express important aspects of the structure of theories in this area, and this author has found it helpful to examine a xeroxed collage of the latter four diagrams.

Information flow over time during model construction. Figure 16.8 unpacks Fig. 16.4 part II in a different way by showing the explanatory modeling processes operating over time from left to right. It emphasizes the many interactions between processes by outlining the flow of information. By showing empirical inputs at the bottom (inputs not actually used in these protocols) and prior knowledge schemas at the top, it shows the potential interplay of top down and bottom up influences. And by showing alternating periods of model generation and model evaluation along the horizontal dimension, it emphasizes an expert's impressive ability to think divergently and convergently in alternate fashion (cf. Kuhn's [1977a] early essay) Multiple sources of ideas feeding the model generation/modification process make it potentially extremely divergent. However, multiple sources of evaluation with possible dissonance are what give the expert the potential to reign in this divergence. These sources of evaluation are depicted in Fig. 16.8, and include: running the model on the target to check whether it explains known behaviors of the target, including accumulated constraints on the target; Gedanken or real experiments; general criteria for scientific models; and compatibility with one's assumed theoretical framework. The vertical dimension depicts top-down influences from above as well as bottom-up influences from observations at the bottom of the figure (although these are shown in parentheses since they were not available in these interviews.) Evaluative cycling means the many heuristics that are "weak" in the sense that they can fail, are still worth trying, since the results can be rejected if faulty. And one initially does not have to find a "perfect" or close analogy or schema–they may be modifiable to a make a contribution to the model.

Thus by setting up a "battle royal" between generation and evaluation, with each side fielding multiple sources of ideas, one creates the possibility for work that is creative and yet valid. This is the rationale for the potential of a spirited dialectic (in this case within one person, but extendible to more). Overall, a substantial part of the reasoning used is nonformal, the outcomes are not predictable,

20.3 Model Construction Findings, Part Two

and a substantial part of the base representation is non verbal. The theory posits that a predominant part of every element in Fig. 16.8 can be imagistic. This is a very different picture than that of a carefully controlled, logical method of model construction, that works within the tight boundaries of linguistically frozen abstractions and rules.

Abductive model generation. But this still leaves open the difficult question of how the system makes an initial model and enlightened improvements in each cycle, the problem of the nature of generative abduction. In reviewing several instances of model construction I argued that the construction process was essentially abductive rather than deductive or inductive. But the problem is that abduction has been an unstable concept in our field, with many meanings. I have tried to clarify its meaning(s) here via diagrams. I distinguished between two meanings for the term "abduction," as shown in Fig. 16.1 (Magnani, 1999):

- "Generative Abduction" (the narrower sense): refers to the formation of explanatory hypotheses during the act of hypothesis generation or revision (but not evaluation).
- "Abductive Model Evolution" (the broader sense) refers to inference to the best explanation, including hypothesis generation, evaluation, and revision cycles plus resolving competition between rival hypotheses.

Abduction in the broad sense is depicted by all of the process in Fig. 16.8, plus model competition. (For some purposes, resolving competition between models can be considered a third, highest level of abduction.) Generative abduction in the narrower sense is depicted inside the dotted box there. (Details were given in steps IIA1–3 in Fig. 16.3.). Figure 16.4 also shows what abduction is not. If, along with many authors, we take abduction as an act of explanation, then that is associated with Cycle II rather than Cycle I. It is not simply prediction from an extreme case or expedient analogy, because those heuristics on their own do not explain a phenomenon in the way that an explanatory model does.

Explanatory modeling inherently difficult. But why bother to go through model construction cycles; why not build the model in one step? The model of scientific investigation shown in Fig. 16.8 responds to a number of inherent difficulties in the problem of constructing a complex explanatory model:

- One lacks resources for either the deductive application of first principles or the inductive formation of a pattern. This leaves abduction as a form of educated guessing as the primary option.
- Abduction is a "weak" strategy in the sense that it is always possible to propose more than one explanatory model for any given phenomenon, and any particular model generation or modification action may later be shown to be invalid.
- Model construction involves the design of a new system to satisfy a large number of constraints at the same time. Working memory limitations may prevent the consideration of all constraints at once. Other essential resources in LTM may only apply in unfamiliar ways and not be activated all at once. This

means one may have to attack only a part of the design problem at a time, and improve the model incrementally over time.

Multiple modeling cycles are needed to complement a weak abduction processes in the learning of complex models. These factors explain the frequent need for an evolutionary approach over several cycles, rather than a single insight. As in writing a complex program or paper or book, by dealing with only a few difficulties at a time, and going through several rounds of revision, a fellow mortal may be able to construct and debug a very complex model. I will expand on this theme in the next chapter in discussing how many of the nonformal strategies being discussed are "weak" – but can actually form a strong system when used together because of their special interactions.

Generative abduction. The process of generative abduction is poorly understood, but I did present some evidence that:

- Subjects were able to generate and evaluate abductions about possible deformations causing stretching in the spring.
- Figure 16.8 makes explicit how multiple constraints from several sources must be kept in mind and how previous results from evaluations can inform the next process of revision.
- A central problem in abduction is accessing the right schemas from prior knowledge to use as building blocks in forming the model, and a major role of analogies in generative abduction was suggesting relevant schemas. In the generative part of the process (within the dotted boxes), analogy can be used, and this has been particularly well documented by Gruber (1974), Millman and Smith (1997) and Nersessian (1984, 2002) in history of science studies. However, the present analysis differs in adding the distinction (adapted from Hesse, 1966) between expedient analogies for behavioral description (used in Process I) and explanatory source analogues for explanatory modeling (used in Process II). The latter often depended on transfer of imagery and runnability in the present protocols and thus played a role in developing the core meanings of a new explanation for subjects.
- There was some subtlety, however in thinking for example, about whether the bending idea should count as part of an abduction. I came to the position that S2 *does not know* early on whether the bending rod is simply an "accidental" expedient analogy, or a deep source analogue for the core mechanism of the spring. It can be viewed as a prototypical exemplar of a schema that forms the core of an abducted, initial explanatory model for the spring. In this view it is part of an abduction, but that abduction is immediately questioned. Much of his energy for attacking the problem appears to come from the tension in that question. Although the bending model is later rejected after a long, intense period of evaluation, S2's engagement in this process sparks many new ideas and leads eventually to the torsion insight. This prompted me to conclude that the worth of an initial model lay not so much in whether it was correct as in whether it had creative heuristic value in starting a productive cycle of evaluation and modification.

20.3 Model Construction Findings, Part Two

- Items within the dotted box in Fig. 16.8 then illustrate the sense in which generative abduction is a nexus or meeting ground for two opposing forces. On the one hand, a number of sources of divergence and creative ideas threaten an explosion of possible ideas for models. In some ways it appears alarmingly open ended like the process of biological evolution under high rates of mutation (that leads only to the generation of monsters!). On the other hand, the number of sources of constraints on such a model could make finding a satisfactory model very difficult, using a system with limited working memory resources. Thus Fig. 16.8 as a whole highlights the potential difficulty of the explanation (generative abduction) process. How it escapes this fate through a structured but flexible control system and contained divergence will be a primary topic in the remainder of this book.

Abductive model evaluation processes: running the model and Gedanken experiments. In the present view, to make up for this fallibility, generative abduction is coupled with a model evaluation process. Together they form a cycle of evaluation and revision that comprises abductive model evolution in the broad sense. A very simple but important way of evaluating a model for basic adequateness was documented as running a new model to produce an imagistic simulation of the target. In one case this process produced an important, dissonant anomaly, and in other cases led to estimates that the contribution of the mechanism would be negligible compared to others. A second very impressive evaluation process was identified in evaluative Gedanken experiments, exemplified by cases such as springs with virtually infinite resistance to bending or zero resistance to torsion. Features of the interaction between generative and evaluative halves of the abduction cycle included:

- Difficulties raised in the evaluation step are passed forward and addressed in the succeeding modification step as constraints.
- Running thought experiments during evaluation can activate schemas and generate unanticipated conflicts and coherences.
- New emergent features of the target can be noticed, even in thought experiments, and these are also passed forward as constraints (e.g. each segment of spring remains horizontal in orientation despite twisting).
- Another dilemma Fig. 16.8 speaks to is how one can maintain some coherence with prior knowledge in science while breaking away from current views to generate new theory. Generative abduction cycles are seen as the focal point that achieves this by using analogies as starting points for models and transformations for improving them, while keeping known constraints and theoretical frameworks in mind. Such techniques build many connections to prior knowledge, in contrast to a random guessing strategy.

The resulting process of generative abduction is fairly complex. Even within the Explanation Process II, there are three types of model refinement that can operate in repeated cycles:

- The basic GEM cycle which proposes, evaluates and modifies an explanatory model M for a relationship R (X, Y) until it attains basic plausibility

- Increasing the level of precision for M by asking more refined questions about details in the imagery and alignment of the model, eventually preparing it for mathematization. Having a model imagistically aligned with a target so that it forms a full spatio-temporally connected model was identified as a new level of precision for a qualitative model that is a prerequisite for developing a mathematical model.
- Increasing the depth of explanation by introducing more intermediate variables such as R_4 and R_5 in Fig. 16.5

Section summary. The processes depicted in Fig. 16.8 appear to take us at least part way toward an imagery based theory of abduction for explanatory model construction. It is a process of assembling a composite image of a mechanism using new combinations of ideas from the resources shown. The mechanism can then be run using the process of compound simulation. It incorporates alternating passes through an abductive generation or modification process (using analogy and imagistic transformations as major subprocesses) and a critical evaluation process (using the running of the model as well as Gedanken experiments as subprocesses). Figure 20.1 illustrates how imagery can be seen as pervading the smaller reasoning modes used within this larger evolutionary model construction process. The issue of an overall control structure for this process that can also account for occasional insights that interrupt and punctuate the evolution process, and the problem of modulating the degree of divergence in different phases, will be discussed in Chapter 21.

20.3.4 Process III: Mathematical Modeling

20.3.4.1 From Qualitative to Quantitative Models

What is the relationship between qualitative and quantitative modeling? Is there a discernible line separating them or are they better thought of as living on the same continuum? There are many sides to this issue and the studies in this book yield a more complex viewpoint that goes beyond a simple yes or no. Intermediate stages between qualitative and fully quantitative were identified. The solution to the spring problem presented in Chapter 14 showed the possibility that nonformal methods could be used in mathematical analysis as well, such as the use of analogies, extreme cases, and imagistic simulation at every level of precision, including those determining mathematical proportions. This implies that although we can separate qualitative and quantitative modeling in theory, they share many common subprocesses.

Intermediate levels between qualitative and quantitative. The presence of five rather than two different stages in Table 14.1 indicates that the progression is somewhat more complicated than a simple dichotomy between qualitative and quantitative. I noted geometric and quantitative stages as the last two of these five stages, and refer to these collectively as mathematical modeling, in contrast to the qualitative modeling occurring in the first three stages. The distance between a completely spatiotemporally connected qualitative model and a model with geometric precision is not great, and these soften the line between qualitative and quantitative. They also illustrate how the imagery of the mathematical model can be constructed as an extension of the qualitative one.

Is mathematical modeling fundamentally different from qualitative scientific modeling? My arguments said yes and no. Mathematical models can use different representations (e.g. equations), and tap different prior knowledge schemas (e.g. functions). Mathematics can make extensive use of deductive arguments, many of which are compressed into standardized algorithmic procedures, and mathematical models can add considerable precision. But, there is perhaps far more overlap between qualitative and mathematical modeling than is commonly realized, specifically with regard to major nonformal reasoning processes and imagistic representations that are used in both. In the *construction* of the mathematical model, both the geometric and quantitative stages made extensive use of nondeductive analogical problem transformations and thought experiments, the same types of plausible reasoning processes that were used in developing the qualitative model. The image developed of the qualitative model can be elaborated and made more precise in the image of a geometric model, and measures and quantitative relationships can be attached to this image in a quantitative model. This leads me to view them as separable but similar processes that can be closely coordinated. Thus a mathematical model for a theory can be thought of as an extension of the qualitative model. While many physicists tend to view an equation as the heart of a theory, in the present view the fully connected qualitative model is the heart, and the mathematics provides the details.

20.4 The Important Role of Imagery in the Expert Investigations

In this section, I will first review some of the potential limitations of imagery as a representation. I will then review evidence that imagery nevertheless played a very important role in the expert protocols analyzed. Two processes will play a central role in this survey: imagistic simulations and spatial transformations.

20.4.1 Limitations of the Imagery and Simulation Systems

I have assembled evidence in this book for the use of imagery in many different aspects of scientific investigation. However, there are theoretical limitations on the imagery system that restrict its use. For scientists working in a complex new area, the imagery can be complex and unfamiliar, and they may be unable to construct complex imagistic models all at once. In particular, I hypothesize that the use of imagery has at least the following limitations:

- I have argued that imagery can be schematic and therefore fairly general – not restricted only to specific cases. However, it is still concrete in the sense of involving perceivable, or potentially perceivable experiences. Therefore its *ability to represent highly abstract material may be limited*. For example, mathematical notations are more efficient for keeping track of and manipulating complex webs of quantitative, deductive relationships at a high level of abstraction.
- The *amount and complexity of spatial information that can be imaged readily is limited*, as is the amount that one can learn to image in a finite time. Thus it is hard to form an accurate mental image of, say, a polygon with 23 sides, or to memorize a map of any significant size in a short time.
- As with propositional representations, the *degree of precision in a particular imagistic representation can vary, and may sometimes be insufficient*. For example, S11 was able to imagine that a heavier cart in motion would have more momentum than a lighter cart, and that it would generate more friction, but was unable to imagine which of these conflicting factors would be more influential. Secondly, the author at first had difficulty imagining the number of twists introduced into a drastically stretched band spring of two coils, and needed to use imagery enhancement techniques to produce cases like those in Fig. 14.4 before becoming confident in the answer.
- Finally, as is documented extensively in the literature, people can harbor certain *intuitive misconceptions* that differ from currently accepted theory, and some of these may be expressed imagistically.

Thus imagery and imagistic simulation have a number of potential weaknesses for a cognitive system. Despite this there is evidence that they were used and were very important for reasoning in the present study.

20.4.2 *Evidence for Imagery Involvement in a Wide Range of Reasoning Processes*

One approach to assessing the *importance* of imagery is to cite the wide range of important reasoning operations where these indicators occurred. Considerable effort went into developing a list of observable indicators that can provide evidence for imagery given in Table 12.3, and many episodes were cited that contain more than one indicator. The large number of examples cited in Chapters 13, 14, 15, 16, 11, 17, and 18. provide evidence that imagery was used by experts and students in important ways. (Note: Unless stated otherwise, in this section, I cite evidence from external subjects only.)

20.4.2.1 Imagistic Simulation and Its Use in Reasoning

Examples with evidence for the use of imagistic simulation were provided for all of the reasoning processes shown in Figs. 15.6 and 15.5, and these comprise the core foundation of reasoning processes used in model construction in the present

20.4 The Important Role of Imagery in the Expert Investigations

theory. These included familiar reasoning patterns of central importance such as analogies and extreme cases.

Evidence for imagistic simulation means evidence for using imagery and making a prediction were observed in the same episode. Elemental schema driven imagistic simulation itself is an underemphasized and poorly understood mode of inference. In the earlier sections of Chapter 15 on imagistic simulation and thought experiments, I provided many examples and an analysis of the sources of conviction in simulations that make plausible their ability to produce new knowledge. Evidence for imagistic simulations was documented in a variety of Untested thought experiments associated with each of the four major benefits of UTEs:

Untested thought experiments:-
- Activating other schemas
- Generating a new coherence relation
- Generating a new dissonance relation
- Generating an emergent global image feature from simulation (e.g. of equal spacing between coils)

However, imagistic simulation was also indicated in examples of other reasoning patterns that have been unrecognized or poorly understood, including:

- Compound simulation
- Projective analogies
- Flexible extended application of a schema (e.g. applying bending and twisting schemas to the square spring)
- New analogy evaluation processes:
  - Bridging analogies
  - Dual simulations
  - (Other analogy subprocesses observed with imagery indicators are listed in Table 20.3)
- Model evaluation processes
  - Running the model as a primary evaluation method
    - Dual imagistic simulations
  - Evaluatory Gedanken experiments as a special type of UTE
    - Minimizing or maximizing a single variable to examine effects

20.4.2.2 Use of Spatial Transformations

When I write, as I have in this chapter and in Chapters 11 and 17, that there was evidence that subjects used imagistic spatial transformations in a wide variety of reasoning processes, I mean that a reference made by external subjects to modifying a system occurred in conjunction with at least one imagery indicator. This was observed during the following processes in science problems:

- Extreme case generation (imagery or simulation enhancement)
- Analogy generation

- Analogy modification
- Explanatory model modification
- Analogy evaluation
- Explanatory model evaluation
- Gedanken experiments (piece of wire into a narrow vs. a wide spring, the vertical band spring)

Virtually all of the above transformations were novel actions that are very unlikely to have been previously performed by the subject, arguing that imagistic transformations were a source of creativity. Spatial transformations were also observed in mathematical modeling and problem solving in:

- Analogy generation
- Analogy evaluation
- Partitioning and reassembly (Chapter 11)

The variety of novel transformations in this book that were generated with imagery indicators, and for different purposes, illustrates the plasticity and creative potential of imagistic transformations for scientific thinking. There is a strong link to the imagery literature here that discusses the availability of perceptual transformations for images, such as Kosslyn (1980). In summary, examples from an extremely wide range of reasoning processes involving (1) imagistic simulation and (2) imagistic transformations, were accompanied by imagery indicators in this study, providing a first type of evidence for the overall importance of imagery.

20.4.3 *Evidence for the Importance to Subjects of Imagistic Simulation*

As a second kind of evidence for importance, in Chapter 13, I cited cases where multiple imagery indicators provided evidence not only that imagistic simulations were used but also that they were important to the subject. The "importance indicators" used were listed separately as part of Table 12.3 and included observations such as displaying an imagery or simulation indicator just after asking a question or before a finding; increased attention or effort indicators such as closing one's eyes (presumably to reduce interference or cognitive load) while displaying another imagery indicator such as giving an imagery report; and intentional efforts to improve and enhance imagery in simulations. Imagery enhancement efforts such as adding paint dots to the imagined spring or using extreme cases to exaggerate perceptual differences in simulations were especially compelling in this regard. Thus evidence was presented suggesting that imagery based processes were very important to subjects themselves in this study. At a more global level, evidence was cited from episodes where imagistic simulations produced by intuition schemas played a central role in the explanations of experts that was more than simply a "start-up" role – imagistic simulations were not just side effects, but were part of, and in some cases acting like axioms for, the central argument.

The pursuit of mechanisms. These findings fit with the proposal by Machamer et al. (2000) that much of the activity of science can be understood as a search for mechanisms. By mechanisms they mean "entities and activities organized such that they are productive of regular changes from start to setup to finish or termination conditions" (p. 3). In their paper, they discuss examples of how biologists value and seek such mechanisms. They suggest that the epistemic adequacy of a mechanism rests on its intelligibility ultimately in terms of base concepts that derive from seeing, kinesthetic and proprioceptive senses, and possibly emotional experiences, as sources of meanings. And "sensory experience with ways of working" is the primitive source of meanings out of which conceptual mechanisms can be constructed. Machamer suggests that all of the hard sciences and perhaps others, with the possible exception of quantum mechanics, have had the pursuit of explanatory mechanisms as an important part of their enterprise (P. Machamer, personal communication 2004.) I believe that this is related to the finding that imagery based processes were important to the subjects in this study.

20.4.4 *Possible Advantages of Imagistic Representations as Knowledge Structures*

In the above two sections, I presented empirical evidence from transcripts for the use and importance of imagery in expert thinking by documenting the wide range of processes involving imagery, and the importance of imagery to the subjects themselves. However, neither of these sections made a case for the advantages of an imagistic representation. In this section, I go beyond the data-based findings above to summarize hypotheses developed at a theoretical level giving reasons that imagistic representations and the processes that utilize them may have certain strengths and or advantages over other types of representation, and in particular, over discrete symbolic representations. Larkin and Simon (1987) cited several possible advantages of diagrammatic over sentential representations as external representations for problem solving:

- Diagrams display implicit information in an easy-to-read form.
- This leads to a reduced cost for perceptual inferences that can be read off of the diagram immediately.
- Diagrams may support perceptual recognition of important elements or applicable principles.
- Diagrams may group together information that is used together, avoiding search efforts. Adjacent locations in a diagram may provide clues to the next step in a problem.

They speculated that mental images may offer some of the same advantages.

The present studies do not make an experimental comparison between thinking with and without imagery and so I cannot reach a definitive data-based conclusion on this issue. However, a theory, grounded in protocol observations, of imagery-based processes in expert model construction has been proposed, and the new elements in that theory should suggest places where one can look for possible new advantages. This issue is as or more difficult than any taken up in this book and some researchers even believe that it is undecidable (Anderson, 1977). I believe that it may be decidable in that the field may eventually prefer one view or a more advanced mix over the other view, based on which exceeds the other on multiple criteria for scientific theories reviewed in Chapter 18, including parsimony. However, more research is required to do that. The present list of possible advantages may provide hints for places to look in such research. I address the possible advantages in sets according to level of processing complexity affected from low to high: basic knowledge representation; elementary imagistic simulations; thought experiments; reasoning operations; abductive model generation and model evaluation; geometric model construction; and scientific benefits from using runnable models. To imagine the underlying processes at these levels, one can construct or envision a vertical column of figures starting at the bottom with Fig. 13.4 at the imagistic simulation level through Fig. 15.3 to Fig. 20.1. The focus here is on possible advantages for developing qualitative and geometric explanatory models. These correspond roughly to developing the mechanisms described above by Machamer et al. (2000).

20.4.4.1 Basic Knowledge Representation – Advantages of Imagery for Representing Spatial Structure: Kosslyn, Shepard

Despite the potential weaknesses of imagistic representations described in an earlier section, there are some general hypothesized properties proposed in the literature on imagery that may make imagistic representations especially useful. Kosslyn (1980) and others have provided evidence that imagery appears to be a natural way to represent perceptual properties such as shape, relative position, and surface texture. Static geometric structure would be a natural extension of this. Shepard (1984), and Farah and Finke have provided some evidence that imagery also appears to be a natural way to represent changes in these properties, and motion trajectories for single objects. These precedents provide a starting point for thinking about areas where static and dynamic imagery may be a natural and efficient representation in science. Others have argued that imagistic representations are a promising approach to addressing problems of meaning, reference, deployment, understanding, and even consciousness. I will not dwell on these more philosophical issues here, but I touched on some of those arguments in Chapter 13.

20.4.4.2 Advantages for Representing Causal Units Via Elementary Imagistic Simulation

Most of the previous work on imagery has concerned static imagery. The studies reported in this book have examined initial evidence for the involvement of dynamic imagery of multiple elements in complex scientific and geometric models. Recent

20.4 The Important Role of Imagery in the Expert Investigations

studies by Hegarty, et al., (2003), Kozhevnikov, et al., (2002) Schwartz and Black (1999), Trickett and Trafton (2002), and Gero (2002) have also begun to examine more dynamic applications. In progressing from static to dynamic imagery, a natural next step is to ask whether imagery would be useful for representing causal relationships. Chapter 13 introduced the concept of elemental schema driven imagistic simulations. The background context for this was that although previous literature had discussed symbolic networks representing multiple causes, relatively little work in cognitive science had been done on the involvement of imagery in representing single causal units.

One of the most central kinds of causal relationship is a change or action causing another change. One conclusion of Chapter 13 was that elemental, schema driven, imagistic simulations are a natural way to represent such causal units. As depicted in Fig. 13.4, most of the schemas modeled took the form of a perceptual recognition of relevance followed by an action A leading to certain expectations B, a form that can embody a primitive concept of A causing B. By projecting one's actions into a situation metaphorically (e.g. "pulling on the spring") one can more generally represent external forces or influences leading to other changes. Such causal units can be chained in sequential simulations, as in Fig. 14.1 for the spring, where Force causes bending, and bending causes stretching. (Later this becomes force causing twisting causing stretching.) For example, S2's description of the operation of the square coil spring in Chapter 6 appeared to involve both of these chains.

Since one of the desired characteristics of a causal explanation is spatiotemporal contiguity between elements involved in actions, it makes sense that a spatiotemporal representation would have advantages as a common representation for such causal chains. Imagery that uses some of the same layers of the perceptual/motor system as those used in real perception of events should qualify as such a representation. In addition if this were linked to spatial reasoning operations that embodied spatiotemporal constraints on any system of objects, such as the constraint that solid objects may not occupy the same space, then those constraints would be automatically built in implicitly to any model created in the representation, avoiding the need to use valuable computational time and effort to compute them at a higher level. I denoted inferences from such operations "spatial reasoning" in Fig. 15.3. Such a system would have strong advantages for thinking about causal relationships.

In Chapter 16, I introduced the idea of a mechanical model in the form of a causal chain or net that is *completely connected* spatially and temporally so that it gives a complete "mechanical explanation" of a system. (this may be related to the concept of the "completeness" of a gap-free explanation (Machamer, et al., [2002]). Although not all explanations in science are able to attain this level, it is valued by scientists when it is attained and is seen here as maximizing the level of sense making and understanding in the subject. A spatiotemporal representation would appear to be ideally suited for developing such models. Thus imagery and schema-driven simulations may provide a very natural representation and reasoning system for causal thinking with advantages for efficiency of operation.

20.4.5 Possible Advantages of Imagistic Representations for Creative Reasoning

20.4.5.1 Level 1: Untested Thought Experiments and Imagistic Simulation

In this section, I consider advantages for processes depicted at the lowest level of Fig. 20.1. The concept of schema driven imagistic simulation is at the center of the present theory of imagery based thinking in science; it is conceived of as relying on perceptual/motor schemas and imagistic representation for efficient operation. These schemas and representations can represent causal knowledge elements as noted above. However, they begin to participate actively in reasoning operations, rather than simply being knowledge elements, when they are applied to unfamiliar circumstances, and when these are intended to lead to predictions we say the subjects are performing thought experiments. Untested thought experiments occurred when a subject predicted the behavior of an untested, concrete, but absent system; this was explained via several possible contributors to imagistic simulation, shown in Fig. 15.3. These include flexible extendibility of the schema to cases outside its normal domain of application, spatial reasoning, and tapping implicit knowledge in the schema. Examples were presented in the case studies where there was evidence that these were imagery-based processes.

Implicit knowledge. Previous research has shown that a distinguishing property of static images is that they can be interrogated for implicit information stored in the image (Kosslyn, 1980) (e.g. the number of windows in one's living room), and I attempted to document this phenomena for the case of causal/dynamic imagery. In cases such as imagining twisting a long and short rod, subjects appeared to tap implicit knowledge about dynamic relations between force and deformation. In Chapter 15 it was argued that this kind of knowledge would be hard to obtain using a discrete symbolic representation.

Flexible extendibility. Flexible extendibility may also contribute to model construction, as recruited schemas are applied within a strange new model in new ways. This means that the action side of a schema must be flexible in adapting to a strange new situation. In the case of the spring problem, the challenge is how to apply schemas representing the standard concepts of torque and torsion to a circular wire. This is a nontrivial problem and requires flexibility in extending the application of these schemas to this case. Given the precedent of flexible tuning parameters within motor schemas, discussed in Chapters 13 and 18, this kind of flexibility or ability to "stretch" to new cases may be easier to envision in a system using perceptual/motor schemas operating on analog images than in a system using algorithms operating on discrete symbols.

Spatial reasoning. Spatial reasoning can come into play when spatial constraints influence the operation and alignment of one or more schemas in a simulation. This includes inferences about how spatial relationships change when one object is moved relative to another. These inferences would appear to be most naturally supported by imagistic representation.

20.4 The Important Role of Imagery in the Expert Investigations

Imagistic simulation allows for the emergence of new image features. A mechanism for generating a new variable, that I call the "emergence of new image features," was suggested by the example of the asymmetrically stretched spring in Fig. 6.3, produced when S2 imagines applying the bending model to the spring. The idea that the coils get farther apart as one moves down the spring is an emergent new image feature. It is unlikely that he has thought about asymmetrically stretched springs before. He later whimsically calls this the "droop" or "flop" effect. Such an effect may emerge before the activation of any familiar schema in memory; in this case, emerging as an interesting novel shape within the image generated by thinking about bending as a mechanism within the helical coil. Here, it appears to require the invention of a new label to name the nonstandard feature. How such new image features arise is poorly understood, but here it would appear to emerge from the application of a schema (bending) to elements of an object outside of its normal domain of application (the spring) and the imagistic "summing" of the effects of these applications via spatial reasoning. As a second example, an unanticipated effect noticed in the thought experiment of the extreme case of the drastically stretched spring was that the spring's curvature disappears as the system approaches the shape of a straight vertical wire. The phenomenon of a novel, unlabeled, spatial feature emerging from an untested thought experiment is harder to explain using a discrete symbolic representation system. A final classic example of emergence would be the recognition of the possibility of jamming in gear systems when one does not possess a prior concept or vocabulary for "jamming" (Schwartz, 1996a; Metz, 1985).

20.4.5.2 Imagistic Representations and Simulations May Enable Perceptual Activation of Useful Schemas

This was the explanation given for discoveries like the recognition of a torque effect a new causal variable in the stretched hexagonal coil. The mechanism by which this takes place is shown in Fig. 15.3 as schema activation.

Flexible activation. For the torque schema to be activated in such a strange case, we need a very flexible activation mechanism. In paradigmatic cases torque is applied via a force at right angles to the axis of rotation of a body, and there are those two perpendicular elements involved. But in the hexagonal coil, there are six elements at odd angles. A schema is assumed to be capable of responding to imagistic activation, but it must be a very flexible one to become active in such a circumstance. In the present theory runnable schemas are assumed to have activation mechanisms that are derived from perceptual ones and that therefore have built in "fuzziness" or flexibility in being able to respond to a wide range of similar objects. This may give them an advantage over rule based categorization schemas that are more rigid.

Spontaneous generation of dissonance and coherence relations from running imagistic simulations was noted as another process that can be imagery based.

Volatility. Taken together, the properties/processes of flexible perceptual activation, the emergence of new image features, and the spontaneous generation of dissonance and coherence relations are called "volatile" processes because they can happen without warning in a way that is outside the control of the scientist. This also means that they can be an important source of divergent creativity. This, then, is a potential benefit of these imagery based processes: they are a way of providing alternative mechanisms for these functions.

Compound simulation. Running two or more schemas in a new compound imagistic simulation or running multiple interacting instantiations of the same schema may also be a common source of emergent unanticipated interactions. This can be another important source of divergent creativity.

20.4.5.3 Level 2: Advantages for Higher Reasoning Operations

In this section, I consider advantages for the higher-level reasoning operations depicted in the middle section of Fig. 20.1.

Extreme case reasoning. Maximizing or minimizing spatial features of a system and then mentally animating the imagery to simulate the result was shown to be an heuristically valuable strategy. Examples were cited where extreme cases appeared to provide "imagery or simulation enhancement" by making it easier for the subject to detect implicit knowledge within a schema being used to run the simulation. This means that after running an initial simulation and getting a result, the extreme case allowed the subject to run the simulation with significantly higher confidence. Here, the subject appears to use an imagistic transformation to change the case image that is the starting point for a simulation. Having this kind of "second chance" to improve dynamic imagery and predictive confidence is again an advantage that would seem hard to obtain using a discrete symbolic representation.

Advantages for analogical reasoning. As was evidenced in Chapter 17, imagistic processes open up alternative ways to generate and evaluate analogies quickly, including:

- Generation via a transformation.
- The flexible perceptual activation of perceptual/motor schemas mentioned in the previous section may also contribute to flexible analogy generation or access by association and this would provide a different route than verbal associations.
- Evaluation via dual simulation (including overlay simulations).
- Evaluation via a conserving imagistic transformation.

Some of these appeared to occur quite rapidly in the protocols and may be quite efficient. They appear to offer important alternatives to analogy access via verbal association and evaluation via mapping of discrete symbols. In particular, the ability to generate an analogy by modifying the target case, via a transformation that is conserving and therefore validating, appears to be a powerful means for simplifying problems.

Flexibility of spatial transformations. Table 15.8 shows a variety of heuristics for how useful extreme cases, Gedanken experiments, and analogies might be generated or improved via imagistic transformations, and this function is indicated by the arrows coming from the lower left in Fig. 20.1. Transformations of a case that enhance the ability to perform imagistic simulations of the case are likely to be useful. The ability of imagistic representations to support this kind of "case improvement" strategy appears to be very valuable.

20.4.5.4 Level 3: Advantages for Abductive Model Generation or Modification

Imagistic representations (third level of Fig. 20.1).may have the following advantages for the generative abduction process in explanatory model generation and modification as one of the most difficult tasks facing an investigator.

Analogy and extreme cases for model construction. These can provocatively activate a relevant schema or principle through imagistic means as an extra alternative to verbal associations for suggesting model elements.

Flexibility in generating new ideas via spatial transformations. This was indicated by the fact that subjects generated many types of novel images, sometimes in a playful and divergent manner. The divergent power of the analog character of an imagistic transformation can be hinted at by the following: A sheet in the shape of an ellipse can be cut into parts in an infinite number of ways. The vast majority of these shapes have no name and cannot even be described well in verbal terms, yet they can be imagined. In particular, constructive spatial transformations were theorized to provide an important source of creative ideas for explanatory model modification, a crucial process in investigating the spring system. For example, once torsion was discovered in the spring in Chapter 14, the position of the "effort point" for applying torque in the spring was transformed several times – from an adjacent side in the hexagonal coil in episode 11a (external subject), to a quarter turn away in the circular coil (episode 17a, external subject), to the center of the coil (episode 18, internal subject), and to a point on the central axis above the coil (episode 19c, internal subject). These spatial transformations enabled the progressive improvement of the model.

Imagistic transformations of various degrees of strength may allow the subject to modulate divergence by producing conserving transformations at a level appropriate to the level of precision at which modeling is taking place at the time, as described in Chapter 16. During mathematical modeling, these may correspond to various levels of geometric transformations such as topological vs. affine.

Imagery can represent multiple constraints efficiently. Imagistic simulation allows one to represent many spatially embedded constraints simultaneously. When a problem with the design is perceived, perhaps via the recognition of a new constraint, the current image of a model operating within constraints can be used as a base in generating a modification that satisfies the old constraints as well as the new one.

Furniture placement example. A simple but instructive example from daily life here is the problem of planning for furniture and partition placement in a large office room with multiple work stations. If there are many pieces of furniture needed in the room, traffic paths will be constrained further by the placement of each piece of furniture. So designing the traffic path is a problem with multiple spatial constraints. In contrast to a set of coordinates for each piece, an image or drawing of the room makes it possible to keep track of these constraints as each piece is placed in the plan. It also provides guidance as to how to adjust or revise the plan if, for example, a traffic path is seen as too circuitous. Adjustments to the position of a piece can be imagined while taking into account the old constraints as well as the new one generated by the criticism. The representation can also represent interconnected constraints, where moving a piece to open up one traffic path can cut off another.

A possible advantage of an imagistic representation is that it allows an expert to "play" with alternative versions of mechanisms within certain chosen constraints, much as one might play with materials on a workbench in trying to design a solution to a new household or industrial problem or play with miniature furniture blocks to solve the traffic pathway problem above. For example, in trying to imagine what ion could be reacting with a certain crystal surface, a chemist might first imagine the configurations of atoms on the surface of the substrate; then imagine the size of an ion that could react with it; then choose a candidate molecule of a known shape; then rotate the image of the molecule in several directions to examine whether the molecule could fit into the shape of the substrate. (Similar processes were described by Watson (1968) with regard to the double helix.) Thus many constraints can be built into the simulation, while a few (in this case the orientation of the molecule in space) are left free. The subject first sets up the problem context in an image; then manipulates the image within the free parameters, possibly quite randomly, until sensing the approach of a solution or match to the goal, when finer adjustments are made to achieve the goal if possible. Similarly, we saw subjects "playing" with various positions for placing a "soft segment" in the spring coil in order to imagine the effects of forces on segments of the spring. Tracking of model modifications and new constraints this way should be especially favored when those involve spatial configurations, as they very often do in explanations via mechanisms in physics, chemistry and biology.

Summary for generative abduction. In this view, at the center of the abduction process is the use of prior knowledge schemas to generate images that are combined in compound simulations in new ways and transformed until they explain the target phenomenon and satisfy required constraints. Abductively generating a really new and successful explanation in science is an extremely difficult task. The strengths mentioned above however, make it possible to conceive of a system that could work by abduction in complex cases. Thus imagistic representations may have important advantages for generative abduction in model construction, because of their ability to:

- Represent analogous cases that trigger relevant schemas imagistically
- Represent target transformations of different "strengths"
- Represent causal chains in models with spatiotemporal contiguity

20.4 The Important Role of Imagery in the Expert Investigations

- Represent model transformations for improving a model
- Represent multiple complex spatial constraints simultaneously
- Allow free play operations on particular variables with feedback

In addition, the idea that imagistic simulation can be ubiquitous in scientific thinking, and that it has the volatile characteristics of generating new image features and activating other schemas, means that imagery opens up additional avenues of divergent thinking as resources for abduction. These are also potential sources for insights that are experienced as Aha phenomena. Spatial representations may be so good for the above purposes, that it becomes desirable to convert nonspatial problems, such as the chain of command in an industry, a computer program, or interconnected processes in cognition, into a spatial one, such as an organization chart, a flow chart, or a diagram of cognitive processes.

20.4.5.5 Advantages for Model Evaluation Phase of Model Construction

Evaluation via simulation. I examined cases in Chapter 15 where the subject appears to run an explanatory model as a way of generating an initial plausibility test for the model. Running the model would be a natural test of the model with respect to spatiotemporal constraints (e.g. the most basic constraints would be whether the model fits physically into the space of the target or predicts obviously false behaviors or trajectories). If a model failed such a simple test it might save a great deal of time that would otherwise be wasted in pursuing the model. Maxwell's criticism of an early model of his of the electromagnetic field, beautifully described by Nersessian (2002), fits this description. That model represented aspects of the magnetic field as discs or gears in contact and rotating in the same direction in space, but there was a conflict in that adjacent gears would have jammed or opposed each other if they were turning in the same direction. Maxwell repaired the theory by adding "idler wheels" between the gears which came to represent another useful feature of the field. This kind of model evaluation on a novel system with spatial properties (e.g. recognizing jamming) could be particularly efficient using an imagistic representation. It is depicted in the present theory as the natural outcome of an imagistic simulation when it spontaneously generates a new conflict relation, as shown in Fig. 15.3. In the present theory this is one of the major benefits of having a runnable model that is imagery based.

Evaluative Gedanken experiments. Gedanken experiments like the torsionless coil or the vertical band spring documented in this study were novel mechanical systems or apparatus constructed in a spatial framework by the subject. As such, they constitute another impressive design problem. Such an experiment can allow the subject to run their new model on a special case and compare the result to a second source–often a simulation using a second schema. The suspected double role of imagery in some of these constructions attests to yet another contribution to creativity.

Formation of coherence and dissonance relations. As discussed above for the Maxwell case, once a model is formed, running it as an imagistic simulation can also enable alternative routes to the formation of coherence and dissonance relations.

20.4.5.6 Advantages for Geometric Modeling as a Step Toward Quantification

A traditional view in opposition to the present theory is that mathematics has highly developed and standardized symbol systems, so the development of a mathematical model should depend only on these. On the contrary Chapter 16 concluded that techniques using imagistic transformations and simulations play a role during model construction, even at mathematical levels of precision. Increasing the precision of description of the model to a geometrically accurate level with relations that can activate symbolic quantitative schemas was seen as a key stepping stone to obtaining a quantitative model. Creative symmetrical cancellation arguments, simplifying analogy transformations, and spatial partitioning and reassembly strategies were used in both the Volume of a Torus Problem in Chapter 11 and the mathematization of the Spring Problem in Chapter 14. One suspects that imagistic representations are invaluable for these.

20.4.5.7 Linguistic and Imagistic Precision

I must admit that one of the heuristics I have used in my career to break out of an exclusively language centered view of reasoning is to try to imagine how much cognition could occur largely independently from language. Therefore the role of language is underemphasized in much of this book (as is mathematical symbolism). It is clear however that careful labeling and formulation in terms of precise language is a powerful part of theory construction/science. It would appear that a viable conception of a schema as a memory unit in science must include verbal labels and descriptions of relations that are attached to an imagery generating, perceptual/motor process that provides core meaning. This kind of representation would allow dual channels for activation, reasoning, and learning, with the linguistic side especially supporting more formal deductions and quantification. Should we also consider language as the primary route to making precise distinctions and general precision in modeling? The present analysis of successive levels of qualitative (unconnected and connected/aligned) and geometric precision suggests resisting this conclusion and argues that imagistic precision is just as crucial, if not more crucial than linguistic precision. It may be that attempts to increase imagistic and linguistic precision can support each other in a synergistic way.

20.4.5.8 Level 4: Models that are Runnable Imagistically Have Significant Advantages for Scientific Theories

Up to now I have been discussing advantages of using an imagistic representation for forming or constructing models. I will now review some of the advantages hypothesized for runnable models after they are formed. Certain subjects in the case studies appeared to be constructing runnable explanatory models capable of generating imagistic simulations. I argued in Chapter 18 that such a model should have some significant advantages. Columns two and three in Fig. 18.7 list some of these cognitive benefits, including:

- Flexible activation
- Extendable, adaptive application
- Imagery-based explanation of dynamic mechanisms
- Projected plausibility
- Novel emergence
- Combinability in compound simulations
- Coherence with experiential knowledge
- A foundation for mathematical modeling

I also argued that these properties can contribute to building an explanatory model that satisfies important criteria for evaluating theories in science, including: generality, explanatory adequacy, predictive accuracy, external coherence with other theories, modularity, heuristic value, clarity, and spatiotemporal coherence.

In summary, building on the processes shown in Fig. 15.3, Fig. 20.1 shows explanatory relations between imagistic simulations and transformations at the lowest level and scientific reasoning processes at a higher level. When the task is such that imagery can be used efficiently at the lower levels, these efficiencies can percolate up to yield benefits for reasoning at the higher levels. These culminate in advantageous properties of the models themselves, shown in Fig. 18.7. Thus imagistic representations are hypothesized to have a considerable number of potential advantages for scientific model construction.

20.5 Transfer of Runnability Leads to Outcomes of Flexible Model Application and Generativity

Chapter 18 put forward the hypothesis that the "runnability" of schemas with perceptual/motor components is a valuable property that can be transferred to models that use those schemas as building blocks. Furthermore it was proposed that the model's "runnability" – its ability to generate new imagistic simulations – is a key source of flexibility in applications to new situations as well as a source of other desirable properties of scientific models.

20.5.1 *Example of Flexible Model Application*

An example of the payoff from runnability transfer was provided by the transcript discussed in Chapter 18 of two students discussing possible sources of friction in a physics class, depicted in Fig. 18.8. In that session, one of the students was able to spontaneously extend her recently learned model of elastic normal forces acting vertically between a static object resting on another object to the frictional force acting in parallel to the surface between two moving objects. She had learned that static objects like tables act as if they are made of very stiff springs so that the downward force of an object placed on the table is met with an equal and opposite force of the "deformed" table on the object. An instructor was drawing and introducing a model for the frictional force that acts between a surface and an object sliding on that surface – a model of bumps on one surface colliding with bumps on the other. She immediately recognized spontaneously that this force could be similar to the upward force from the "elastic" table if one thinks in the friction case of the bumps (on the table) bending and pushing back. As evidenced by her similar hand motions and similarity description she was perceiving the new friction case as an extended application of her recently learned elastic normal force model. Since the subjects are naive students, this application of a model of elastic normal forces for static cases, to a microscopic model of frictional parallel forces of a different kind for dynamic cases, is an impressive case of model flexibility.

More sophisticated examples of this kind from the history of science are the extension of the germ theory of fermentation to the germ theory of infection or the extension of field theory from fluid mechanics to electromagnetism and later to gravity. As opposed to "inert" knowledge that is compartmentalized and limited in scope to previous applications, flexible knowledge is more likely to be activated and adapted when needed in new situations. I now review a historically opposed point of view that claimed that models were of only marginal importance in science.

A philosophical viewpoint directly opposed to the view of runnable models providing flexibility is the following: Strategies such as running simulations and making analogies may have some value as "tricks" for provoking or suggesting the right "moves" in constructing an abstract model, but once the model has been built, these tricks can be discarded; a mature scientific theory becomes completely divorced from imagery and analogies, thereby achieving greater power and generality. (This argument is inherited from Duhem, who opposed the arguments of Campbell in the early part of the 20th century that analogies and models were of continuing importance in science, as described in Hesse (1966).)

In contrast, Mary Hesse supported and extended the Campbellian view by describing analogies as providing concrete elements for explanatory model construction that have "material similarity" to the model. (See section on Explanatory Models vs. Expedient Analogies near the end of Chapter 14.) The present transfer of runnability idea extends her view; I have argued that runnable schemas can constitute such concrete elements. Grounding the models in this way has important advantages for the flexibility and generativity of the model. To support this view I have asked: Why did the subjects studied in Chapter 14 bother to pursue "understanding" for problems where they already had a confident answer? I argued that this pursuit of understanding in these ways ultimately gave them an ability to "run" the model to provide an explanation of the target behavior. This corresponds to the scientist's interest in finding mechanisms (Machamer et al., 2000) described earlier, mechanisms with material parts thought of as having material similarity to nature's elements.

20.5.1.1 Summary of Support for the Transfer (or Inheritance) of Runnability Hypothesis

The "transfer of runnability" theory presented in Chapter 18 was supported by its ability to explain:

- The example above showing a similarity in hand motions during the running of a source schema and the running of a model constructed from that schema.
- Similar depictive hand motions in S2's discussion of a source analogue of, for example, the twisting rod, and the square coil model of the spring.
- Similar data on parallel hand motions from a separate case study of electricity learning at the high school level (Clement and Steinberg, 2002).
- Why expert subjects in the present protocols bothered to pursue "understanding" when they already had confident predictions and went to the trouble to anchor their arguments in concrete, imageable, analogies, using strategies like "imagery enhancement."
- Why the anchoring strategy was successful in the tutoring studies described in Chapter 10.
- An initial cognitive model of how scientists attain many of the properties they seek in a scientific model, such as flexible application and heuristic value.
- What is according to Nagel (1961), the long-standing preference of many scientists for visualizable models.

Thus I consider two of the points in Nagle's early discussion to be particularly prescient where he theorized that a visual model can add to a scientific theory by: (a) allowing the application of theories to concrete physical problems by suggesting points of correspondence between theoretical elements and observable variables and (b) suggesting key questions for the refinement and extension of theories. However, in his writing there is still a sense in which visualizable models are seen as appendages to a theory. In the present view imageable models are seen as highly prized as the foundation of a theory.

20.5.2 Role of Runnable Intuitions in Conceptual Understanding and Recursive Runnability

Most of the examples of elemental imagistic simulation in this book involved the use of physical intuitions where the runnable schema being used carries some self-evaluated conviction, often presumably because it is a schema with perceptual/motor components that has been constructed largely from direct experience with the world. In addition to transfer of runnability, cases like these motivate the idea of transfer of some level of conviction as one source of confident knowledge for an explanatory model. This would provide confidence about how one element would operate in such a model. This would not be a complete source of confidence in the model, since the conviction that is transferred would not constitute support for whether the model is part of Nature's mechanism. That support is only developed when the whole model survives repeated evaluation and criticism cycles.

20.5.2.1 Recursive Runnability of Models as Thought Experiments

A major integrating connection between the present view of the benefits of grounding models in intuitions and the earlier theory of thought experiments was made in Chapter 18. It was proposed that once a model inherits runnability and conviction, it is eligible to participate recursively as a runnable schema, as in Fig. 15.3, with the ability to run simulations in untested thought experiments, utilize spatial reasoning, and participate in compound simulations. These properties make the model robust in the sense of being flexible and general for application to unfamiliar situations and generating new theory. (The episode in Fig. 18.8 was interpreted in this way.) This constitutes the recursive pattern of largest scope discussed in this book. Essentially, most of the relations in columns 1, 2, and 3 of Fig. 18.7 can be inferred by the analysis of benefits of thought experiments, as unpacked in Fig. 15.3, if one makes the assumption that a model assembled from runnable schemas becomes a runnable schema assembly or macro-schema. A runnable model should then be able to participate recursively as a schema in the plausible reasoning operations in Fig. 15.6, giving it a form of compatible modularity with other runnable structures. When this happens, the model (such as the twisting model of the spring) begins to act as though it were the schema at the left side of Fig. 15.3. Theoretically, it can therefore proceed to act in the role of any of the Sc schemas in Fig. 15.6 to create larger models or inferences. These possibilities create the potential for an impressive degree of flexibility and generativity (what Darden calls fruitfulness) in a scientific model.

20.5.2.2 Conceptual Understanding as a Fully Connected Explanatory Model of Grounded Schemas

A possible infinite regress is avoided in the present case studies because certain physical intuitions (about bending and twisting in the case of the spring problem) were taken as primitives. That is, they were seen by the subjects as not needing further justification. It is plausible that these intuitions consist of perceptual/motor schemas that correspond to a primitive level of experience and meaning for the subject because they were built up out of many perceptual/motor experiences with manipulating objects (Clement, 1994b). They do not need further grounding because they are already grounded in this sense. This avoids the axiomatic grounding problem and this is related to the symbol grounding or basis of meaning problem faced by certain other representations.

Indeed, in the case study of S2 in Chapter 6, the bending and twisting ideas played a role parallel to that of axioms in the final argument structure shown in Fig. 6.7c; and the schema for twisting played the same role in the more complete explanation developed in Chapter 14. "Bottoming out" in physical intuitions that carry necessity or at least conviction may be the most satisfying way there is for a theory to be grounded.

In section 16.1.3 I referred to a model constructed from several schemas as a schema assembly. When these schemas are runnable and are aligned so that their causal interactions are fully imageable (connected spatio-temporally with no gaps) I referred (in Section 14.3.1.2) to a fully connected model or mechanism. An explanatory model that is a fully connected assembly of schemas, each grounded in an intuition, *may be the most satisfying form of conceptual understanding* of a scientific system. Since such a mechanism is grounded in intuitions, in this case one can say that *science is an extension of intuition*.

20.5.3 Comparison to Lakoff and Nunez's Embodied Mathematics

In one way this book can be seen as responding to a challenge posed by the controversial and widely noted book, *Where Mathematics Comes From: How the Embodied Mind Brings Mathematics into Being* (Lakoff and Nunez, 2000). They propose that the meaning of *mathematical ideas*, even sophisticated ones like limit, are grounded in elemental perceptual/motor ideas like "putting an object in a container" and "moving on a path from place A to place B." Grounding for them takes place via "Conceptual metaphor" whereby abstract ideas are understood via metaphor in terms of these more concrete concepts. One of the purposes of the present book is to show how *scientific ideas*, even sophisticated ones, can be grounded in perceptual/motor ideas. This means that some of the concepts underlying science are "embodied" in the perceptual and motor systems of the thinker. Although Lakoff has cited evidence from linguistic patterns in earlier books for the idea of conceptual metaphors in general, Lakoff and Nunez's intriguing theory of mathematical thinking itself has yet to be supported by evidence.

Findings in the present book are based on evidence from protocols of scientists thinking aloud about the strategies they are using to solve problems, including their statements, drawings, and depictive gestures. In this book there has been an attempt to unpack the idea of elemental intuition schemas more deeply by analyzing the way in which they are expressed in imagistic simulations and thought experiments, and connected to spatial reasoning, as depicted in Fig. 15.3. This view was extended to argue that certain useful, intuitive, perceptual/motor preconceptions in students can be utilized as analogical anchors for constructing imageable models in science instruction, and this finding also has received some empirical support from case studies.

The parallel to Lakoff and Nunez's "conceptual metaphor" in the present work is the description of "analogical grounding" of scientific ideas in intuitions, which I sometimes refer to as "anchoring" a new concept (Clement, 1993b). I attempted to unpack this idea further by discussing several purposes for analogies, and distinguishing between expedient analogies, and actual source analogues that become part of an explanatory model being constructed abductively by the subject. The evidence for transfer of imagery and runnability from source analogues to explanatory models justifies consideration of a theory that goes beyond symbolic mappings. I believe this touches on the heart of the issue. The book goes on to describe how many types of nonformal scientific reasoning operate on images and are spatial in character. These include analogical reasoning, thought experiments, and scientific model construction. In this way it presents evidence for a view of science as being at least partly embodied by being grounded in perceptual/motor primitives and imagistic reasoning.

20.5.4 Payoffs from Transfer of Runnability

Figure 18.7 presents the outline of a theory that culminates the argument presented in this book for the importance of imagistic simulations in expert cognition. These are outcomes or "payoffs" of runnability accrued after a model is constructed. Personally, I find that these ideas point to the importance of grounding models in runnable intuitions whenever possible.

As discussed elsewhere in the concluding chapters, model runnability also can be seen to have strategic value during the model construction process itself. For example, running a tentative model may help to evaluate it by triggering new emergent features and conflicts, or by producing a prediction in a Gedanken experiment. Positive educational effects of these features could be imagined in a fifth column to the right in Fig. 18.7. For example, retention or likelihood of appropriate activation of a model would be affected positively by generality, coherence with other ideas, and clarity, from the fourth column. These ideas will be discussed further in Chapter 21.

20.6 Comments on Methodology

The observation and analysis strategies used in this study were described in Chapter 1. Here, I reflect on the status of the method used and the varying relation between observation and theory in different parts of the study.

20.6.1 Small Samples

It has been argued that qualitative scientific model construction can occur through a process in which an initial model is constructed abductively and then successively refined through a cycle of criticism and modification. One can apply this view of model construction to the methodology of investigation used in this book. The process of early model evolution does not necessarily depend on large sample sizes, unlike induction by enumeration and methods of statistical pattern generation. Abduction can occur quite readily from a single important example. Of course, when triangulation is possible from more than one example that is preferred. This was possible for some behaviors, but it is very difficult to capture multiple instances of unusually creative behavior. Even within a single rich protocol however, the challenge of producing a coherent model that accounts for most episodes in the protocol is an enormous challenge, and if one has to account for say, 20 or more episodes in the protocol, the space of alternative models that will do this is heavily constrained.

The "depth of precision" levels in Table 14.2 give us a way to characterize this study itself as being at a *qualitative level* of precision in modeling. Figure 16.4 gives us another way to characterize this study: as spending most of its energy within the *Explanation Process* in section II of that figure, as different mechanisms were proposed for plausible reasoning processes. The study also spent time in the Description Process in section I. However, as noted in Chapter 16, descriptive observations can carry the solver only so far in developing a scientific investigation. Without an understanding of the hidden qualitative mechanisms operating within the spring system (developed by the Explanation Process), one could not progress to higher levels of detail. Similarly, for an exploratory study of scientific reasoning processes that are poorly understood, it seems appropriate that much of the present study concentrated on developing initial models of hidden processes, and supporting lines of connection to the descriptions of case study data that they explain.

20.6.2 Links Between Data and Theory

One can ask which parts of the present theory are more and less closely tied to transcript data. Theory represented in figures such as 15.3 and 20.1, showing knowledge and reasoning structures, were most closely linked – there was an attempt to provide one or more examples from transcripts of external subjects for every element and relation shown in these figures to provide an existence proof for that aspect of the model.

For the larger overall investigation process in Figs. 16.8 and 16.4 (and the more detailed Fig. 16.3) I have been working from the somewhat idealized composite protocol in Chapter 14 and therefore the model of investigation processes is also idealized. It serves as an initial prescriptive model, and any one subject would be expected to exhibit only pieces of it. Many of the items have support in transcripts of external subjects, but others have been added from internal subject moves or to make the theory more complete, especially for the mathematical modeling and higher-level control portions, so the overall model at those levels is more speculative. Figure 18.7 is the most speculative of all, although the core idea of transfer of runnability is documented by transcripts in Chapter 18. That figure is included to propose possible connections to philosophy of science issues and to data from the history of science.

In summary, of the figures depicting the core of the present theory, Figs. 15.3, 15.6, and 20.1 represent data-based models, based on and constrained by case studies of external subjects. These then served as a base for constructing more speculative hypotheses for higher level investigation processes that were partially grounded in observations. It is hoped that these models will lead to investigations that can further evaluate and improve them.

Chapter 21
Creativity in Experts, Nonformal Reasoning, and Educational Applications

In this chapter I first give a summary of the overall theoretical framework developed followed by a section on how experts use creativity effectively. That section discusses how such apparently weak nonformal methods could possibly produce successful theories. I then consider educational issues by starting from a summary of expert–novice similarities and differences identified here and examining their educational implications. I end by discussing whether creative processes in experts appear to be ordinary or unnatural and the extent to which they are explainable.

21.1 Summary of the Overall Framework

21.1.1 View from Multiple Diagrams

An introductory summary of the theoretical framework developed in this book was given at the beginning of Chapter 19. The overall view of expert investigation processes was summarized in the section on model construction in Chapter 20, with qualitative explanatory model construction playing a key role. Figure 21.1 shows perhaps the most concise summary in the book of key elements in that process. It emphasizes the dependence of the process on cycles of generative abduction plus model evaluation via running the model. These in turn depend on schema-driven imagistic simulations. The resulting potential for transfer of runability from source schemas to the model can yield a runable model that generates its own simulations. This evolutionary view contrasts with a one-step Eureka model. But it also allows for intermittent insights, as unexpected activations or dissonance relations are triggered by simulations. It also allows for deep grounding in perceptual motor schemas such as physical intuitions.

A more detailed view of the overall framework can be organized around Fig. 20.1, which depicts the subprocesses used in a single cycle from Fig. 21.1. Figure 20.1 is constructed in four levels, which correspond to increasing complexity of the modeling processes described in the overall theory. Table 21.1 uses these levels of complexity as an organizing principle for a number of the diagrams in this book.

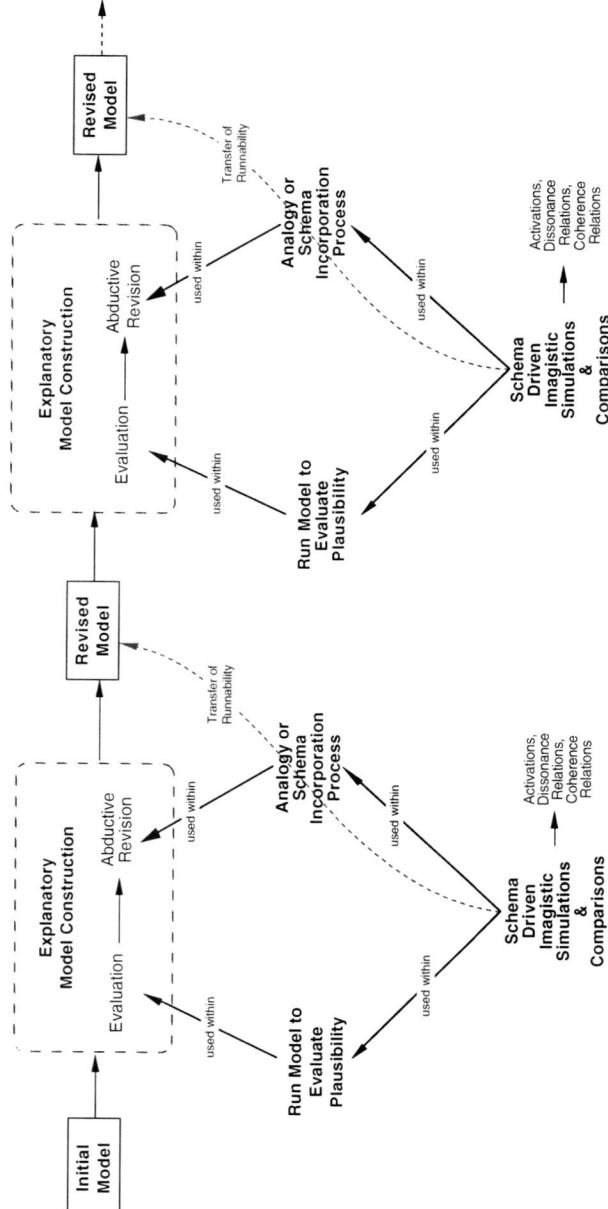

Fig. 21.1 Some major nonempirical subprocesses involved in explanatory model construction

21.1 Summary of the Overall Framework

Table 21.1 Summary of the book using four levels of modeling

	Coarse-grained analysis	Finer-grained analysis
Scientific knowledge outcome: Runnable models with flexibility (result of scientific investigation process below) *Top level of Fig. 20.1*	Four types of knowledge in science (Fig. 6.10)	Special benefits of inherited runnability for scientific models (Fig. 18.7)
Scientific investigation process (utilizes nonformal reasoning processes below) *Third level of Fig. 20.1*	Cycles I–III in scientific investigation process (Fig. 16.4) (elaborated in 16.3)	Flow of information over time for explanatory modeling cycle II (Fig. 16.8)
Nonformal reasoning processes (utilize imagistic processes below) *Second level of Fig. 20.1*	Schemas generate imagistic simulations in four types of nonformal reasoning (Fig. 15.6)	Four types of nonformal reasoning, details (Table 19.1)
Basic knowledge structures and imagistic processes by which they are "run" *Bottom level of Fig. 20.1*	Imagistic simulation process with possible benefits on the right and possible origins of conviction on the left (Fig. 15.3)	Perceptual motor schema driving an imagistic simulation (Fig. 13.3)

Although there is not space to actually display it here, the author has found it helpful to photocopy reduced copies of these diagrams and tape them together to provide an overview of the theory. (If one were to do this with only four of the drawings instead of nine, I would choose 15.3, 15.6, 16.8, and 18.7.)

The *bottom row* of Table 21.1 contains diagrams of the most basic knowledge structures and imagistic processes in the theory. The first, Fig. 15.3, expands the process of running a single imagistic simulation. This is the smallest "atomic" level of modeling in the present theory. Figure 13.3 is a more detailed depiction of the schema and simulation ovals in Fig. 15.3. The process of imagistic simulation can be used by a number of nonformal reasoning processes, such as analogies and Gedanken experiments, which are shown on the *second row* from the bottom. Figure 15.6 details four types of nonformal reasoning processes, moving from simple to more complex, and correspondingly moving from level 1 to level 2 within Fig. 20.1. A finer-grained analysis of these four processes is given in Table 19.1.

These reasoning processes are in turn utilized by the Scientific Investigation Process, summarized on the *third row*. Figure 16.4 illustrates cycles I to III in this process. Figure 16.8 shows a finer-grained analysis of Cycle II, including how it can be run repeatedly to achieve greater explanatory depth (indicated by the looping arrow in Fig. 16.4). It shows an abductive design/assembly process that generates or modifies a model as an assembly of more primitive runnable schemas. The design process alternates with evaluation processes in order to evolve new models. It also corresponds to level 3 in Fig. 20.1.

The explanatory model produced by the scientific investigation process is shown on the *top row* of Table 21.1 along with processes that apply it. The model is seen as fitting into a framework of observations, models, and formal principles, as shown in Fig. 6.10. This figure shows a range of types of knowledge used in science, from the empirical to the theoretical. The four columns of Fig. 18.7 track the inheritance of flexibility and runnability from the elemental schemas that drive imagistic simulations, to the models that run them, and to a set of desirable properties of an explanatory model produced as the outcome. These properties form a basis for comparing the model to other rival models.

21.1.2 Central Role of Imagery

One feature of the vertical levels of processes presented is the case it makes for the possibility that the entire qualitative model construction process can be based on imagery as a major form of representation. Undoubtedly, some key features have verbal labels and propositions associated with them, but the sequence of levels indicates how construction can happen in a way where the central part of the process does not depend on deductions or inferences from strings of symbols. The sequence from levels one to three also indicates how runnable, scientific, explanatory models can be grounded with respect to meaning, runnability, and self-evaluated plausibility in concrete, runnable, perceptual motor schemas.

Thus by working from protocol data from individual think-aloud statements, drawings, and hand motions, these findings complement those from historical studies of experts by formulating finer-grained models of imagery and analogy use than would be possible there. These models were then in turn used to formulate finer-grained models of subprocesses involved in model generation and evaluation. Figure 18.7 at the highest level outlines hypotheses for how imagistic simulations may have interesting advantages as a representation for finished scientific models, in terms of desirable characteristics for scientific theories from history and philosophy of science studies (e.g. Darden, 1991). In sum, this hidden world of qualitative and nonformal, but powerful reasoning processes contrasts with the image of scientists as abstract thinkers who use only formal logic and mathematics.

21.1.3 Highlighted Findings

In what follows I will list some selected findings that for me were the most interesting ones in the expert studies, again organized according to the four levels in Fig. 20.1. The case studies provided initial grounding in transcript observations for the following hypotheses.

21.1.3.1 Level 1: Schema-driven Imagistic Simulation, Intuitions, and Thought Experiments

Basic Knowledge Structures

- *Imagery*. A large number of observable indicators were defined in Table 12.3 that add to our ability to detect the use of imagery. A large number of these were detected in protocols, supporting the case for the use of imagery in advanced cognition. Although they have limitations imposed by prior knowledge and memory capacity, two important operations using imagery are: (1) spatial transformations that can shape new cases and mechanisms and (2) schema-driven imagistic simulations that can utilize spatial reasoning in "running" these cases and mechanisms.
- *Perceptual motor schemas*: Experts used very concrete perceptual motor schemas in powerful ways, challenging the idea of expertise being distinguished primarily by the use of abstract knowledge in science. I used these to explain the presence of self-evaluated intuitions that were seen as the primary units of causal knowledge.
- *Knowing as action*: For some types of expert knowledge, knowledge is action oriented, as discussed in Chapter 13. An active state of knowing can involve the readiness to perform an action or the mental performance of it in a simulation. In this view, motoric images of actions were equally important to, if not more important than, perceptual images. Along with the central role of imagery, this view of knowledge contrasts with discrete symbolic knowledge, and connects the domain of higher-level cognition strongly with motor and perceptual theory.
- The finding that experts engage in *imagery enhancement strategies* was a surprise for this author that supports the claim that imagistic thinking was important to expert subjects.

Thought Experiments

- A surprisingly large number of *untested thought experiments, in the broad sense*, occurred, wherein a subject predicts the behavior of an untested, concrete, but absent system. One way these were conceived of operating was via *schema-driven imagistic simulations* that are "extended" to apply to an unfamiliar case.
- *Knowledge-based reasoning*: Observations of these thought experiments, where perceptual motor schemas adapt in order to apply to unfamiliar cases, led to the view that such *knowledge has a certain amount of reasoning power built into the knowledge*. This means that knowledge is not conceived of simply in terms of static data structures or procedural rules. This is expressed in: the activation of a schema by an unfamiliar case in extended application; and the adaptivity of the

schema to details of the case in generating an imagistic simulation for it. A major precedent here is the adaptivity of generalized motor schemas with tuning parameters as described by Schmidt (1982). This provides one possible avenue for beginning to explain how people are able to make predictions from an unfamiliar thought experiment. The focus here on action-oriented knowledge and perceptual motor schemas as sources of simulations is the sense in which knowledge elements can be *embodied* in the present framework.

- A surprising variety of other *inference sources for thought experiments*, in addition to the extended application of a schema in imagistic simulation, were inferred from transcripts, as expressed in the left half of Fig. 15.3, including tapping implicit knowledge, spatial reasoning, and unanticipated interactions from compound simulations.
- *Thought experiment paradox*: These sources can begin to explain the fundamental paradox of thought experiments, expressed as: How can findings that carry conviction result from a new experiment conducted entirely within the head? The sources utilize prior knowledge but extend it via flexible imagistic reasoning.
- A variety of *multiple outcomes from thought experiments* using imagistic simulation were suggested, as expressed in the right half of Fig. 15.3, in addition to predicting future states of a system.
- *Volatility*: These properties make such thought experiments *volatile* in the sense that they can trigger a surprising number of divergent, unplanned-for processes in many ways, such as new: recognitions of image features, coherence relations, dissonance relations, and schema activations. An unexpected schema activation was central in the explanation of a major insight episode in S2's protocol.
- *Spatial reasoning* can make spatially embedded inferences and imagistic representations can keep track of constraints at a very low cost in terms of mental resources.
- Sometimes thought experiments lead to a feeling of "getting the *physics for free*." This may be virtually true in one sense when one is able to tap implicit knowledge or spatial reasoning very efficiently, but the phrase can be deceiving, since preparing for successful simulations or spatial reasoning can take considerable effort.

The references to "knowledge-based reasoning" above reveal an interesting fuzziness in the line between knowledge and reasoning in the present framework.

21.1.3.2 Level 2: Advanced Nonformal Reasoning Processes

The Foundational Role of Simulations

- *Schema-driven imagistic simulations* can be involved not only in elemental thought experiments as described above but also serve as a basis for a surprising number of critical higher-order reasoning processes such as analogy, extreme cases, model evaluation, and Gedanken experiments, as depicted in Fig. 15.6 and summarized below.

21.1 Summary of the Overall Framework

Compound Simulations

- *Compound simulations* were identified as a fundamental means by which multiple schemas can be run in combination to produce predictions for novel systems.

Analogy

- *Invented analogies.* Analogies are often invented creatively, not just retrieved. This can take place via an imagistic transformation of the target, rather than only accessed via a symbolic association. The spring problem led 10 subjects to generate 31 significant analogies. 18 of these were generated via a transformation versus 8 via association.
- *Improvement cycles*: Analogous cases can be revised and improved.
- *Intuitively grounded perceptual motor schemas* were seen as primitives that could play the role of *anchors as the base of an analogy*.
- *Imagistic evaluation subprocesses*: In addition to the commonly described mapping process, analogy relations can be evaluated via imagery in processes such as conserving transformations, dual simulations, and bridging. These raise the question as to whether discrete mapping is a fundamental, complimentary, or derivative process (Clement, 2004).
- *Transfer of runnability*: In a projective analogy, it is not just a prediction as a discrete symbolic predicate, but imagery and/or a simulation-generating schema element that is inference from a source schema to the target or model.
- *Widely different types of analogy*: As shown in Table 20.2, there are a surprising number of purposes and modes of application for analogies that involve different kinds of inference or provocation in scientific investigation, not just one kind. Some analogies appear to activate latent schemas rather than being a source of direct inference. These different types of analogy make the relation between analogy and explanatory model construction more complex than described previously.

Since associative access, evaluation by mapping, and direct inference are standard features in classical models of analogical reasoning, it was interesting that the present imagery-based view of analogy can differ from these processes in many ways.

Gedanken Experiments

- *Evaluative Gedanken experiments* in the narrow sense were seen as a special type of thought experiment in the broad sense. Describing them as cases designed to evaluate the current explanatory model allows one to separate them from analogies as another form of advanced reasoning. They can utilize imagistic simulation in the same way as all thought experiments described in level 1 above.

- *Several types of Gedanken experiments*: There were at least two functions for qualitative Gedanken experiments documented: making a hidden mechanism perceivable; and evaluating the existing model by attempting to generate conflicts (the latter function was cited by Kuhn, 1977d). The latter function is impressive from the point of view of the fundamental paradox in that a Gedanken experiment can play a powerful role in contributing to the elimination of a scientific theory. However, one also sees Gedankens that provide support for a model. Some principles of design used in lab experiments (e.g. control of variables) appeared to also be used for Gedanken experiments.
- Gedanken experiments can be designed to evaluate quantitative models as well.

21.1.3.3 Level 3: Explanatory Model Construction

The following model construction processes are supported by the nonformal reasoning processes above.

Explanatory Model Generation and Modification

- The *distinction between expedient analogies and explanatory models* emphasizes the sense in which the latter are hidden mechanisms projected into a target.
- A new corresponding *distinction was made between a description cycle that increases confidence in the presence of a relationship and an explanatory model construction cycle* that deepens the explanation of a relationship.
- In contrast to much of the literature on mapping theory, protocols provided evidence for Harre's *three-part view of explanatory model construction* instead of a two-part view of an analogy modeling a target: (1) a source analogue which supplies material for (2) a more complex explanatory model of (3) the target (Harre, 1972).
- Intuitively grounded perceptual motor schemas were seen as primitives that could play the role of *source analogues that provide concrete elements of the model*.
- *Types of abduction*: The analysis differentiated between two important concepts of abduction: "generative abduction" (narrow sense) and "abductive model evolution" (broad sense) (Magnani, 1999).
  - As a way of generating or modify models within a set of constraints, generative abduction can play a central role in scientific investigation in addition to processes of induction or deduction.
  - Experts can use a *source analogue* or *provocative analogy* or *extreme case* to initiate generative abduction as explanatory model generation or modification.
- *Relatively sudden Insight (Aha!) phenomena* were documented, but these were intermittent, and there were also periods of stagnation. This was explained by a "punctuated evolution" model of discovery where Einstellung effects can be strong but are then broken by "volatile" processes such as activations or dissonances created by running simulations.

Explanatory Model Evaluation and Evolution

- Experts were observed engaging in dialectic *GEM (generation, evaluation, and modification) cycles* in constructing models. In constructing a model, experts are "bipolar" in that they can alternate between very divergent generation processes and very convergent evaluation processes in a way that appears to utilize very different skills.
- *Running the current model* on the target case in a simulation can be a major, basic form of model evaluation for plausibility.
- *Intelligent model evolution*: The metaphorical parallels between biological evolution mechanisms and model construction processes are substantial. However, there are important ways that abductive expert model construction can be more "intelligent" than biological evolution:
  - *Passing forward of constraints* from criticisms of a model and from highlighted features in the target, to guide the revision process.
  - This is facilitated when an imagistic representation allows one to *represent multiple spatial constraints simultaneously*; and allows free play operations on particular variables with feedback on how they affect multiple spatial constraints
- *The process of evaluating* a new model by running it *can actually contribute to the generative* process of model revision by activating schemas.
- Multiple ways to achieve "contained" or *modulated divergence* via associations and conserving transformations at different solution stages were hypothesized in Chapter 16.
- Five observed stages of modeling in an investigation process were described in Table 14.1. Intermediate stages in *levels of precision* between qualitative and quantitative stages were found in *seeking in a completely connected spatial model* and then a model with *geometric precision*.
- The importance of the connected spatial model stage and the detailed *spatial alignment* of a model with the target was driven home by the fact that once the correct qualitative model of the spring had been identified (torsion), it still led to a paradoxical result (zero stretching) because the model was aligned to the target incorrectly. Attaining spatial alignment is necessary before one can reach for higher (mathematical) levels of precision in modeling. Very little research has been done on techniques for spatial alignment.
  - *Imagery alignment analogies* were seen as a special technique that can help with this process.
- The many possible functions of analogy listed in Table 20.2 summarize the finding that the relationship between analogy and explanatory model construction is more complex than commonly realized.

21.1.3.4 Level 4: Runnable Models

- *Transfer of Runnability*: The transfer of imagery and runnability from a source schema to a model gives models the valuable capability of generating and running simulations. In the present view, this is a very *different process than simply inferring or transferring symbolic relations from the source.*

More Advanced Processes

Elements in the four levels of the framework above had initial grounding in transcript episodes from external subjects. In addition, the following more speculative hypotheses were formulated based on (1) episodes from the author's self protocol on quantitative aspects of the spring problem and (2) connections to findings in history of science.

- The runnability of a model can help explain many desirable properties of successful scientific models, as described in Fig. 18.7.
- An intermediate level of *partially centralized control* over the entire investigation process can maintain focus while allowing for interruptability and multiple methods of model generation and evaluation, seen as key features for fostering effective creativity.
- *Mathematical Modeling.* Mathematical models were seen to be built upon qualitative models that have been aligned imagistically.
  - There was a progression from qualitative alignment to geometric description to quantitative description. The latter two types of mathematical modeling *can share a surprising number of subprocesses* with qualitative explanatory model construction, including analogy and extreme case generation, Gedanken experiments, and modulated divergence. However, the source schemas used and standards for evaluation are different, requiring a higher level of precision.
  - Nevertheless it took a *surprisingly large amount of qualitative preparation* to understand the spring problem at a level that could be quantified, reinforcing the importance of qualitative analysis.
  - Mathematical modeling did not simply consist of translating models into equations. Moving to a geometric/mathematical level required *expanded precision in imagery* in the model to support the assignment of equations to the model.

21.1.3.5 Different Metaphors Contribute to Modeling at the Four Levels

The four levels discussed above use ideas and metaphors from different disciplines. For example:

Level 4: Abduction, multiple criteria for theory assessment—drawing on history and philosophy of science

Level 3: Procedural, recursive, and parallel programming metaphors—drawing on AI; Einstellung and insight phenomena—drawing on Gestalt psychology; punctuated evolution—from biology

Level 2: Image coordination and projection, image comparison and transformation —drawing on imagery theory: dissonance, coherence, and analogy processes—from psychology

Level 1: Perceptual motor schemas and imagistic simulation—drawing on motor control and imagery theory

One of the themes of this work is that explanatory models can draw on multiple analogies or metaphors, and that theme also applies to this study. I believe that scientific creativity issues are too layered and too complex to attack with a single existing theoretical framework, and that it is important to draw on ideas from different frameworks for different levels.

21.1.4 Larger Integrating Processes

In addition to the four levels of findings above, there were a number of large-scale integrating processes that cut across the four levels:

- *Overall investigation process*. The coherence and continuity between qualitative and quantitative representations was reinforced in Chapters 14 and 16 by showing how the overall investigation process in Fig. 16.4 could gradually produce a series of five closely linked stages or levels of precision in model development, ending with a quantitative model.
- *Common imagistic representation*. In Chapter 20, I summarized evidence that imagery was used in most of the processes discussed in this book. Imagistic representations were of primary importance for the inputs to and outputs from processes at all the different levels (with the possible exception of algebraic inference).
- *Recursive integration from transfer of runnability*. Transfer of runnability from source schema to model (usually through an analogy) is depicted in Fig. 20.1 as a dotted arrow, and serves to tie all four levels of modeling shown in that figure together by showing how grounded runnability at the schema (bottom) level can affect runnability at the model (top) level. Additional integration was achieved by thinking of the model produced at the end of the process as recursively possessing all the properties of a runnable schema, leading to the following benefits:
  - *Model flexibility*: The idea that a scientific model is runnable allows the model to be thought of recursively as having the properties of flexibility of imagistic simulations implied by the left side of Fig. 15.3. Although this figure was introduced initially to portray mechanisms for the role of an elemental schema, since a model is also a schema it can participate in the same relationships.
  - *Model volatility*: This act of running a new model on the target situation then also shares the "volatile" properties on the right-hand side of Fig. 15.3 and is a basic form of model evaluation that can lead to surprising conflicts as well as confirmations. In this view, runnability not only has advantages for applying a final model, but also for improving an intermediate model.
  - *Model generativity*. Runnability also allows a model to participate as a schema element recursively in any of the plausible reasoning operations in Fig. 15.6 and the model can become a runnable building block for grounding and assembling even more sophisticated theories.

- *Organizing framework of separate but interacting model modification and evaluation processes.* Figure 16.8 provides an organizing framework by associating certain reasoning processes primarily with model generation or modification and others primarily with model evaluation. These alternate in a GEM cycle. In a later section of this chapter, I will try to expand this layer of integration by discussing how a set of divergent generation and modification processes can be pitted against a set of convergent, critical evaluation processes in a cycle to produce the "survival of the fittest model" in a complementary and synergistic way.

21.1.5   Position on Concrete vs. Abstract Thinking

Experts were found to seek both abstraction and concreteness. Their use of diagrams is characteristic of this result. They used concrete but schematic drawings (and I hypothesize, concrete but schematic images) to represent target systems and general models. The trick here is how to conceive of explanatory models as being both concrete and general at the same time. In this regard, the subjects' drawings and inferred images were generic in the sense that they did not appear to commit the subject to a particular thickness or type of material or size for the spring. Yet they were concrete in the sense of being perceivable (in principle), spatial representations of a physical system, and when necessary, subjects were capable of imagining a more specific system of a size that one could imagine manipulating with one's hands. Therefore I hypothesized the use of schematic imagery stripped of detail, that could have the property of being concretely perception-like but also general (of wide scope) at the same time.

In particular, protocol data suggested that experts sought *concreteness* by:

- Seeking to build concretely imageable models as theories.
- Seeking intuitive anchoring in experiential knowledge schemas used to build these models.
- Seeking to align their model imagistically with the target.
- This means they sought an increase in precision from rough qualitative to more precisely qualitative, which I interpreted as seeking increased imagistic precision.
- This appeared to be part of seeking a *completely connected model* – with spatio-temporal connectedness from microscopic model elements to macroscopic target behaviors.
- As described in Chapter 15, subjects sometimes used "imagery enhancement" strategies, in an apparent effort to make cases they were thinking about more concretely imageable or simulatable.
- Subjects transferred material, concrete, runnable elements from source analogues to an explanatory model (e.g. incorporating twisting and torsion into the spring), thereby *adding* to it and enriching it with new features. (This presents an alterna-

21.1 Summary of the Overall Framework

tive to a common view of theory formation as an abstraction from one or more analogies or cases (Gentner, 1989; Gick and Holyoak, 1983), during which one *removes* or strips features from cases to obtain a more abstract model. It also contrasts with Nersessian's (2002) view of Maxwell's eventual process of "generic abstraction" in symbolizing and removing concreteness from his model of the electromagnetic field. She describes Maxwell as eventually having to in some sense leave behind his mechanical models of gears and rotational fluid flows when he finally formulated the symbolic field equations via generic abstraction. Because Nersessian includes generating symbolic equations within her process of generic abstraction, for her this is something more abstract than just making imagery more schematic.

On the other hand, experts in the present study sought *abstraction* when:

- Within qualitative modeling, they sought to generalize results from one part of the system to others or to the whole system. Here, I need to emphasize that this kind of generalization was not seen as a reformulation in abstract language or mathematics but toward more schematic imagery or toward applying a property of an image to a wider array of images. Thus this is a more concrete type of generalization than Maxwell's move to the field equations.
- Given the opportunity, the subjects might have continued this trend and sought to generalize results to other systems, but this was not encouraged by the protocol question or format.
- I assume there were parallel attempts during the solution at all stages to re-represent imagistic actions and results symbolically by verbalizing them more or less precisely.
- Eventually, some subjects attempted to use symbolic equations or graphs while mathematizing solutions. This occurred late in the game in the spring solution in Chapter 14, but early on in some solutions to the doughnut problem in Chapter 11.

21.1.5.1 Present Position vs. Position of Others on Abstraction

Studies by Gooding (1990, 1992) and Tweney (1985, 1996) have shown how deeply the thinking of scientists like Faraday was situated in the concrete world of their experiments. They argue convincingly that the next step in theory growth can sometimes be intertwined with and strongly influenced by particulars of the apparatus manipulations a scientist is working with. In considering the qualitative model construction done by external subjects, the present study appears to occupy a middle ground between Nersessian's emphasis on generic abstraction and Gooding and Tweney's emphasis on experimental concreteness. It contrasts with the latter in revealing the extent to which scientists can make progress via thought experiments when they do not have access to real experiments. They were able to "run" many of these experiments successfully, often giving evidence for using imagery and/or drawings that were schematic. The thought experiments were concrete enough to yield "perceivable" results but schematic enough to yield somewhat general findings.

Thus the thinking in these protocols, especially during the running of explanatory models and thought experiments, appears to have taken place at a fairly concrete but intermediate level of abstraction. So to those who would emphasize abstraction as central, I would argue that grounding an explanatory model in concrete perceptual motor schemas can be central for attaining explanatory power and a satisfying understanding. And to those who would emphasize concrete, hands-on experimental manipulation and observation as central, I would argue that imageable but schematic and therefore general models and thought experiments involving imagined actions can be central to theory development. This places the present study in a curiously intermediary position with respect to other current research on this issue.

The above is stated somewhat too strongly since Gooding and Tweney, myself, and Nersessian all agree on the abductive construction of qualitative models as a step in the process. So the differences are a matter of emphasis. If we think of these three positions on a vertical spectrum from lower to higher abstraction, Nersessian's emphasis on what she calls generic abstraction puts her 'above' the other two. But in fact she must be seen also as well 'below' classical views in philosophy of science, because of her equal emphasis on Maxwell's analogical thought experiments in contrast to formal deduction. And while Tweney and Gooding emphasize the role of actual experiments, each of them have also explored the role of thought experiments and their connection to real experiments (Gooding, 1994; Ippolito and Tweney, 1995). This suggests that what is needed is a combination of perspectives.

In the case of Maxwell's field theory, removal of concreteness was necessary (1) because of the need to go to a highly refined, quantitative level of detail in the model and (2) because the modern physics of electromagnetic fields, relativity, and quantum mechanics is an exception in science in that it extends to realms so far from our world of experience that scientists have not been able to model these realms fully in terms of concrete mechanisms. Maxwell therefore had to give up certain concrete properties of his qualitative, mechanical model of the field in moving to equations. But the need to leave major schematic aspects of a mechanism behind may not be necessary in the development of all or even most models in science (Machamer et al., 2000). Mathematical models may be developed "on top of" full qualitative models that provide an interpretive foundational base. Most scientists prefer to have a lucid, visualizable model as part of their theory if possible (Nagel, 1961). In the present framework, perceptual motor imagery rather than symbolic equation manipulation is seen as the central foundation of scientific thinking. These considerations suggest that future efforts may need to distinguish between at least two very different kinds of abstraction: imagistic generalization where a more and more schematic and general image of a scientific model is formed; and symbolization where observations or images are described in the form of verbal principles or mathematical representations.

In comparison with Faraday's playful experimental explorations with apparatus, the present cases of thought experiments are at least one step removed from real experiments, since no such manipulations were allowed. However, when one contemplates S2's hand motions as he imagines twisting long and short rods to examine their

21.2 How Experts Used Creativity Effectively

resistances to twisting, it does not seem *so* far removed, and my acknowledgment of perceptual motor engagement spans from the mental manipulation of very generic geometric objects, down to imagined manipulations that are very close to Gooding's concrete experiments. This suggests that there may be a spectrum of possible analog representations available for scientific thinking, from "thinking via external object manipulations," to manipulating images of specific apparatus, to manipulating very schematic images of generic objects.

The present study, by documenting thinking via schematic diagrams, thought experiments, and, especially, imageable but schematic internal models, indicates that a scientist's mode of thinking via qualitative and geometric models can be concrete and general at the same time – that is, concretely imageable and experientially grounded, but also schematic so as to have wide applicability and generality within the domain being modeled. Therefore results of the present study lie somewhere in the middle between Nersessian's generic abstraction on the one hand, and Gooding and Tweney's situated experimentation on the other. How this intermediate mode that is both concrete and general complements or interacts with the modes documented by these authors is an important area for future research.

21.2 How Experts Used Creativity Effectively

In this section, I first discuss the idea of nonempirical processes in science and then go on to consider how such weak nonformal methods are able to overcome the dilemma of fostering both creativity and validity in the protocols.

21.2.1 Do Expert Discovery Processes in Science Always Have an Empirical Focus?

21.2.1.1 Argument Against

Model evaluation and modification. The model of investigation in Fig. 16.8 certainly recognizes the possible bottom-up influence of scientific observations (shown in the bottom row) as well as the top-down influence of prior knowledge. However, a "no" answer to the above question for modest timescales is emphasized throughout the book by showing how much processing can occur on an issue in the absence of new data. This statement applies to the great majority of protocols in the book – exceptions are the "launching a cart" interview for S10 and S12 in Chapter 18 (where the students had limited access to manipulating a real cart) and Episode 13a in Chapter 14. In the last case subjects were given a statement confirming that torsion had been detected in springs by engineers, but the presence of torsion had already been theorized by the relevant subjects. Subsequent issues in that protocol go far beyond the presence of torsion issue, and were not informed by other empirical information. Thus new empirical information played a role in almost none of the protocol events.

Instead, subjects used nonformal reasoning and thought experiments to make discoveries about the target cases presented to them. How is this possible? The analysis suggested that by using new combinations of old prior knowledge schemas and running them in simulations on new cases, many new possible relations and explanations could be found. This suggests that significant episodes of scientific thinking can occur without using new empirical information.

Perhaps more surprisingly, this finding extended to model evaluation processes as well, in contrast to experimental testing as the traditional form of model evaluation. These nonempirical evaluation processes included:

- Running a new model to evaluate it, with the possibility of the spontaneous recognition of new dissonance relations with prior knowledge
- Evaluative Gedanken experiments
- Mathematical Gedanken experiments
- Evaluation via more esoteric scientific criteria such as symmetry, linearity, and simplicity

That is, productive cycles of model criticism and revision can take place entirely within the head without making new observations or subjecting those models to empirical testing. This of course does not deny the ultimate importance of empirical observation and testing in science, as it will often occur in separate cycles (perhaps done by different researchers). Rather, this suggests that there are other mechanisms for theory development and testing as well.

Model generation. It is hard to think of generative abduction of a new model as empirical because the generative abduction process can involve the creative invention of new mechanisms, in contrast to induction by enumeration. Such an abduction allows for the strong influence of prior knowledge schemas on the model that is developed, so the system is not just bottom up. In fact abductions may suggest new observation descriptors, and this is another important sense in which the system is top down. In the spring protocol we saw at least the following new potentially observable descriptors emerge from the nonempirical deliberations: slope of the spring wire; changes in slope; inward curvature of the wire in the horizontal plane of a loop; changes in curvature; torque on the wire, twist and torsion in the wire, and angular orientation of an element of the wire.

21.2.1.2 Argument for and Rapprochement

On the other hand, from the perspective of long developmental timescales, one might answer yes to the question of ubiquitous empirical involvement. This is because the subject's explanations were found to be grounded in schemas that were concrete and that were assumed to be experiential in origin (such as physical intuitions). That is, the ultimate origin in the past of the prior knowledge being used was undoubtedly partly empirical in nature. The abductive investigation process was constrained by at least a small number of prior observations and practical experiences with springs and with other objects in the source analogues, such as the twisting rod, so it is both top down and bottom up, from a longer developmental perspective. This same issue can apply microscopically to individual thought experiments; e.g. some authors (Norton, 1996) have tried to explain away the fundamental paradox of thought experiments by claiming that they are simply based on hidden empirical premises and therefore are empirical in nature. A more even-handed view of thought experiments has been articulated with concise clarity by Gooding (1994): "What is needed is a combination of empirical knowledge and the ability to reason with it" (p. 1041). The present theory fleshes out this statement in more detail by proposing that thought experiments and explanatory model construction processes can use schemas with an empirical history of experiential grounding; however, they can also go way beyond prior experience by applying those schemas to cases well outside their normal domains of application, or by using multiple schemas in new and novel combinations along with spatial reasoning to produce new emergent results. Thus the theory takes the position that subjects can use both empirical and non empirical resources together to produce creative products.

In summary, significant discoveries can be made during protocols where a subject receives no new empirical information. However, viewed from a developmental perspective, the use of perceptual motor schemas in these protocols suggests that the background knowledge used has a strong empirical component in its history. These schemas can be reasoned with very creatively in new combinations and extensions. This suggests the view that expert thinking is both empirical and nonempirical in its origins, but that it can be much more nonempirical at times than is commonly realized.

21.2.2 How a Coalition of Weak Nonformal Methods are Able to Overcome the Dilemma of Fostering Both Creativity and Validity

21.2.2.1 Weak Processes

Individually, the processes shown in Fig. 20.1 would appear to be rather weak, compared to using an "established" method like making deductions from first principles. Newell and Simon (1972) used the term "weak methods" to describe processes that were more general problem solving heuristics, and meant they were weak only in comparison to more specialized, knowledge based, "strong methods" that are designed and tailored to apply to a problem in a particular domain. The idea is that strong methods should be tried first if available, because they are more efficient. In this section I will first take a devil's advocate position and go beyond Newell's meaning to say how each individual heuristic on its own might be weak in the different sense of not being equal to the difficult task posed by serious creative modeling. I will then discuss how they might still be used successfully by using them together. The weaknesses of these processes are summarized in Table 21.2, which shows that there are many

Table 21.2 Potential weaknesses of "weak methods"

Abduction: Abductions are underdetermined and open up the possibility of generating too many alternate hypotheses as answers to the same question; also we do not seem to have agreed-upon procedures for successful abduction.

Flexible, weak, overall control process: The system is not as tightly controlled as in a routinized, algorithmic procedure, and surprising activations, interruptions, conflicts, inferences, and emergent images are possible. This makes the system potentially unmanageable.

Einstellung effects: Humans are subject to becoming "stuck" in a particular viewpoint or approach to a problem, so that once a misconceived model is generated, it may be difficult to break away from. Divergent strategies are needed in order to break away from such models.

Divergence leads to misconceptions: Approaches that are too divergent can generate misleading models that are not useful. The subject may not always be able to detect that a model is invalid and may be misled by it.

Transformations: In particular an object can be transformed spatially in an infinite number of ways. This generates a potentially explosive number of non-useable possibilities to consider that could exhaust the investigator.

Analogies: Because of their informal, approximate, and divergent nature, analogies have long been considered in some circles to be useful only as ornamental metaphors or pedagogical strategies, but not as important components of scientific discovery. And they suffer from the same danger described above for divergent processes in general.

Nonempirical Evaluation: Most model evaluation processes used here were nonformal and many involved the use of thought experiments outside of the realm of direct experience; subjects were not able to obtain more objective, reliable knowledge from conducting real experiments.

Limitations on Imagery Systems: Imagery can be limited in its ability to represent highly non-concrete material, and can be imprecise. And humans are limited in the amount (and complexity) of spatial information they can imagine at one time.

Use of intuitive knowledge structures as starting points: Intuitions can also be wrong or imprecise, as documented by the literature on misconceptions.

possible reasons for why science based on these processes might never get off the ground. This feeling is compounded by the fact that scientific investigation is one of the most difficult enterprises attempted by mankind. The enterprise is extremely open ended and lacks standardized procedures in many areas. As depicted in Fig. 16.8, there are a large number of different sources of multiple constraints on the design of a model that make successful theorizing extremely difficult. So we are left with the question of how any system can achieve progress, much less insights, by using such weak processes.

21.2.2.2 Overcoming the Dilemma of Fostering Both Creativity and Validity

Despite the weaknesses listed in Table 21.2, the examples presented in this book provide evidence that these processes can be quite powerful when used together in an investigation system. In this section, I go beyond this initial data-based finding in an attempt to explain it. Creating new theories is a hard problem, but perhaps the first strategy is to work in small steps that only solve a part of the

21.2 How Experts Used Creativity Effectively

problem at a time. This is represented in Fig. 16.3. In addition to the strategy of breaking the problem into parts, the separate kinds of cycles in the figure indicate an incremental approach. They are: repeatedly evaluate and modify ideas and models; slowly increase the level of precision of modeling (until finally it is quantitative); and slowly increase the depth of explanation being sought. These cycles indicate a sense in which the investigation process could be thought of as a pristinely smooth evolution process where each step in each cycle takes the growing model one step closer to the desired state. However, the protocols indicate that things do not always go smoothly, and that the system can become "stuck" by generating invalid models, being subject to Einstellung effects, and being swamped by an excessive number of divergent choices, as shown at the bottom of Fig. 21.2. These problems appeared to limit experts to punctuated evolution at best.

A coalition of processes that are individually "weak" but that collectively can overcome these problems is depicted above them in that figure, which summarizes certain features of the model presented in Figs. 16.8 and 16.3. The dilemma of divergent creativity vs. convergent validity is represented in the bottom two corners of Fig. 21.2 (cf. Kuhn, 1977a). Between them is shown the third problem, the open-endedness of abduction – the number of possible models is theoretically limitless, and if the system is too divergent there will be a combinatoric quagmire of possible directions. Figure 21.2 shows key characteristics of Figs. 16.8 and 20.1 that allow the "weak" processes to work together to resolve these problems, as follows:

- The central organizing cycle across the middle shown with bold arrows and type depicts model construction as a dialectic process that alternates between *model generation (or modification)* on the right and *model evaluation* on the left (GEM cycles). Thus the diagram shows more divergent processes on the right, and more convergent processes on the left. The figure is organized so that processes that help solve each problem at the bottom appear roughly above it, as follows.
- On the right side, a major problem is that even when there are known difficulties with a model, the subject may be "stuck" in the present way of viewing the model due to Einstellung effects (or simply have no model at all). Countering this are three divergent sources for Generative Abduction. These generation and modification processes were described in Chapter 16: associative activations, analogies that can tap runnable schemas; and transformations for model modification. The open-endedness of these processes may be critical for producing and exploring enough creative ideas to be successful in overcoming Einstellung effects.
- A major problem is that invalid models can be generated or held initially, so that strong sources of criticism are required. On the left side three major subprocesses for model evaluation are shown: evaluation via the current theoretical framework and criteria for models, evaluation via simulation (running the model as a thought experiment) and evaluation via more specialized real or Gedanken experiments.
- This figure is a "hybrid" in that it shows both control relations and information flow and is not intended to specify the structure of a system in detail. Rather it is integrative in intent as a way to show how the many processes involved in creative modeling influence each other.

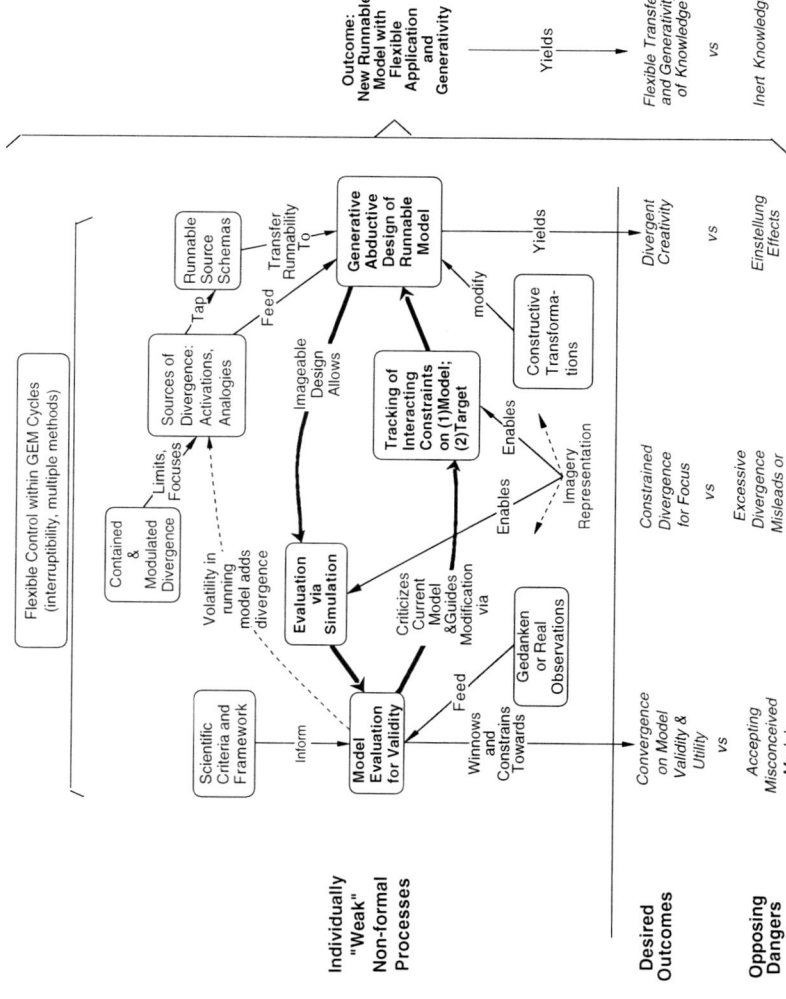

Fig. 21.2 A balanced system for productive creativity: how a coalition of seemingly "weak" nonformal methods can overcome the dilemma of fostering both creativity and validity during model construction

21.2.2.3 A Set of Mutually Supporting Processes Utilizing Imagery

Back on the generative abduction side, another problem is that of the multiple interactive constraints that a successful model must satisfy. These can accumulate each time a model evaluation is done. It was hypothesized that representing the model as a dynamic image could provide advantages for representing accumulated spatially linked constraints simultaneously, shown in the diagram as "Tracking of Interacting Constraints." The same holds for newly apprehended features and constraints on the target phenomenon. It was also hypothesized that the effects of spatially linked model transformations on these imaged constraints could be apprehended simultaneously during model repair. These are important potential advantages that can make generative abduction feasible during model revision.

Thus the most basic systemic factor in overcoming the dilemma of divergence vs. validity is the codependency of the abductive model design process and the model evaluation process. It has been hypothesized that this cycle can use imagery to great advantage. In fact, imagery was seen as a representation serving almost all of the processes in the figure. First, the formation of an imageable, runnable model makes possible evaluation via simulation, the most basic form of criticism of a model for viability. Secondly, this evaluation process can in turn feed multiple constraints on images to the generative abduction process and can activate new schemas by verbal or imagistic association. This makes possible an "intelligent" rather than simply a guess and check form of evolution. Analogies can keep the process from having to start "from scratch" – it can use image generating schemas from prior knowledge, where appropriate, in the design of the model. When imagery or runnability is transferred from a source analogue to the model, this makes the model runnable so that it can be evaluated via simulation, and so on. In addition, imagistic transformations were seen as the most controlled or intentional means of modifying a model. Assuming transfer of runnability from runnable source schemas, the outcome shown on the right in the figure is a runnable model with inherent flexibility and generativity, as argued in Chapter 18. Thus we have a somewhat complicated set of feedback loops with processes that are mutually supporting and that appear to be capable of bootstrapping to an explanatory model. This provides another perspective on how creative theory construction can gain important benefits from the use of imagery. I reviewed other possible advantages of processes using imagistic representations in the section on "How Imagery May Contribute to Creativity" in Chapter 20, so I will not repeat them here. Thus multiple generation and evaluation processes can feed each other in very important ways in order to overcome the dilemma of fostering both creativity and validity, creating the potential for converging on a viable scientific explanatory model.

21.2.2.4 Contained Divergence Helps Prevent Combinatorial Swamping

However, a third problem listed at the bottom of Fig. 21.2, that of excessive divergence and combinatorial swamping, stems from the fact that some of the above moves can involve imagistic simulations. It was proposed that such simulations are *volatile*, in that they can at any time, in an involuntary way, produce new activations, dissonance relations, or emergent image features. This is potentially helpful when one needs a new idea, but unfortunately the number of sources for divergence and multiple activation possibilities gives the potential for (1) unproductive associations that are misleading and go in unproductive directions or (2) simply generating too many choices in the system to explore. In addition there are a limitless number of combinations of transformations that could be used to change a model. Focusing strategies or control processes to counter excessive divergence are shown roughly in the middle of the figure. Whereas the problem dealt with on the right-hand side is the need for *sufficient divergence* to get new solutions, the problem in the middle is the need to *limit divergence* to avoid misleading and swamping the system.

One hypothesized solution to this problem was the idea of "controlled" or "contained" divergence that limits the degree of divergence. Identified sources of contained divergence include analogy generation under constraints (e.g. bridging analogies), adjusted analogies, conserving transformations, and associative "temperature" control. The latter is somewhat analogous to a nuclear reactor, where a volatile chain reaction is controlled by strategic damping, except that here it is controlled by idea evaluation and channeled activation and modification. In addition, the ideal level of divergence changes from high at the beginning of the investigation, to much lower near the end, suggesting that the level of divergence needs to be *modulated* in this way. Using this strategy, later stages of the solution use divergent processes modulated for less divergence and more focused evaluation according to stricter criteria. In the protocol in Chapter 14, associative analogies were deployed early on for wide divergence, whereas smaller conserving transformations tended to be used to generate analogies late in the solution. And within the latter strategy the severity of the transformations used was modulated further so as not to distort the system at the level of precision being described. This method is a generalization to the science domain from the well-known pattern in mathematics of using tightly constrained conserving transformations to produce "strong" equivalencies (e.g. at the affine level) and weakly constrained transformations to produce "weaker" equivalencies (e.g. at the topological level). More generally, the process described in Fig. 16.3 also has modulation built into it by using different kinds of analogies for the different purposes listed in Table 20.2 at different times. The modulation of divergence is one of the most sophisticated strategies described in this theory and speaks directly to the solution of the dilemma of fostering both creativity and validity at the same time. The strategy only makes sense within a higher level strategy of model evolution that moves from a rough model to more precise ones.

21.2.2.5 Flexible Control Structure with Intermediate Level of Control

Another problem closely connected to that of optimizing divergence is the problem of whether to specify a tightly controlled order for using the methods shown in Fig. 21.2. A persistent question in science studies is: Is scientific investigation a highly structured procedure worthy of the label Scientific Method writ large; or is there really no method at all, and anarchy reigns, with the scientist casting about for ideas, and using "whatever works?" Feyerabend (1975) has argued for the latter position, while many other historians of science have attempted to find some pattern or structuring in the scientific investigation process. Cognitive scientists engage in a milder version of this debate in arguing about central, hierarchical control structures vs. distributed ones with less rigidity.

The procedural structuring represented in Fig. 16.3, organized around a GEM cycle, as shown by bold arrows in the center of Fig. 21.2, expresses one model for how scientific investigation can be different from an extreme anarchistic position. That procedure serves to initiate, evaluate, and winnow divergent ideas. This structure is an important part of the explanation for how the process can be creative and yet convergent enough to be valid. On the other hand, the protocols do not support the positing of too much structure; they suggest that multiple methods and flexible sequencing allowed the exploration of many pathways on the way to an explanation. They also suggest an *interruptability* of the investigation process that allowed insights to be pursued when desirable, e.g. when a process activated a new schema that could be useful for another goal. The torsion discovery protocol provides examples of a number of the processes just discussed: The sequence begins with an apparent conscious decision to attempt to increase the level of divergence when S2 says in line 44: "I need to think of it in a radically different way". (2) S2 created a bridging analogy, the hexagonal coil case, as a way of evaluating the bending model, with (3) some evidence that it was created via an imagistic transformation in line 117. (4) The subsequent torsion insight itself apparently sprang from an involuntary process of volatile activation of the torsion schema (described in Fig. 15.3) stemming from S2 running a simulation of the hexagonal coil while staring at a drawing of it. (5) The insight interrupted the comparison being made and led to a flood of ideas that created the foundation for a deep understanding of the system. Apparently the strategy that worked in this case was to increase the level of divergence, but not to diverge so much that one loses the key constraints inherent in the problem. It gives one possible answer for the questions raised earlier about the cause of insight episodes and about how it is possible to break out of an Einstellung trap. Most of the above processes are represented in Fig. 21.2. Interruptability is modeled by a control system having multiple parallel goals operating at the same time rather than a single active goal. This feature is not captured well by the procedural notation used in Fig. 16.3, therefore an alternative model of higher-level control was sketched in Chapter 16 as a production system. In addition, imagistic mechanisms for activation and application of schemas are fuzzy, allowing schemas to be "stretched" to novel cases (Chapter 18).

This is not to say that creative thinking only occurs during breakthrough insights. Others have theorized that processes used during normal science can be creative (Sternberg, 2006). The documentation of extreme cases, bridging analogies, imagistic alignment analogies, and Gedanken experiments that were not considered to be breakthrough insight episodes in this study also argues in this direction.

In summary, the system described is less than fully structured in at least four senses from the presence of: a set of actions that is incompletely ordered; multiple parallel goals, fuzziness of activation and application conditions, and multiple sources of divergent and parallel activation of structures that can lead to interruptions. On the other hand, the system described is not anarchistic. Essentially the system has an evolutionary structure. Although the individual strategies are heuristic rather than guaranteed, Fig. 16.3 presents more structure than Polya's (1954) original four part list of heuristics for problem solving, for example. Also the system is more structured than a blind variation version of evolution via trial and error. The additional structuring comes from abduction within constraints, intelligent revision processes, partial ordering of strategies represented by the structure of Fig. 16.3 and founded on GEM cycles, the specialized use of certain types of analogies at certain times, and other specialized strategies for modulating divergence according to different stages of the solution.

In addition, not shown in Fig. 21.2 is the *focusing center of attention*: an image of the target system's behavior and constraints and an image of the current partial explanatory model and its behavior and constraints, and an overriding goal of having the latter explain (cause) the former in a spatiotemporally contiguous way. In an advanced investigation where a subject is implementing section IIA1 of Fig. 16.3, every new activation or transformation could be damped or retained based on its potential contribution to this goal. I have conjectured that this can be done via processes somewhat like compound simulations and the dual simulations discussed in Chapter 17 (of the model and the target), but exactly how this is done is not well understood. However, experts who set a goal of developing an explanatory model of the system were remarkably persistent in their pursuit of this goal.

The present model then reflects an *intermediate level of control* that gives it focused purpose but also flexibility, and this again helps explain how it is able to produce creative new ideas without becoming lost or permanently misled. If one were to add empirical inputs in the form of exploratory observations and experiments to this system as shown in Fig. 16.8 that would give it another tremendous resource for both creative stimulus and critical constraints in the dialectic between observation and theory.

21.2.3 *Overlap Between the Context of Discovery and Context of Evaluation*

In closing this section, I want to reexamine the way in which the context of discovery and context of evaluation can be mixed in the present model of investigation. Earlier I stated that a basic means of model evaluation was to run the model on a target case and inspect the effects. However, on occasion, new variables or key features may also be discovered from running a model. Thus the very act of evaluating a model (or analogy) for the first time can provide further ideas for abduction, in the prepared mind, as shown in Fig. 16.8. In the torsion insight, the discovery of the new variable of torsion interrupted the evaluation process that was occurring and sent the investigator suddenly back to the generation/discovery process. This challenges the distinction between the context of discovery and context of evaluation in science when one is examining thinking at this level of detail.

However, the distinction may still be useful in describing an effective planning strategy, as in Fig. 16.4, where there are separate steps for evaluation and modification. The saying is, "some plan is better than no plan." Even though the plan may generate surprises that ultimately interrupt and change the plan, it can still be a useful heuristic tool. Figure 21.2 expresses this intermediate position on the issue of separation vs. integration of the context of discovery and the context of evaluation. It is organized so that they are separated at the right and left sides of the picture, respectively. But the net of interconnections between them shows how they can interact strongly, sometimes involuntarily. Similarly, Fig. 16.8 separates the context of evaluation from the context of discovery (there called generative abduction), but shows how they can feed each other in repeated cycles over a relatively short period of time. In this view, over a short timescale, the two contexts can be thought about separately, but over a longer timescale, they can be heavily intertwined.

21.2.4 Section Conclusion

The theory of dialectic learning cycles of abductive generation and evaluation developed in Chapter 16 was proposed as part of the explanation of how scientists can construct complex new imagistic models. In section 22.2.2, I listed some severe difficulties faced by any system that attempts to construct really new explanatory models. The processes that appear to be used in the protocol in Chapter 14 are individually quite weak. However in this section, I endeavored to show how a coalition of ostensibly weak processes, when used together within a GEM cycle, can be powerful enough to overcome the dilemma of fostering both creativity and validity simultaneously. I discussed a number of key process interactions used to both generate and contain or modulate divergence as outlined in Fig. 21.2. As shown there it is possible to envision a system that is unconstrained enough to overcome Einstellung effects, but is guided enough to converge on solutions and avoid a variety of invalid models. However, the individual weakness of the processes involved still helps explain why the investigation process can take many false starts, and take many iterations of the model construction cycles, as was illustrated in the case studies of the spring problem. These strategies are not guaranteed to work, because it is fundamentally still partly a "guess and test" approach. But the coalition pictured appears to be a way to maximize the chances for progress or insight by using imagistic representations to advantage and attempting to optimize the degree of divergence at different stages.

21.3 Educational Applications: Needed Additions to the Classical Theory of Conceptual Change in Education

In this section, I will discuss gaps in the classical theory of conceptual change in education, summarize the expert–novice similarities identified in this book, and use these to suggest aspects of an expanded model of instruction for conceptual change in science that builds on student's intuitions and natural reasoning processes.

21.3.1 Uses and Criticisms of Kuhn

In their classic paper, Posner et al. (1982) drew on Thomas Kuhn's description of scientific paradigms and revolutions as a metaphor in developing the outline of a theory of conceptual change in science education. This likened the problem of persistent misconceptions to that of persistent paradigms in science, and the role of anomalies in fostering paradigm shifts to the role of discrepant events in fostering conceptual change in the classroom. I refer to their outline here as the classical theory of conceptual change in science education.

Kuhn's ideas about anomalies and the persistence of paradigms, were reviewed in Chapter 7. Kuhn's ideas have been exceedingly influential but have also drawn criticisms. Some of the most common are: there is vagueness in what is meant by the term "paradigm"; science progresses more often by "evolution" than "revolution," progressing in a more continuous way. Kuhn also introduces the idea that concepts in different paradigms can be "incommensurable"; this means that discussion on crucial points cannot take place between members of different paradigms due to their possession of different concepts and languages. This idea of strong, global incommensurability has also been criticized as overly pessimistic. So an important question concerns which of Kuhn's ideas can serve as useful metaphors or starting points for a theory of conceptual change and how they should be revised or augmented.

21.3.2 Criticisms of Classical Conceptual Change Theory

Limitations and extensions of the classical conceptual change theory have also been identified by others. I divide these criticisms into two categories, external and internal, as shown in Table 21.4. External limitations point to larger issues that a cognitive theory will not cover, and these should be handled by extending the theory to encompass them as complementary issues. Some might argue that these larger systemic or situated factors make cognitive theories irrelevant, but that would be like saying that molecular structure theory makes atomic structure irrelevant. Internal critiques on the other hand claim that elements of classical conceptual change theory were wrong and must be changed. These include the idea that paradigms and conceptual changes go through large, relatively sudden changes in a revolutionary phase, since many aspects of the paradigm must be changed all at once for the new paradigm to make sense. Critics claim that neither science nor students can typically do that, and that both often go through changes that are smaller and more gradual. In this view, a theory is often not "replaced" in a revolution so much as it is modified during its evolution. This would allow instruction to start from a student's ideas and modify them gradually. Another criticism is that the use of strong dissonance may discourage certain students.

21.3.3 Need for an Expanded Theory of Conceptual Change for Education

I would argue that the identified flaws internal to the classical theory can be repaired by appropriate modifications to the theory. I believe a more serious problem not shown in Table 21.4 was that the theory was not only flawed, but seriously underdeveloped internally; that is, it merely provided an outline for a theory of learning and some *conditions* for learning rather than a set of learning *mechanisms*. In the next section, I summarize the evidence from this book on whether there are similarities between expert and student learning mechanisms. The hope is that this analysis of learning mechanisms will contribute to expanding and informing the evaluation of classical conceptual change theory, including the issue of whether there are valid analogies between expert processes such as those described by Kuhn and student learning processes. Then at the end of the education sections, I will use this analysis to suggest that certain of Kuhn's ideas should be retained, avoided, or extended in an expanded theory of conceptual change that includes detail on learning mechanisms.

21.4 Expert–Novice Similarities in Nonformal Reasoning and Learning

Many educators have assumed that there are sharp differences between experts and students in their approach to scientific thinking. There is some research that highlights the curiosity of young children and draws some parallels to investigatory processes in scientists (Metz, 1998; Brewer and Samarapungavan, 1991; Hewson and Hennessey, 1992). However, less work has been done with older students on the question of whether they can reason about models of scientific concepts at the secondary level and higher. In this section, I summarize findings on students' spontaneous reasoning from the present studies, and attempt to give a balanced picture by pointing to both similarities as well as differences with the findings on experts.

21.4.1 Similarities Concerning Resistance to Change

21.4.1.1 Experts

Certain aspects of Kuhn's theory receive some support from the expert case studies in this book: that Kuhn was partially right in identifying real *resistances to change in science* that were not only social but cognitive/individual. As discussed in Chapter 7, analysis of protocols of individual scientific thinking allows one to see real barriers to change in the form of Einstellung effects and the coherence that can be present in a wrong theory. In particular, S2's long period of little progress where he is "stuck" in the view of the spring as bending vertically and in which he expresses his frustration with not being able to get new ideas illustrates this. The role of recognized inconsistencies that cause dissonance as an analogy to the role of anomalies in Kuhn's description of what drives change in science was also evident in that example. These factors are shown in column two of Table 21.3. I take this finding as very consistent with Kuhn's point that there can be considerable resistance to theory change in science. If people have difficulty breaking out of procedures or ways of viewing problems when they have just developed those procedures, then they should have even more difficulty breaking out of older procedures they have learned and even more out of historically sanctioned ones. As discussed in Chapter 19, science may progress in a pattern of punctuated evolution, as depicted in Fig. 19.1a, graph 2. This is an intermediate position between Eurekaist and accretionist points of view.

21.4.1.2 Students

In the "Book on the Table" protocols in Chapter 10, I also presented evidence of *resistances to change in science students* where they did not respond quickly to instruction for change, but had to be taken through numerous intermediate steps in order to begin to give up their own point of view. By selecting concrete examples carefully however, the tutor was able to promote enough *dissonance* with this belief to make progress.

In fact, the instructional topic in this instance was of intermediate difficulty; there are some that meet less resistance – and some that meet more resistance, such as relative motion. Some offer much more resistance such as the relationship between force and motion, where misconceptions can take weeks or months to overcome in advanced high school and college classes. In all these cases the subject is "stuck" in a previous way of thinking.

These factors are shown in row 2 of Table 21.3.

21.4.1.3 Incommensurability

Kuhn raises the question of whether scientists undergoing a paradigm change must make major changes in fundamental concepts (such as moving from classical mass to relativistic mass) as well as changes in larger theories (see also Wiser and Carey, 1983). There is no change of this magnitude in the expert protocols

21.4 Expert–Novice Similarities in Nonformal Reasoning and Learning 535

Table 21.3 Kuhnian issues in the protocols, classical conceptual change theory, and extensions to education (T = Teacher; S = Student)

T. Kuhn	Present expert findings	Recent student findings	Classical conceptual change (CC) theory	Recommended expansion of CC theory for education
Incommensurability	New concept differentiations required	Difficulties from T and S using different key concepts		Need concept differentiation or integration strategies
Cognitive (and social) persistence of paradigms	Resistance to change – Einstellung effects	Resistance to change from pre-conceptions and Einstellung effects	Resistance to change from persistent pre-conceptions	Resistance to change from Einstellung effects as well as preconceptions
Paradigm shifts triggered by multiple strong anomalies	Role of anomaly or inconsistency as source of dissonance	Role of anomaly or inconsistency, as a source of dissonance	Role of anomalies producing dissonance	Additional role of discrepant questions and non-empirical inconsistencies as important sources of dissonance
Role of thought experiments in disconfirming theories	Role of thought experiments in both supporting and disconfirming models	Students can run models in thought experiments to test them		Role of thought experiments in both supporting and disconfirming models
Theory as a coherent system	Import of spatio-temporally connected models	Nonformal reasoning abilities but instruction ignores them & isn't coherent	Instructional objectives as coherent concept clusters	Researched learning pathways; coherently connected spatial models
(Model evolution not emphasized)	Punctuated model evolution cycles: successive model modifications			Scaffolded model evolution cycles; expect periods of faster and slower progress depending on resistance
	Analogy to intuitions; multiple roles for analogy; transfer of runnability	Analogy to intuitions; multiple roles for analogy; transfer of runnability	Analogies may help	Theory of analogy formation and evaluation; multiple roles for analogy; transfer of runnability
	Imagistic simulation; grounding in perceptual motor schemas; transformations	Imagistic simulation; grounding in perceptual motor schemas; transformations		Theory of imagistic simulation; grounding in perceptual motor schemas; transformations

Table 21.4 Common criticisms of the classical theory of conceptual change

External criticisms claim that the classic theory of CC is incomplete:
Motivation
Situational context of application
Social learning
Metacognition
Internal criticisms claim that the classic theory of CC is flawed:
Big changes only
Sudden change
Learning as replacement
Sharp dissonance discourages students

discussed here. The closest case is probably the new differentiation and definition of "curvature" required to move from a "vertical curvature introduced by bending" view of the drastically stretched spring to a "horizontal curvature reduced by unbending" view. However, there is no evidence that these two views are strongly incommensurable, although lack of this differentiation in the definitions of curvature could presumably held up progress.

Readjustment and redefinition of concepts were definitely issues brought out by the "Book on the Table" instructional interviews since it was recognized that the student's concept of "force" and the tutor's were not the same. This again could cause fatal communication problems if unrecognized, but students appeared to be capable of making progress with a savvy instructor on this topic despite that difficulty, although this might take considerably more time and energy than most would expect (see also Leander and Brown, 1999).

Thus there are at least partial similarities on the dimensions of resistance to change, the role of anomalies, and local, partial concept incommensurability between processes in expert think-aloud data and in Kuhn's theory of change in science. These factors also appear to be important in the case of students.

21.4.2 Similarities in the Use of Intuition and Imagery

21.4.2.1 Students' Knowledge

Observations such as the depictive hand motions described in Chapters 9, 12, 13, 15, and 18 suggested that both students and experts were using physical intuitions to ground their ideas in perceptual motor schemas as they solved unfamiliar problems. Most of the spontaneous analogies students reported in Chapter 8 were to familiar situations where the student was likely to have an intuition from prior experience. Many of these were classified as personal analogies where the analogous case involved some bodily action, suggesting that perceptual motor intuitions were preferred. And for problems involving the conception of force, students often gave evidence for grounding conceptions about moving objects in ideas about actions exerted by their own muscles. Evidence provided in Chapters 9, 13, and 18 supported the interpretation of these as imagistic simulations involving perceptual motor schemas.

21.4.2.2   Expert Knowledge

What is perhaps more surprising is that experts thinking about the spring and wheel problems appeared to use these same processes. In unfamiliar problems they used very concrete, often kinesthetic knowledge as axioms on which their arguments were founded. Thus, both experts and naive students were observed to use physical intuitions spontaneously in anchoring cases in answering problems when these problems were outside the domain of their formal science knowledge. And these intuitions were often of a perceptual motor type.

21.4.2.3   Potential Value of Anchoring Intuitions

It is also significant that examples which tap useful anchoring intuitions have been found for many topics in our previous instructional research efforts (Clement et al., 1989; see also Fischbein, 1987 on intuition in mathematics). There are examples where 75% or more of the students in physics classes will make a prediction that agrees with the scientist's prediction PRIOR to instruction. These indicate that not all preconceptions are misconceptions; students not only have physical intuitions that conflict with current theory, but also ones that are in agreement – an important expert/student similarity. Such anchoring intuitions are a valuable untapped resource for science education.

In summary, students are not blank slates. They have relevant prior knowledge in the science areas studied here, both in the form of misconceptions that are largely incompatible with currently accepted theory, and in the form of anchoring conceptions that are largely in agreement. Students and experts were both observed to use physical intuitions as important resources in their thinking and these were modeled here as involving concrete perceptual motor schemas which generate imagistic simulations.

21.4.3   Use of Analogies by Students

By documenting a large number of spontaneous analogies generated during problem solving, Chapter 8 indicated that many engineering students have a natural ability to engage spontaneously in analogical reasoning. However, on the negative side, it was noted that not all of the analogies were correct from a physicist's point of view and that only a few students gave evidence of criticizing their own analogies. In contrast, for the experts studied here, evaluating the validity of analogies appeared to play an important role, and they rejected many of the analogies that occurred to them as possibly misleading. Thus it is unlikely that students will generate completely successful analogies nearly as often as experts.

However, in some cases the student analogies led to significant conceptual progress. For example, students were able to access helpful analogies for a relative motion problem. And some of the analogous cases were invented creatively by students rather than simply accessed as an existing case in memory. As described in Chapter 9, S20 was able to overcome a difficult and commonly held misconception by using analogies. And in a separate case study, the women quoted for the friction example in Figure 18.8 were able to tie together two rather different theories via a spontaneous analogy.

Overall, these findings suggests that although most of these students exhibited spontaneous analogical reasoning, analogy criticism and evaluation appears to be an area where scaffolding and training is needed to complement natural analogical reasoning capabilities. With that support, there is reason to believe that students are capable of using analogical reasoning productively during learning. This is also supported by the finding that children are good analogists and are good at exploiting usable partial analogical relationships, particularly often in naive biology (Inagaki and Hatano, 1987). Teaching studies by Johsua and Dupin (1987), Clement (1993b), Harrison and Treagust (1996), Mason (1994) and others support the value of analogies in science instruction.

21.4.4 *Model Construction by Students*

Experts were described in this study as engaging in a model construction process of generation, criticism and revision, often by starting from a rough analogy and refining it. Did students display this pattern? Previous research on such questions has generated controversy. D. Kuhn et al. (1988) has argued that nonscientists do not use deductive reasoning patterns reliably enough to succeed in making inferences from scientific experiments. On the other hand, Metz (1998) and Brewer and Samarapungavan (1991) have found that even primary school students can learn to make such inferences under supportive conditions. But little previous work has been done in the area of nonempirical model construction processes.

Among the 11 students who generated articulated analogies in the study in Chapter 8, three generated a series where an analogy was refined or improved. This constitutes a process of model evaluation and improvement (where model is used in the broad sense) and prefigures the process of criticizing and refining an explanatory model in the narrow sense. However, on the negative side, the low frequency of spontaneous evaluations cited for students in the previous section suggests that few students are likely to generate precise scientific models on their

21.4 Expert-Novice Similarities in Nonformal Reasoning and Learning

own by criticizing and revising an initial analogy or model during problem solving. In addition, there were few or no examples of very sophisticated strategies seen in experts, such as evaluative Gedanken experiments or strategies for modulating divergence.

An exception to this on the positive side is S20's protocol analyzed in Chapter 9. His thinking was remarkably similar to processes of model construction seen in expert protocols: he was able to criticize and modify a whole series of analogous cases on his way to evaluating a model of the resistance of objects to acceleration in outer space. In doing this he was able to overcome a common misconception on his own by engaging spontaneously in many of the basic nonformal reasoning processes. In addition, one of his analogies appeared to be an effective bridging analogy. His solution demonstrates that it is not only experts that can use multiple forms of nonformal reasoning in sophisticated cycles. This episode, along with the "elastic friction forces" episode of student learning described in Chapter 18, also contain substantial evidence that fits the hypothesized expert pattern of seeking to "transfer runnability" from existing intuitions to the problem of constructing a runnable explanatory model. Almost all of the 34 articulated analogies in the study of spontaneous analogies in Chapter 8 fit a somewhat parallel pattern, in that the student seeks to transfer intuitions from a dynamic situation that is familiar to the dynamics of the target problem. And although we did not attempt to record hand motions or other imagery indicators in that early study, 53% of the analogies involved personal action projections, consistent with a transfer of runnability mechanism.

We therefore have a mixed picture in the area of student model construction. Most students studied did not exhibit spontaneous model construction cycles. However, the context of solving short but unfamiliar physics problems may not have been conducive to this. On the other hand, occasionally a student did generate a series where an analogy was evaluated and improved, and many did exhibit important subprocesses, such as analogy generation and the use of personal action projections. Many show aptitude for the process of transferring runnable intuitions from a familiar situation to the dynamics of the target problem. Thus, many of the students we have studied show a potential for using basic subprocesses of model construction described in this book.

The broader literature on student model construction is also mixed. Driver (1973) did document model construction cycles in middle school students in a university lab school participating in open ended laboratory investigations in mechanics, with very minimal scaffolding by the teacher. Many of these cycles showed students generating modifying, and expanding impressive, reasonable theories that made sense to them, although they had much more difficulty dealing with misconceptions and designing experiments to test their theories thoroughly so as to lead to well-argued conclusions. Roth and Roychoudhury (1993), White (1993), Metz (1998), Hewson and Hennessey (1992), Easley (1990), Johnson and Stewart (1990), Linn (1992), and Hammer (1995) have documented very impressive student model construction cycles, but they have reported various levels of scaffolding by the teacher to support it, so that they lead to target models.

21.4.5 Summary: Expert–Novice Comparisons

In summary, there are differences between the expert and novice protocols in this work, but there are also important similarities, and these are summarized in columns 2 and 3 of Table 21.3. On the negative side, experts and students face similar challenges in unfamiliar areas in the form of persistent misconceptions and Einstellung effects. On the positive side, expert–novice similarities in the four areas of intuitive grounding, imagistic simulation, analogical reasoning, and runnable models suggest that there exist promising resources for developing scientific models in students, if only these cognitive resources can be tapped in the right way. These findings complement the prior focus in the literature on expert–novice differences with a focus here on expert–novice similarities.

Science education researchers are in agreement on the need for active involvement in reasoning on the student's part for lasting learning. This raises the question of whether students are capable of such reasoning, which goes beyond rote memorization. Older perspectives on science education have asked whether students possess the deductive reasoning skills and the mathematical skills to learn science content. Teachers and researchers vary in their opinions about what the answer is. However, the present framework shifts the emphasis to what I believe is the more central question of whether students have the necessary nonformal reasoning and model based learning skills to learn science content at the secondary level or higher. I believe the answer to the last question is "Yes, with support" and "Rarely, without support." Placed in a problem context without training, most students studied did not spontaneously exhibit a coordinated use of many processes together in a sustained cycle of theory development, criticism and revision such as that shown in Fig. 21.2. However, they did exhibit the use of a significant number of individual subprocesses such as analogy generation and simulation using physical intuitions. This suggests that novices do not reason just like experts, but there are significant overlaps in reasoning subprocesses used. This does not mean that all novice processes are strong, or that they are executed with the same degree of precision or skill as in experts. But it does suggest that, to a surprising degree, the foundations are present for model based learning, and that with the right context and proper scaffolding and support from an instructor, students may be able to participate in model construction and criticism processes that are similar in many ways to expert processes. In the next section, I will concentrate on ideas for tapping these valuable resources. Such an approach builds on the student's intuitions and natural reasoning processes.

21.5 Implications for Instructional Strategies and Theory

I will break the discussion of instructional implications of these findings down into two parts: strategies suggested by Part One of the book, including the chapters on teaching, and strategies suggested by more recent work on the role of imagery in model construction in Part Two of the book.

21.5.1 Strategies Suggested by Initial Studies of Analogy and Model Construction in Part One of the Book

21.5.1.1 Dissonance

The use of dissonance in conceptual change approaches is widespread but somewhat controversial. Stavy (1991) argues that dissonance may discourage certain types of learners. However, successful uses of dissonance have been described by Zietsman and Hewson (1986), Chan et al. (1997), Dreyfus et al. (1990), Jensen and Finley (1995), Niedderer (1987), Rea-Ramirez and Nunez-Oviedo (2008b), Tsai and Chang (2005), Clement and Steinberg (2002), and others. A simplistic model of the role of dissonance in conceptual change is shown in Fig. 21.3a. A discrepant event from a demonstration or experiment DE conflicts with the student's model M_1, leading the student to adopt the target model M_T. While this model provides a reason to reject or revise M_1, it says nothing about the source of the new model M_T. The use of analogies speaks to this gap as follows.

21.5.1.2 Basic Strategy of Learning by Analogy to an Intuition

Content Goals

Anchoring. Figure 2.4 depicted the elements of a "basic" strategy of analogical reasoning, derived from expert protocols. Efforts are made to find a case B that can be seen as analogous to the target case A. There are then three major requirements for comprehending the intended analogy: (1) subjects must understand the anchoring case with some degree of conviction; (2) they must confirm the plausibility of the analogy relation; and (3) they must apply findings from case B to case A. There is a fairly exact match between the basic subprocesses of expert analogical reasoning (also shown in Table 2.1) and those used in the instructional anchoring strategy described in Chapter 10. In this strategy the first step, generation of the analogous case, is often done by the teacher, but also sometimes by the student. Then understanding of the behavior of the anchoring case is evaluated, usually through "voting" in class, and refined if necessary. After this, the question is raised as to whether the suggested anchoring case "really works in the same way" as the target case (evaluating the validity of the analogy relation). Finally students practice transferring their intuition from the anchor to the target and to other similar examples.

Research needed on anchors. With respect to the first of the three requirements listed above, I would emphasize the need to search for anchoring intuitions. There has now been a considerable effort in the educational research community to identify students' common misconceptions. An effort to search for anchoring intuitions would open up a large field of research that would complement the ongoing research effort on misconceptions. Potential anchoring examples can be listed by

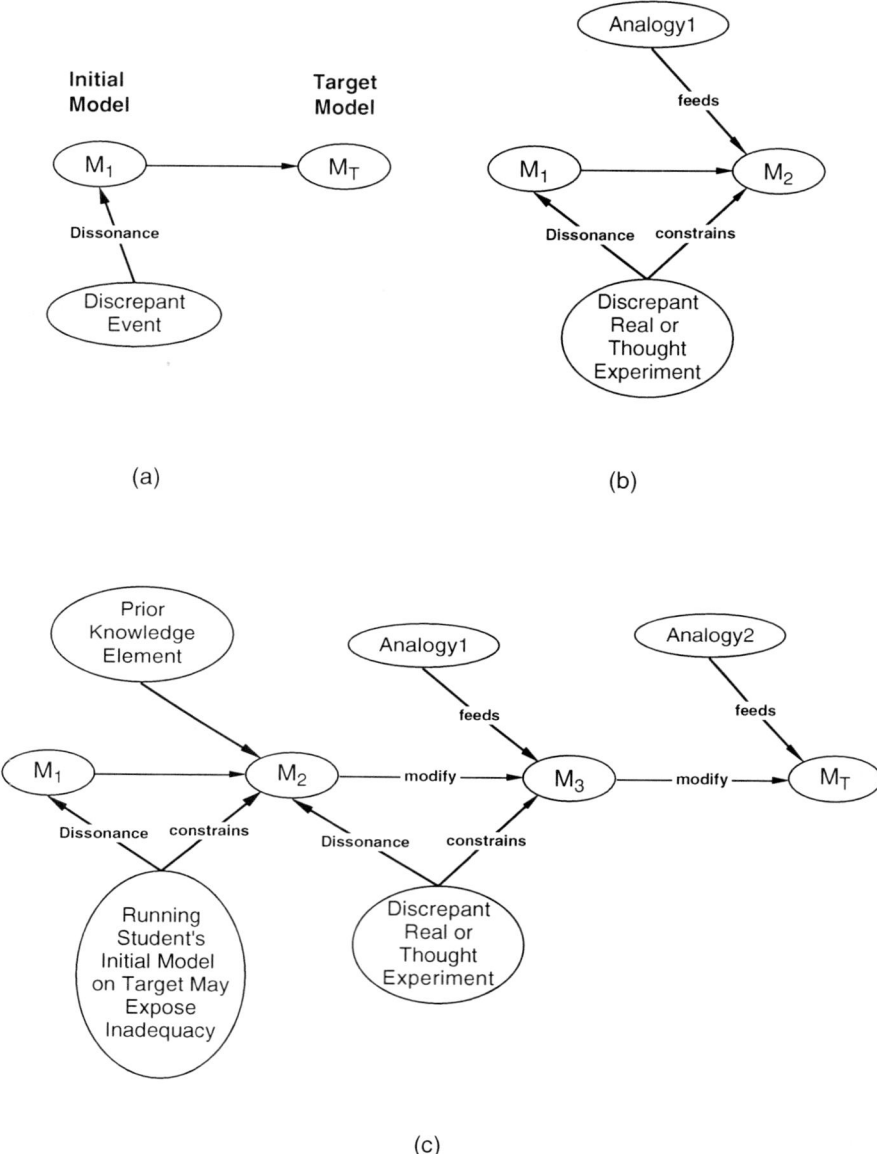

Fig. 21.3 Possible models of conceptual change learning. (a) Simplistic model of the role of dissonance in conceptual change; (b) Analogy can play the role of a proto model as one source of material to help build a new model; (c) Series of student models produced by GEM cycles

21.5 Implications for Instructional Strategies and Theory

skilled teachers, but they require empirical confirmation. For example, our team confidently predicted that hitting a wall with one's fist would be an excellent anchoring example for the idea that a static object can exert a force. Surprisingly however, only 41% of pre-physics students tested agreed that the wall would exert a force on one's hand.This should forewarn lesson designers using analogy from basing their work only on an armchair analysis of a topic. Anchors can sometimes be suggested by students as well. For example, my own group has utilized the anchor generated by S20 (in Chapter 9) of imagining pushing on something very large, like a rocket, in space. It was tested in trials and used in the teacher guide by Camp et al. (1994).

In areas where students have insufficiently developed anchoring intuitions they may need to be developed by real or simulated experiences such as Arons' (1990) activity of having students push large objects (e.g. a block of dry ice) in a low friction environment, McDermott's (1984) use of air hoses to accelerate dry ice pucks, or diSessa, Horwitz, and White's use of dynaturtle (White, 1993). I noted expert S3's use of an extension analogy in Chapter 10 as another technique for developing a weak anchor.

Analogy evaluation and bridging. The major educational implication of this work with regard to analogy evaluation is that much more effort than is usually allocated should be focused on helping students to make sense of an analogy. In classrooms this requires careful instructional planning that promotes discussion of these qualitative issues. In particular there is an excellent match between the expert strategy of evaluating an analogy by using intermediate bridging analogies (e.g. Fig. 4.3) and instructional uses of bridging to increase the plausibility of an analogy. As exemplified by the tutoring studies in Chapter 10, since students often have difficulty in the crucial step of seeing the plausibility of an instructional analogy, the use of bridging analogies may be very important. In a full classroom this often has the effect of increasing the intensity of the discussion and eventually of making the analogy make sense to more students. The hypothesized effect of such bridging strategies, as used in the book on the table problem for example, is to expand the domain of application of the anchoring schema M_1 – in this instance of something pushing back when it is deformed (elasticity). Students use this schema naturally for the hand on the spring case (anchor A). When a successful bridging case of, say, the book placed on a bending yardstick is introduced, the elasticity schema's domain of application expands to form model M_1'. Finally the domain expands more in applying the elasticity model M_1'' to the book on a solid table. This conflicts with their previous conception of the table as rigid and incapable of generating forces. Starting from the student's model of easily deformable objects like springs, in this case it is the model's domain of application that changes during learning rather than the basic imagery of the model.

It is hypothesized that students are more likely to change an intuitively grounded misconception if they work through the reasoning involved in confirming an analogy to a different grounded intuition that can compete with the misconception. This has support from Brown (1992), who compared a group of students who learned about Newton's third law from multiple examples in an inductive strategy to a group who were given a sequence of bridging analogies that was grounded in an anchoring intuition, showing significant pre-post gain differences in favor of the latter group.

21.5.1.3 2nd Function of Analogy as Proto Model

In response to the questions raised earlier about the instructional strategy in Fig. 21.3a, Fig. 21.3b depicts an improved strategy by showing how analogy can play the role of a proto model as one source of material to help build a new model. For example, the concept of colliding balls (as in billiards) can be incorporated into a particle model of gasses as a prototype that will be developed further. Figure 21.3b also contrasts this role for analogies with that of a "domain stretcher" described above. There the analogies are cases that are explained by the developing model. In Fig. 21.3b, the analogy becomes a subschema that is incorporated into the model itself. It was noted in Chapter 10 that this last process conflicts with a widespread view of science learning as always seeking higher levels of abstraction. Rather, analogies used as proto-models are designed to help *enrich* the students' conceptions by adding imagery to the models. This enrichment process was shown schematically in Fig. 10.5.

Dissonance. One possible emerging theme from these studies is that dissonance is unlikely to work on its own as a strategy but that it can be useful early on in conjunction with other strategies. Therefore I have retained the use of dissonance producing strategies as a suggested part of a theory of conceptual change (shown in the last column of Table 21.3). Figure 21.3b shows the integrated use of analogy and discrepant situations as combined methods. There the discrepant situation (a real experiment or a thought experiment) plays a negative role in casting doubt on the initial model, while the analogy plays a positive role in providing material for the new or revised model. However, a number of questions about dissonance remain. Whether there is a way to foster conceptual change with minimal cognitive conflict is an empirical question. Vivid discrepant events may produce the strongest dissonance (but not always – see Chinn and Brewer (1998)). Bridging analogies used in Chapter 10 represent a "softer" approach to dissonance but can still produce it. The idea that dissonance in small manageable doses may be an optimal approach has been discussed by Clement and Rea-Ramirez (1998) and Clement and Steinberg (2002).

Success of lessons using these expert learning strategies. Interviews in Chapter 10 provided evidence for the persistence of students' preconceptions in some areas of science, as does a great deal of literature mentioned there. A major theme of this book is that nonformal expert reasoning methods such as analogy may be useful in instruction for dealing with such preconceptions by producing conceptual change. Evidence for change was present in the protocols in Chapter 10 as well as from larger classroom studies reported there (Minstrell, 1982; Brown and Clement, 1992; Clement, 1993b). The last two studies indicated large significant gain differences over control groups on the order of one standard deviation in size. These studies have shown that positive widespread effects are possible to attain in real classrooms as measured by multiple choice tests. Tutoring interviews such as those in Chapter 10 included explanations of each answer in the student's own words as well as statements about how much sense the ideas made; this gave stronger evidence for real understanding. Having these two types of findings point in the same direction is very encouraging.

21.5.1.4 General Theoretical Implications of Model Construction Cycle Findings

In this section, I will propose that the *expert* model construction cycle in Fig. 6.8 may also be useful as a description of processes which need to take place when *students'* learn and apply scientific models. This figure is a simplified version of Fig. 6.9.

As shown in Fig. 6.10, explanatory models are seen here as a separate type of knowledge from either empirical laws or formal quantitative principles. Easley (1978), Gilbert, et al. (1998), and others have noted the unfortunate tendency of educators to associate "real" scientific thinking with only the latter two types of knowledge. Unfortunately, physics students for example also tend to assume that the priority for learning is on physics formulas. *They need to become aware of the central role of qualitative models as an underlying foundation.*

At the most general level, the present framework suggests that students need to learn complex models via an internal construction process, not via a direct transmission process during lecture. I cannot support this assumption fully here, but current research in science education is providing an increasing amount of evidence in this direction. As illustrated by the protocols in Chapter 10, the problem is that having one's current model criticized by the teacher does not seem to be enough. The complex, tacit, nonobservable, and sometimes counterintuitive nature of scientific models means that misconceptions or "bugs" (from the expert's point of view) will be the rule rather than the exception during instruction, requiring critical feedback and correction processes. If students are to understand *why* an established model has certain advantages over their preconceptions, they must be able to take an active role in criticizing their own conceptions and evaluating which make the most sense. In other words a major part of *what it means to be involved in the process of making sense of the topic is being involved in the self-evaluation of their own understandings.* This means that the learning of complex, unfamiliar, or counterintuitive models in science requires a kind of learning by doing and by construction and criticism rather than by listening alone.

The generation, evaluation, and modification or GEM process in Fig. 6.9, is seen as a core idea for the learning of scientific concepts via such an internal construction process. The term "knowledge construction via active learning" has been much used in discussions of education by educators originally inspired by Piaget and others, using concepts such as knowledge construction, disequilibrium, and accommodation. The view developed here offers a explicit description of the knowledge construction process as it occurs in science at a higher level than the cycles in infants originally described by Piaget. Fig. 21.3c shows the resulting series of student models that can be produced by such a GEM cycle. This figure is both an extension of Fig. 21.2b and a greatly simplified version of the expert model construction process in Fig. 16.8. Successful examples of such a process in instructional settings have been documented by Clement and Rea-Ramirez (2008), and Clement and Steinberg (2002). Figure 21.3c shows both "top down" sources of material for models from analogies to prior knowledge, and "bottom up" sources from real or thought experiments. It indicates a suggested direction for expanding and deepening our notion of conceptual change processes in education. More details on this synthesis are discussed in Nunez-Oviedo (2008b).

21.5.1.5 Model Evolution Cycles in Instruction

In this section, I discuss a number of factors that are important for fostering such model evolution cycles in students, organized according to the three aspects of the GEM cycle.

Model generation. Chapter 10 provided evidence that teacher guided cycles of more and more challenging analogies could be effective in tutoring. In normal classrooms, balancing individual and *social* learning is a key factor. The larger classroom lessons in studies that were referenced in Chapter 10 use large-group and small-group discussions extensively to promote the joint construction of scientific models. Very valuable nonempirical generation, criticism, and modification processes may take place if students attempt to *give explanations and discuss their strengths and weaknesses* in large- or small-group discussions (Brown and Clement, 1992; Nunez et al., 2008a,b; Khan, 2003; Duschl and Osborne, 2002; Wells et al., 1995; Driver, 1973). This simple implication is probably greatly under emphasized in instruction.

Educators differ on the degree of input and support that a teacher should provide for an inquiry process like the GEM cycle. In general, those emphasizing process goals favor less explicit input and those emphasizing content goals favor more. This makes sense. However, even when content goals are foremost, the degree of support question is still difficult and important. One approach is to use a process of *teacher–student co-construction* where both teacher and students contribute ideas to model construction, as in Nunez et al. (2002), Williams and Clement (2007a, b). In this approach it is argued that students cannot be expected to invent all standard models of complex hidden processes on their own in a reasonable amount of time. The teacher may share equally with students the responsibility for generating ideas for models in such discussions as a first step in promoting model evolution cycles in instruction. When students introduce poorly articulated models, the teacher can play a clarifying, summarizing, and focusing role during discussions that is very important (Nunez et al., 2008b; van Zee and Minstrell, 1997; Clement, 2008c). When students are unable to generate a relevant model, the teacher can introduce an analogy based on physical intuitions, as discussed above.

Model evaluation. Demonstrations or laboratories may be arranged with the goal of providing empirical dissonance with students' misconceptions. But a danger here is the false expectation that such a strategy on its own will lead automatically to the modification of the student's model in a positive direction. Classroom discussion that incorporates students' interpretations of laboratory results are seen as essential. Teachers can also ask students to evaluate models on the basis of whether they explain the phenomenon. One way our group has done this is via "Explanation Criticism Sheets" which ask small groups to discuss, evaluate, and rank written alternative explanations of a phenomenon they have seen in lab (Camp et al., 1994). One of the explanations is the physicist's point of view, but the students do not find out which one until after the exercise. This can make model evaluation an active process for the student, even in those cases where the model is generated by the teacher. The fostering of dialectic discussions requires the careful development of a spirit of inquiry in the classroom where students' ideas are valued. But on the other hand, ideas in science cannot be immune to criticism. Minstrell (1982) discusses strategies for distancing ownership of ideas from individual students to make criticism less threatening.

Model modification and evolution. Explicit strategies for fostering modification are rare, but I will mention two examples. One is to simply ask students for modifications but to only *consider small criticisms of the model one at a time*; this makes it simpler for students to generate a "fix," as suggested in Fig. 16.9. Rea-Ramirez, et al. (2004) have used this approach in an experimental middle school biology curriculum. Another is to list several possible modifications, one of which goes in the correct direction, and use them to foster class discussion.

Evidence from curriculum trials. Steinberg (1992) documented large gains over control groups using a curriculum that takes students through a series of model revisions in electric circuits, with critical experiments triggering the need for revision of each intermediate model. Teachers and students using such approaches need to become comfortable with the idea of discussing intermediate models that are not entirely correct, prior to students developing a more sophisticated model. At an even more fine-grained level than intermediate *models*, Brown and Clement (1992) describe large gains over controls for mechanics lessons which teach students a set of "intermediate *concepts*" of inertia such as "keeps going tendency" and "holdback tendency," before modifying and combining them into a single expert concept. Other studies have shown that model evaluation and modification cycles can be used effectively in: biology (Rea-Ramirez and Nunez-Oviedo, 2008a; Johnson and Stewart, 1990); chemistry (Khan et al., 2002), heat (Linn, 1992), and mechanics (White and Frederiksen, 2000; Clement, 1993b).

Concept diagrams: a representation for instructional planning. Figure 10.6 shows a concept diagram from the latter study used in the design of a physics lesson on static normal forces in a mechanics curriculum (Camp and Clement, 1994). In the past the units of analysis for instructional design have been postulated elements of an expert's knowledge structure (mostly mathematical elements in the case of physics) and logical relationships between them. The approach suggested here is to focus on the student's available prior knowledge structures and reasoning processes such as those shown in Fig. 10.6, namely, key anchoring analogies, bridges, explanatory models, and the qualitative observations which support them. This involves the heavy use of concrete examples either in demonstrations, labs, or thought experiments with familiar objects, and avoids the introduction of mathematics and formal notation until qualitative conceptual models are in place. Thus concept diagrams like Fig. 10.6 are a useful planning tool for outlining the structure of a lesson that utilizes nonformal reasoning and learning processes.

21.5.1.6 Section Conclusion

I conclude that there are some promising initial findings suggesting that the reasoning processes discussed in Part One of this book, including the use of anchoring analogies (Chapter 10), and model construction cycles, can be used very fruitfully in educational settings. One indicator of the extent to which these lessons are taking advantage of expert–novice similarities in nonformal reasoning and learning is the comparison of Fig. 10.6, showing a successful teaching strategy, and Fig. 21.4, showing an expert bridging and modeling strategy. The forms are virtually the same, incorporating anchoring examples, analogy confirmation via bridging, and construction of an imageable, dynamic explanatory model.

21.5.2 Strategies and Implications Suggested by Findings on Imagistic Knowledge Representations in Part Two of the Book

Figure 20.1 illustrates the present view of the centrality of imagery-based processes in scientific thinking. The introduction of the role of imagery and simulation in Part Two of the book not only led to the analysis of new reasoning operations but also expanded the view of what knowledge is. This has basic implications for instruction but little research has been done to date. Therefore in the first part of this section, I speculate on

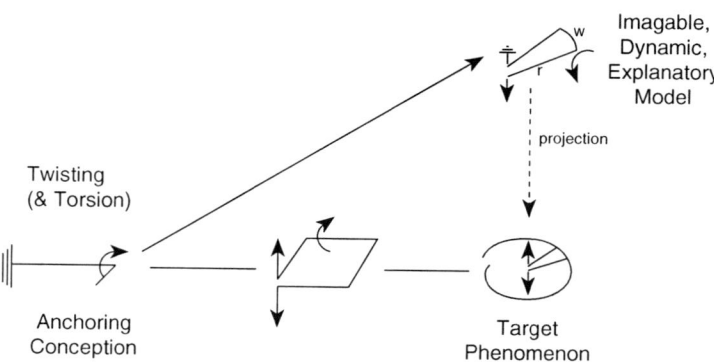

Fig. 21.4 Model construction in an expert

the implications of an imagery-centered view of knowledge for education, while in the second part, I discuss implications of imagistic reasoning and learning processes.

21.5.2.1 Runnable Explanatory Models

Given the imagery-centered view of knowledge developed in Part Two of this book, instead of aiming only for students to acquire static symbolic structures, the goal of instruction becomes the development of dynamically runnable mental models (or some mix of the two). This goal specifies not only that students can generate simulations of previously observed phenomena as a kind of "mental movie," but also that they can generate simulations of a hidden mechanism operating "underneath" the observable phenomenon being studied. The ability to generate a *simulation from a runnable explanatory model* constitutes a deep form of scientific explanation and an important content goal. This is when the model becomes a *tool for thought*. This is so different from the view of knowledge as memorized verbal symbols that its importance for teaching cannot be overemphasized. It is essentially an elaborated descendant of Piaget's theme of knowledge as action.

Hand motions and other imagery indicators as a means of communication in classrooms. One immediate implication of this goal is that it raises the question of whether teachers are setting up effective communications channels in the classroom to support imagery. It is only recently that imagery indicators have been used in classroom research to detect mental simulations (Clement et al., 2005). The spoken language is instantly available to all and is used ubiquitously in instruction, whereas visual/depictive communication channels are used much less, and in many classrooms very rarely. Depictive hand motions, as an indicator of mental simulation processes, may be an important and undervalued means of communication that can contribute to both process and content goals (Alibali and Nathan, in press; Givry and Roth, 2006) and to the task of diagnosing students' models at the beginning, middle, and end of a unit. Another challenge, for teachers who want to stimulate discussions of conceptual issues in their classrooms, is that of helping students understand each other, given that their articulation of tentative ideas is sometimes difficult to follow. Drawings are an obvious means of visual communication, but they are underutilized in classrooms. And when students point to and gesture over a drawing this can be a way of animating it and communicating that to others at the same time. I believe this is a simple but powerful mode of communication that is undervalued and underfostered. Gestures may help students do more explaining and running of their models and comprehending of each other as well as allowing us as teachers to comprehend them, contributing to both process and content goals. It may be that gesturing and hand motions on the part of teachers and students during classroom discussion (over drawings or alone) is a more "direct" and more effective communication system for reflecting dynamic imagery than is language. For this reason I believe gesturing should be encouraged by request, by example, and by verbal praise, as a means for making classroom discussions more coherent. Educators may be able to generalize this point by examining whether sensitizing teachers to other imagery indicators in Table 12.3 would aid in diagnosis and communication.

A central problem for ongoing assessment that can inform instruction is: "how can one tell whether imageable models are being learned?" It can be argued that this problem calls for the addition of less conservative imagery indicators to the list in Table 12.3 for use in the classroom. For example, given the question, "Which takes more force to stop in a 3 s period, a 500 lb boat or a 700 lb car, each of which is going 5 mph?" a student who says "the car, because we learned that F = ma, and the mass of the car is greater" is indicating a kind of formula-centered knowledge, in contrast to a student who says "the car, because the car moving toward you on the road has greater mass and momentum, and therefore you must push harder in the direction opposing the motion than you would if the boat were moving toward you at the end of a dock." The latter is highly suggestive of imagery use because it (1) describes concrete pictures or scenes, (2) the scenes are depicted from a particular point of view, and (3) the inferences made fit the form of a spatially embedded reasoning inference. In addition the student may (4) draw a picture of the explanation with an arrow diagram. Teachers may recognize these indicators as ones they use instinctively to detect student understanding at the level of sense making, and from the point of view of cognitive science they fit well with an imagery hypothesis. I believe the day may come when some of these indicators can be used along with depictive hand motions and personal action projection (also present in the above example) to detect imagery in research, but, because they are somewhat less directly implicative than those in Table 12.3, as a conservative measure none of these indicators were used in the present studies. Nevertheless teachers, in order to make on the spot decisions, need to use "every kind of antenna they have" to detect imagery-based models and these indicators may help.

The finding that explanatory models are schematic may have important implications for the design of visual supports in instruction. The distinction between rich experiential examples and the much sparser schematic models they help to form is important for educators to heed. Although they are extremely important, too often the examples are assumed to carry the message, whereas the present view argues that helping the student to construct a more general schematic, runnable model is central. Using the richest, high resolution computer simulation in the world may not on its own lead the student to develop a sparse model that they can run on their own mentally without computer support. Thus one's vision of the goal of instruction may cause one to evaluate the proper use of such instructional tools differently. Their usefulness as visual supports needs to be balanced to avoid the potential problem that they may in some cases become crutches the keep students from learning to run their own mental models (Monaghan and Clement, 1999).

21.5.2.2 Why Anchoring in Runnable Schemas is Important: It "Strengthens" a Model and Makes It More Competitive with Other Preconceptions

Chapter 13 also documented the sense in which a student's prior knowledge structures could take the form of dynamically runnable intuitions or schemas. This leads to the idea that one may be able to transfer imagery and runnability from an aptly

21.5 Implications for Instructional Strategies and Theory 551

chosen analogy to a model under development. In Chapter 18, I discussed a considerable number of possible scientific benefits stemming from anchoring a scientific model in one or more runnable intuition schemas. Here, I will argue by analogy that many of the same benefits (shown in the right-hand column of Fig. 18.7) can serve to make a runnable model more powerful for students as well as experts. As a result, there should be a sense in which an anchored model is "stronger" and more competitive with persistent preconceptions than one that is not anchored. That is, it will be more likely to be activated appropriately, to remain active in interpreting and responding to a situation, and to be able to displace a competing preconception in the same area. Since competing preconceptions are often based in intuitions, and these are self-evaluated and appear self-evident, they can be difficult to overcome. Anchoring in other intuitions that are compatible with accepted theory may be one of the best ways to make new models competitive with preconceptions, so that they attain a strong level of self-evaluated plausibility, and eventually conviction.

Benefits of runnability. Cognitive benefits and desiderata in science in Fig. 18.7 that can also be seen as educational benefits are: flexible transfer, plausibility, generality, clarity, explanatory satisfaction and predictive adequacy, and coherence. Further educational benefits that can be hypothesized to accrue from anchoring a model in a runnable intuition are: retention, appropriate application, and "strength" or overall plausibility. The latter benefits could be listed in a *fifth column* to the right in Fig. 18.7, as follows. Retention, or likelihood of activation, should be affected positively by (have arrows coming to it from the fourth column items of) generality, coherence with other ideas and clarity. Appropriate application should be positively influenced by clarity. And "strength" of the conceptions (their ability to compete with other conceptions) would accrue from explanatory satisfaction, projected plausibility, coherence with other models, and spatiotemporal coherence. If this theory has merit, it is clear that anchoring a model in runnable intuition schemas should be very beneficial to a student. In fact, in this framework runnable explanatory models are a central component of conceptual understanding in science. The usefulness to the student is enhanced if the models provide a mechanical explanation with a fully imageable and *runnable causal chain* for explaining patterns in the observations, when possible.

Science as extension of intuition. Intuitive grounding appears to be highly valued in science but it is not always fully attained. Quantum mechanics and relativity appear to be at least partially incompatible with basic notions of space, time, and causality. That is not unreasonable given that they deal with realms that are extremely far from our everyday experience. These may be viewed as exceptions since in other areas compatibility with intuitions is highly valued. However, Cohen (1975) has argued that visual models are still used in these domains, even though some of our intuitions must be given up or modified. In domains of science that are closer to our experience, visual models, and intuitive grounding are still highly valued because they offer valuable ways of thinking about those domains. This argues that they are also important for science education.

Can intuitions be learned or changed or taught? In some circles, use of the term "intuition" signifies an unteachable concept. In contrast the concept of intuition as developed in this book is designed to allow for intuitive knowledge structures that are changeable, learnable, and to some extent teachable. The key notion for the concept of intuition developed here is "self-evaluated plausibility." This makes it possible to talk about changing one's intuition, for example, when someone overcomes misconceptions in mechanics and constructs a more complex set of schemas to understand force and motion. In that case one must change intuitions by removing, modifying, or at least reducing the priority of certain intuitive preconceptions. And if one develops a "feel for" the Newtonian system – that is, if one can image the way force causes objects to accelerate and decelerate, then the newly learned conceptions are intuitive because they are plausible under self evaluation, not just because the textbook says so. When one evaluates them by running them as simulations to explain test cases, they give a satisfying explanation. In this view conceptual understanding depends on developing a runnable set of new schemas that are coherent with each other, with observations, and with some, but not all of the older schemas. Ideally they should be more coherent and predictive than was the subject's original set of preconceptions. Their self-evaluated plausibility then allows us to refer to them as having become "intuitive" as well. Thus grounding a model in anchoring intuitions may have a considerable educational payoff. This helps explain the positive results in learning studies cited in Chapter 10, where there is some evidence of learning that changes intuitions, or at least builds up a competing and hopefully more valued intuition.

It was also claimed that the transfer of runnability achieved by grounding a new model in a runnable prior knowledge schema as a source analogue may foster a type of model flexibility that aids the use of the model in new contexts. This has implications for the choice of source analogue. Brown and Clement (1989) recommended that they be selected for transfer of the appropriate kind of runnability to the model, not just for high confidence and mapability to the target.

21.5.3 Educational Implications of Imagistic Learning Processes in Part Two of the Book

21.5.3.1 Process Skill Benefits of Model Construction

The previous sections on educational benefits have pointed to *content* objectives, such as improved conceptual understanding of specific topics that can be aided by newly identified model construction processes. But there can also be benefits for *process* objectives, defined as the development of scientific investigation or thinking skills. It can be argued that the experts analyzed in this book have demonstrated "adaptive expertise" as opposed to the expertise used to solve routine problems. I agree with Hatano and Oura (2003) that: "While basic schools cannot make students real experts, they can place students on a trajectory toward expertise or prepare them for future learning (Bransford and Schwartz, 1999). In this sense, an important goal of basic schooling is to make each student a "baby adaptive expert" of the

21.5 Implications for Instructional Strategies and Theory

domain or topic of choice." Since scientific investigation is a skill of knowing how to learn systematically, it can be argued that all students can benefit from it. I will discuss this at the level of (1) problem-solving skills and (2) inquiry skills.

Problem solving. Figure 6.8 can also be thought of as a model of the process of constructing a representation for an ill-structured problem. (Here, memories of prior experiences can play a role in empirical testing if no new empirical information is available). For the content goals discussed above, considerable support might be given by the teacher in guiding students through such a cycle. But, in order to learn problem-solving skills, students eventually need to be able to generate, evaluate, and modify problem representations by going through construction cycles without teacher support. Despite this difference, Fig. 6.8 provides the basis for seeing some significant overlap in the strategies for achieving content and problem solving goals in science and mathematics education. Clement and Konold (1989) describe a program where pair problem solving was used in conjunction with minimal training in heuristics to encourage model construction and revision cycles in mathematics courses.

Inquiry skills. The most ambitious goal in science education is that of teaching scientific investigation or inquiry skills. In fact, classes in which students are asked to propose scientific hypotheses for phenomena are not common. It is even more unusual for teachers to explicitly recognize criticism and revision processes as desirable goals in science classes. I certainly have difficulty recalling ever being asked to criticize a model in my own educational experience in science. Here again, it seems important not to assume that learning by bottom-up induction from data is the dominant inquiry process in the scientific method. The nonempirical model generation, criticism and modification processes identified in this book would also seem to be important to the design of inquiry activities.

Johnson and Stewart (1990) describe a program in which high school biology students learn to run computer simulations of fruit fly experiments in genetics. They first teach the students Mendelian genetics and practice simulating classical genetics experiments. However, they then provide the students with anomalous data, that is, data that does not fit the Mendelian theory, and ask them to invent an explanation for the data. The students are being asked to modify or replace the classical models that they have learned and are thus participating fully in model criticism and revision processes. Stewart and his colleagues have shown that many high school students can participate in these processes successfully and in some cases they have reproduced discoveries of modern theories of genetics such as multiple alleles, sex-linked characteristics, etc. (Hafner and Stewart, 1995). Thus there is hope that students can participate fully in model construction cycles in classrooms. The studies by Metz (1998), Hewson and Hennessey (1992), and Duckworth et al. (1990) document the ability to foster scientific thinking processes at the elementary level, as do White and Frederiksen (2000) at the middle school level.

In summary, there have been a few initial indicators of success in designing learning experiences where some of the model construction processes described here are treated not only as a means toward the end of achieving content goals, but as an end in themselves in the classroom. More work in this area is very much needed.

Imagistic processes that could expand science process goals in the future. The original list of process skill goals for science education identified by Padilla (1991) was primarily oriented to empirical processes of science. Hammer (1995) later added a variety of nonempirical process skills. The expanded additional set of nonempirical model construction processes documented in this book and Fig. 21.2 may provide a description of a number of new categories of nonempirical process skills for science education. These can be used to evaluate whether students are engaging in processes used in scientific discovery, in classrooms that pursue such process goals. My guess is that analogy evaluation strategies, the role of imagistic transformations and compound simulations in model design, running an explanatory model to evaluate it, engaging in multiple modeling cycles, and spatially aligning a model with the target are some of the under-recognized processes that are of primary importance in student inquiry that develops scientific explanations.

Section summary. For both content and process goals, instructional strategies that tap into plausible reasoning modes observed in experts and students can be designed, and some promising initial successes have been cited.

21.5.3.2 Using Imagistic Learning Processes in Pursuit of Content Goals

Fig. 20.1 emphasizes the idea that learning processes take place at several different levels, and unpacking the processes for visual model construction in these levels opens up exciting challenges for education. Having a conceptual change theory that is structured in this way may allow us to organize teaching strategies by levels, something that may help make them more understandable and learnable (Clement, 2008). Processes in the figure include: imagery enhancement, evaluatory Gedanken experiments and other tests of model plausibility, imagistic conserving transformations, and analogies with multiple purposes. Since very few educational studies have been conducted as yet in these areas, in this section I will speculate on educational implications and promising future directions for educational research. I consider these according to the levels in Fig. 20.1, moving from bottom to top.

Imagery enhancement. At the lowest level in Fig. 20.1, the use of *imagery enhancement* techniques may be a powerful instructional tool. To give just one example, one curriculum we have worked on includes the use of imaginary "saran wrap" boundary around a swimmer in a river to imagine how the block of water the swimmer is in is moving with the current. This may help with chronically difficult learning problems in the area of relative motion in transparent mediums (Camp et al., 1994). Educational applications of other specific heuristics using imagery enhancement listed in Tables 15.7 and 15.8 should also be explored (Stephens and Clement, 2007).

Rationalistic criticism of models in discussion. At level 2 and as illustrated in Chapter 10, with proper support, students can run thought experiments to criticize models under discussion. A full analysis of the use of imagery during model construction by students is beyond the scope of this book and an area in great need

21.5 Implications for Instructional Strategies and Theory

of further study. Some initial efforts in this area have been made by Schwartz and Black (1996a,b), Hegarty, et al. (2003), Zietsman and Clement (1997), Monaghan and Clement (1999), Clement and Steinberg (2002), Clement (2003), and Stephens and Clement (2006a,b).

Imagery-based analogy. As suggested by Chapter 17, *overlay diagrams* for analogies, such as the lever drawn on top of the wheel, may help students perform internal dual simulations of the base and target in order to comprehend and evaluate analogies. Imagistic alignment analogies may help in the same way. Given the multiple purposes detected for analogy use in Table 19.1, it may very well be that different teaching strategies should be designed for each purpose. These are important areas for further research.

Subprocesses in model construction. As shown in Figs. 21.2c and 16.1, educators need to distinguish between abductive processes aimed at forming explanatory models, and processes aimed at forming empirical law hypotheses or formal quantitative principles, since the cognitive processes involved may be quite different. Nor are students likely to construct qualitative models deductively from the study of formal quantitative principles alone. The present view of this process for experts was outlined in Fig. 16.8 and its simplified parallel for education in Fig. 21.3c. The focus on imagery-based model construction adds another layer of depth to that figure. Using Fig. 21.3c as a starting point, one can then outline a direction in which more sophisticated models of conceptual change in instruction can be developed by incorporating more of the features in Figs. 16.8 and 16.4.

Stages in model construction. As suggested in Chapter 14, there may be *more stages involved in model construction than is commonly recognized*. There were five different stages identified in the spring protocol in Chapter 14. Different levels of precision and to some extent, different processes, are used in each stage. Asking whether sufficient attention is given to each stage may be an important guide for curriculum developers and teachers. Hestenes (1997) has proposed a theory of modeling in physics and suggests using different types of notation systems to represent models at different levels of precision. The vision of building an explanatory model that goes beyond initial analogies, refining and aligning it imagistically, building a geometric model on top of that, and finally using this as a foundation for a grounded mathematical model contrasts sharply with the vision of simply memorizing facts or formulas to learn science.

In addition, the distinction between *three cycles involved in scientific investigation processes* shown in Fig. 16.4 may suggest a set of important instructional processes for progressing through the stages. These were initial description, explanatory model construction, and mathematical model construction. These considerations suggest that previous models of instruction containing only a single learning cycle may be oversimplified (see related views by Lawson (1988)).

The idea of *modulating the divergence or "distance" of analogies* according to the stage (level of precision) of modeling may have important instructional implications. It was proposed that distant analogies may be more useful early in the model development process, with close analogies generated by transformations of the target situation being used at later stages. This may provide a useful heuristic for curriculum developers attempting to generate effective analogies at different stages of modeling. More generally, the several *different types or purposes of analogies* identified in Chapter 20 may give important clues as to how to use analogies more effectively in instruction. Thus Analogy 1 in Fig. 21.3c may be aimed at developing a rough model, whereas Analogy 2 may add fine structure to the model. This is a higher level of sophistication in teaching, and may require advanced planning by curriculum developers, although some teachers may do it intuitively or learn to do it. Similarly, designing Gedanken experiments is a sophisticated teaching strategy, but these can have an important place as shown in Fig. 21.3c (Reiner, 1998; Reiner and Gilbert, 2000; Stephens and Clement, 2006a).

Other application ideas may come from considering the educational implications of an overall view of modeling like that in Fig. 16.8. This figure captures the idea of GEM cycles with small step sizes, but also shows a potentially large number of processes involved in trying to balance the need for divergence and convergence in scientific modeling. Attempts by our own group to promote model construction in classrooms indicate that if one can get the teacher to hand over some responsibility to the students for model generation (usually done in small groups over a desk-sized whiteboard) that there is no trouble getting them to generate lots of divergent ideas for, say, how the digestive or circulatory system works, even at the sixth-grade level. A harder task is to manage that diversity and to guide model competition, evaluation, and revision that needs to ensue, but this can also be done (Clement and Rea-Ramirez, 2008). Some of the most sophisticated processes in Fig. 16.8 may often have to be initiated by the teacher, such as (1) generating Gedanken experiments and (2) fostering constrained divergence leading to productive modifications. Successfully managing the second process is an interesting way to describe what the teacher with content goals in a model-based constructivist classroom does. *Thus managing learning can be a balancing act for expert scientist and teacher alike (Price, 2007); and in considering how to adapt Fig. 16.8 (or 21.2) as one possible model for constructivist teaching, one might title the figure "A Balanced System for Productive Teaching and Learning."*

The task of applying the more advanced concepts in this section to the problems of education opens up important territory for future work.

21.5.4 Conclusion– Educational Applications

21.5.4.1 Ubiquitous Nature of the GEM Cycle and Active Learning

We concluded in the first part of this chapter that even though they use different subprocesses, the analogy formation and model construction processes use, at the topmost level, the same basic cycle of generation, evaluation and modification. This GEM cycle, shown in Fig. 6.8, can also be seen in the construction of a representation for an ill-structured problem; thus it is very widely used. Because it is also fairly simple to describe, it is therefore a high-priority item for dissemination into the teaching community. For me it goes a long way in explaining the value of constructivist methods of instruction such as small-group projects where students are asked to design, assess, and debug a product, solution, or model. I have argued that the cycle is quite general and is relevant to three major educational goals: the content goal of comprehending established scientific models; the process goal of learning to solve ill-structured problems; and the even more ambitious process goal of learning scientific method or scientific inquiry skills. The cycle is relevant even down to the level of hands on projects in the primary grades. Decisions concerning when such approaches are worth continuing might be informed by using student engagement in GEM cycles as an explicit goal and criterion.

The GEM cycle provides some of the ideas that provide a more detailed view of the concept of "active learning." They are student participation in: the abductive assembly of an explanatory model from prior knowledge, evaluating the model by recognizing consistencies or anomalies, using those to guide modification of the model, engaging in repeated passes through this (GEM) cycle, transferring runnability from prior knowledge schemas to the model, and applying the model to generate imagistic simulations of new situations.

If knowledge is thought of as static propositions, then the most obvious learning models are "learning by memorizing" or "learning by deducing." But if knowledge is thought of as dynamic models capable of generating imagistic simulations that require the coordination, modification, and tuning of several dynamic schemas, then "learning by doing" emerges as a more appropriate model for learning. Such coordinations must be tried, practiced, debugged and refined while doing them.

21.5.4.2 Need for a Newly Expanded Theory of Conceptual Change that Goes Beyond Kuhn and Classical Conceptual Change Theory

I can now summarize the present view of modifications that are needed to work toward an expanded theory of conceptual change for education. The last column in Table 21.3 shows how ideas from Kuhn's theory and the classical theory of conceptual change would be retained, modified, or expanded, as follows. The idea of global, insurmountable incommensurability has been dropped, but better theory is needed on how to help students work through conceptual change with teachers in the face of locally incompatible basic concepts, which can present a significant barrier to communication. Cognitive resistance to change of an established model is retained as a key concept. The common view of anomalies as empirical discrepant events is retained, but augmented by the possibility of nonempirical thought experiments or inconsistencies with prior knowledge, as a second major source of dissonance for motivating change (which was also recognized by Kuhn, 1977d). The role of large structure-replacing events is discounted as the pace of change, but the role of thought experiments is retained and extended to include those that support as well as disconfirm models. Table 21.3 indicates that Kuhn's idea of theories as coherent systems can be elaborated via the ideas of spatio-temporally connected models and utilized in an image of target theories in a curriculum. Learning pathways that take advantage of natural student reasoning processes and gradually build such coherent models need to be designed. Thus, in an improved conceptual change theory, some of Kuhn's ideas would be retained in some form and others would be modified.

However, there are also large areas of expansion needed as shown in the bottom four rows. Those entries provide only a hint of the complex set of additional learning mechanisms that need to be understood in order to inform instruction. Among them are evolution *cycles* that produce a sequence of model modifications. When the cycle gets "stuck" at difficult change points, it may involve unlearning and model element replacement or downgrading, rather than accretion, with resistance to unlearning, so knowledge is not just cumulative. We therefore can expect unevenly progressing, sometimes fitful, punctuated evolution, shown in Fig. 19.2. Multiple roles for analogy and a theory of runnable models with grounding in schemas generating imagistic simulations are needed. Much attention has shifted in recent years to issues at levels *outside* of the cognitive level, such as social interactions, to address important areas where conceptual change theory was incomplete. The present findings however speak to what I believe was a large gap *within* the cognitive level – the lack of any detail in the mechanisms of higher-level learning and the structure of scientific knowledge. Work on filling this gap is still proceeding.

Teaching strategy levels. The view of model-based learning developed in this book and outlined in Fig. 20.1 is described at several levels: (1) the perceptual (and often motor) processing that makes imagistic simulations possible; (2) the nonformal reasoning operations that utilize imagistic simulations; and (3) abductive model evolution (GEM) cycles of model generation, evaluation, and modification. The levels describe processes at shorter and longer timescales, and this may be a useful way to think about student learning and teaching processes. As we assemble larger numbers of teaching strategies for fostering model construction we will need a way to organize them. Clement (2008c) proposes placing teaching strategies in six levels roughly according to the timescales of the learning processes being promoted, and three of these strategy levels correspond to the three process levels shown in Fig. 20.1. For example these include strategies for fostering GEM cycles at level 3, for promoting model evaluation at level 2, and for promoting imagistic simulation at level 1. The goal is to help build an improved model of conceptual change teaching.

In summary, analyses of expert protocols have led to an expanded model of conceptual change processes in science. Students were shown to have natural abilities for many aspects of these processes, and to be able to do others with scaffolding from a teacher. Still others have not yet been investigated with students. But the results so far can be used to envision a variety of new instructional strategies. Promising initial evaluations have been done for several of these strategies, but much work remains to be done.

21.6 Are Creative Processes in Experts a Natural Extension of Everyday Thinking?

In this section, I will assess the expert–novice similarities and differences found in this study, as a contribution to the debate about how ordinary or extraordinary creative scientific thinking is. Here, I will argue that some expert processes are ordinary, and others are not – there are some expert processes that are rarely observed in novices – but that the protocols lend support to the idea that much of expert thinking can be seen as an extension of (1) "natural knowledge" in the form of intuitions and (2) "natural reasoning" in the form of nonformal, simulation-based reasoning processes.

21.6.1 Expert–Novice Comparisons for Knowledge Structures: Science as an Extension of Intuition?

With regard to the use of knowledge, it is obvious that experts possess considerably more knowledge than students and that it is more highly structured and interrelated. On the other hand, the book has provided examples where both students and experts grounded their solutions on runnable, self-evaluated intuitions rather than on formal or mathematical principles. Experts were portrayed as relying more on concrete, imageable source schemas than is commonly recognized, especially when they were working in what was a "frontier area" for them. There is evidence from history of science for this statement as well in the cases of Maxwell, Watson, Darwin, and Faraday (Nersessian, 2002; Watson, 1968; Millman and Smith, 1997; Gooding, 1990; Tweney, 1985). This supports the idea of science as an extension of intuition and supports the teacher's impulse to make science knowledge coherent with everyday knowledge wherever possible.

The precise extent to which everyday knowledge structures are similar to expert structures is still debated, and it varies across subject matter topics (diSessa, 1985; Vosniadou, 2002). But both the latter two researchers and others agree on the desirability of developmental continuity in the learning pathway from naive to expert concepts. I have tried to express this idea here in the notion of "intuitive grounding" of models gained by using intuitive source analogues as anchors. Historically, these ideas are derivatives of Piaget's (1955) grand constructivist scheme advocating that an important part of our knowledge can unfold from a few extremely primitive prior knowledge schemas interacting with the world and with each other, beginning in a newborn child. The relationship between practical knowledge, implicit knowledge, and developing expertise remains an important area for future research (Cianciolo et al., 2006).

21.6.2 Expert–Novice Differences in Reasoning

Is expert reasoning similar in type to student reasoning? On the one hand, there were some sophisticated reasoning processes used by certain experts that illustrated a level of thinking hard to imagine in a novice. For example, the use of evaluative Gedanken experiments and certain scientific theory evaluation criteria such as symmetry may be candidates for processes that are more sophisticated than those of most lay problem solvers. And certain forms of modulated divergence, such as using transformations that are conserving at the level of precision needed, or choosing to use near vs. far analogies, may be rare in novices.

Figures 16.4 and 21.2 are attempts to depict adaptive and successful scientists as capable of engaging in a very sophisticated balancing act that is highly refined and that coordinates multiple strategies. Really creative expert thought, as pictured there, involves a complex orchestration of many resources. In addition, Darden (1991) has chronicled a large set of sometimes interacting or conflicting criteria by which scientists evaluated theories in the history of genetics, and this requires advanced judgment in weighting conflicting criteria. I am doubtful that we will find many young students who are capable of such orchestration or weighting of criteria on their own; this is another issue that could use more research.

21.6.3 Expert–Novice Similarities in Reasoning

Despite the above, when I began my own studies of expert reasoning, I seriously underestimated the extent of the overlap with student reasoning. Students can use far more expert-like reasoning processes than one might anticipate, including tapping positive physical intuitions, engaging in intelligent discussions of proposed analogies, models, and thought experiments by running imagistic simulations within them, and in some cases being flexible and creative in their activation or construction of analogous cases and models of their own. This opens up an array of possibilities for fostering active learning in science via these reasoning processes. It also makes it more plausible that many expert processes grow out of natural reasoning processes as an extension of everyday thinking.

21.6.4 Some Expert Processes are Neither Extraordinary Nor Ordinary

Very recently, Weisberg (2006) has provided a useful and interesting collection of case studies of creativity from masters in the fields of science, engineering, and art. This is especially welcome since it is time consuming and rare for researchers to

21.6 Are Creative Processes in Experts a Natural Extension of Everyday Thinking? 561

investigate creative processes as opposed to creative characteristics of individuals. Weisberg strives to build a case for the view that creative methods are just methods of ordinary thinking used in especially opportune circumstances. He argues that this is true because both ordinary and creative thinking exhibit a connection to prior thinking and use top-down as well as bottom-up processing from environmental events: "we have seen much evidence that creative achievement is not based on breaking away from the past" (p. 599).

One limitation of the Weisberg study is that none of the expert case studies are think-aloud studies, so there is an upper bound on the grain size with which he can describe thinking processes. Nevertheless, the first criterion for "normal" thinking, some connection to prior thoughts, dovetails with a criterion I used in Chapter 7, to claim that S2's thinking, even in strong "Aha" insights, was not necessarily an episode of "extra-ordinary" thinking (Clement, 1989b).

However, in countering the "genius" theory of "extraordinary" thought in creative thinking, Weisberg runs the risk of going to the opposite extreme in reducing creative thinking to the ordinary. This leads him to make statements in the conclusion such as: "there is no compelling reason to introduce anything beyond ordinary processes into explanations of creativity" (p. 591).

It is true that all of the thinking processes discussed in the present book were interpreted as being connected to a previous process in some way (as opposed to being a "bolt from the blue"). This is one reasonable criterion for calling the process a rational one that is not "extraordinary." However, as was emphasized in Chapter 7, it is important to distinguish between a break in the train of thought, and a break away from the subject's (or the field's) currently assumed model. S2's discovery of torsion, which provided the long sought for break from the persistent dominance of the bending model of the spring, was one such breakthrough. And in Chapter 14 another subject's sudden rejection of his shear model of the spring is another example of a break from previous thinking. And thirdly, coming to see the spring as "unbending" rather than bending qualifies as a 180° turn. These subjects were unusually successful in breaking away from their previous views.

We saw that this was not at all an easy process, but rather could entail a considerable amount of effort, frustration, perseverance, ability to try multiple divergent methods strategically, and motivation. Because this was a rare kind of event in the expert protocols, I suspect that such breakthroughs may take special talents and/or special coordinations of processes to carry out. The trouble with calling them 'ordinary' is that that position would imply, wrongly I think, that someone who revolutionizes 80% of continent formation theory, for example, has not broken with the past because of the remaining similarity in 20% of the theory. This leaves me wanting to argue that the best creative thinking is both ordinary (in the sense of being "not extraordinary") and not ordinary (in the sense of taking special skills). I believe the resolution to this dilemma lies in being more precise about what one means by the term "ordinary" (there may be several different meanings) and in moving away from a false dichotomy of ordinary vs. extraordinary thinking to a more nuanced spectrum.

21.6.5 A Spectrum from Ordinary Thinking to Unusually Effective Creative Thinking to Extraordinary Thinking

In order to make such differentiations, I propose the spectrum shown in Fig. 21.5. The *x*-axis shows four Domains in a spectrum from ordinary to extraordinary thinking instead of a simple dichotomy. Domain 1 is ordinary thinking such as lay problem-solving methods and methods of explanation. Domain 2 consists of basic model construction processes such as abductive guessing about hidden causes, the use of analogy, running a model to generate predictions and examine plausibility, and model modification. Both experts and student utilize processes in Domains 1 and 2. Domain 3 consists of more unusual and creative processes used by the best experts such as pushing to form precise spatiotemporal connections in a model, generating Gedanken experiments, and modulating divergence according to the stage of development of a model. These are advanced innovation processes used to make theoretical breakthroughs in real science. Domain 4 consists of extraordinary processes that go beyond cognitive/motivational explanations in science such as incubation, unconscious influences, or unexplained insight processes.

The curve labeled "student processes" indicates frequency of use of a number of processes proportional to the area below it in each of the domains. It naturally shows the largest area of processes encompassed below it in the ordinary thinking domain. However, the present study argues that students also share significant number of processes in Domain 2 with experts, even though the curve labeled "expert processes" shows a larger region of competence in Domain 2 than for students. Spontaneous student processing in Domain 3 is rare at best, although Stephens and Clement (2007) have documented one or two instances of proposed Gedanken experiments in advanced high school physics discussions). The region between the student and expert curves in Domain 1 represents less precise or faulty forms of natural reasoning used by students, forms that experts avoid in scientific thinking (see D. Kuhn et al., 1988). Of course, I do not really have data to support an accurate curve for each

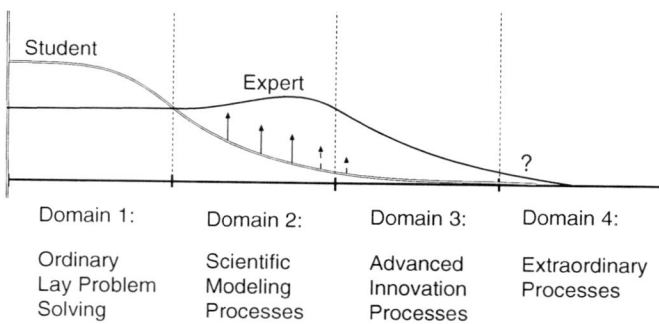

Fig. 21.5 Possible expert and student profiles on spectrum from ordinary to extraordinary thinking in science

group, but the curves are not inconsistent with present data and allow one to express a theoretical proposal. Whether experts possess skills in Domain 4 is controversial, and it may or may not contain documentable processes, but there was no evidence for them in the data discussed in this book. The vertical arrows mostly in Domain 2 represent the expected effects of teacher scaffolding to support and boost the level of scientific reasoning that students can use in the classroom. Given that help, students may not need advanced innovation processes in Domain 3 to learn science content with flexible understanding (although it would still be quite interesting to see if students could develop some of those skills). For example, whether some students can be taught to generate evaluative Gedanken experiments in Domain 3 is unknown.

In sum, this figure allows me to simultaneously articulate the positions that bold, successful innovation in complex scientific thinking via Domain 3 processes is not simply ordinary thinking; but to also argue that it is not extraordinary in the sense of being incapable of being explained as a rational act. Furthermore, students share with experts a significant number of processes as natural reasoning resources. But experts have extended their natural reasoning processes in the direction of the upward arrows to develop specialized ones. So there are both expert–novice differences as well as similarities.

21.6.5.1 Beyond the Ordinary I: Flexible Processes in Scientific Modeling

It is important to tease out some different meanings of the term "ordinary." If "ordinary" means "could potentially be explained in cognitive and motivational terms without reference to the unconscious or to unexplainable processes," then yes, our experts were ordinary. But if "ordinary" means" following traditional, rule-bound methods," then many of the expert processes studied here were not ordinary. The following methods were deemed to lie outside the set of traditional, rule-bound methods of knowledge generation in science because of their flexibility or divergent nature. Many of these methods are in Domain 2 in Fig. 21.5. So I consider the divergent portion of the processes in Domain 2 to be somewhat unordinary in the sense that they are not following rigid procedural methods or traditional techniques of "normal science."

- Generation of analogous cases via
  - Free association
  - Associations under constraints
  - Playful transformations of the problem
- Generation of extreme cases via
  - Extreme transformations
- Generation of new models via abduction – that involves divergent guessing

Figure 21.2 represents the idea that by setting up a "battle" between model generation and evaluation (often within a single individual), with each side fielding multiple sources of ideas, one creates the possibility for creative work through an intense dialectic. In this battle, a substantial part of the reasoning used is heuristic in that there is no guarantee that it will work, the abductions are, in part, guesses, and a substantial part of the base representation is imagistic, meaning that assimilations and applications are a matter of judgment rather than unambiguous decision points. This is a very different picture than that of following traditional, rule-bound procedures. In this milder meaning of "beyond the ordinary," students may also sometimes go beyond the ordinary in their thinking, as we have documented instances of the above processes in students in Chapters 8, 9, and 18.

21.6.5.2 Beyond the Ordinary II: Advanced Innovation Processes

Alternatively, if "ordinary" is taken in the broader meaning of "natural reasoning processes used by most laymen in everyday thought" then the answer given is yes and no: experts do use such lay processes, but some use advanced innovation processes (Domain 3 in Fig. 21.5) that go beyond these and that are "unordinary" in a stronger sense. Particularly skillful creative processes identified were:

- During model generation and revision:
  - Abductive conserving transformations of a scientific analogy or explanatory model that conserve many constraints
  - Ability and judgment to penetrate and focus on only the essential constraints so that one can diverge radically around them if necessary
  - Modulated divergence of associations, transformations, and analogies–varies with model development stage
  - Flexible control: orchestrating multiple goals and methods in 3 cycles for model construction (as shown in Figs. 16.3 and 16.4)

- During model evaluation:
  - Of six subjects who thought about bending, S2 was the only one who noticed the discrepancy with uniform distances between coils. This attention to detail in criticism is suggestive of a highly developed skill of "problem finding" (Getzels and Csikszentmihalyi, 1976).
  - Generation of an evaluative Gedanken experiment as a way to evaluate a scientific model.
  - Ultimately, when two models compete, scientists favor one model by making complex judgment calls based on multiple, sometimes conflicting, criteria discussed in Chapter 18.
  - Ability to maintain and adjust a balance of divergent generation and convergent criticism and evaluation; the complexity of the overall process depicted in Fig. 21.2 is high, especially when carried out successfully on a hard problem.

I suspect that the ability to execute the above processes well in a particular area is not "ordinary," even in science; rather it takes an unusual combination of developed skills. This view is consistent with the general finding that outstanding expertise takes years of concerted effort and training in order to develop specialized skills (Ericsson, 2006).

21.6.5.3 Beyond the Ordinary III: Breaking with the Past

Probably Thomas Kuhn's most famous contribution was to highlight the depth of the cognitive and social obstacles facing scientists trying to break away from ways of thinking of the past, and the present study has argued that there can certainly even be significant obstacles to breaking away from the cognitive framework one is currently using to think about a problem. Long periods where no progress occurs in spite of sensed anomalies can occur in protocols and this can be extremely frustrating for the subject. Kuhn showed that the obstacles can be even greater in the case of social pressures from the traditional past.

Garry Knox Bennett is an artist with works in the Smithsonian who has created some very interesting pieces (among others) by sawing chairs along various planes. In one piece the chair looks as if it is tucked under a red, semicircular side table whose flat side rests against the wall (see Fig. 21.6). But the back of the chair is actually underneath and glued to the round side of the table and serves as a third leg. And the front half of the chair is missing as if the front of the chair had been magically slid halfway through the wall under the table. This piece is interesting in

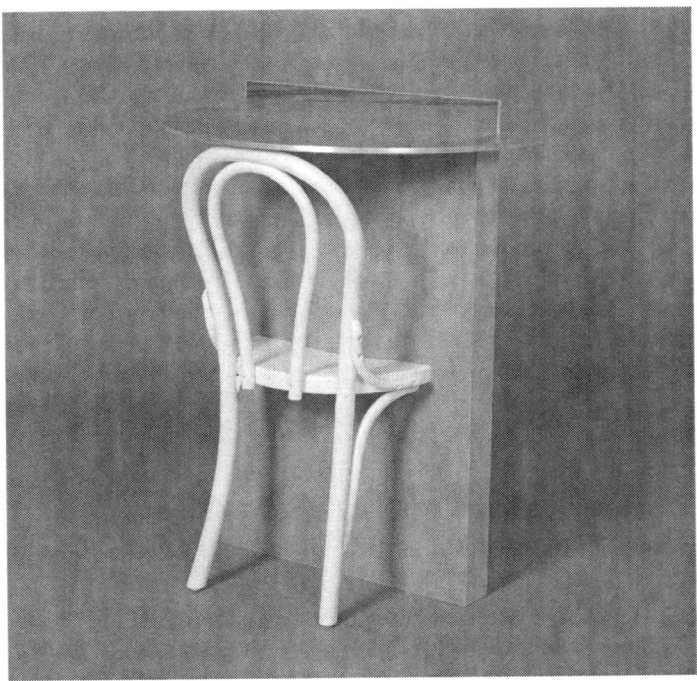

Fig. 21.6 *Hall Table and Chair* by Garry Knox Bennett
Copyright: Garry Knox Bennett, used with permission; Photo: Alison McLennan

large part because it violates our schema expectations for what a table and chair are. The artist has done something really novel because he was able to saw through traditional categories that we are all used to taking for granted. The transformations are radical, not because they are complex or hard to see but because of Einstellung effects associated with standard perceptual motor categories that are not easy for the innovator to think of breaking.

When a scientist uses a playful transformation to generate a surprising analogous or contrasting case (e.g. a square spring or helical molecule or torsionless coil), it can be unordinary in the same way. Or when a scientist invents an experiment (in thought or in the lab) that exposes a problem and threatens a familiar theory, this can also be a creative act that challenges the ordinary. The present protocols documented both kinds of thinking. One's thinking can fall into very deep grooves that are difficult to break out of. In model evaluation processes the most successful experts have a striking degree of persistence and willingness to criticize their own ideas stringently in the search for consistent and coherent models. These can motivate creative theorizing, but one also needs a playful approach to generating new theories, such as using analogies and extreme cases.

These processes resulted in some relatively sudden breakthroughs that I called scientific insights in the present study. Here again, instead of a strict dichotomy between a complete break with prior ideas vs. continuity with all prior ideas, we need to allow for intermediate cases where thinking is still connected in some ways to the past, but also makes a particularly difficult break with strong habits of the past. There are multiple sides to this story because the emphasis on GEM cycles in this book focused on the modification of a previous model, and this is very dependent on prior thinking. However, it was recognized that the method of small changes via GEM cycles does not always work; sometimes a radical break is needed. This led to an overall picture of punctuated evolution: more or less steady evolution much of the time, but broken by an occasional impasse requiring a difficult shift, as shown in Fig. 19.2. Thus I consider persevering against strong Einstellung forces and eventually breaking with one's prior views or assumptions to be another kind of thinking that is not ordinary. I have been able to argue this position using data from problem-solving efforts that were very short compared to those in real science innovations; arguments along these lines for real science innovations may be even stronger.

21.6.6 Summary: How Creative Expert Reasoning is not Ordinary

In summary, being able to say how highly innovative thinking is different from everyday thinking is an important challenge. On the one hand, I have argued that expert reasoning may grow out of natural reasoning as an extension and that expert theories can grow out of intuitive ideas. It would seem unfair to call these trajectories "extraordinary" in the sense of being outside the realm of rational processes. But on the other hand, it seems unfair to call expert innovation methods "ordinary" because of their ability to:

- Generate novel transformations and flexible case generations that go beyond traditional, rule-bound methods
- Maintain complex, specialized processes for advanced innovation used by some experts
- Apply persistent, stringent criticism to one's own ideas or to traditional ideas in the face of persistent Einstellung effects in order to break away from a previous model

In this sense it seems preferable to describe creative reasoning in talented experts as lying between the ordinary and the extraordinary; in some cases it is sophisticated and remarkable, even though it can be seen as a developed skill that has its origins in natural reasoning.

21.6.7 *Implications for Instruction: Utilizing Natural Reasoning Processes*

I have referred to work showing that successful instructional strategies can be designed that tap into the plausible reasoning modes observed in both experts and students. Unfortunately, many students may not be accustomed to learning in this way in school, so they must be acclimated to the approach. However, student transcripts have supported the idea that when students' reasoning is scaffolded by a teacher's support, students become capable of understanding and playing a significant role in more sophisticated reasoning. Given these findings, it is essential that we not destroy the student's natural ability to use imageable mental models, model-based reasoning, and intuition-based grounding for new meanings in science, which is what can happen if students perceive science as an exercise in memorization. If we are going to have any chance of producing independent thinkers – and have some counter to the mindless conformism that is so easily manipulated and weakens a society – then we need to take seriously the job of encouraging and developing students', natural reasoning processes. Such plausible reasoning methods appear to be used quite naturally by students in other domains such as the assignment of unobservable, unstated motives to other individuals, or the assignment of hidden causes to devices needing repair, and it would be unfortunate if we do not take advantage of the power of such reasoning processes in science instruction. If this can be done, one can envision courses helping students experience the power of scientific reasoning, the feeling of sense-making, and the acquisition of models that are flexible and powerful, where learning science is seen as an extension of one's natural impulse to understand and make sense of the world, rather than as an isolated academic exercise.

21.7 Assessing the Potential for a Model of Creative Theory Construction in Science

21.7.1 *Expertise and Domain Specificity*

The findings presented here also have a bearing on the nature of adaptive expertise and the question of whether there are general, interdomain processes in science. To what extent are the processes identified in this book domain specific to mechanics? Or are they general scientific investigation skills? Examining the processes summarized by the figures in Table 19.1, they appear to be applicable to other domains as well, such as chemistry, geology, and biology. It would seem reasonable that reasoning via analogy, model generation and evaluation, and imagistic simulation, should apply to other sciences (and aspects of everyday life). Significant connections to the heuristic strategies identified in Dunbar's (1997) analysis of genetics labs, Darden's (1983) analysis of the history of genetics, and the descriptions of Machamer et al. (2000) of the pursuit of qualitative visualizable mechanisms in biochemistry as a central goal, suggest that some central high-level nonformal reasoning processes described here, such as the use of analogies and extreme cases within model construction cycles, are domain general and not specific to physics. And for those processes that were found similar to novices' everyday reasoning, it makes sense that they are most likely domain general. On the other hand, the aspects in the present study that I consider to be obviously domain specific are the particular schemas used, such as schemas for force, deformation, torque, etc. and associated references to "physical intuitions" about these derived from practical experience with them. Here, the expert will have the advantage of having selected and honed more specific intuitions than the novice. Also, it is possible that the grounding of biology and chemistry concepts directly in confident intuitions based on experience may be present but less common or extensive than in mechanics. And some biology concepts will essentially be grounded in chemistry ideas, and some chemistry concepts in physics. Thus the chain of grounding may be longer for some concepts in the latter fields. Markman and Gentner (2001) consider analogy to be in an intermediate position – not as domain general as deduction, since analogies are sensitive to domain content, but more general than specialized knowledge. In summary, more work needs to be done, but the initial indication through Darden's work is that some of the major higher-level reasoning and model construction processes discussed herein appear to be domain general science investigation processes; however individual applications of reasoning via these processes also appear to be grounded in the concrete content and meaning of the material they are working with.

21.7.2 Can Creative Behavior be Explained?

Related to the question of whether creative processes in experts are a natural extension of everyday reasoning is the question of whether creative behaviors are possible to explain or whether some are too difficult to explain. If a significant part of expert thinking is extraordinary, then the answer could be no to both questions. Here, I list aspects that are still largely unexplained and then speculate on the extent to which we may eventually be able to explain creative behaviors.

21.7.2.1 Aspects Still Poorly Understood in the Present Model

It must be said that there are many processes that are still poorly understood at present. Some of these are:

- Detailed nature of imagery representations and imagistic simulation, as expressed in the thought experiment paradox. The unpacking of such processes in Fig. 15.3 is a beginning, but details are needed on the fundamental question of how a perceptual motor schema produces a simulation for a new case, and how it "stretches" to do this for a case outside its normal domain of application.
- Nature of spatial reasoning and image transformations, and how these are connected to the navigation and hand-eye manipulation systems.
- Exactly how the large number of processes described in this book are sequenced and coordinated with neither too much nor too little control. Here, more work with models such as production systems may be useful.
- How processes identified herein are coordinated with the design and implementation of real experiments.
- The nature of incubation and other long-term effects on the "network of enterprise" in the creative process.
- How these processes can be aided, impeded, or complemented by those from social interactions; and by external representations.
- How aesthetics and emotions can aid in or interfere with the creative process.
- How questions are formulated which drive the inquiry process. Initial analyses of how dissonance occurs have been discussed, but this may be only one source of questions.
- The nature of generative abduction under multiple constraints. This has been described as an imagery-based design process central to theory construction, involving combining schemas to generate compound simulations, but more detail is needed.
- How the subject can exhibit an "Aha" reaction that something important has been discovered, even before its implications have been developed and articulated. For example, the Aha episode upon considering the square and polygonal coils in line 117 in Chapter 6 is of this form.

The above phenomena are not well understood, indicating that we have not formulated adequate explanations for many aspects of creative processes. For me, many of these processes still inspire awe.

21.7.2.2 Toward a Model of Creative Theory Construction in Science

Even though the above issues have not been resolved, I would argue that enough of a beginning has been made on a model of creative theory construction in science to make further progress seem possible, and to make it more difficult for someone to claim that creative processes are unexplainable in principle. I will first discuss reasons to be optimistic about this, then reasons to be pessimistic, and then speculate on what is possible within probable limits.

Optimistic view. Viewing Fig. 20.1 provides some optimism, from the different levels of explanation: elemental simulations and transformations at the bottom supporting nonformal reasoning, model construction, and model application processes. In such a hierarchy one can begin to see volatility at the bottom level percolating up and providing divergence for creative theory generation. And one can see runnability being transferred upward in the same way. Each level can be unpacked into more detailed processes. This and Figs. 16.4 and 16.8 (as well as other figures in the composite Table 20.1 of figures) outline at least partial explanations, given in the text, of analogy, thought experiments, transfer of runnability, and a control structure that harnesses these to produce the major stages of creative model construction described in Chapter 14.

In particular, a flexible control process was cited in explaining how Einstellung effects could lead to a punctuated evolution pattern that included insights. This was based on a foundation of ongoing, more gradual, GEM cycles, but with the possibility of Einstellung effects leading to the subject being trapped in a "rut" and analogies or associations generating sudden insights that allow an escape from the "rut."

Pessimistic view. On the other hand, there are reasons to be pessimistic. The construction process in Fig. 21.2 utilizes an impressive variety of nonformal reasoning processes that have been described. When one thinks of the central GEM cycle as the eye of the vortex with its feeder arms in that figure – operating as only one stage within Fig. 16.4, followed by further stages at finer levels of detail and depths of explanation, it suggests a somewhat daunting multitude of nested and sometime competing actions to be sorted out by the analyst.

Previous studies of advanced cognition have tended to focus on "close" inferences that rely heavily on standard or given problem representations. Within these representations subjects can make deductions or at least probable inferences in a relatively systematic and stepwise manner. Some of the episodes in the present study appear to exemplify a more difficult form of cognition where the representations themselves are being invented rather than simply manipulated. For example, when analogies are explored the subject appears to intentionally leave the original problem space, not solving the problem on its own terms, but in effect rejecting the original problem in favor of finding a "better" one. This strategy must eventually come full circle to apply back to the original problem, but nevertheless there is a

21.7 Potential for a Model of Creative Theory Construction

period where one has the sense of the subject "leaving terra firma" and launching into a meta-space of problem spaces. It is as if suddenly any schema in prior knowledge is eligible to serve as a basis for reinterpreting the problem states and operators, and that means the number of possibilities is exceedingly large (a problem that has also been described by Anderson [1990], p233).

But the situation is even worse. Subjects did not restrict themselves to familiar analogous cases that activated prior knowledge schemas. They also invented new cases that were clearly novel and not previously seen. Cases have been described where the subject invents a "torsion less spring coil made of elements that can turn with perfect ease," or "bar of rectangular cross section that was constructed by reassembling pieces of a doughnut that was cut into apple rings." Seeing such flexibility leads one to appreciate the highly creative but effortful processes some scientists are capable of. The task of analyzing and systematizing the process by which subjects do this is daunting because there would seem to be no algorithm for inventing such diverse and far fetched creative inventions that form new problem representations. Like a sculptor facing the options for molding blocks of clay, the number of options is infinite. How can we hope to find any consistent processes underlying such diverse and creative behavior?

Invention and our ability to attain a predictive theory. My view is that this can be done to some extent. Although we may not be able to *predict* the exact path that a creative thinker will go down next, I still believe that we can uncover patterns at a somewhat larger grain size which allow us to explain such thinking more deeply than was done in the last century. Each of the processes discussed in this book, including the ones shown in Fig. 21.2, is an example of such a pattern. They essentially provide a set of units, images and vocabulary for describing or explaining subjects' reasoning. After more cases are analyzed to refine the theory, one should be able to use this vocabulary to make predictions over multiple subjects. A possible example is: "On average, radical problem transformations should tend to happen early in the process of model construction (unless a late contradiction arises), whereas smaller problem transformations should happen late in the process and in the final mathematical stages of development." Such statements are predictive at a certain level of explanation, but allow for individual variation and creativity in the details. As in the case of the weather, we may be able to theorize and predict trends down to a certain grain size of explanation, but not further.

Another reason that I am pessimistic about deterministic theories is that I have given up some of the certainty and stability associated with physical symbol systems. The units of knowledge (schemas) here have dynamic and flexible properties (like motor schemas) of "becoming activated," "stretching to attend to and assimilate an unfamiliar object in the world," "tracking an image of a moving object over time," "having momentum in the sense of staying active for a while," "acting over time in continuous coordination with another action schema" and "adapting action output dynamically to accommodate new constraints." The very units of knowledge are less predictable because they have some flexible dynamic autonomy, in contrast to static symbol data structures. This kind of plasticity in knowledge, discernable even in infants, is presumably what led Piaget to search for biological metaphors for knowledge and reasoning rather than mechanical ones, and it is another reason for being skeptical about the possibility of a fully deterministic theory.

21.8 Conclusion

21.8.1 Creative Thinking

Scientists may have a public face that is quite different, but think-aloud protocols allow one access to a hidden world of nonformal reasoning on images of both perceptions and actions that contrasts with the view of scientists as abstract thinkers who use only formal logic or mathematics. Reflecting on examples of creative constructions from the protocols highlights how different the thinking there is from more mundane cognitive processes. Thinking of a spring as a sponge or polygon, a wheel as a lever, a torus as "apple rings," and comparing a slingshot to a meteor are creative ideas of the first order. So is the invention of specially designed Gedanken experiments for the spring like a "torsionless coil" or "vertical band spring." These are cases where the thinker leaves tradition and algorithmic methods behind and launches into a riskier creative venture for the sake of model construction. To then be able to "run" a model internally in an imagistic simulation even though one has never seen such a model is just as remarkable. This gives one the ability to do a very rapid evaluation of the initial plausibility of the model. That subjects can use imagery enhancement techniques to improve these images and make these internal tests "easier to see" is another remarkable tool. The scientists appear to be creative inventors of these imageable models rather than simply being logical manipulators of linguistic symbols. A major goal of this book was to better understand this "hidden world" of expert nonformal reasoning by outlining a theory for each process that has initial grounding in evidence from case studies.

21.8.2 The Model Construction Process Portrayed Here in Contrast with Oversimplified Models

The image of expertise we are left with is one that is complex and sophisticated and yet one that has primitive roots in its dependence on concrete prior knowledge and non-formal reasoning. Given its potential for producing too many scattered ideas that could swamp the system, the collection of processes in Fig. 21.2 highlights the impressive balancing act that a successful creative expert performs. Balancing influences are needed to avoid the pitfalls described at the bottom of the figure. The need for balance is reflected in the present framework taking an intermediate position on a number of historical issues. These can be expressed most compactly in pairs of seeming opposites that are nevertheless both used by a scientific modeler, such as divergent vs. convergent thinking. Other pairs are: algorithmic control vs. anarchistic flexibility;

21.8 Conclusion

change vs. conservation; and evolution vs. Eureka events ("mini-revolution"). In addition, a central juxtaposition that is largely outside the scope of this book is: observation vs. theory. With respect to knowledge structures, other pairs are: old knowledge used in new ways; concrete but general; and stable but flexible. Thus Fig. 21.2 is intended to portray a delicately balanced system that is negotiating and accommodating for different and sometimes competing pressures shown at the bottom of the figure. Another way to say this is that the study provides counterexamples to a number of overly simplistic views of scientific thinking, as follows:

- *Growth in theories only from inductive or deductive truth transmission.* Subjects were restricted to using nonempirical methods, and their progress in spite of this was used to argue against an overly inductivist view of scientific investigation. Model construction was seen as an inventive, abductive, hypothesis generation and evaluation process vs. one of truth transmission either from statistically strong patterns in data in bottom-up fashion or from deduction from axioms in top-down fashion. In this process model evaluation can also originate from nonempirical as well as empirical sources. Although this book must be seen as concentrating on providing evidence for the importance of nonempirical processes, the intended image of science that results is a balanced one that uses both empirical observations and complex plausible reasoning operations.
- *Eurekaism.* The protocols on the spring problem also argue against either a pure gradualist or a pure Eurekaist view of the pace of change in scientific hypothesis formation. They revealed a process of punctuated evolution: many small-sized changes (evolution), periods of stagnation (Einstellung effects), and a few medium-sized and rational, but powerful, insights.
- *Divergent thinking.* The hallmark of effective creativity in these protocols was not simply divergent thinking, but rather the ability to alternate between generative divergent production and evaluative convergent criticism as part of the above cycle, and to use "contained or focused divergence" strategies with increased odds for success.
- *Algorithmic vs. anarchistic control.* Rather than either of these extremes, the present view includes a focus on goals and structured cycles of generation and evaluation appropriate to different stages in an investigation complemented by a loosely ordered collection of heuristics used within those cycles, and the flexibility to switch goals when interrupted by a good idea. This provides an intermediate degree of control.
- *Scientific knowledge as abstract.* Although some types of scientific knowledge are highly abstract, the present findings also emphasize the role of concrete experiential knowledge, including perceptual motor schemas and kinesthetic intuitions. The role of concretely visualizable (but at the same time schematic and general) imagery at an intermediate level of abstraction was highlighted. Qualitative models were seen as providing a foundation for quantitative ones, with intermediate levels in between.

Thus the present findings suggest that the methods used by scientists are more varied, balanced, and complex than the more one-sided views above, and suggest that in addition to induction by enumeration or the hypothetico-deductive method – creative abduction, analogy, and rationalistic evaluation methods that use imagistic processes can each play important roles at different times in scientific thought.

21.8.3 Questions About Scientific Thinking

The conclusion section of this book began with a hierarchy of questions about scientific thinking in Table 19.3. The framework that has been developed allows a way of seeing how the questions are interrelated. Starting from the bottom of the table and working up, basing the theory on perceptual motor schemas capable of generating imagistic simulations paves the way for understanding analogies as a way of projecting imagery selectively from one context to another. Compound simulations offer a way to think about how schemas can be combined during the abduction of a new model, as an alternative to induction or deduction. Since model runnability is inherited from the runnability and flexibility of its constituent schemas, this enables some major modes of model evaluation: via running the model on the target phenomenon (this first mode seems widely underappreciated) or on Gedanken experiment cases. These properties also give the model flexibility to complement its stability. Repeated evaluation and modification cycles can produce model evolution. When this process reaches an impasse, more divergent processes can be used: transforming the model image in a radical way or activating additional schemas that provide a new way of imagining the mechanism. A sequence of stages of increasing precision in the imagery and accompanying language can eventually enable attaching mathematical schemas to the model.

Individually these processes do not seem powerful, and because they are nonformal and build on prior knowledge that includes concrete intuitions, they appear to be extensions of natural thinking processes to a considerable extent. But harnessed together, this coalition of nonformal processes appears to be capable of real scientific creativity in constructing models that are both new and viable. Protocols in the present study give evidence that this can include scientific insights, and yet they do not contain episodes that are "extraordinary" in the sense that we cannot begin to explain them using models of nonformal reasoning processes.

The overall picture is that these scientists appear to start from natural forms of knowledge and reasoning, although they also extend them to become very powerful tools. Their natural nature means that there is a large potential for engaging students in these processes. The examples of instructional applications presented suggest that the analysis of nonformal reasoning processes in science can open up many new ideas for science instruction, indicating an important area for further work.

We have only begun to develop models of creative scientific behavior and there may be limits to how detailed such a theory can be. However, the author is optimistic that by evaluating and modifying these explanations in repeated cycles, we can continue to build progressively better models of creative theory formation in science that include the roles of imagery, analogy, and mental simulation.

References

Achinstein, P. (1970). Inference to scientific laws. In Stuewer, R. (Ed.), *Minnesota studies in philosophy of science,* Vol.5: *Historical and philosophical perspectives of science* (pp. 87–104). Minneapolis: University of Minnesota Press.

Alibali, M. W. (2005). Gesture in spatial cognition: Expressing, communicating, and thinking about spatial information. *Spatial Cognition and Computation,* 5(4), 307–331.

Alibali, M. W., Bassok, M., Olseth-Solomon, K., Syc, S. E., & Goldin-Meadow, S. (1999). Illuminating mental representations through speech and gesture. *Psychological Science,* 10, 327–333.

Alibali, M. W, & Goldin-Meadow, S. (1993). Gesture-speech mismatch and mechanisms of learning: What the hands reveal about a child's state of mind. *Cognitive Psychology,* 25(4), 468–523.

Alibali, M. W., & Nathan, M. J. (in press). Teachers' gestures as a means of scaffolding students' understanding: Evidence from an early algebra lesson. In R. Goldman, R. Pea, B. Barron, & S. J. Derry (Eds). *Video research in the learning sciences.* Mahwah, NJ: Erlbaum.

Anderson, J. R. (1977). Arguments concerning representations for mental imagery. *Psychology Review,* 85, 249–277.

Anderson, J. R. (1990). *The adaptive character of thought.* Hillsdale, NJ: Lawrence Erlbaum.

Anzai, Y., & Simon, H. A. (1979). The theory of learning by doing. *Psychological Review,* 86, 124–140.

Arbib, M. A. (1981). Perceptual structures and distributed motor control. In V.B. Brooks (Ed.), *Handbook of physiology* (Section 2: The nervous system, Vol. II, Motor control, Part 1, pp. 1449-1480). American Physiological Society.

Arons, A. B. (1990). *Teaching introductory physics.* New York: Wiley.

Barsalou, L. W. (1999). Perceptual symbol systems. *Behavioral and Brain Sciences,* 22, 577–660.

Barwise, J., & Shimojima, A. (1995). Surrogate reasoning. *Cognitive Studies: Bulletin of the Japanese Cognitive Science Society,* 4(2), 7–27.

Black, M. (1979). More about metaphor. In A. Ortony (Ed.), *Metaphor and thought* (pp. 19–45). Cambridge, England: Cambridge University Press.

Bransford, J. D. & Schwartz, D. (1999). Rethinking transfer: A simple proposal with multiple implications. In A. Iran-Nejad & P. D. Pearson (Eds.), *Review of Research in Education* (Vol. 24 pp. 61–100). Washington, DC: American Educational Research Association.

Brewer, W. F., & Samarapungavan, A. (1991). Children's theories vs. scientific theories: Differences in reasoning or differences in knowledge?. In R. R. Hoffman & D. S. Palermo (Eds.), *Cognition and the symbolic processes: Applied and ecological perspectives* (pp. 209–232). Hillsdale, NJ: Erlbaum.

Brown, D. E. (1987). Using analogies and examples to help students overcome misconceptions in physics: A comparison between two teaching strategies (Dissertation Abstract International No. 473A). University of Massachusetts.

Brown, D. E. (1992). *Teaching electricity with capacitors and causal models: Preliminary results from diagnostic and Tutoring study data examining the CASTLE project.* Paper presented at the meeting of American Educational Research Association, San Francisco, CA.

Brown, D. E. (1993). Refocusing core intuitions: A concretizing role for analogy in conceptual change. *Journal of Research in Science Teaching, 30*(10), 1273–1290.

Brown, D. E. (1994a). Facilitating conceptual change using analogies and explanatory models. *International Journal of Science Education, 16*(2), 201–214.

Brown, D. E., (1994b). A statistical study of CASTLE curriculum effectiveness in 14 high schools for U.S. National Science Foundation (analysis available by request).

Brown, D. and Clement, J. (1989). Overcoming misconceptions via analogical reasoning: Factors influencing understanding in a teaching experiment. *Instructional Science, 18*, 237–261.

Brown, D. and Clement, J. (1992). Classroom teaching experiments in mechanics. In R. Duit, F. Goldberg, and H. Niedderer (Eds.), *Research in physics learning: theoretical issues and empirical studies.* San Diego, CA: San Diego State University.

Brown, J. R. (1991). *Laboratory of the mind: Thought experiments in the natural sciences.* London: Routledge.

Brown, J. S., & VanLehn K. (1980). Repair theory: A generative therory of bugs in procedural skills. *Cognitive science, 4*, 379–426.

Burstein, M. H. (1986). Concept formation in incremental analogical reasoning and debugging. In R. S. Michalski, J. G. Carbonell, & T. M. Mitchell (Eds.), *Machine learning: An artificial intelligence approach* (pp. 351–370). Palo Alto, CA: Tioga.

Burstein, M. H. (1988). Incremental learning from multiple analogies. In A. Prieditis (Ed.), *Analogica* (pp. 37–62). Los Altos, CA: Morgan Kaufmann.

Butterworth, B. & Hadar, U. (1997). Iconic gestures, imagery, and word retrieval in speech. *Semiotica, 115*(2), 147–172.

Camp, C., Clement, J., Brown, D., Gonzalez, K., Kudukey, J., Minstrell, J., Schultz, K., Steinberg, M., Veneman, V., and Zietsman, A. (1994). *Preconceptions in mechanics: Lessons dealing with conceptual difficulties.* Dubuque, IA: Kendall Hunt.

Campbell, D. (1960). Blind variation and selective retention in creative thought as in other knowledge processes. *Psychological Review, 67*(6), 380–400.

Campbell, D. (1979). Degrees of freedom and the case study. In T. Cook & C. Reichardt (Eds.), *Qualitative and quantitative methods in evaluation research.* Beverly Hills, CA: Sage.

Campbell, N. (1920). *Physics: The elements.* Cambridge: Cambridge University Press. Republished in 1957 as *The foundations of science.* New York: Dover.

Casakin, H, & Goldschmidt, G. (1999). Expertise and the use of visual analogy: Implications for design education. *Design Studies, 20*, 153–175.

Catrambone, R., & Holyoak, K.J. (1989). Overcoming contextual limitations on problem-solving transfer. *Journal of Experimental Psychology: Learning, Memory, and Cognition, 15*(6), 147–1156.

Chan, C., Burtis, J., and Bereiter, C. (1997). Knowledge building as a mediator of conflict in conceptual change. *Cognition and Instruction, 15*(1), 1–40.

Chi, M., Feltovich, P. J., and Glaser, R. (1981). Categorization and representation of physics problems by experts and novices. *Cognitive Science, 5*, 121–152.

Chinn, C. A., & Brewer, W. F. (1998). An empirical test of taxonomy of responses to anomalous data in science. *Journal of Research in Science Teaching, 35*(6), 623–654.

Cianciolo, A. T., Matthew, C., Sternberg, R. J., & Wagner, R. K. (2006). Tacit knowledge, practical intelligence, and expertise. In K. A. Ericsson, N. Charness, P. J. Feltovich, and R. R. Hoffman (Eds.), *The Cambridge handbook of expertise and expert performance* (pp. 613–632). Cambridge: Cambridge University Press.

Clancey, W. J. (1997). *Situated cognition.* Cambridge: Cambridge University Press.

Clement, J. (1979). Mapping a student's casual conceptions from a problem solving protocol. In J. Lochhead. and J. Clement (Eds.), *Cognitive process instruction* (pp. 143–146). Philadelphia, PA: Franklin Institute Press.

References

Clement, J. (1981). Analogy generation in scientific problem solving. *Proceedings of the Third Annual Conference of the Cognitive Science Society.* Berkeley, California.

Clement, J. (1982a). Analogical reasoning patterns in expert problem solving. *Proceedings of the Fourth Annual Meeting of the Cognitive Science Society, 4*, 79–81.

Clement, J. (1982b). Students' preconceptions in introductory mechanics. *American Journal of Physics, 50*, 66–71.

Clement, J. (1983). Use of analogies and spatial transformations by experts in solving mathematics problems. *Proceedings of the Fifth Annual Meeting of the International Group for the Psychology of Mathematics Education, North American Chapter, 2*(5), 102–111.

Clement, J. (1986). Methods used to evaluate the validity of hypothesized analogies. *Proceedings of the Ninth Annual Meeting of the Cognitive Science Society.* Hillsdale, NJ: Lawrence Erlbaum.

Clement, J. (1988). Observed methods for generating analogies in scientific problem solving. *Cognitive Science, 12*, 563–586.

Clement, J. (1989). Generation of spontaneous analogies by students solving science problems. In D. Topping, D. Crowell, and V. Kobayashi (Eds.), *Thinking across cultures: Proceedings of the Third International Conference, Honolulu, HI* (pp. 303–308). Hillsdale, NJ: Lawrence Erlbaum.

Clement, J. (1989). Learning via model construction and criticism: Protocol evidence on sources of creativity in science. In J. Glover, R. Ronning, and C. Reynolds, (Eds.), *Handbook of creativity: assessment, theory and research* (pp. 341–381). NY: Plenum.

Clement, J. (1991). Nonformal reasoning in experts and in science students: The use of analogies, extreme cases, and physical intuition. In J. Voss, D. Perkins, and J. Siegel (Eds.), *Informal reasoning and education* (pp. 345–362). Hillsdale, NJ: Lawrence Erlbaum.

Clement, J. (1993a). Model construction and criticism cycles in expert Reasoning. In *Proceedings of the Fifteenth Annual Conference of the Cognitive Science Society.* Hillsdale, NJ: Lawrence Erlbaum.

Clement, J. (1993b). Using bridging analogies and anchoring intuitions to deal with students' preconceptions in physics. *Journal of Research in Science Teaching, 30*(10), 1241–1257.

Clement, J. (1994a). Imagistic simulation and physical intuition in expert problem solving. *The Sixteenth Annual Meeting of the Cognitive Science Society.* Hillsdale, NJ: Lawrence Erlbaum.

Clement, J. (1994b). Use of physical intuition and imagistic simulation in expert problem solving. In D. Tirosh (Ed.), *Implicit and explicit knowledge* (pp. 204–244). Norwood, NJ: Ablex.

Clement, J. (1998). Expert novice similarities and instruction using analogies. *International Journal of Science Education, 20*(10), 1271–1286.

Clement, J. (2000). Analysis of clinical interviews: Foundations and model viability. In R. Lesh and A. Kelly (Eds.), *Handbook of research methodologies for science and mathematics education* (pp. 341–385). Hillsdale, NJ: Lawrence Erlbaum.

Clement, J. (2002). Protocol evidence on thought experiments used by experts. In Wayne Gray and Christian Schunn (Eds.), *Proceedings of the Twenty-fourth Annual Conference of the Cognitive Science Society 22*, 32. Mahwah, NJ: Erlbaum.

Clement, J. (2003). Imagistic simulation in scientific model construction. In *Proceedings of the Twenty-fifth Annual Conference of the Cognitive Science Society.* Mahwah, NJ: Erlbaum.

Clement, J. (2004). Imagistic processes in analogical reasoning: Conserving transformations and dual simulations. In K. Forbus, D. Gentner, and T. Regier, (Eds.), *Proceedings of the Twenty-sixth Annual Conference of the Cognitive Science Society 26* (pp. 233–238). Mahwah, NJ: Erlbaum.

Clement, J. (2006). Thought experiments and imagery in expert protocols. In L. Magnani, (Ed.), *Model-based reasoning in science and engineering*. pp. 151–166. London: College Publications.

Clement, J. (2008a), Six levels of organization for curriculum design and teaching. In J. Clement & M. A. Rea-Ramirez (Eds.), *Model based learning and instruction in science* (pp. 255–272). Dordrecht: Springer.

Clement, J. (2008b). Student-teacher co-construction of visualizable models in large group discussion. In J. Clement and M. A. Rea-Ramirez (Eds.), *Model based learning and instruction in science.* (pp. 11–22). Dordrecht: Springer.

Clement, J. (2008c). The role of explanatory models in teaching for conceptual change. In S. Vosniadou (Ed.), *Handbook of research on conceptual change*. Mahwah, NJ: Lawrence Erlbaum.

Clement, J., & Konold, C. (1989). Fostering problem-solving skills in mathematics. *For the Learning of Mathematics, 9*(3), 26–30.

Clement, J., and Rea-Ramirez, M.A. (Eds). (2008). *Model based learning and instruction in science*. Dordrecht: Springer.

Clement, J. and Rea-Ramirez, M. A. (1998). The role of dissonance in conceptual change. *Proceedings of the 1998 Annual Conference of the National Association for Research in Science Teaching*, San Diego, CA.

Clement, J. and Steinberg, M. (2002). Step-wise evolution of models of electric circuits: A "learning-aloud" case study. *Journal of the Learning Sciences, 11*, 389–452.

Clement, J., Brown, D., Camp, C., Kudukey, J., Minstrell, J., Schultz, K., Steinberg, M., and Veneman, V. (1987). Overcoming students' misconceptions in physics: The role of anchoring intuitions and analogical validity. In J. Novak (Ed.), *The second international seminar on misconceptions and educational strategies in science and mathematics, 3* (pp. 84–97). Ithaca, NY: Cornell University.

Clement, J., Brown, D., and Zietsman, A. (1989). Not all preconceptions are misconceptions: Finding "anchoring conceptions" for grounding instruction on students' intuitions. *International Journal of Science Education, 11*, 554–565.

Clement, J., Zietsman, A., and Monaghan, J. (2005). Imagery in science learning in students and experts. In J. Gilbert (Ed.), *Visualization in science education* (pp. 169–184). Dordrecht, The Netherlands: Springer.

Cohen, H. (1975). The art of snaring dragons. Massachusetts Institute of Technology Artificial Intelligence Laboratory A.I. Memo 338 November, 1974 (LOGO Memo No. 18) Revised May, 1975 Reprinted March 2002. http://64.233.167.104/search?q = cache:r_DK9SeLG1AJ: homepage.cs.latrobe.edu.au/image/dragons/TheArtDragonHunting.pdf + papert + cohen + physics&hl = en&ie = UTF-8

Collins, A. and Gentner, D. (1987) How people construct mental models. In D. Holland and N. Quinn (Eds.), *Cultural models in language and thought* (pp. 243–265). Cambridge: Cambridge University Press.

Confrey, J. (1990). A review of the research on student conceptions in mathematics, science, and programming. In C. B. Cazden (Ed.), *Review of research in education* (pp. 3–56). Washington, D.C.: American Educational Research Association.

Cooper, L. A., & Shepard, R. N. (1973). Chronometric studies of the rotation of mental images. In W. G. Chase (Ed.), *Visual information processing* (pp. 75–176). New York: Academic Press.

Craig, D. L., Nersessian, N. J., & Catrambone, R. (2002). Perceptual simulation in analogical problem solving. *Model-based reasoning: Science, technology, & values*. (pp. 167–191). New York: Kluwer Academic / Plenum Publishers.

Croft, D. and Thagard, P. (2002). Dynamic imagery: A computational model of motion and visual analogy. In L. Magnani and N. Nersessian (Eds.), *Model-based reasoning: science, technology, values* (pp. 259–274). New York: Kluwer/Plenum.

Cronbach, L. J. (1975). Beyond the two disciplines of scientific psychology. *American Psychologist, 30*, 116–127.

Crowder, E. M. & Newman, D. (1993). Telling what they know: The role of gesture and language in children's science explanations. *Pragmatics & Cognition, 1*(2), 339–374.

Darwin, C. (1892). 1958. In F. Darwin (Ed.), *The autobiography of Charles Darwin and selected letters*. New York: Dover.

Darden, L. (1983). Artificial intelligence and philosophy of science: Reasoning by analogy in theory construction. *Philosophy of Science Association 1982, 2*, 147–165.

Darden, L. (1991). *Theory change in science: Strategies from Mendelian genetics*. New York: Oxford.

References

Darden, L. and Rada, R. (1988). Hypothesis formation via interrelations. In A. Prieditis, (Ed.), *Analogica*. Los Altos, CA: Kaufmann.

Decety, J. and Ingvar, D. H. (1990). Brain structures participating in mental simulation of motor behavior: A neuropsychological interpretation. *Acta Psychologica, 73*, 13–34.

Decety, J. (1996). Do imagined and executed actions share the same neural substrate? *Cognitive Brain Research, 3*, 87–93.

Decety, J. (2002). Is there such a thing as functional equivalence between imagined, observed, and executed action? In M. A. Meltzoff & W. Prinz (Eds.), *The imitative mind*. Cambridge: Cambridge University Press.

de Kleer, J., & Brown, J.S. (1983). Assumptions and ambiguities in mechanistic mental models. In D. Gentner & A. Stevens (Eds.), *Mental models* (155–190). Hillsdale, NJ: Lawrence Erlbaum.

diSessa, A. (1983). Phenomenology and the evolution of intuition. In D. Gentner and A. Stevens (Eds.), *Mental models*. Hillsdale, NJ: Lawrence Erlbaum.

diSessa, A. (1985). Knowledge in pieces. In G. Forman and P. Pufall (Eds.), *Constructivism in the computer age* (pp. 49–70). Hillsdale, NJ: Lawrence Erlbaum.

Dijksterhuis, E. J. (1987). *Archimedes* (C. D Ikshoorn, Trans.). Princeton, NJ: Princeton University Press. (Original work published 1938)

Drake, S. & Drabkin, I.E. (Eds). (1969). *Mechanics in sixteenth century Italy: Selections from Tartaglia, Benedetti, Guido Ubaldo, & Galileo*. Wisconsin: University of Wisconsin Press.

Dreistadt, R. (1969). The use of analogies and incubation in obtaining insights in creative problem solving. *Journal of Psychology, 71*, 159–175.

Dreyfus, A., Jungwirth, E., & Eliovitch, R. (1990). Applying the "cognitive conflict" strategy for conceptual change -some implications, difficulties, and problems. *Science Education, 74*(5), 555–569.

Driver, R. (1973). The representation of conceptual frameworks in young adolescent science students. Doctoral dissertation, University of Illinois, Urbana-Champaign, IL.

Driver, R. and Easley, J. (1978). Pupils and paradigms: A review of literature related to concept development in adolescent science students. *Studies in Science Education, 5*, 61–84.

Driver, R. and Erickson, G. (1983). Theories-in-action: Some theoretical and empirical issues in the study of students: Conceptual frameworks in science. *Studies in Science Education, 10*, 37–60.

Duckworth, E., Easley, J., Hawkins, D., & Henriques, A. (1990). *Science education: A minds-on approach for the elementary years*. Hillsdale, NJ: Lawrence Erlbaum and Associates.

Duit, R. (1987). Research on students' alternative frameworks in science topics, theoretical frameworks, consequences for science teaching. In J. D. Novak (Ed.), *Second International Seminar Misconceptions and Educational Strategies in Science and Mathematics*. Cornell U.

Dunbar, K. (1997). How scientists really reason: Scientific reasoning in real-world laboratories. In R. J. Sternberg and J. Davidson (Eds.), *Mechanisms of insight*. Cambridge, MA: MIT Press.

Duncker, K. (1945) On problem solving. Psychological Monographs, vol. 58 (whole No 270).

Duschl, R. A. and Osborne, J. (2002). Supporting and promoting argumentation discourse in science education. *Studies in Science Education, 38*, 39–72.

Easley, J. (1974). The structural paradigm in protocol analysis. *Journal of Research in Science Teaching, 11* (3), 281–290.

Easley, J. (1978). Symbol manipulation reexamined: An approach to bridging a chasm. In B. Presseisen, D. Goldstein, and M. Appel(Eds.), *Topics in cognitive development, 2* (pp. 99–112). New York, NY: Plenum.

Easley, J. (1990). Stressing dialogic skill. In E. Duckworth, J. Easley, D. Hawkins, & A. Henriques (Eds.), *Science education: A minds-on approach for the elementary years* (pp. 61–95). Hillsdale, NJ: Erlbaum.

Einstein, A., & Infeld, L. (1967). *The evolution of physics*. New York: Simon and Shuster.

Ericsson, K. A. (2006). An introduction to *The Cambridge handbook of expertise and expert performance*: Its development, organization, and content. In K. A. Ericsson, N. Charness, P. Feltovich, and R. R. Hoffman (Eds.), *Cambridge handbook of expertise and expert performance* (pp. 3–20). Cambridge, UK: Cambridge University Press.

Falkenhainer, B. (1989) *Learning from physical analogies: A study in analogy and the explanation process.* PhD thesis, University of Illinois, Department of Computer Science, Urbana, IL. ERIC document ED310166.

Falkenhainer, B. (1990). A unified approach to explanation and theory formation. In Jeff Shrager and Pat Langley (Eds.), *Computational models of scientific discovery and theory formation* (pp. 157–195). San Mateo, CA: Morgan Kaufmann.

Feyerabend, P. K. (1975). *Against method: Outline of an anarchistic theory of knowledge.* London: Verso.

Feyereisen, P. and Havard, I. (1999). Mental imagery and production of hand gestures while speaking in younger and older adults. *Journal of Nonverbal Behavior*, 23(2), 153–171.

Finke, R. A. (1989). *Principles of mental imagery.* Cambridge, MA: MIT Press.

Finke, R. A. (1990). *Creative imagery: Discoveries and inventions in visualizations.* Hillsdale, NJ: Lawrence Erlbaum.

Fischbein, E. (1987). *Intuition in science and mathematics: An educational approach.* Dordrecht, Reidel.

Forbus, K. (1984). Qualitative process theory. *Artificial Intelligence*, 24, 85–168.

Forbus, K. D. & Falkenhainer, B. (1990). Self-explanatory simulations: An integration of qualitative and quantitative knowledge. *Proceedings of the Eighth National Conference on Artificial Intelligence, Boston*, pp. 380–387. Cambridge, MA: AAAI Press/The MIT Press.

Forbus, K., Gentner, D., Everett, J., and Wu, M. (1997). Towards a computational model of evaluating and using analogical inferences. *Proceedings of the 19th Annual Conference of the Cognitive Science Society* (pp. 229–234). LEA.

Freedman, N. (1972). The analysis of movement behavior during the clinical interview. In A. W. Siegman & B. Pope (Eds.), *Studies in dyadic communication.* New York: Pergamon.

Freyd, J., (1987). Dynamic mental representations. *Psychological Review*, 94, 427–438

Freyd, J. and Finke, R. (1984). Representational momentum. *Journal of Experimental Psychology: Learning, Memory, and Cognition*, 10, 126–132.

Galileo Galilei (1638/1974). *Two new sciences* (S. Drake, Trans.). Madison, Wisconsin: University of Wisconsin Press.

Galileo Galilei (1976). *Dialogue concerning the two chief world systems* (S. Drake, Trans.). Berkeley: University of California.

Gelerntner, H. (1959). Realization of a geometry theorem-proving machine. In *Proceedings of the International Conference on Information Processing, Paris*, 273–282, UNESCO House.

Gendler, T. S. (1998). Galileo and the indispensability of scientific thought experiment. The British Journal for the Philosophy of Science, 49, 397–424.

Gentner, D. (1982). Are scientific analogies metaphors? In D. Miall (Ed.), *Metaphor: problems and perspectives.* Bridgton, Sussex, England: Harvester Press.

Gentner, D. (1983). Structure-mapping a theoretical framework for analogy. *Cognitive Science*, 7, 155–170.

Gentner, D. (1989). The mechanisms of analogical learning. In S. Vosniadou and A. Ortony (Eds.), *Similarity and analogical reasoning* (pp. 199–241). New York: Cambridge University Press.

Gentner, D., Brem, S., Ferguson, R. W., Markman, A. Levidow, B., Wolff, P., and Forbus, K. (1997). Analogical reasoning and conceptual change: a case study of Johannes Kepler. *Journal of the Learning Sciences, Special Issue: Conceptual Change*, 6(1), 3–40.

Gero, J. S. (2002). Computational models of creative designing based on situated cognition. In T. Hewett and T. Kavanagh (Eds.), *Creativity and Cognition 2002*, (pp. 3–10). New York, NY: ACM Press.

Getzels, J., & Csikszentmihalyi, M. (1976). *The creative vision: A longitudinal study of problem finding in art.* New York: John Wiley.

Gick, M. L. and Holyoak, K. J. (1980). Analogical problem solving. *Cognitive Psychology*, 12, 306–355.

References

Gick, M. L. and Holyoak, K. J. (1983). Schema induction and analogical transfer. *Cognitive Psychology, 15*, 1–38.

Giere, R. (1988). *Explaining science: A cognitive approach.* Chicago: Chicago University Press.

Gilbert, J. K., Boulter, C., and Rutherford, M. (1998). Models in explanations, part 2: Whose voice? Whose ears? *International Journal of Science Education, 20*, 187–203.

Givry, D. & Roth, W.-M. (2006). Toward a new conception of conceptions: Interplay of talk, gestures, and structures in the setting. *Journal of Research in Science Teaching, 43*(10), 1086–1109.

Glaser, B. G. & Straus, A. L. (1967). *The discovery of grounded theory: Strategies for qualitative research.* Chicago: Aldine Pub. Co.

Glynn, S. M. (1991). Explaining science concepts: A teaching-with-analogies model. In S. M. Glynn, R. H. Yeay, and B. K. Britton (Eds.), *The psychology of learning science* (pp. 219–240). Hillsdale, NJ: Erlbaum.

Glynn, S. M. (2003). Teaching science concepts: Research on analogies that improve learning. In D. F. Berlin and A. L. White (Eds.), *Improving science and mathematics education: insights for a global community* (pp. 179–192). Columbus, OH: International Consortium for Research in Science and Mathematics Education.

Glynn, S. M., & Duit, R. (1995). Learning science meaningfully: Constructing conceptual models. In S. M. Glynn & R. Duit (Eds.), *Learning science in the schools: Research reforming practice* (pp. 3–33). Mahwah, NJ: Erlbaum.

Goldin-Meadow, S. (1999). The role of gesture in communication and thinking. *Trends in Cognitive Sciences, 3*(11), 419–429.

Gooding, D. (1990). *Experiment and the making of meaning: Human agency in scientific observation and experiment.* Dordrecht, The Netherlands: Kluwer.

Gooding, D. (1992). The procedural turn: Or, why do thought experiments work? In R. Giere (Ed.), *Cognitive models of science, Minnesota Studies in the philosophy of science, 15* (pp. 46–78). Minneapolis: University of Minnesota Press.

Gooding, D. C. (1994). Imaginary science. *British Journal for the Philosophy of Science, 45*(4), 1029–1045.

Gorman, M. (2006). Technological thinking and moral imagination. In L. Magnani, (Ed.), *Model-based reasoning in science and engineering.* London: College Publications.

Gould, S. (1980). Darwin's middle road. In S. Gould (Ed.), *The panda's thumb: More reflections in natural history.* New York, NY: W. W. Norton.

Gould, S. J. and Eldredge, N. (1977). Punctuated equilibira: The tempo and mode of evolution reconsidered. *Paleobiology, 3*, 115–151.

Gregory, R. L. (1981). *Mind in science.* Cambridge, England: Cambridge University Press.

Greeno, J. G. (1997). On claims that answer the wrong questions. *Educational Researcher, 26*(1), 5–17.

Griffith, T. (1999). A computational theory of generative modeling in scientific reasoning. Unpublished Doctoral dissertation, Georgia Institute of Technology.

Griffith, T. W., Nersessian, N. J., Goel, A. K., and Clement, J. (1997). Exploratory problem-solving in scientific reasoning. *The Nineteenth Annual Conference of the Cognitive Science Society,* Stanford University, Lawrence Erlbaum.

Griffith, T. W., Nersessian, N. J., and Goel, A. (2000). Function-follows-form transformations in scientific problem solving. In *Proceedings of the Cognitive Science Society, 22* (pp. 196–201). Mahwah, NJ: Lawrence Erlbaum.

Gruber, H. (1974). *Darwin on man. A psychological study of scientific creativity.* New York, NY: E. P. Dutton.

Hadamard, J. (1945). *The psychology of invention in the mathematical field.* Princeton, NJ: Princeton University Press.

Hafner, R. and Stewart, J. (1995). Revising explanatory models to accommodate anomalous genetic phenomena: Problem solving in the "context of discovery." *Science Education, 79*(2), 111–146.

Hammer, D. (1995). Student inquiry in a physics class discussion. *Cognition and Instruction*, *13*(3), 401–430.

Hanson, N. R. (1958). *Patterns of discovery*. Cambridge: Cambridge University Press.

Harre, R. (1961). *Theories and things*. London: Newman History and Philosophy of Science Series.

Harre, R. (1967). Philosophy of science, history of. In P. Edwards (Ed.), *The encyclopedia of philosophy* (pp. 289–296). New York: Macmillan and The Free Press.

Harre, R. (1972). *The philosophies of science*. New York, NY: Oxford University Press.

Harre, R. (1983). *An introduction to the logic of the sciences*. (2nd edition). New York: St. Martin's Press.

Harrison, A. G., & Treagust, D. F. (1996). Secondary students mental models of atoms and molecules: Implications for teaching science. *Science Education*, *80*, 509–534.

Hatano, G. I., & Inagaki, K. (1986). Two courses of expertise. In H. Azuma, K. Hakuta, and H. W. Stevenson (Eds.), *Center for Advanced Study in the Behavioral Child Development and Education in Japan* (pp. x, 315 p. ill., 324 cm). New York: W. H. Freeman.

Hatano, G. and Oura, Y. (2003). Reconceptualizing school learning using insight from expertise research. *Educational Researcher*, *32*(8), 26–29.

Hegarty, M., Kriz, S., and Cate, C. (2003). The roles of mental animations and external animations in understanding mechanical systems. *Cognition and Instruction*, *21*(4), 325–360.

Hegarty, M., Mayer, S., Kriz, S., and Keehner, M. (2005). The role of gestures in mental animation. *Spatial Cognition and Computation*, *5*(4), 333–356.

Hesse, M. (1966). *Models and analogies in science*. South Bend, IN: Notre Dame University Press.

Hesse, M. (1967). Models and analogies in science. In P. Edwards (Ed.), *The encyclopedia of philosophy* (pp. 354–359). New York: Macmillan/Free Press.

Hestenes, D. (1997). Modeling methodology for physics teachers. In E. Redish & J. Rigden (Eds.), *The changing role of the physics department in modern universities*, American Institute of Physics Part II. (pp. 935–957).

Hewson, P. W., Hennessey. (1992). Making status explicit: A case study of conceptual change. In R. Duit, F. Goldberg, H. Niedderer (Eds.), *Research in physics learning: Theoretical issues and empirical studies: Proceedings of an International Workshop at the University of Bremen* (pp. 176–187). Kiel: IPN.

Holland, J. H., Holyoak, K. J., Nisbett, R. E., & Thagard, P. R. (1986). *Induction: Processes of inference, learning, and discovery*. Cambridge, MA: MIT Press.

Holyoak, K. J., & Thagard, P. (1989). Analogical mapping by constraint satisfaction. *Cognitive Science*, *13*, 295–355.

Holyoak, K. J., and Hummel, J. E. (2001). Toward an understanding of analogy within a biological symbol system. In D. Gentner, K. Holyoak, & B. N. Kokinov, (Eds.), *Analogical mind: Perspectives from cognitive science*, (pp. 23–58). Cambridge, MA: The MIT press.

Hostetter, A. and Alibali, M. (2004). On the tip of the mind: Gesture as a key to conceptualization. In K. Forbus, S. Gentner, and T. Regier (Eds.), *Proceedings of the Twenty-sixth Annual Conference of the Cognitive Science Society* (pp. 589–594). August 4–7, Chicago, IL, Mahwah, NJ: Lawrence Erlbaum.

Ippolito, M. F. and Tweney, R. D. (1995). The inception of insight. In R. J. Sternberg and J. E. Davidson (Eds.), *The nature of insight* (pp. 433–462). Cambridge, MA: The MIT Press.

Inagaki, K., & Hatano, G. (1987). Young children's spontaneous personification as analogy. *Child Development*, *58*(4), 1013–1020.

Inhelder, B. & Piaget, J. (1958). *The growth of logical thinking from childhood to adolescence*. New York: Basic Books.

Ippolito, M. F., & Tweney, R. D. (1995). The inception of insight. In R. J. Steinberg & J. E. Davidson (Eds.), *The nature of insight*. Cambridge, MA: The MIT Press.

Iverson, J. M. and Goldin-Meadow, S. (1997). What's communication got to do with it? Gesture in children blind from birth. *Developmental Psychology*, *33*(3), 453–467.

Iverson, J. M. and Goldin-Meadow, S. (1998). Why people gesture when they speak. *Nature*, *396*, 228.

References

Jacobsen, E. (1930). Electrical measurement of neuromuscular states during mental activities: II. Imagination and recollection of various muscular acts, *American Journal of Physiology*, 94, 22–34.

Jeannerod, M. (2001). Neural simulation of action. *NeuroImage*, 14, S103–S109.

Jensen, M. S., & Finley, F. N. (1995). Teaching evolution using historical arguments in a conceptual change strategy. *Science Education*, 79(2), 147–166.

Jervis, K. and Tobier, A (Eds.) (1998). *Education for Democracy, Proceedings from the Cambridge School Conference on Progressive Education*. Weston, MA: The Cambridge School.

Johnson, M. (1987). *The body in the mind: The bodily basis of meaning, imagination, and reason*. Chicago: University of Chicago Press.

Johnson, S. and Stewart, J. (1990). Using philosophy of science in curriculum development: An example from high school genetics. *International Journal of Science Education*, 12, 297–307.

Johsua, S. and Dupin, J. J. (1987) Taking into account student conceptions in an instructional strategy: An example in physics. *Cognition and Instruction*, 4(2) (pp. 117–135). Hillsdale, NJ: Erlbaum.

Karmiloff-Smith, A. and Inhelder, B. (1975). If you want to get ahead, get a theory. *Cognition*, 3(3), 195–212.

Kendon, A. (1972). Some relationships between body motion and speech. In A. W. Siegman & B. Pope (Eds.), *Studies in dyadic communication*. New York: Pergamon.

Khan, S., Stillings, N., and Clement, J. (2002). Developing inquiry skills in chemistry students using compact simulations. Presented at the 2002 Annual Meeting of the National Association for Research in Science Teaching, April 7–10, New Orleans.

Klatzky, R. L., Pellegrino, J. W., McCloskey, B. P., and Doherty, S. (1989). Can you squeeze a tomato? The role of motor representations in semantic sensibility judgments. *Journal of Memory and Language*, 28, 56–77.

Koestler, A. (1964). *The act of creation*. New York, NY: Macmillan.

Kosslyn, S. (1980). *Image and mind*. Cambridge, MA: Harvard University Press.

Kosslyn, S. (1994). *Image and brain: The resolution of the imagery debate*. Cambridge, MA: MIT Press.

Koyré, A. (1968). *Metaphysics and measurement*. London: Chapman and Hall.

Kozhevnikov, M., Hegarty, M., & Mayer, R. (2002). Revising the visualizer-verbalizer dimension: Evidence for two types of visualizers. *Cognition and Instruction*, 20(1), 47–77.

Krauss, R. M. (1998). Why do we gesture when we speak? *Current Directions in Psychological Science*, 7, 54–59.

Krist, H., Fieberg, E. & Wilkening, F. (1993). Intuitive physics in action and judgment: The development of knowledge about projectile motion. *Journal of Experimental Psychology: Learning, Memory, and Cognition*, 19, 952–966.

Kuhn, T. (1970). The structure of scientific revolutions. (2nd edition). Chicago, IL: University of Chicago Press.

Kuhn, T. (1977a). The essential tension: Tradition and innovation in scientific research. In T. Kuhn (Ed.), *The essential tension*. Chicago, IL: University of Chicago Press.

Kuhn, T. (1977b) Objectivity, value judgment, and theory choice. Reprinted in Kuhn, *The essential tension* (pp. 320–339). Chicago, IL: University of Chicago Press.

Kuhn, T. (1977c). Concepts of cause in the development of physics. In T. Kuhn (Ed.), *The essential tension* (pp. 21–30). Chicago, IL: University of Chicago Press.

Kuhn, T. (1977d). A function for thought experiments, reprinted in Kuhn, *The essential tension* (pp. 240–265). Chicago, IL: University of Chicago Press.

Kuhn, T. (1977e). Second thoughts on paradigms. In F. Suppe (Ed.), *The structure of scientific theories* (pp. 459–482). Urbana, IL: University of Illinois Press.

Kuhn, D., Amsel, E., & O'Loughlin, M. (1988). *The development of scientific thinking skills*. New York: Academic Press.

Kurz, E. M. & Tweney, R. D. (1998). The practice of mathematics and science: From calculus to the clothesline problem. In M. Oaksford & N. Chater (Eds.), *Rational models of cognition* (pp. 415–438). Oxford: Oxford University Press.

Lakatos, I. (1976). *Proofs and refutations : The logic of mathematical discovery*. New York: Cambridge University Press.

Lakatos, I. (1978). The methodology of scientific research programmes. *Philosophical Papers, 1*. Cambridge: Cambridge University Press.

Lakoff, G. & Nunez, R. (2000). *Where mathematics comes from*. New York: Basic Books.

Lakoff, G., & Johnson, M. (1980). *Metaphors we live by*. Chicago: University of Chicago Press.

Langley, P. (1981). Data-driven discovery of physical laws. *Cognitive Science 5*, 31–54.

Langley, P., Shiran, O., Shrager, J., Todorovski, L., & Pohorille, A. (2006). Constructing explanatory process models from biological data and knowledge. AI in Medicine, *37*, 191–201.

Langley, P., Simon, H. A., Bradshaw, G. L., & Zytkow, J. M. (1987). *Scientific discovery*. Cambridge, MA: MIT Press.

Larkin, J. (1983). The role of problem representation in physics. In D. Gentner and A. Stevens (Eds.), *Mental models* (pp. 75–98). Hillsdale, NJ: Lawrence Erlbaum.

Larkin, J. and Simon, H. A. (1987). Why a diagram is (sometimes) worth ten thousand words. *Cognitive Science, 10*, 65–100.

Lawson, A. E. (1988). Three types of learning cycles: A better way to teach science. *Proceedings of the National Association for Research in Science Teaching Annual Meeting, USA, 61*.

Leander, K. M., & Brown, D. E. (1999). "You understand, but you don't believe it": Tracing the stabilities and instabilities of interaction in a physics classroom through a multidimensional framework. *Cognition and Instruction, 17*(1), 93–135.

Leibniz, G. W. (1898). Monadology and other philosophical essays. R. Latta (Trans.). Oxford, England: Oxford University Press. (Original work written 1714, published 1940 in French).

Lenat, D. B. (1977). Automated theory formation in mathematics. *Proceedings of the Fifth International Joint Conference on Artificial Intelligence, 5*, 833–842.

Lenat, D. B., (1983). The role of heuristics in learning by discovery. In T. Mitchell (Ed.), *Machine learning* (pp. 243–306). Palo Alto, CA: Tioga.

Lindsay, R. K. (1988). Images and reference. *Cognition, 29*, 229–250.

Linn, M. C. (1992). The computer as learning partner: Can computer tools teach science? In K. Sheingold, L. G. Roberts, and S. M. Malcolm (Eds.), *This year in school science 1991: technology for teaching and learning* (pp. 31–69). Washington, DC: American Association for the Advancement of Science.

Lozano, S. and Tversky, B. (2005). Gestures convey semantic content for self and others. Talk given at the *Twenty-Seventh Annual Conference of the Cognitive Science Society*. July 21–23, Stresa, Italy. Accessed 11/20/2006. http://www.psych.unito.it/csc/cogsci05/frame/talk/f707-lozano.pdf

Luchins, A. S. (1942). Mechanization in problem solving. *Psychological Monographs, 54*, 248.

Machamer, P., Darden, L. & Craver, C.F. (2000). Thinking about mechanisms. *Philosophy of Science, 67*, 1–25.

Magnani, L. (1999). Model-based creative abduction. In L. Magnani, N. J. Nersessian, & R. Thagard (Eds.), *Model-based reasoning in scientific discovery* (pp. 219–238). New York, NY: Kluwer Academic/Plenum Publishers.

Maier, N. R. F. (1931). Reasoning in humans: The solution of a problem and its appearance in consciousness. *Journal of Comparative Psychology, 12*, 181–194.

Markman, A. B. and Gentner, D. (2001). Thinking. *Annual Review of Psychology, 52*, 223–247.

Mason, L. (1994). Cognitive and metacognitive aspects in conceptual change by analogy. *Instructional Science, 22*(3), 157–187.

Mayer, R. E. (1999). Fifty years of creativity research. In R. J. Sternberg (Ed.), *Handbook of creativity* (pp. 449–460). New York: Cambridge University Press.

McDermott, L. C. & Redish, E. F. (1999). Resource Letter: PER-1: *Physics Education Research, American Journal of Physics, 67*(9), 755–767.

McNeill, D. (1985). So you think gestures are nonverbal? *Psychological Review, 92*, 350–371.

McNeill, D. (1992). *Hand and mind: What gestures reveal about thought*. Chicago: University of Chicago Press.

Metz, K. E. (1985). The development of children's problem solving in a gears task: A problem space perspective. *Cognitive Science, 9*, 431–471.

References

Metz, K. E. (1998). Scientific inquiry within reach of young children. In B. J. Fraser & K. G. Tobin (Eds.), *International handbook of science education* (pp. 81–96). Dordrecht, Netherlands: Kluwer Academic Press.

Miller, A. I. (1984). *Imagery in scientific thought: Creating 20th century physics*. Boston, MA: Birkhäuser.

Millman, A. B. & Smith, C. L. (1997). Darwin's use of analogical reasoning in theory construction. *Metaphor and symbol, 12*(3), 159–187.

Minstrell, J. (1982). Explaining the 'at rest' condition of an object. *The Physics Teacher, 20*, 10.

Mitchell, M., and Hofstadter, D. R. 1989. The role of computational temperature in a computer model of concepts and analogy-making. In *Proceedings of the Eleventh Annual Conference of the Cognitive Science Society*. Lawrence Erlbaum Associates.

Miyake, N. (1986). Constructive interaction and the iterative process of understanding. *Cognitive Science, 10*(2), 151–177.

Monaghan, J. M. and Clement, J. (1999). Use of a computer simulation to develop mental simulations for understanding relative motion concepts. *International Journal of Science Education, 21*(9), 921–944.

Mounoud, P. and Bower, T. G. R. (1974). Conservation of weight in infants. *Cognition, 3*(1), 29–40.

Murray T., Schultz, K., Brown, D., and Clement, J. (1990). An analogy-based computer tutor for remediating physics misconceptions. *Interactive Learning Environment, 1*(2).

Nagel, E. (1961). *The structure of science*. New York, NY: Harcourt, Brace, & World.

Neisser, U. (1976). *Cognition and reality*. San Francisco, CA: W. H. Freeman.

Nersessian, N. J. (1984). *Faraday to Einstein: Constructing meaning in scientific theories*. Dordrecht: Martinus Nijhoff/Kluwer Academic Publishers.

Nersessian, N. J. and Greeno, J. G. (1990). Multiple abstracted representations in problem solving and discovery in physics. In *Twelfth Annual Proceedings of the Cognitive Science Society* (pp. 77–84). Hillsdale, NJ: Lawrence Erlbaum.

Nersessian, N. J. (1991). Why do thought experiments work? In *Thirteenth Annual Proceedings of the Cognitive Science.Society*. Hillsdale, NJ: Lawrence Erlbaum.

Nersessian, N. J. (1992). How do scientists think? Capturing the dynamics of conceptual change in science. In R. N. Giere (Ed.), *Cognitive models of science: Vol. 15*. (pp. 3–44). Minneapolis: University of Minnesota Press.

Nersessian, N. J. (2002). Maxwell and the method of physical analogy: Model-based reasoning, generic abstraction, and conceptual change. In D. Malamet (Ed.), *Reading natural philosophy: essays in history and philosophy of science and mathematics in honor of Howard Stein on his 70th birthday*. LaSalle, IL: Open Court.

Newell, A., & Simon, H. A. (1972). *Human problem solving*. Englewood Cliffs, New Jersey: Prentice-Hall.

Newell, A. (1973). You can't play 20 questions with nature and win. In Chase (Ed.), *Visual information processing* (PSY BC:C 39).

Niedderer, H. (1987) A teaching strategy based on students' alternative frameworks -theoretical conceptions and examples. In J. D. Novak (Ed.), *Proceedings of the Second International Seminar in Misconceptions and Educational Strategies in Science and Mathematics* (pp. 360–367). Ithaca: Cornell University.

Niedderer, H. (2001). Physics learning as cognitive development. In R. H. Evans, A. M. Andersen, and H. Sørensen (Eds.), *Bridging research methodology and research aims* (pp. 397–414). The Danish University of Education. (ISBN: 87-7701-875-3). http://didaktik.physik.uni-bremen.de/niedderer/personal.pages/niedderer/Pubs.html#lpipt. Last accessed January 27, 2007.

Norton, J. (1996). Are thought experiments just what you always thought? *Canadian Journal of Philosophy, 26*, 333–366.

Nunez-Oviedo, M. C., Clement, J., and Rea-Ramirez, M. A. (2008a). A competition strategy and other discussion modes for developing mental models in large group discussion. In J. Clement & M. A. Rea-Ramirez, (Eds.), *Model based learning and instruction in science* (pp. 117–138). Dordrecht: Springer.

Nunez-Oviedo, M. C., Clement, J., and Rea-Ramirez, M. A. (2008b). Developing complex mental models in biology through model evolution. In J. Clement and Mary Anne Rea-Ramirez (Eds.), *Model based learning and instruction in science* (pp. 173–194). Dordrecht: Springer.

Padilla, M. J. (1991). Science Activities, process Skills, and thinking. In S. M. Glynn, R. H. Yeany & B. K. Britton (Eds.), *The psychology of learning science* (pp. 205–217). Hillsdale, NJ: Erlbaum.

Peirce, C. S. (1958). *Collected papers*, 8 Vols. C. Hartshorne, P. Weiss, and A. Burks (Eds.). Cambridge, MA: Harvard University Press.

Perkins, D. (1981). *The mind's best work*. Cambridge, MA: Harvard University Press.

Piaget, J. (1930). *The childs conception of physical causality*. London: Kegan Paul.

Piaget, J. (1955). *The child's construction of reality*. London: Routledge/Kegan Paul.

Piaget, J. (1976). *The grasp of consciousness*. Cambridge, MA: Harvard University Press.

Pinker, S. (1984). Visual cognition: An introduction. *Cognition*, *18*, 1–63.

Plucker, J. and Renzulli, J. S. (1999). Psychometric approaches to the study of human creativity. In R. J. Sternberg (Ed.), *Handbook of creativity* (pp. 35–60). New York: Cambridge University Press.

Polanyi, M. (1966). The tacit dimension. Garden City, NY: Doubleday.

Polya, G. (1954). *Mathematics and plausible reasoning, vol. 1: Induction and analogy in mathematics*, 155. Princeton, NJ: Princeton University Press.

Popper, K. (1959). *The logic of scientific discovery*. London: Hutchinson.

Posner, G., Strike, K., Hewson, P., and Gertzog, W. (1982). Accommodations of scientific conceptions: Toward a theory of conceptual change. *Science Education*, *66*, 211–227.

Price, N. (2007). Self-study of the evolution of a "deferred judgment questioning" discussion mode in a middle school science teacher. *Proceedings of the National Association for Research in Science Teaching Annual Meeting*.

Pylyshyn, Z. W. (1981). The imagery debate: Analogue media versus tacit knowledge. *Psychological Review*, *88*, 16–88.

Qin, Y. and Simon, H. (1990). Imagery and problem solving. *Twelfth Annual Proceedings of the Cognitive Science Society* (pp. 646–653). Hillsdale, NJ: Lawrence Erlbaum.

Rada, R. (1985). Gradualness facilitates knowledge refinement. I.E.E.E. Transactions on Pattern Analysis and Machine Intelligence, *7*(5), 523–530.

Raghavan, K. and Glaser, R. (1995). Model-based analysis and reasoning in science: The MARS curriculum. *Science Education*, *79*(1), 37–61.

Rea-Ramirez, M. A., Nunez-Oviedo, M. C., Clement, J., and Else, M. J. (2004). *Energy in the human body curriculum*. Amherst, MA: University of Massachusetts.

Rea-Ramirez, M. A. (2008) Determining target models and effective learning pathways for developing understanding of biological topics. In J. Clement and M. A. Rea-Ramirez, (Eds.), *Model based learning and instruction in science*. Dordrecht: Springer.

Rea-Ramirez, M. A., & Nunez-Oviedo, M. C., (2008a). Model based reasoning among inner city middle school students. In J. Clement & M. A. Rea-Ramirez (Eds.), *Model based learning and instruction in science* (pp. 233–254). Dordrecht: Springer.

Rea-Ramirez, M., & Nunez-Oviedo, M. C., (2008b). The role of discrepant questioning leading to model element modification. In J. Clement and M. A. Rea-Ramirez (Eds.), *Model based learning and instruction in science* (pp. 195–213). Dordrecht: Springer.

Reiner, M. (1998). Thought experiments and collaborative learning in physics. *International Journal of Science Education*, *20*(9), 1043–1058.

Reiner, M. and Gilbert, J. (2000). Epistemological resources for thought experimentation in science learning. *International Journal of Science Education*, *22*(5), 489–506.

Reiner, M. & Burko, L. M. (2003). On the limitations of thought experiments in physics and the consequences for physics education. *Science & Education*, *12*, 365–385.

Rime, B. (1982). The elimination of visible behavior from social interactions: Effects of verbal, nonverbal and interpersonal variables. *European Journal of Social Psychology*, *12*, 113–129.

Rimé, B., & Schiaratura, L. (1991). Gesture and speech. In R. S. Feldman & B. Rimé (Eds.), *Fundamentals of nonverbal behavior*. New York: Press Syndicate of the University of Cambridge.

References

Roland, P. E., Larsen, B., Lassen, N. A., & Skinhoj, E. (1980). Supplementary motor area and other cortical areas in organization of voluntary movements in man, *Journal of Neurophysiology*, *43*, 118–136.

Romero, D. H., Lacoursea, M. G., Lawrencea, K. E., Schandlera, S., & Cohena, M. J. (2000). Event-related potentials as a function of movement parameter variations during motor imagery and isometric action. *Behavioral Brain Research*, *117*, 83–96.

Rosenbaum, D. A. (1991). *Human motor control*. San Diego, CA: Academic Press.

Roth, W-M. & Roychoudhury, A. (1993). The development of science process skills in authentic contexts. *Journal of Research in Science Teaching*, *30*(2), 127–152.

Roth, W-M. (2001). Gestures: Their role in teaching and learning. *Review of Educational Research 71*(3), 365–392.

Roth, W-M. (in press). From gesture to scientific language. *Journal of Pragmatics*.

Rothenberg, A. (1979). *The emerging goddess*. Chicago, IL: University of Chicago Press.

Rumelhart, D. E., & Norman, D. A. (1981). Analogical processes in learning. In J. R. Anderson (Ed.), *Cognitive skills and their acquisition*. Hillsdale, NJ: Lawrence Erlbaum.

Shepard, R. N., & Metzler, J. (1971). Mental rotation of three-dimensional objects. *Science*, *171*, 701–703.

Schmidt, R. A. (1982). *Motor control and learning*. Champaign, IL: Human Kinetics.

Schon, D. (1981). Intuitive thinking? A metaphor underlying some ideas of educational reform. D.S.R.E. Working paper WP-8, Massachusetts Institute of Technology, Cambridge, MA.

Schoenfeld, A. H. (1985). *Mathematical problem solving*. Orlando, FL: Academic Press.

Schrager, J. (1990). Commonsense perception and the psychology of theory formation. In Shrager, J. & Langley, P. (Eds.) Computational models of scientific discovery and theory formation (pp. 437–470). Morgan Kaufman, San Mateo, CA.

Schwartz, D. L. and Black, J. B. (1996a). Shuttling between depictive models and abstract rules: Induction and fall-back. *Cognitive Science*, *20*(4), 457–497.

Schwartz, D. L. and Black, J. B. (1996b). Analog imagery in mental model reasoning: Depictive models. *Cognitive Psychology*, *30*, 154–219.

Schwartz D. L. and Black, T. (1999). Inferences through imagined actions: Knowing by simulated doing. *Journal of Experimental Psychology Learning, Memory, and Cognition*, *25*, 116–136.

Schweber, S. (1977). The origin of the origin revisited. *Journal of the History of Biology*, *10*, 229–316.

Shepard, R. N. (1984). Ecological constraints on internal representation: Resonant kinematics of perceiving, imagining, thinking, and dreaming. *Psychological Review*, *91*, 417–447.

Shepard, R. N., & Metzler, J. (1991). Mental rotation of three-dimensional objects. *Science 171*, 701–703.

Shepard, R. and Cooper, L. (1982). *Mental images and their transformations*. Cambridge, MA: MIT Press.

Shepard, R. (2006). Thought experiments in scientific discovery: What emergent mental capabilities underlie their efficacy? *Proceedings of the 28th Annual Conference of the Cognitive Science Society*.

Smith, C. L. & Millman, A. B. (1987). Understanding conceptual structures: A case study of Darwin's early thinking. In J. Lochhead & J. C. Bishop (Eds.), *Thinking* (pp. 197–211). Hillsdale, NJ: Erlbaum.

Sorensen, R. (1992). *Thought experiments*. Oxford: Oxford University Press.

Spiro, R. J., Feltovich, P. J., Coulson, R. L., and Anderson, D. K. (1989). Multiple analogies for complex concepts: Antidotes for analogy-induced misconceptions in advanced knowledge acquisitions. In S. Vosniadou and A. Ortony (Eds.), *Similarity and analogical reasoning* (pp. 498–530). New York: Cambridge University Press.

Stavy, R. (1991). Using analogy to overcome misconceptions about conservation of matter. *Journal of Research in Science Teaching*, *28*(4), 305–313.

Steinberg, M. (1992). *Electricity visualized – the CASTLE project*. Roseville, CA: Edmund Scientific.

Steinberg, M. (2006). The nature of creativity. *Creativity Research Journal 18*(1), 87–98.

Steinberg, M. (2008). Target model sequence and critical learning pathway for an electricity curriculum based on model evolution. In J. Clement and M. A. Rea-Ramirez, (Eds.), *Model based learning and instruction in science* (pp. 79–102). Dordrecht: Springer.

Stephens, L. and Clement, J. (2006a). Designing classroom thought experiments: What we can learn from imagery indicators and expert protocols. *Proceedings of the 2006 Annual Meeting of the National Association for Research in Science Teaching*. San Francisco, CA.

Stephens, L. and Clement, J. (2006b). Using expert heuristics for the design of imagery-rich mental simulations for the science class. *Proceedings of the 2006 Annual Meeting of the National Association for Research in Science Teaching*. San Francisco, CA.

Stephens, L. and Clement, J. (2007). Depictive gestures as evidence for dynamic mental imagery in four types of student reasoning. In L. McCullough, L. Hsu, and P. Heron (Eds.), *2006 Physics Education Research Conference: AIP Conference Proceedings, 883* (pp. 89–92). Melville, NY: American Institute of Physics.

Sternberg R. J. (1977). Intelligence, information processing, and analogical reasoning: *The componential analysis of human abilities*. Hillsdale, NJ: Erlbaum.

Sternberg, R. J. (2006). The nature of creativity. *Creativity Research Journal* 18(1), pp. 87–98.

Suppe, F. (1974). The search for philosophic understanding of scientific theories. In F. Suppe (Ed.), *The structure of scientific theories*. Urbana, IL: University of Illinois Press. Thirteenth Annual Proceedings of the Cognitive Science.

Thagard, P. (1989). Explanatory coherence. *Behavioral and Brain Sciences, 12*, 435–502.

Thagard, P. (1992). *Conceptual revolutions*. Princeton, NJ: Princeton University Press.

Thagard, P. and Shelley, C. P. (1997). Abductive reasoning: Logic, visual thinking, and coherence. In M. -L. Dalla Chiara et al. (Eds.), *Logic and scientific methods* (pp. 413–427). Dordrecht, The Netherlands: Kluwer.

Thiele, R. B. & Treagust, D. F. (1994). An interpretive examination of high school chemistry teachers' analogical explanations. *Journal of Research in Science Teaching, 31*, 227–242.

Toulmin, S. (1972). *Human understanding, 1*. Oxford, England: Oxford University Press.

Trafton, G., Trickett, S., and Stitzlein, C. (2004). Spatial transformations and gestures. In the interrelationship between spatial cognition and gestures. *Proceedings of the Twenty-sixth Annual Conference of the Cognitive Science Society* (p. 20). K. Forbus, S. Gentner, and T. Regier (Eds.). August 4–7, Chicago, IL. Mahwah, NJ: Lawrence Erlbaum.

Trickett, S. and Trafton, J. G. (2002) The instantiation and use of conceptual simulations in evaluating hypotheses: Movies-in-the-mind in scientific reasoning. In Wayne Gray and Christian Schunn (Eds.), *Proceedings of the Twenty-fourth Annual Conference of the Cognitive Science Society, 22* (pp. 878–883). Mahwah, NJ: Erlbaum.

Tsai, C. -C. and Chang, C. -Y. (2005). Lasting effects of instruction guided by the conflict map: Experimental study of learning about the causes of the seasons. *Journal of Research in Science Teaching, 42*(10), 1089–1111.

Tweney, R. (1985). Faraday's discovery of induction: A cognitive approach. In D. Goodling and F. James (Eds.), *Faraday rediscovered: Essays on the life and work of Michael Faraday, 1791–1867* (pp. 189–209). New York, NY: Stockton Press.

Tweney, R. (1986). Procedural representation in Michael Faraday's scientific thought. *PSA 1986*, Volume 2.

Tweney, R. D. (1991). Faraday's notebooks: the active organization of creative science. *Physics Education, 26*, 301–306.

Tweney, R. D. (1996). Presymbolic processes in scientific creativity. *Creativity Research Journal, 9*(2&3), 163–172.

Tweney, R. D., Doherty, M. E., Worner, W. J., Pliske, D. B., Mynatt, C. R., Gross, K. A., & Arkkelin, D. L. (1980). *Strategies of rule discovery on an inference task*. Quarterly Journal of Experimental Psychology, 32, 109–123.

Tweney, R. D. & Hoffner, C. E. (1987). Understanding the microstructure of science: An example. In *Program of the Ninth Annual Conference of the Cognitive Science Society*, pp. 677–681. Hillsdale, NJ: Lawrence Erlbaum.

References

Varela, F. J., Thompson, E. & Rosch, E. (1991). *The embodied mind: Cognitive science and human experience* (pp. 172–179). Cambridge, MA: The MIT Press.

van Zee, E. H., & Minstrell, J. (1997). Reflective discourse: Developing shared understanding in a physics classroom. *International Journal of Science Education, 19*(2), 209–228.

VanLehn, K. & Brown, J. S. (1980). Planning nets: A representation for formalizing analogies and semantic models of procedural skills. In R. E. Snow, P. A. Frederico, & W. E. Montague (Eds.). *Aptitude learning and instruction volume 2: Cognitive process analyses of learning and problem solving*. Hillsdale, NJ: Lawrence Erlbaum Associates, Inc.

Van Meel, J. M. (1984). Kinesic strategies in representation and attention-deployment. In *Relation of language, thought and gesture*. The Netherlands: Tilbulg.

Vosniadou, S. (2002). On the nature of naive physics. In M. Limon & L. Mason (Eds.), *Reconsidering conceptual change: Issues in theory and practice* (pp. 61–67). Amsterdam: Kluwer.

Vosniadou, S., & Ortony, A. (1989). Similarity and analogical reasoning: A synthesis. In S. Vosniadou & A. Ortony (Eds.), *Similarity and analogical reasoning* (pp. 1–17). Cambridge: Cambridge University Press.

Watson, J. (1968). *The double helix: A personal account of the discovery of the structure of DNA*. London: Weidenfeld and Nicolson.

Weisberg, R. W. (1993). *Creativity: Beyond the myth of genius*. New York: W. H. Freeman.

Weisberg, R. W. (2006). *Creativity: Understanding innovation in problem solving science invention and the arts*. Hoboken, NJ: John Wiley & Sons, Inc.

Weld, D. (1990) Exaggeration. *Artificial Intelligence, 43*(3), 311–368.

Wells, M., Hestenes, D., and Swackhamer, G. (1995). A modeling method for high school physics instruction. *American Journal of Physics, 63*, 606.

Wertheimer, M. (1959). Productive thinking. In White, B. (Ed.) (1983). *Sources of difficulty in understanding Newtonian*. New York: Harper & Row.

White, B. Y. (1993) Thinkertools: Causal models, conceptual change, and science education. *Cognition and Instruction, 10*, 1–100.

White, B., & Frederiksen, J. (2000). Metacognitive facilitation: An approach to making scientific inquiry accessible to all. In J. Minstrell and E. van Zee (Eds.), Inquiring into inquiry learning and teaching in science (pp. 331–370). Washington, DC: American Association for the Advancement of Science.

Williams, E. G. & Clement, J. (2007a). Identifying model-based teaching strategies: A case study of two high school physics teachers. *Proceedings of the National Association for Research in Science Teaching Annual Meeting, USA, 80*.

Williams, G. & Clement, J. (2007b). Strategy levels for guiding discussion to promote explanatory model construction in circuit electricity. In L. McCullough, L. Hsu, & P. Heron (Eds.), *2006 Physics Education Research Conference: AIP Conference Proceedings*, Vol. *883* (pp. 169–172). Melville, NY: American Institute of Physics.

Wiser, M. and Carey, S. (1983). When heat and temperature were one. In D. Gentner and A. Stevens (Eds.), *Mental models* (pp. 267–297). Mahwah, NJ: Lawrence Erlbaum.

Witz, K. (1975). Activity structures in four year olds. In Scandura, J. (Ed.), *Research in structural learning*, Vol. II. New York: Gordon & Breach.

Zietsman, A. I. and Clement, J. (1997). The role of extreme case reasoning in instruction for conceptual change. *Journal of the Learning Sciences, 6*(1), 61–89.

Zietsman, A. I. and Hewson, P. W. (1986). Effect of instruction using microcomputer simulations and conceptual change strategies on science learning. *Journal of Research in Science Teaching, 23*(1), 27–39.

Name Index

A

Achinstein, P., 72
Alibali, M. W., 174, 194, 196–198, 549
Anderson, J. R., 172 , 490, 571
Anzai, Y., 13
Arbib, M. A., 216
Aristotle, 54, 90
Arons, A. B., 543

B

Barsalou, L. W., 176, 217, 452
Barwise, J., 474
Bennett, G. K., 565
Black, J. B., 9, 177, 197, 555
Black, M., 14, 74, 109
Black, T., 216, 491
Bower, T. G. R., 226
Bransford, J. D., 552
Brewer, W. F., 6, 366, 533, 538, 544
Brown, D., 129, 139, 141, 151, 153, 418, 474, 543, 544, 546, 547, 552
Brown, D. E., 139–155, 417, 536
Brown, J. R., 471
Brown, J. S., 9, 21, 58, 206, 226, 227
Burstein, M. H., 57, 459
Butterworth, B., 174

C

Campbell, D., 11, 110
Campbell, N., 14, 21, 73, 88, 89, 270, 468, 500
Camp, C., 123, 150, 151, 543, 547, 554
Carey, S., 534
Casakin, H., 383
Catrambone, R., 25, 383
Chan, C., 541
Chang, C.-Y., 541

Chi, M., 119, 206, 230, 450
Chinn, C. A., 6, 366, 544
Cianciolo, A. T., 559
Clancey, W. J., 9
Clement, J., 3, 6, 12, 13, 21, 23, 25, 29, 33, 36, 43–45, 47, 56–58, 60, 67, 82, 97, 109, 119, 123, 129, 139, 140, 151, 153, 163, 170, 174, 175, 177, 205, 210, 217, 220, 222, 225, 233, 247, 272, 279, 289, 294, 322, 331, 342, 349, 353, 357, 367, 373, 383, 405, 406, 417, 418, 423, 429, 438, 451, 454, 455, 459, 462, 463, 469, 473, 474, 501, 502, 504, 513, 537, 538, 541, 544, 546, 547, 549, 550, 552–556, 558, 561, 562
Cohen, H., 4, 551
Collins, A., 176, 188, 210, 211, 448
Confrey, J., 139
Cooper, L. A., 187, 464
Craig, D. L., 177, 383
Croft, D., 383
Cronbach, L. J., 12
Crowder, E. M., 174
Csikszentmihalyi, M., 564

D

Darden, L., 6, 85, 111, 112, 331, 357, 370, 426, 427, 502, 510, 560, 568
Decety, J., 175, 176, 195, 216
De Kleer, J., 9, 58, 206, 226, 227
Dijksterhuis, E. J., 170
DiSessa, A., 9, 21, 119, 208, 453, 543, 559
Drabkin, I. E., 54
Drake, S., 54
Dreistadt, R., 21
Dreyfus, A., 541
Driver, R., 58, 139, 153, 247, 539, 546

Duckworth, E., 553
Duit, R., 139
Dunbar, K., 7, 8, 25, 372, 437, 568
Duncker, K., 21, 457
Dupin, J. J., 538
Duschl, R. A., 546

E

Easley, J., 5, 13, 139, 181, 539, 545
Einstein, A., 19, 21, 429, 431, 441, 468
Eldredge, N., 108, 444
Erickson, G., 139
Ericsson, K. A., 564

F

Falkenhainer, B., 25, 85, 227, 240, 357, 462
Faraday, M., 454, 519, 559
Feyerabend, P. K., 529
Feyereisen, P., 196
Finke, R. A., 11, 176, 228, 291, 293, 400, 464, 446
Finley, F. N., 541
Fischbein, E., 537
Forbus, K., 9, 23, 47, 58, 206, 227, 240, 383
Frederiksen, J., 547, 553
Freedman, N., 174
Freyd, J., 176, 228

G

Galileo, G., 54, 59, 235, 439
Gelerntner, H., 226
Gendler, T. S., 279
Gentner, D., 6, 7, 21–25, 33, 43, 85, 149, 176, 188, 210, 211, 438, 448, 457, 462, 468, 519, 568
Gero, J. S., 491
Getzels, J., 564
Gick, M. L., 21, 149, 462, 519
Giere, R., 6, 426, 429
Gilbert, J., 471, 545, 556
Givry, D., 174, 549
Glaser, B. G., 11, 13
Glover, J., 67, 97
Glynn, S. M., 469
Goel, A., 352
Goldin-Meadow, S., 174, 195, 197
Goldschmidt, G., 383
Gooding, D., 6, 9, 206, 240, 318, 322, 454, 471, 473, 519, 520, 521, 523, 559

Gorman, M., 85
Gould, S., 69, 108, 444
Greeno, J. G., 9, 206
Gregory, R. L., 72
Griffith, T., 348–350, 352–354, 478
Gruber, H., 4, 6, 112, 482

H

Hadamard, J., 161, 206, 429
Hadar, U., 174
Hafner, R., 153, 553
Hammer, D., 539, 554
Hanson, N. R., 70, 71, 87, 88, 186, 326, 329
Harre, R., 11, 14, 72, 73, 89, 90, 93, 148, 271, 272, 422, 426, 429, 468, 514
Harrison, A. G., 538
Hatano, G., 6, 538, 552
Havard, I., 196
Hegarty, M., 177, 197, 491, 555
Hennessey, 533, 553
Hesse, M., 14, 21, 73, 88–90, 93, 148, 153, 271, 358, 417, 426, 428, 429, 468, 482, 500
Hestenes, D., 555
Hewson, P. W., 533, 539, 541, 553
Hoffner, C. E., 454
Hofstadter, D. R., 374
Holland, J. H., 25, 85, 111, 149, 357, 462
Holyoak, K. J., 7, 21, 23–25, 33, 149, 438, 457, 462, 519
Hooke, R., 55, 389
Horwitz, P., 543
Hostetter, A., 197

I

Inagaki, K., 6, 538
Infeld, L., 19, 21, 432, 468
Ingvar, D. H., 175, 176, 195, 216
Inhelder, B., 5, 91, 106, 358, 477
Ippolito, M. F., 9, 293, 520
Iverson, J. M., 195, 197

J

Jacobsen, E., 175, 195, 216
Jeannerod, M., 175, 216
Jensen, M. S., 541
Jervis, K., 117
Johnson, M., 21, 174

Name Index

Johnson, S., 539, 547, 553
Johsua, S., 538

K
Karmiloff Smith, A., 91, 106,
    358, 477
Kendon, A., 174
Kepler, J., 6, 462
Khan, S., 546, 547
Klatzky, 173, 176
Koestler, A., 73
Konold, C., 553
Kosslyn, S., 172, 173, 176, 177, 195, 252, 291,
    293, 303, 391, 400, 446, 464, 465, 488,
    490, 492
Koyre, A., 240, 471
Kozhevnikov, M., 491
Krauss, R. M., 196, 197
Krist, H., 226
Krueger, 175, 177
Kuhn, D., 155, 538, 562
Kuhn, T., 70, 71, 87, 90, 99, 107, 108, 235,
    240, 270, 370, 426, 427, 441, 471,
    473, 474, 480, 514, 525, 532–536,
    557, 558, 565

L
Lakatos, I., 87, 238, 239, 357
Lakoff, G., 21, 233, 402,
    503, 504
Langley, P., 5, 6, 73
Larkin, J., 5, 226, 240, 489
Lawson, A. E., 555
Leander, K. M., 536
Lenat, D. B., 111
Lindsay, R. K., 226, 291
Linn, M. C., 539, 547
Lozano, S., 196
Luchins, A. S., 106

M
Machamer, P., 489–491, 500, 520, 568
Magnani, L., 326, 481, 514
Maier, N. R. F., 106
Markman, A. B., 568
Mason, L., 538
Maxwell, J., 6, 367
Mayer, R. E., 5
McDermott, L. C., 139, 543
McNeill, D., 174, 175, 183, 194–196, 200

Metz, K. E., 493, 533, 538, 539, 553
Metzler, J., 176, 228, 400
Miller, A. I., 206, 429
Millman, A. B., 85, 112, 482, 559
Minstrell, J., 50, 544, 546, 547
Mitchell, M., 374
Miyake, N., 85
Monaghan, J., 174, 175, 550, 555
Mounoud, P., 226
Murray, T., 155, 351

N
Nagel, E., 14, 61, 73, 74, 89, 426, 429,
    501, 520
Neisser, U., 216
Nersessian, N. J., 6–8, 85, 206, 240, 272, 278,
    279, 312, 322, 331, 348, 352, 357,
    367, 383, 462, 468, 471, 482, 497,
    519–521, 559
Newell, A., 5, 10, 12, 523
Newman, D., 174
Newton, I., 55, 90, 389, 439
Niedderer, H., 541
Norman, D. A., 21
Norton, J., 523
Nunez-Oviedo, M. C., 153, 541,
    546, 547
Nunez, R., 233, 402, 503

O
Ortony, A., 21
Osborne, J., 546
Oura, Y., 552

P
Padilla, M. J., 554
Papert, 4
Pavio, 173, 446
Peirce, C. S., 88, 326, 329
Perkins, D., 21, 73, 102
Piaget, J., 5, 75, 117, 216, 222, 226, 386, 387,
    420, 451, 545, 549, 559, 571
Pierce, 186
Pinker, S., 159, 451
Plucker, J., 5
Polanyi, M., 73, 441
Polya, G., 3, 4, 21, 127, 133, 161, 170, 237,
    238, 468, 469, 478, 530
Popper, K., 70–72, 441
Posner, G., 532

Price, N., 556
Pylyshyn, Z. W., 172

Q
Qin, Y., 206

R
Rada, R., 85, 111
Rea-Ramirez, M., 153, 541, 544, 546, 547, 556
Redish, 139
Reiner, M., 471, 556
Renzulli, J. S., 5
Reynolds, C., 67, 97
Rime, B., 174, 194, 195
Roland, P. E., 175, 195
Romero, D. H., 216
Ronning, R., 67, 97
Rosenbaum, D. A., 420
Rothenberg, A., 73
Roth, W.-M., 174, 539, 549
Roychoudhury, A., 539
Rumelhart, D. E., 21

S
Samarapungavan, A., 533, 538
Schiaratura, L., 174
Schmidt, R. A., 216, 226, 420, 512
Schoenfeld, A. H., 161
Schon, D., 21
Schrager, J., 240
Schwartz, D. L., 9, 177, 197, 216, 491, 493, 552, 555
Schweber, S., 112
Shelley, C. P., 331
Shepard, R., 11, 65, 176, 187, 206, 217, 228, 279, 400, 464, 490
Shimojima, A., 474
Siegel, J., 21
Simon, H., 5, 10, 12, 13, 206, 226, 489, 523
Smith, A., 112
Smith, C. L., 85, 112, 482, 559
Sorensen, R., 471
Spiro, R. J., 469
Stavy, R., 541

Steinberg, M., 153, 174, 406, 417, 429, 469, 501, 541, 544, 546, 547, 555
Stephens, L., 554–556, 562
Sternberg, R. J., 21, 529
Stewart, J., 153, 539, 547, 553
Straus, A. L., 11, 13
Suppe, F., 70, 440

T
Thagard, P., 7, 23, 33, 149, 331, 342, 370, 383, 438, 457
Thiele, R. B., 140
Tirosh, D., 205
Tobier, A., 117
Toulmin, S., 71
Trafton, J., 175, 177, 197, 383, 491
Treagust, D. F., 140, 538
Trickett, S., 175, 177, 383, 491
Tsai, C.-C., 541
Tversky, B., 196
Tweeney, R., 6, 8, 9, 102, 206, 240, 293, 454, 519, 520, 521, 559

V
VanLehn, K., 21
Van Meel, J. M., 174, 196
Van Zee, E. H., 546
Vosniadou, S., 21, 559
Voss, J., 21

W
Watson, J., 496, 559
Weisberg, R. W., 73, 560, 561
Weld, D., 298
Wells, M., 546
Wertheimer, M., 3, 21, 161, 163, 167, 384, 385, 442, 457, 460, 464, 468
White, B. Y., 153, 539, 547, 553
Williams, E. G., 546
Wiser, M., 534
Witz, K., 5

Z
Zietsman, A., 175
Zietsman, A. I., 153, 373, 541, 555

Subject Index

A

Abduction
 analogies and extreme cases used in, 152, 331, 340, 367, 371, 374, 377, 378, 445, 487, 566, 568
 defined, 326
 evolutionary view, 71, 95, 325, 326, 507
 generative abduction, 326–328, 330–332, 335, 363–365, 371, 378–381, 470, 481–483, 495–497, 507, 514, 522, 525, 527, 531, 569
 model generation, 84, 86, 93, 108, 112, 319, 321, 326, 329, 330, 332, 333, 355, 358, 364, 366, 370, 371, 377, 378, 381, 434, 435, 445, 478, 480, 481, 490, 495, 510, 514, 516, 518, 522, 525, 546, 553, 556, 558, 563, 564, 568
Abstract *vs.* concrete thinking, 87, 206, 230, 271, 299, 417, 441, 510–511, 518–521
Accretion, 97, 98, 104, 112, 441, 443, 558
Adaptive tuning, 419, 420
Algorithmic control, 572
Analogical reasoning
 classical views of, 24–26, 457–460, 467, 470, 485, 513
Analogy
 analogous case, development of, 57–60, 62
  breaking problems into parts, 58
  extension analogies, 59–60
  extreme cases, 58–59
 characteristics of, 57, 119, 121, 122
 "close" analogies, 23, 43, 269, 374, 438
 definition of, 34–36, 43, 74, 438
 educational implications, 61, 119, 433, 445, 507, 552, 554, 556
 evaluation of, 47–56, 58, 62, 64, 86, 383–405, 436, 458, 459, 464–467, 538, 543, 555
 expedient analogies *vs.* explanatory models, 89–91, 148–149, 269–274, 308–310, 333–334, 339–340, 353–354, 436, 462–463, 468–470, 478, 481–482, 514
 generation methods, 13, 37–42, 44, 111, 163, 164, 438
  from a formal principle, 37
  via an association, 38, 39, 41, 42, 44, 45, 64, 109, 136, 163, 164, 374
  via a transformation, 33, 38, 39, 41, 42, 44, 45, 63, 109, 134, 136, 163, 164, 374, 376, 387, 391, 392, 438, 460, 464, 494, 513
 and imagery, 8, 15, 24, 115, 173, 185, 206, 212, 218, 221, 229, 247, 249, 299, 312, 314, 378, 383–405, 448, 450, 463, 487–488, 494, 517, 536
 invented, 72, 92, 121, 122, 440, 461, 471, 513
 vs. literal similarity, 22, 23, 43, 438
 and model construction, 2, 8, 15, 325, 434, 455, 457, 468, 470, 471, 515, 541, 548, 556, 568
 vs. persistent misconceptions, 445
 personal *vs.* physical, 121
 presented *vs.* spontaneous, 22
 progressive refinement of, 85
 roles of, in scientific thinking, 24, 63, 153–154, 461–463, 469
 significant *vs.* decorative, 33, 36
 spontaneous *vs.* provoked, 64
 structural mapping, 7, 22–24, 49, 56, 58, 353, 383–385, 398, 400, 403–405, 438, 457–458, 464
Anarchistic control, 573
Anchoring case, 140, 141, 144, 152, 315, 399, 537, 541

595

Subject Index

Anomalies
   analogy to persistence of a paradigm, 107, 445

B

Blind variation, 110, 111, 530
Bridging analogies, 24, 48–50, 52, 53, 55, 56, 60, 64, 82, 83, 139, 141, 148, 150, 152, 246, 375, 377, 389, 397, 398, 402, 403, 407, 439, 440, 462, 463, 466, 467, 470, 487, 528, 529, 543, 544

C

Causal mechanisms, 55, 99, 238, 268, 271, 340–342, 400, 427–429, *see also* Spatio-temporally connected causal model
Classroom teaching
   analogies, use of, 7, 26, 27, 48, 53, 114, 120, 139, 163, 237, 434, 445, 484, 537, 541, 568
   thought experiments, use of, 524
Coherence
   between models, 369–370, 418, 428, 480
Collective abduction, 186–187
Combinatorial swamping, 527–528
Composite model, 273, 358, 359
Computational model, 348, 350, 352, 353
Compound simulation, 246, 280, 284, 287, 289, 290, 292–294, 304–311, 397–398, 434, 451, 469, 472, 474, 484, 494, 502, 512–513, 530, 574, 582
Conceptual change theory
   classical, criticisms of, 532
   need for an expanded theory of, for education, 533
Conceptual understanding, 5, 14, 149, 152, 435, 501–503, 551, 552
Concrete *vs.* abstract thinking. See Abstract vs. concrete thinking
Connected causal model. See Spatio-temporally connected causal model
Context of discovery, 93, 530, 531
Context of justification or evaluation, 93, 530
Conviction, sources of
   compound simulation, 246, 284, 285–287, 289–293, 304–307, 472, 474, 484, 487, 494, 496, 499, 512, 574
   extended application of schema, 287, 289, 301, 472
   implicit knowledge, 289, 290, 294, 296, 297, 299–301, 305, 432, 472
   perceptual motor schemas, 287, 288, 291, 293, 321, 322, 434, 446, 448–455, 471, 472, 492
   spatial reasoning, 286, 289–292, 294, 295, 301–305, 312, 314, 322, 434, 446, 472–476
   symmetry, 290, 292, 294, 301, 303–305, 322, 472, 473
Core meaning, 14, 174, 195, 197, 469, 498
Creative reasoning, 1, 4, 22, 104–111, 113–114, 123, 136, 137, 168, 446, 492, 521–531, 567
Creative theory construction
   domain specificity in, 568
   expertise and, 4, 568
   and imagistic representations, 485, 512
   model of, 521–531, 568–572
"Cross talk"
   between major investigation processes from an analogy, 354

D

Darwin's theory of natural selection, 67, 112–113
Deduction, 15, 72, 85, 87, 110, 279, 299, 323, 326–330, 364, 378–380, 413, 434, 441, 451, 474, 498, 510, 514, 520, 523, 568, 570, 573, 574
Deductive truth transmission, 573
Depictive gestures
   not translated from sentences, 195
Description cycle (Descriptive prediction cycle), 348, 350, 354, 378, 478, 514
Dialectic tension, 105–106
Discrete symbolic methods, 403–404
Dissonance, 6, 106, 107, 151, 153, 155, 281, 283, 289, 319, 369, 372, 443–445, 472, 480, 487, 493, 494, 497, 507, 508, 512, 514, 516, 522, 528, 533–536, 541, 542, 544, 546, 557, 569
Distributed control, 358–359
Divergent thinking
   analogies and extreme cases as sources of, 371
   contained divergence, 372, 373, 376, 478, 483, 527, 528
   dangers of, 372
   modulating divergence, 380, 516, 528, 560, 564
      conserving transformations, 374
      constrained analogy generation, 374, 375
      temperature control, 374
   optimal level of, 370–377
Doughnut problem, 52–53, 161, 163, 165, 453, 519

Subject Index

Dual simulation, 389–400, 492–495, 513, 530
   vs. compound simulation, 397–398
   for square and circular coils, 392
Dynamic imagery, 9, 16, 75, 169, 171, 176–178, 180, 184, 187–188, 191, 192, 194–199, 205, 206, 209–214, 218, 221, 228–231, 263, 282, 298, 321, 369, 383, 387, 391, 407, 410, 415, 446, 447, 449, 465, 471, 490–492, 494, 549

E

Einstellung effects, 107, 268, 370, 372, 377, 469, 514, 524–525, 531, 534, 540, 567, 570, 573
   defined, 106
   reducing, 372–374
Elemental simulation, 9, 206, 209, 230, 231, 308, 309, 436, 448, 450, 480, 570
Embedding, 165, 168, 337
Embodied mathematics, 503–504
Embodied thought, 229–233, 206, 226, 322, 387, 430, 446, 449–451, 453, 455, 512
Empiricism vs. rationalism, 70–72, 114, 288, 292, 307, 382, 441, 454, 473, 479, 521–524, 546, 553, 557, 573
Eureka event, 71, 97–100, 102–104, 111, 442, 443, 445
Eurekaism, 68, 69, 113, 444, 445, 573
Evaluation functions, 350–352
Everyday reasoning, 444–446, 568, 569
Evolution of models, 12, 93, 113, 326, 329, 346, 348, 349, 364, 372, 379, 481, 483, 505, 514, 515, 528, 535, 546–548, 558, 574
Expedient analogies, 89–91, 148, 149, 269–274, 308–310, 325, 333, 334, 339, 340, 343, 351, 353, 354, 378, 436, 462, 463, 468–470, 478, 481, 482, 500, 504, 514
Expedient models, 88–94
Expert methods
   expertise, 6, 9, 511, 552, 568
   transfer of findings via an inverse transformation, 63
Expert-novice comparisons, 533–540
   analogy as proto model, 544–545
   concept diagrams, 547–548
   concerning resistance to change, 534–536
   instructional strategies dissonance, 541

   learning by analogy, 541–544
   model evolution cycles
      evidence from curriculum trials, 547
      model evaluation, 547
      model generation, 546
      model modification, 546–547
      theoretical implications of, 545–546
      use of intuition and imagery, 536–537
Expert-novice similarities, 9, 136, 137, 155, 435, 507, 532, 533, 540, 548, 559, 590
Explanatory model
   defined, 74, 148–149
Explanatory modeling cycle, 338, 363–368, 405–406
   information flow during, 367, 480
   inherent difficulty of, 478
External vs internal subjects, defined, 241
Extraordinary reasoning processes, 101–104, 441–443, 559–563
Extreme case reasoning, 49–50, 95, 132, 164–165, 169, 242–244, 247, 251, 253, 264, 267, 295–305, 310, 320, 323, 331–332, 334, 335, 338–340, 342, 361, 369, 371, 373, 374, 381, 390, 393, 405–407, 436, 479, 488, 493–495, 514, 563
   defined, 58

F

Flexible activation, 418, 419, 493, 499
Flexible model, 2, 411, 417, 422, 479, 499–501, 526–527, 574
   Free ride. *See* "Physics for free"

G

Galileo, 54, 59, 67, 235, 439
Gedanken experiment. *See* Thought experiment: evaluative Gedanken experiment
GEM cycle, 84–86, 346, 365, 367
   and active learning, 556–557
   defined, 84
   ubiquitous nature of, 556–557
Generative abduction, 326–332
Geometric model
   advantages of, 498
   stage of modeling, 476, 477
Gestures, depictive. *See* Hand motions
Grounding in runnable schemas, 409–428

H

Hand motions, 171–203
  previous research on
    internal motor representation, 173–174
    language and, 174
  as a source of information about imagery, 197–198
  as a source of information about mental simulations, 210–211
  and thinking, 172, 173
Hierarchical control, 358–359, 529
History of science, 9, 13, 48, 53–54, 93, 95, 99, 240, 278, 389, 418, 473, 482, 500, 506, 516, 559
Hypothesis formation, mechanisms of, 68, 71–73, 115, 441, 442, 444–445

I

Ill-structured problem, 370–371, 553, 556, 557
Imagery
  evidence for importance of, 180, 184–186, 191–192, 219–221, 298, 305, 429
  evidence for, in wide range of reasoning processes, 486–488
  and gesture, 171–202
  importance of, in expert investigations, 485–499
  indicators giving evidence for, 179–181, 298
  limitations of, 485–486, 524
  visual supports, in instruction, 548–555
Imagery enhancement, 179–180, 185, 203, 220–221, 228–229, 249, 251–252, 294–305, 314, 317, 319, 322, 361, 447, 472, 486, 488, 511, 518, 554, 572
  compound (or "linearity") enhancement, 304
  effort and, 488
  symmetry enhancement, 303–305
  and teaching, 544
Imagery report, 115, 169, 173, 177–180, 185, 188, 202, 209, 210, 212, 217–219, 227–229, 250, 251, 280, 282, 283, 298, 299, 310, 312, 388, 390, 391, 403, 406, 407, 410, 446, 447, 465, 488
Imagistic alignment, 240, 342, 351, 405, 436, 459, 463–467
  imagistic alignment analogy, 337, 347, 360, 376, 380, 396–397, 405, 459, 462, 466, 529, 555
  imagistic alignment analogy, defined, 345
Imagistic representations, possible advantages of, 489–490, 492–499

Imagistic simulation
  concrete intuitions and, 219–220
  definition and terminology for, 176, 289, 448–449
  indicators for, 210, 218–220
  intuitive anchors in, 453
  mechanisms for, 210–217, 286–293 447–448
  perceptual motor schemas, 171, 177, 232, 321, 434, 446, 448, 451, 455, 471, 472, 492, 511, 517, 537, 574
  schema-driven, 211–213, 215, 279, 282, 289, 310, 323, 410, 434, 446, 450, 453, 454, 479, 507, 511, 512
  situated action, connection to, 454
  sources of new knowledge in, 286–293, 451–455
  use of motor system in, 205, 206, 211, 452–453
Implicit knowledge, 221–226, 230–232, 290, 296–301, 423, 472
  defined, levels of, 222–223
  as source in thought experiments, 300–301, 305, 327–329, 423, 474, 492, 512, 559
  unconscious, 225–226
  undescribed, 224, 228, 231
Induction, 6, 68–70, 72–74, 85–87, 102, 110, 112, 207, 288, 326, 327, 329, 330, 364, 371, 379, 380, 434, 436, 451, 461, 474, 505, 514, 522, 553, 574
Inquiry skills, 553, 557
Insight
  vs. accretion, 441–442
  definition of, 102–104
  insight behavior, 102–104, 355, 422
  vs. natural reasoning, 440–441
  and punctuated evolution, 443–444
  pure Eureka event, 97–100, 102–104, 111, 442, 443, 445
  torsion insight, 81–83, 87, 90, 98, 100, 101, 110, 248, 267, 285, 304, 344, 350, 375, 416, 460, 468, 482, 529, 531
Instruction, 139–154, 532–538, 540–558
  analogy, 15, 121, 123, 137, 139–154, 417
  bridging, 49–56, 58, 60, 63, 141–154
  imagery, 15–16, 549–550
  and natural reasoning, 15–16, 567
  schema driven imagistic simulation, 507, 508, 511–512
  successful strategies 150, 151, 541, 544, 546, 553
  see also; Classroom teaching
Integrating processes
  recursive integration, 517

Subject Index

Internal vs. external subjects, defined, 241
Intuition. *See* Physical intuition
Investigation processes, evolutionary model of, 476–485

K

Kinesthetic imagery
    existence of, 195, 198
    motor imagery, physiological evidence for, 175
Kinesthetic intuition, 243, 297, 452
Knowledge
    dynamic, 226–229
    ordinal, 264, 266
    prior, 9, 14, 69, 79, 87, 88, 111, 419
Kuhn, Thomas
    anomaly, 106, 107
    causality, 562, 565
    commensurability, 532, 534–536
    criticisms of, 532
    gestalt switch, 71, 108
    models, 557, 558
    paradigms, 105–107, 532, 534
    social processes, 107, 534, 565

L

Limit, concept of, 53, 402–403
Literal similarity, 22–23, 43, 438

M

Mapping in analogies, 7, 8, 383, 384, 389, 405, 440
Mathematical modeling, 161–170, 240, 258–269, 333–342, 346–347, 356–357, 359–363
    fundamentally different from qualitative? 485
    intermediate levels of, 484
    mathematical Gedanken experiments, 361–362
    mathematics and explanation, 362–363
Mental simulation, 8, 9, 11, 15, 57, 58, 75, 108, 171, 176, 187, 188, 192, 199, 206, 208–211, 229–231, 240, 241, 289, 308, 309, 350, 423, 436, 446, 448, 450, 471, 472, 478, 480, 549, 570, 574
    definition of, 176, 289, 448–449
Meteor analogy, 130–136
Methodology of descriptive case studies, 10–13, 214–215, 504–505

Misconceptions, persistent
    teaching strategies for, 139–154, 543
Model construction
    precision levels in, 265, 266, 336–340, 505
    quantitative, 238–239
    triangular relation in, 272–273, 275
Model construction cycles, 84, 85, 93, 104, 110, 434, 514, 545
    in expert reasoning, 84–88
    in students, 539, 553
    see also GEM cycle
Model evaluation, 93, 95, 317–321, 344
    and running a model, 364–367, 368, 369, 378, 435, 479, 483, 497, 515, 521, 530, 546, 564
Motor control theory, 173–188, 193–197, 213–228, 289
Multiple constraints
    and imagery, 331, 373, 376, 495, 527, 569

N

Natural reasoning, 5, 14, 15, 137, 139, 440, 441, 532, 540, 559, 560, 562–564, 566, 567
Newton
    Newton's canon, 389
Non-explanatory models. *See* "Expedient models"
Novel emergence
    novel image feature, 285, 321, 335, 337, 355, 418, 423, 428, 493, 499

O

Overlay simulation, 383, 392–400, 403, 405, 407, 459, 465, 470, 494

P

Pace of change, and insight, 99–100, 113, 445
Partitioning, 165–169, 261, 263, 267, 330, 335, 337, 347, 348, 380, 406, 470, 488, 498
Perceptual motor schemas, 9, 171, 177, 215, 216, 225, 226, 229, 231, 232, 234, 287, 288, 291, 293, 321, 322, 344, 353, 397, 417, 420, 423, 434, 446, 448–455, 471–472, 492, 494, 502, 507, 510–514, 517, 520, 523, 535–537, 573, 574

Physical intuition, 205–234, 242–245, 247, 279, 282, 289, 295, 309, 313, 351, 353, 409, 416, 417, 420, 429, 434, 446–450, 455, 501–503, 507, 522, 536, 537, 540, 546, 560, 568
  features of, 207–209
  intuition reports, 207, 212, 218
  observable behaviors associated with, 208–209
"Physics for free", 459
  caveats to, 475
  "virtually for free" or "very inexpensively", 474–475
Plausibility, 250, 283, 314, 351, 360, 418
  defined, 351
  projected, 422–423
  spatio-temporal coherence and, 422–423, 551
Precision, 266, 269
  linguistic and imagistic, 498
  levels of in modeling cycles, 336–342, 351, 355–357, 362–363, 484–486, 498, 515–518, 525, 554, 555
Prior knowledge, building on, 9, 139, 419
Process goals, 152, 546, 554, 557
Production rules, 352

Q

Quantitative models, 7, 238–240, 260–265, 269, 270, 337, 353, 359, 380, 476, 484–485, 498, 514, 517

R

Rod case, 84, 108, 246, 295, 388, 462
Rules, memorized, 386
Runnability
  as desiderata for scientific theories, 331, 418, 426, 551
  educational benefits, 233, 412, 429, 455, 551, 552
  recursive runnability
    and thought experimentation, 424, 428, 502
  role of
    Nagle's functions, 426
    in scientific knowledge, 426
    for visualizable models, 74, 426, 501
  See also Transfer of runnability
Runnable models
  in mathematical modeling, 516

S

Schema
  activation, 285, 335, 373, 379, 443, 445, 472, 493, 512
  grounded, 215, 233, 284, 289, 416, 502–503
  schema based simulation, 323, 449
Science as extension of intuition, 503, 551, 559
Scientific model, defined, 74, 75
Simulation. *See* Imagistic simulation, Mental simulation
Sisyphys problem. *See* Wheel problem
Source analogues, 24, 25, 58, 60, 63, 94, 95, 154, 271–273, 275, 316, 332, 335, 345, 376, 396, 407, 412, 413, 434, 436, 460, 462, 468, 469, 482, 501, 504, 514, 518, 522, 527, 552, 559
Spatial reasoning
  vs. constructive transformations, 368, 369
  defined, 368–370
  spatial reasoning operators, 291, 368, 369, 416, 452
Spatio-temporally connected causal model, 255–258, 268, 335, 345, 491, 503
Spring problem, 26–30, 33–37, 40, 45, 48, 49, 59, 63, 76, 79, 82, 109, 162, 206, 207, 212, 220, 237, 239, 241, 243, 277, 328, 333, 339, 340, 351, 359, 370, 387, 398, 406, 409, 427, 442, 458, 460, 465, 468, 476, 478, 484, 492, 498, 502, 513, 516, 531, 573
Symbolic relations, 47, 350, 353, 385, 399, 403, 465, 467, 516

T

Targeted relationship, 238, 241, 266, 267, 334, 343, 385, 387, 389, 391, 399–401, 464, 476
Think aloud protocols, 10, 21, 22, 26, 161, 171, 177, 185, 194, 205, 229, 239, 240, 280, 407, 433, 435, 471, 473, 572
Thought experiments, 277–304, 317–322
  coherence relations, spontaneous generation of, 281–284, 286–287, 493, 494
  compound simulations in, 280, 285, 292–294, 301, 306, 309, 312
  defined, 278
  dissonance, spontaneous generation of, 281, 283, 493, 494

Subject Index

evaluative Gedanken experiment, 280, 283, 306, 312–315, 317–318, 321, 361, 472–474, 479, 483, 487, 512–513, 514, 522
flexible extendibility of, 492
fundamental paradox of, 231, 277–278, 322–323, 434, 471–473, 512, 523
and imagistic simulation, 277–323, 475, 476, 492–493
new image features from, 321, 355, 472, 493, 494, 497
perceptual activation of useful schemas in, 285, 321, 493–494
real experiments, connection to, 319, 321, 520, 524
schema based imagistic simulation and, 281–285
spatial reasoning in, 286, 289–292, 301–305, 424, 472–476, 492, 493, 502, 503, 511, 512
volatility of, 379, 512
Torsion definition, 95–96
Torsion, 80–85, 87, 88, 90–93, 95, 96, 98–102, 104, 107, 109–111, 113, 211, 219, 246–250, 253–257, 260, 261, 263–265, 267–269, 277, 281, 285, 302, 304, 306, 307, 312–315, 318, 327, 328, 330, 340, 344–346, 350, 354, 367, 371, 373–375, 379, 397, 416, 442, 443, 460, 461, 465, 468, 474, 482, 483, 492, 495, 497, 515, 518, 521, 522, 529, 531, 548, 561, 566, 571, 572

Torus problem, 26, 27, 52, 402, 406, 498
Transfer of conviction, 424–425
Transfer of runnability
  in instruction, 416–418
  mechanism of, 415
  transfer of runnability hypothesis, 410–411, 415, 429, 501
Transformations
  conserving, 24, 56, 169–170, 270, 292, 334, 335, 337, 374, 376, 377, 380, 383–389, 398, 400–405, 436, 464, 466, 467, 495, 513, 515, 528, 554, 564
  constructive, defined, 368 (footnote)
  constructive, 320, 332, 365, 367, 464
  vs. mapping, 401
  as a source of creativity, 108–109, 488
  spatial
    flexibility of, 495
Triangulation, 178, 184, 214, 505

U

Unconscious
  as extraordinary source of insight, 444–445
  as type of implicit knowledge, 221–226

W

Wertheimer's parallelogram, 167, 384–385, 464
Wheel problem, 48, 50–52, 310, 375, 386, 387, 392, 394, 396, 398, 403, 404, 406, 537

Printed in the United States
124440LV00001B/34-39/P